Communications
in Computer and Information Science   355

Kang Li   Shaoyuan Li   Dewei Li
Qun Niu (Eds.)

Intelligent Computing for Sustainable Energy and Environment

Second International Conference, ICSEE 2012
Shanghai, China, September 12-13, 2012
Revised Selected Papers

 Springer

Volume Editors

Kang Li
Queen's University Belfast
School of Electronics, Electrical Engineering and Computer Science
Stranmillis Road, Belfast BT9 5AH, UK
E-mail: k.li@qub.ac.uk

Shaoyuan Li
Shanghai Jiao Tong University, Department of Automation
800 Dongchuan Road, Shanghai 200240, China
E-mail: syli@sjtu.edu.cn

Dewei Li
Shanghai Jiao Tong University, Department of Automation
800 Dongchuan Road, Shanghai 200240, China
E-mail: dwli@sjtu.edu.cn

Qun Niu
Shanghai University, School of Mechatronic Engineering and Automation
149 Yanchang Road, Shanghai 200072, China
E-mail: nq@shu.edu.cn

ISSN 1865-0929   e-ISSN 1865-0937
ISBN 978-3-642-37104-2   e-ISBN 978-3-642-37105-9
DOI 10.1007/978-3-642-37105-9
Springer Heidelberg Dordrecht London New York

Library of Congress Control Number: 2013935375

CR Subject Classification (1998): J.2, H.2.8, F.2.m, G.1.10, I.2.1, I.5.4

Typesetting: Camera-ready by author, data conversion by Scientific Publishing Services, Chennai, India

Printed on acid-free paper

Springer is part of Springer Science+Business Media (www.springer.com)

Preface

The International Conference on Intelligent Computing for Sustainable Energy and Environment (ICSEE) has been formed to provide a platform dedicated to the emerging and challenging topics in artificial intelligence, machine learning, data mining and their applications in tackling energy and environment problems for sustainable development. It aims to bring together researchers and practitioners to share ideas, problems, and solutions related to the multifaceted aspects of intelligent computing with its applications to sustainable energy and environment issues. It also aims to bring together key figures from the public sector, academia, and industry, in order to discuss the crucial role of science, innovation, and technology in driving economic growth and tackling the energy and environment challenges facing mankind.

ICSEE 2012, which was held in Shanghai, China, during September 12–13, 2012, was the second International Conference on Intelligent Computing for Sustainable Energy and Environment, building on the success of ICSEE 2010 held in Wuxi, 2010.

This year, the conference mainly concentrated on the theories and methodologies as well as the emerging applications of intelligent computing in sustainable energy and environments. It intended to unify the contemporary intelligent computing techniques within an integral framework highlighting the trends in advanced computational intelligence and bridging the theoretical research with key applications in sustainable energy and environment. In particular, bio-inspired computing emerges as a key player in pursuing for novel technology in recent years to deal with challenging issues such as the curse of the dimensionality problem in handling imbalanced, large-scale, multi-mode, heterogeneous, inconsistent, and temporal and spatial data in energy and environment applications, such as smart grids. In light of this trend, the theme for this conference was the "Emerging Intelligent Computing Technology and Applications in Sustainable Power and Environment." Papers related to this theme were especially solicited, including theories, methodologies, and applications.

These proceedings of ICSEE 2012 contain 61 contributions from different parts of the world, especially from leading UK and China Science Bridge partners. Each paper received three or more reviews. There were five plenary keynote speeches and nine sessions. Each session was consisted of five to seven papers. Papers were assigned with the sole purpose of forming coherent sessions. The authors have further revised their papers according to the feedback from the audience after the conference before inculsion in this volume of CCIS.

Shanghai is one of the most populous and fascinating cities in the world, which has contributed greatly to technology, education, finance, commerce, fashion, and culture. Before or after the conference, delegates were able to attend various networking and high-profile events, and one of the highlights was the UK-China Science Bridge forum that brought together key figures from the public sector, academia, and industry to review some of the large-scale key collaboration projects between partner institutions from UK and to China, and to discuss issues to sustain international collaborative research and innovation.

The organizers of ICSEE 2012 would like to acknowledge the enormous contributions made by the following: the Advisory Committee for their guidance and advice, the Program Committee and the numerous referees for their efforts in reviewing and soliciting the papers, and the Publication Committee for their editorial work. We would also like to thank the editorial team at Springer for their support and guidance. Particular thanks are of course due to all the authors, as without their high-quality submissions and presentations, the ICSEE 2012 conference would not have been successful.

Finally, we would like to express our gratitude to our sponsors and technical co-sponsors of the UK-China Science Bridge forum and ICSEE conference: Research Councils UK, National Natural Science Foundation of China, IEEE Communication and Control Chapter Ireland, and IEEE Shanghai Section. The support of the organizers including Queen's University Belfast and Shanghai Jiaotong University, and, co-organizers including Tsinghua University, Zhejiang University, Shanghai University, Chongqing University, South-eastern University, Hunan University, Harbin Institute of Technology, Shanghai Research Center for Wireless Communications, Herriot Watt University, Lancaster University, and Bradford University, is also acknowledged.

October 2012

George W. Irwin
Yugeng Xi
Kang Li
Shaoyuan Li
Dewei Li
Qun Niu

Organization

The Second International Conference on Intelligent Computing for Sustainable Energy and Environment (ICSEE 2012) was organized by Shanghai Jiao Tong University and Queen's University Belfast.

Advisory Committee

Cao, Xiren   China
Ge, Shuzhi   Singapore
Irwin, George   UK
Jin, Weiliang   China
McCanny, John   UK
Wu, Cheng   China
Scott, Stan   UK
Heskes, Tom   The Netherlands
Thompson, Stephen   UK

General Chairs

Honorary Chairs: Irwin, George (UK)
                          Xi, Yugeng (China)
General Chairs: Li, Kang (UK)
                          Li, Shaoyuan (China)

International Program Committee

Program Chairman: Li, Shaoyuan (China)

IPC members:

Bai, Yun (UK)   Du, Dajun (China)
Bai, Erwei (USA)   Fei, Minrui (China)
Basheer, Muhammed (UK)   Gan, Deqiang (China)
Chen, Sheng (UK)   Gu, Xingsheng (China)
Chen, Weidong (China)   Hong, Xia (China)
Deng, Jing (UK)   He, Haibo (USA)
Ding, YongSheng (China)   Heskes, Tom (The Netherlands)
Duan, Guangren (China)   Huang, Dan (China)

Huang, Guangbin (Singapore)
Hussain, Amir (UK)
Infield, David (UK)
Jia, Li (China)
Laverty, David (UK)
Li, Dewei (China)
Li, Kang (UK)
Li, Ning (China)
Li, Xue (China)
Lin, Wenfeng (UK)
Littler, Tim (UK)
Long, Chengnian (China)
McAfee, Marion (Ireland)
Mei, Shenwei (China)
Maione, Guido (Italy)
McLoone, Seán (Ireland)

Naeem, Wasif (UK)
Niu, Qun (China)
Song, Shiji (China)
Thompson, Stephen (UK)
Wang, Chengxiang (UK)
Wang, Dainhui (Australia)
Wang, Haifeng (UK)
Wang, Lin (China)
Wang, Shujuan (China)
Weng, Zhenxin (China)
Xu, Lie (UK)
Yang, Zheng (China)
Zhai, Guofu (China)
Zhang, Shaohua (China)
Zhao, Guangzhou (China)
Zhou, Donghua (China)

Organizing Committee

Organizing Chairs:	Li, Dewei (China)
	Du, Dajun (UK)
Organizing Co-Chairs:	Li, Ning (China)
	Huang, Dan (China)
Finance Chair:	Wu, Yijun (China)
Publication Chairs:	Naeem, Wasif (UK)
	Long, Chengnian (China)
	Niu, Qun (China)
Special Session Chairs:	Hong, Xia (UK)
	Wu, Jing (China)
Secretariat:	Yuan, Lijuan (China)
	Li, Xia (China)

Sponsoring Institutions

National Natural Science Foundation of China
Research Councils UK

Table of Contents

Study on Controlling Power Quality
Based on Thermodynamics Modeling Method
for 6.3kV Power System

Liguo Wang[*,**], Jian Liu, Shibo Zhang, and Dianguo Xu

Dept.of Electrical Engineering, Harbin Institute of Technology,
Harbin 150001, Heilongjiang Province, China
wlg2001@hit.edu.cn

Abstract. Based on controlling the power quality of the 6.3kV power system of a coal mine in North China, a topology consist of Passive Power Filter (PPF), Thyristor Switched Capacitor (TSC) and Fixed Compensation Capacitor (FC) has been studied. Combining the thermodynamics modeling theory with network topology analysis method, the parameters of PPF, TSC and FC was proposed and optimized. Though there is a 5.th single tuning filter only, the measured results show that the average harmonic current injected into the power system completely accord with Chinese national standard GB/T-14549-93 and the power factor is increased to 0.99 without any possibility of over-compensation or under-compensation. Due to the action of the PPF the negative effect that caused by zero-voltage-switch of TSC was eliminated by proposed topology. The study of this paper gives the corner stone of theory and method for realizing smart grid system with Chinese characteristics.

Keywords: power quality, thermodynamics modeling, PPF, TSC.

1   Introduction

The concept of smart power grid, which was presented by Mark Kerbel in 2005, is essentially obtaining the feedback from the consumers via intelligent meters, and taking advantage of the advanced measuring, telecommunication, computer and automation technology, in order to achieve the closed-cycle control among the electricity transmission, transforming and using, and enhance the robustness, self-healing, adaptability, efficiency and compatibility of the power systems. Until now, the Smart Power Grid is still in its primary stage all over the world. Council of the European Union pointed out explicitly in 2006 that Smart Power Grid is the key to guarantee a sustainable, competitive and secure power quality. After the Boulder, Colorado becoming the first Smart Grid City in April 2008, the Google and the GE

---

[*] Manuscript received July 30, 2012. This work was supported by the "Natural Science Foundation of China (51177028)". Liguo Wang is with the Department of Electrical and Electronics Engineering, Harbin Institute of Technology, Harbin, 86-451-86403145, China.
[**] Corresponding author.

K. Li et al. (Eds.): ICSEE 2013, CCIS 355, pp. 1–9, 2013.

put forward plans at the same year which boosted the development of Smart Grid. Barack Obama issued American Recovery and Reinvestment Plan in January 2009, indicating that US was planning to intelligentize its power grid. All the events mentioned above reveals the urgent demand for the smart grid [1-4].

The Smart Grids in Europe and US are characterized by self-adapt, self-optimize, and self-heal. They have achieved the power distribution intelligent measurement, controllable load and the smart building, which are based on Ethernet and intelligent bidirectional meters [5-6]. In China, some institutes such as China Electric Power Research Institute, Tsinghua University, Harbin Institute of Technology, North China Electric Power University, Central-South University, Xian Jiaotong University, are researching on the Smart Grid [7-9]. Compare to the intelligent power distribution of US, Europe, China should pay particular emphasis on intelligentizing power transmission, in order to meet the increasing demand of electrical energy and enhance the security and reliability of the power grid, and further, promoting the development of coal-fired power, hydro-electricity, nuclear electricity and large scale of renewable energy resource in intensive way [10-11]. It is necessary to explore a new way to achieve smart grid under certain power distribution circumstances, gathering the early experience of smart grid transformation of China [12-13].

As the preliminary work on developing the smart grid, we focus on controlling power quality of a coal-mine in Heilongjiang Province of northern China. According to analyzing topology of network thermodynamics, the harmonic suppression and reactive power compensation device was proposed based on analyzing the stability of PPF, TSC and FC to equivalent reactance of the power system. By analyzing combinatorial topology of filter and compensator, optimization capacity and parameters of PPF and TSC was given based on FC which capacity of 6600kVar and 4 channels. Considering terrible 5.th harmonic current that measured by HIOKI power quality analyzer 3196 the network topology that results in an infinite response at resonance was analyzed. While this work remains to be seen how this is received in other field, there's no reason why it shouldn't be effective for suppressing harmonic current and compensating reactive power in 6.3kV power system. Moreover, the negative effect that caused by zero-voltage-switch of TSC was eliminated by proposed topology. The overheat phenomena of the filter reactor was analyzed also. This work can provide reference for building up the smart grid in line with the modern power quality in China.

2   The Research Background

The topology of the distribution substation of a coal mine is shown in Figure 1. The loads include rectifier and converter device such as service incline winch with 272kW DC motor, main mine lifter converter of capacity of 1000kVA, pumps, tunnel fan, belt conveyor, excavation area winch and so forth. There is a main transformer which the capacity of 20,000kVA and the short-circuit impedance is 9.06%. The voltage of primary side and secondary side of the transformer is 66kV and 6.3kV respectively.

There were 2 banks of fixed capacitor which capacity of 6600kVar.

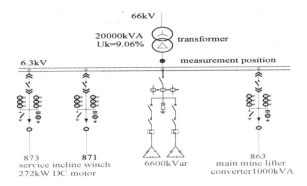

Fig. 1. Topology of distribution substation of a Coal Mine

In order to solve the power quality problems in power system it is necessary to monitor and analyze actual harmonics and reactive power firstly. For this purpose, the reactive power, harmonic current of the secondary side of the transformer are obtained by HIOKI power quality analyzer 3196 in July 14th, 2009.

Figure2 and Table 1 show the max current of 5th, 7th, 11th and 13th harmonic order of the 6.3kV power system. It indicates the 5th harmonic current reached 111.60A which far exceeds 63.04A that allowed by the national standard GB/T-14549-93.

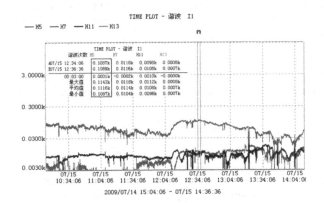

Fig. 2. The harmonic current of 6.3KV bus

Table 1. Measured harmonic current of 6.3KV bus (A)

	harmonic current			
harmonic order	5	7	11	13
allowed value	63.04	44.50	29.66	29.66
average values	111.60	11.40	10.80	7.00

Figure2 to Figure4 show that there are more surge current and voltage flash of 6.3kV bus. These phenomena can be illustrated by Figure4. Due to there were 2 banks of fixed capacitor which capacity of 6600kVar maximum power factor of 6.3kV bus reached 0.98 almost.

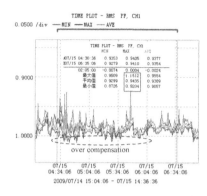

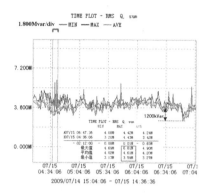

Fig. 3. Power factor of 6.3KV bus   **Fig.4.** Reactive power of 6.3KV bus

After all, it must not be overlooked that there are some over compensations and shocks of power reactor that caused by high power device such as service incline winch. Figure2 to Figure4 show that there are not only some distortions of harmonic current but also shocks of power reactor which must be considered for improving power supply quality of a Coal Mine.

Due to cost, efficiency, power, and form factor constraints, a device that consists of PPF, TSC and FC was put forward in this paper. By comparing figure3 to figure4 the maximum capacity of PPF and remained FC can determined by following expression

$$Q_{FC} + Q_{PPF} = 6600 - 1200 = 5400(kVar) \tag{1}$$

There 6600kVar and 1200kVar is original compensation capacity and over compensation capacity of secondary side of the transformer respectively. Considering the compensation effect of original fixed capacitor the capacity of TSC can expressed as follows

$$Q_{TSC} = 150 + 300 + 600 = 1050(kVar) \tag{2}$$

The major difficulty is that how to avoid the resonance of PPF, TSC and FC occurs. It needs to analyze all impossible combination between equivalent reactance of the transformer, every channel of TSC, FC, and PPF under the harmonic current that caused by harmonic loads. For this purpose, the thermodynamics modeling method that consists of network topology and statistics was used to analyze impossible resonance of proposed device.

3   Thermodynamics Modeling Method

The network topology [14-15] of equivalent inductance, TSC, FC and PPF is shown in Figure5. Based on the capacity and the short-circuit impedance of the transformer the equivalent inductance of 6.3kV power system can be calculated by following expression

$$L_S = ku_{second}^2 / S_N = (9.06/100) \times 6300^2 / (2000 \times 10^3) = 1.80(mH) \tag{3}$$

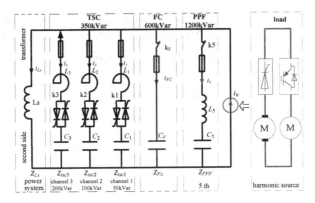

Fig.5. Equivalent phase circuit of proposed device

There symbols S_N, k and u_{second} denote capacity, the short-circuit impedance and secondary voltage of the transformer respectively. The Eq.(3) is calculated at $S_N = 2000\,\text{kVA}$, $k = 9.06\%$ and $u_{second} = 6300\,\text{V}$.

TSC consists of 3 channels which capacity of 150kVar, 300kVar and 600kVar respectively. In order to avoid terrible 5th harmonic current there is a reactance ratio of 6% of the TSC. AS shown in figure5 the compensation capacitor of every channel was denoted by C_1, C_2 and C_3, corresponding smoothing reactor is L_1, L_2 and L_3.

There is a single tuning filter with the capacity of 3600kVar in PPF based on analyzing harmonic current and original compensation capacitor. According to classical calculation method the parameters of the PPF can be given.

Considering the economic benefits of the project and space constraint there is remained fixed capacitor which capacity of 1800kVar. The symbol C_F denotes equivalent capacitor of the remained FC.

The harmonic source denotes total harmonic current that caused by corresponding rectifier and converter device of the secondary side of the transformer. Detail harmonic currents were illustrated by TABLE I. According to Thevenin and Norton theorem the repercussion effect of every harmonic current need to analyze.

As shown in Figure5 due to the equivalent inductance fall into a part of the 6.3kV bus, FC and PPF could be switched individually or in a flotilla, and there are 3 channel of TSC, so their manner of working may be expressed as C_1^1, $\left(C_2^1 + C_2^2\right)$ and $\left(C_3^1 + C_3^2 + C_3^3\right)$ respectively. The total combination of equivalent inductance, TSC, FC and PPF can be theoretically is presented as,

$$n_1 = C_1^1\left(C_2^1 + C_2^2\right)\left(C_3^1 + C_3^2 + C_3^3\right) = 21 \qquad (4)$$

Equation (4) denotes that there are 21 network topologies of proposed device. Every circuit topology need to analyze in order to avoid the resonance occurs. Equivalent impedance of every above part is defined as Z_{Ls}, Z_{Fc}, Z_{PPF} and $Z_{TSCi}(i=1,2,3)$ which are illustrated with Figure5. Above symbols are written as: $Z_{Ls} = \omega Ls$, $Z_{Fc} = -1.0/(\omega C_F)$, $Z_{PPF} = \omega L_5 - 1.0/(\omega C_5)$ $Z_{TSCi} = \omega L_i - 1.0/(\omega C_i)$ $(i=1,2,3)$.

Based on above analysis the impossible combination was expressed in Table 2. The symbols $FC_i (i = 1,2,\cdots,20,21)$ denote combination type of equivalent inductance, TSC, FC and PPF. For example the symbol FC_{18} in Table 2 denote following topology sets

$$FC_{18} = (Z_{TSC1}, Z_{TSC2}, Z_{Ls}, Z_{Fc}, Z_{PPF}) \tag{5}$$

Table 2. parameters of a capacitor

term	$Z_{Ls} + Z_{Fc}$	$Z_{Ls} + Z_{PPF}$	$Z_{Ls} + Z_{Fc} + Z_{PPF}$
Z_{TSC1}	FC_1	FC_8	FC_{15}
Z_{TSC2}	FC_2	FC_9	FC_{16}
Z_{TSC3}	FC_3	FC_{10}	FC_{17}
$Z_{TSC1} + Z_{TSC2}$	FC_4	FC_{11}	FC_{18}
$Z_{TSC1} + Z_{TSC3}$	FC_5	FC_{12}	FC_{19}
$Z_{TSC2} + Z_{TSC3}$	FC_6	FC_{13}	FC_{20}
$Z_{TSC1} + Z_{TSC2} + Z_{TSC3}$	FC_7	FC_{14}	FC_{21}

The difficulty of proposed device is how to avoid the parallel resonance of variables of Figure 5 occur in condition of the harmonic currents in Figure2 and Table I. Considering the Ls is fixed it is necessary to determine the other parameters such C_F and (L_i, C_i) $(i = 1,2,3,5)$.

By the thermodynamics modeling method the PPF consists of 5 factors which set is expressed as

$$S'_{PPF} = \{contactor 5, fuse 5, L_5, C_5\} \tag{6}$$

In Eq.(6), considering the factors such as contactor5、 fuse5 are linear, their voltage and current are determined by the L_5 and C_5 by the Kirchhoff law. When calculating, they can be seen as linear node of directed edge. So the set of the Eq.(6) can be simplified as

$$S_{PPF} = \{L_5, C_5\} \tag{7}$$

Similarly, the basic factors of TSC thermodynamics modeling such as breaker, fuse$i$ $(i = 1,2,3)$ and ki $(i = 1,2,3)$. The factor breaker fuse$i$ $(i = 1,2,3)$ can also be seen as linear node of directed edge. The set of the TSC can be written as

$$S_{TSC} = \{L_i, C_i\} \ (i = 1,2,3) \tag{8}$$

The basic factors of thermodynamics modeling of the proposed compensation devices can be expressed as

$$S = \{L_s, S_{PPF}, S_{TSC}, C_F\} = \{L_s, L_i, C_i, C_F\} \ (i = 1,2,3,5) \tag{9}$$

According to thermodynamics, under the power constraint E , the set S can be characterized by the following equations

$$E = \{(u_{Ls}, i_{C_F}, u_{Lj}, i_{Cj}) | j = 1,2,3,5\} \tag{10}$$

The factors Eq. (10) can be denoted by follows: $u_{Ls} = L_s di_{Ls} / dt$, $u_{Lj} = L_j di_j / dt$ $(j = 1,2,3,5)$, $i_{Cj} = C_j du_j / dt$ $(j = 1,2,3,5)$, $i_{C_F} = C_F du_{C_F} / dt$.

The correlation function of those factors in set S can be rewritten as

$$\Psi = \{i_h, i_{PPF}, i_{FC}, i_{TSC}, i_{Ls}\} \tag{11}$$

There are following relationship in Eq.(11) ,

$$i_h = i_{PPF} + i_{FC} + i_{TSC} + i_{Ls} \tag{12}$$

$$i_{PPF} = i_5 \tag{13}$$

$$i_{TSC} \in \{i_1, i_2, i_3, (i_1 + i_2), (i_2 + i_3), (i_1 + i_3), (i_1 + i_2 + i_3)\} \tag{14}$$

Based on above analysis the maximum harmonic current that injected into the PPF, FC, TSC can be obtained. For example the i_5 and i_{Ls} can be described as follows

$$i_5 = (Z_{Ls} // Z_{Fc} // Z_{TSC}) / (Z_{Ls} + Z_{Fc} + Z_{TSC} + Z_{PPF}) i_h \tag{15}$$

$$i_{Ls} = (Z_{PPF} // Z_{Fc} // Z_{TSC}) / (Z_{Ls} + Z_{Fc} + Z_{TSC} + Z_{PPF}) i_h \tag{16}$$

There Z_{TSC} belong to following set

$$Z_{TSC} \in \{Z_{TSC1}, Z_{TSC2}, Z_{TSC3}, (Z_{TSC1} + Z_{TSC2}), (Z_{TSC2} + Z_{TSC3}), (Z_{TSC1} + Z_{TSC3}), (Z_{TSC1} + Z_{TSC2} + Z_{TSC3})\} \tag{17}$$

Table 3. parameters of a capacitor

PPF	Capacity (kVar)	Connection mode	rated voltage (kV)	capacitor (μF)
C_5	300	2 series and 8 parallel	$6.600/\sqrt{3}$	65.766

Table 4. parameters of a reactor

PPF	rated voltage (kV)	fundamental voltage (V)	rated current (A)	reactor (mH)
L_5	$6.600/\sqrt{3}$	161.994	314.916	1.637

Considering different Z_{TSC} the maximum i_5 and i_{Ls} can be obtained. In order to improve security it is not necessary to take the resonance point is 4.85. Following parameters such as (L_i, C_i, C_F) $(i = 1,2,3,5)$ can be calculated. Giving attention to evidently decrease the cost and analyze parallel resonance of every combination of Table 2, the parameters of the PPF was derived in Table 3 and Table 4.

4   Actual Measurement Results Analysis

The device that consists of PPF, FC and TSC has been serviced in a coal mine of northern China since November 24th 2009. Figure6 to Figure9 show the result that obtained by proposed device. Figure6, Figure7 and Table 5 show that the 5th, 7th, 11th and 13th harmonic current was reduced significantly and far less than the national standard GB/T-14549-93. Compare Figure6 to Figure7 denotes that the 7th harmonic current was magnified by the action of FC and TSC only. Figure 7 indicates that the resonance of 7th harmonic current, i.e. the negative effect of the TSC was suppressed by PPF.

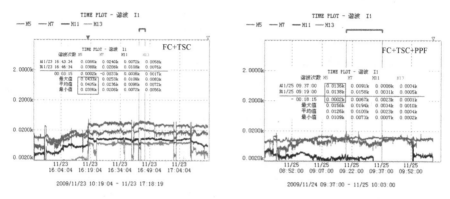

Fig. 6. The harmonic current by FC and TSC   **Fig.7.** The harmonic current by FC, TSC and PPF

Table 5. Harmonic current of 6.3KV bus by fc, tsc and ppf （A）

harmonic current				
harmonic order	5	7	11	13
allowed value	63.04	44.50	29.66	29.66
average values	**111.60**	11.40	10.80	7.00
FC+TSC	40.20	**23.10**	10.30	**7.50**
FC+TSC+ PPF	12.60	10.10	2.40	4.00

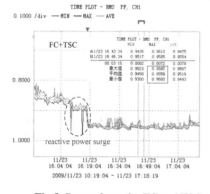
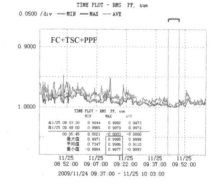

Fig.8. Power factor by FC and TSC   **Fig.9.** Power factor by FC, TSC and PPF

Figure8 and Figure9 show that though the maximum power factors reached 0.96 there still were some surges of reactive power by the action of FC and TSC only. Figure 13 indicates that the negative effect can be eliminated by the PPF.

5   Conclusions

This paper aims at promoting development of the smart grid by controlling the power quality of the power system. For this purpose, a thermodynamics modeling method was used to analyze the stability and performance of typical filter and reactive power compensation device in order to improve the power quality of the distribution systems of the coal mine.

This work emphasizes the smart grid projects and development experience based on initial base of the power system. Combining FC, TSC with PPF a cost-effectively method that suppresses harmonic current and compensates reactive power was given based on analyzing the topology of the power system. The effect of proposed method was validated by an actual engineering which can offer a useful theoretical and practical reference for the smart grid with Chinese characteristics.

References

[1] Zhang, W., Liu, Z., Wang, M., Yang, X.: Research Status and Development Trend of Smart Grid. Power System Technology 33(13), 1–11 (2009)

[2] Chen, S., Song, S., Li, L., Shen, J.: Survey on Smart Grid Technology. Power System Technology 33(8), 1–7 (2009)

[3] Zhang, B., Sun, H., Wu, W., Guo, Q.: Future Development of Control Center Technologies for Smart Grid. Automation of Electric Power Systems 33(17), 21–28 (2009)

[4] Zhang, Q., Wang, X., Fu, M., Wang, J.: Smart Grid from the Perspective of Demand Response. Automation of Electric Power Systems 33(17), 48–54 (2009)

[5] Lu, J., Xie, D., Ai, Q.: Research on Smart Grid in China. In: IEEE T&D Asia, pp. 1–4 (2009)

[6] Ma, Y., Zhou, L., Tse, N., Osman, A., Lai, L.L.: An Initial Study on Computational Intelligence for Smart Grid. In: Proceedings of the Eighth International Conference on Machine Learning and Cybernetics, Baoding, July 12-15, pp. 3425–3429 (2009)

[7] Yu, F., Yi, T., Xike, W.: Parameter Estimation Method of Power Quality Disturbances Based on Electrical Parameters Analysis. Proceedings of the CSEE 29(16), 100–107 (2009)

[8] Zhang, W., Zhou, X., Bai, X., Tang, Y.: Countermeasures Against Sudden Events to Ensure the Security of Urban Power Supply Systems. Proceedings of the CSEE 28(22), 1–7 (2008)

[9] Zhou, Z., Bai, X., Li, W., Li, Z., et al.: A Novel Smart On-line Fault Diagnosis and Analysis Approach of Power Grid Based on WAMS. Proceedings of the CSEE 29(13), 1–7 (2009)

[10] Zhao, J., He, Z., Qian, Q.: Detection of Power Quality Disturbances Utilizing Generalized Morphological Filter and Difference-entropy. Proceedings of the CSEE 29(7), 121–127 (2009)

[11] Zhao, W., Luo, A., Cao, Y., Yu, L.: Hybrid Var and Harmonic Dynamic Compensator and Application to Three-two Phase Traction Substation. Proceedings of the CSEE 29(28), 79–82 (2009)

[12] Luo, A., Ou, J., Tang, J., Rong, F.: Research on Control Method of STATCOM for Grid Voltage Unbalance Compensation. Proceedings of the CSEE 29(6), 55–60 (2009)

[13] Guo, W., Wu, J., Xu, D., Wang, L.: Hybrid Shunt Active Power Filter Based on Novel Sliding Mode Control. Proceedings of the CSEE 29(27), 29–35 (2009)

[14] Wang, L., Xu, D., Miao, L., Guan, B.-L.: A Modeling Analysis of Passive Power Device and Influence Parameter Perturbation on Power System 25(10), 70–74 (2005)

[15] Wang, L., Wang, Y., Xu, D., Fang, B., Liu, Q., Zou, J.: Application of HHT for Online Detecting the Inter-area Short Circuit of Rotor Windings of Wind-Generator Based on Thermodynamics Modeling Method. Journal of Power Electronics 11(5), 759–766 (2011)

Digital Control Strategy for Harmonic Compensation of Dynamic Voltage Restorer

Jianwei Wang, Xiaoguang Hu, Guoqing Hu, and Yi Wang

School of Automation Science and Electrical Engineering, Beihang University,
100191, Beijing, China
wjwjian@163.com

Abstract. To make dynamic voltage restorer (DVR) compensate the low-order harmonic voltage, overcome the effects on system performance of digital control, and improve the voltage compensation effect, a dual closed-loop digital PR control strategy, consisting of voltage outer loop fundamental proportional resonant (PR) control and current inner loop low-order harmonic PR control is proposed. The discretization methods of PR controller are analysed, a detailed analysis and parameter design of the fundamental PR controller and the 3rd, 5th and 7th harmonic PR controller in the discrete domain are carried out. The analysis shows that the harmonic PR controller can effectively compensate for the 3rd, 5th and 7th harmonic voltage, the fundamental PR controlled can achieve zero steady-state error tracking and inhibit the effects on output voltage of load current, and digital control system has good dynamic response characteristics. Based on the theoretical analysis, 11KVA DVR prototype is developed and tested. The effectiveness and feasibility of the proposed control strategy are verified by experimental results.

Keywords: dynamic voltage restorer, harmonic compensation, digital control, fundamental PR controller, harmonic PR controller.

1 Introduction

When the dynamic voltage restorer detects the system voltage, if the system voltage drops, the invert unit will instantaneously produce dynamic offset voltage to compensate the dropping voltage to a rate level. That will ensure normal operation for the system sensitive load [1–3].

As the DVR is a new power quality series compensation device, the domestic and foreign relevant researchers have made enormous study on its control strategy. But few DVR control strategy considers the distortion of the voltage waveform caused by low order harmonics which are produced by the dead time of the inverter unit switching devices and the nonlinear load [8], and calculation delay of digital control causing the adverse effects on the dynamic characteristics of the DVR.

In order to improve the DVR inverter unit's ability to restrain the low order harmonics, overcome digital control's effect on the system and have better

K. Li et al. (Eds.): ICSEE 2013, CCIS 355, pp. 10–18, 2013.

voltage compensation effect, this paper presents a double closed loop digital PR control strategy by using the fundamental PR to control voltage outer loop and low order harmonics PR to control current inner loop electrical. At the resonance frequency, the open loop gain of the PR controller is infinite, which can achieve no static error control to particular frequency, and can provide enough attenuation to prevent interference between adjacent frequencies in other frequencies. Because the bandwidth of the voltage outer loop is narrow, the low order harmonic PR control can't be applied in the outer loop. The paper can use fundamental PR control in voltage outer loop and low order harmonic PR control in current feedback inner loop to eliminate the steady state error and compensate the low order harmonics. To avoid the influence produced by the sampling, calculation delay and other discrete processes on the steady-state error and dynamic response characteristics, the paper use a digital control method. The paper develops a 11KVA DVR prototype based on the theory and do some corresponding test. The theory research and test result show that the proposed control scheme is effective and feasible.

2   Inverter Unit Digital Proportion Resonance Control Strategy

When DVR inverter unit uses a double closed loop PR control, since the bandwidth of the voltage of the outer ring is small, and that of the voltage of the inner ring is big, if the outer loop uses the harmonic PR control, though the PR controller has a very big gain at the resonance frequency, the system will be instable and it is still difficult to realize the satisfied compensation effect in condition that the resonant frequency exceeds the bandwidth of the voltage of the outer ring, and the reduction of the gain of the outer ring will greatly offset the resonance gain. Thus, the fundamental wave PR controller is applied in the outside loop and the harmonic PR controller in the inside current loop.

As the digital processor has a sampling and a calculation time delay, PWM pulse update generally lags behind for a beat of sampling time. And for the DVR emphasizes the dynamic characteristics, this influence can't be ignored. The key to continuous domain discretization digital control is how to choose an appropriate discretization method. And this paper uses the step response invariant method of zero-order holder combined to discrete the consecutive time model of the DVR inverter unit, which makes the system stable and the gain unchanged. The high harmonics produced by the inverter unit switching device in switch action can be filtered,while 3rd, 5th and 7th low order harmonics produced by nonlinear factors such as the nonlinear load can only be compensated by designing a reasonable controller. Therefore, as discussed and analyzed in Fig. 1, the paper presents a double loop PR control strategy, that the fundamental wave PR controller is applied in the outer loop and the 3rd, 5th, 7th harmonic low pass filters are applied in the inner loop. Among them, K_{p1} and $R_1(z)$ respectively stand for the fundamental wave PR controller, $K_{ph}, R_3(z), R_5(z)$ and $R_7(z)$ respectively stand for 3rd, 5th and 7th harmonic PR controllers.

The low order harmonic current of the load current i_o of the inverter unit of the H Bridge is a disturbance in the continuous time model. It will distort the output voltage u_o when it passes through $1/Cs$. The action of current inner loop harmonic PR controller on the inverter unit inhibits the influence of low order harmonic current on the system. That ensures the DVR to track and compensate the voltage dips and harmonic voltage quickly and accurately.

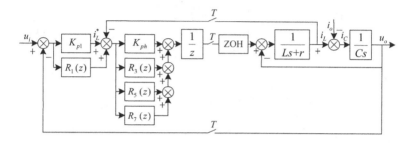

Fig. 1. Dual-loop digital resonant controller

3   Discretization of Proportion Resonant Controller

According to the principle of reference [11–13], the transfer function of PR controller can be shown as formula (1).

$$PR(s) = K_p + \frac{K_r s}{s^2 + \omega_n^2} \tag{1}$$

Where: K_p is the radio coefficient of the proportional control link, K_r the resonant coefficient, and ω_n the resonant angle frequency.

That is, the discussion about the discrete method of the PR controllers is converted into that about the discrete method of resonant controllers, shown as formula (2).

$$R(s) = \frac{K_r s}{s^2 + \omega_n^2} \tag{2}$$

However, in the resonant frequency formula (2) will introduce the 90° lagging phase angle, and an compensation for lagging phase angle is necessary in order to increase the stability of phase margin [11–13]. Resonant coefficient K_r can be neglected during the discussion of the discrete method of PR controller. So formula (2) can be rewritten as:

$$R(s) = \frac{s \cos \theta_n - \omega_n \sin \theta_n}{s^2 + \omega_n^2} \tag{3}$$

Where: θ_n in above formula refers to the compensation angle of phase angle lag. Formula (3) can be achieved by the virtual LC methods in reference [13], its output equation can be expressed as:

$$R_o(s) = [I_L(s) \cos \theta_n - U_C(s) \sin \theta_n]/\omega_n \tag{4}$$

Where: $I_L(s)$ and $U_C(s)$, respectively, refer to the inductor currents and capacitor voltage of virtual LC circuit. The input signal of virtual LC circuit's control function towards inductance current and capacitor voltage can push forward that the control function of the resonant controller to sinusoidal signal is equivalent to the control function of integral controller to DC signal [11–13].

As ω_n is the resonance angular frequency, $1/\omega_n$ can be regarded as a factor of resonant coefficient K_r, the output equation of the virtual LC circuit (4) can be rewritten for:

$$R_o(s) = I_L(s)\cos\theta_n - U_C(s)\sin\theta_n \tag{5}$$

To discrete above continuous-time output equation may have:

$$R_o(k) = C\begin{bmatrix} I_L(k) \\ U_C(k) \end{bmatrix} \tag{6}$$

Where: $C = [\cos\theta_n - \sin\theta_n]$.

The continuous time state equation of virtual LC circuit is:

$$\begin{bmatrix} \dfrac{dI_L(t)}{dt} \\ \dfrac{dU_C(t)}{dt} \end{bmatrix} = \begin{bmatrix} 0 & -\omega_n \\ \omega_n & 0 \end{bmatrix}\begin{bmatrix} I_L(t) \\ U_C(t) \end{bmatrix} + \begin{bmatrix} \omega_n \\ 0 \end{bmatrix} R_i(t) \tag{7}$$

To discrete above continuous-time state equation may have:

$$\begin{bmatrix} I_L(k+1) \\ U_C(k+1) \end{bmatrix} = A\begin{bmatrix} I_L(k) \\ U_C(k) \end{bmatrix} + BR_i(k) \tag{8}$$

Where: $A = \begin{bmatrix} \cos(\omega_n T) & -\sin(\omega_n T) \\ \sin(\omega_n T) & \cos(\omega_n T) \end{bmatrix}$; $B = \begin{bmatrix} \sin(\omega_n T) \\ 1 - \cos(\omega_n T) \end{bmatrix}$.

Take inductor current and capacitor voltage of virtual LC circuit as output, the Z-domain transfer function of input and output can be obtained by formulas (6) and (8) as:

$$R(z) = C(zI - A)^{-1}B = \frac{(k_{1n} - k_{2n})z - k_{1n} - k_{2n}}{z^2 - 2z\cos(\omega_n T) + 1} \tag{9}$$

Where: $k_{1n} = \cos\theta_n \sin(\omega_n T)$; $k_{2n} = \sin\theta_n [1 - \cos(\omega_n T)]$.

If directly discrete the formula (3) by step response invariant method can get the following formula (10):

$$R(z) = C(zI - A)^{-1}B = \frac{(k_{1n} - k_{2n})z - k_{1n} - k_{2n}}{z^2 - 2z\cos(\omega_n T) + 1} \cdot \frac{1}{\omega_n} \tag{10}$$

Compared discrete time transfer function (9) with (10), (9) expands ω_n times, this is because in formula (4), $1/\omega_n$ is treated as a factor of resonant coefficient K_r in order to facilitate the discussion of virtual LC discretization method.

From the above analysis and discussion, it is known that the discretization effect on discrete resonance controller is exactly the same by using virtual LC method and step response invariant method. However reference [14] pointed out that virtual LC method makes the transfer function of step response invariant method expand w_n times in number, which obviously ignores the influence on discrete-time transfer function of rewriting formula (4) into formula (5). Using the discrete-time transfer function of reference [10] to analyze the stability of the system is very desirable. Therefore, the method of reference [7–10] will be employed to the design of PR controller discretization.

4   Analysis of Control Strategy and Design of Parameters

4.1   Current Inner Loop

The transfer function of inverter unit can be deduced from Fig. 2, which is shown as follows:

$$G(s) = \frac{1}{LCs^2 + Crs + 1} \tag{11}$$

The specific parameters of the experimental prototype are as the following: $L = 680\ \mu H$, $C = 100\ \mu F$, $r = 0.2\ \Omega$, and sampling period $T = 0.1$ ms. Using the ZOH with the step response invariant method to discrete transfer function $G(s)$, which can be transformed to $G(z)$ in Z – domain.

$$G(z) = Z\left[\frac{1 - e^{-Ts}}{s} \cdot \frac{1}{LCs^2 + Crs + 1}\right] = \frac{0.1448z + 0.1419}{z^2 - 1.655z + 0.9412} \tag{12}$$

The design parameters of current inner loop's harmonic PR controller are based upon discrete method in formula (10). After a beat delay and LC filter, the 3rd, 5th and 7th harmonics phase angle lag respectively are about 42.5° 45° and 47.5, then $\theta_3 = 42.5°$, $\theta_5 = 45°$ and $\theta_7 = 47.5°$. Transfer functions of the harmonic resonant controller respectively are $R_3(z)$, $R_5(z)$ and $R_7(z)$ in the Z-domain as follows:

$$R_3(z) = \frac{0.0664z - 0.0724}{z^2 - 1.9912z + 1} \cdot \frac{1}{942.4778} \tag{13}$$

$$R_5(z) = \frac{0.1019z - 0.1193}{z^2 - 1.9754z + 1} \cdot \frac{1}{1570.8} \tag{14}$$

$$R_7(z) = \frac{0.1296z + 0.1652}{z^2 - 1.9518z + 1} \cdot \frac{1}{2199.1} \tag{15}$$

References [15, 16] brought forth the principle of priority to system stability, DVR control system has a good stability at $K_{ph} = 0.9$.

4.2   Voltage Outer Loop

Similar to the current inner loop, the parameters design of voltage outer loop adopt the virtual LC discrete method in formula (10). After a beat delay and LC filter, phase angle lag of fundamental wave is approximately $40°$, then can calculate $\theta_1 = 40°$. The transfer function $R_1(z)$ of fundamental resonant controller in the Z-domain is:

$$R_1(z) = \frac{0.0238z - 0.0244}{z^2 - 1.9990z + 1} \cdot \frac{1}{314.1593} \tag{16}$$

In accordance with the principle of priority to the stability, taking $K_{p1} = 1.4$, then transfer function $PR_1(z)$ of fundamental PR controller in the Z-domain is:

$$PR_1(z) = K_{p1} + R_1(z) = 1.4 + \frac{0.0238z - 0.0244}{z^2 - 1.9990z + 1} \cdot \frac{1}{314.1593} \tag{17}$$

If taking load current i_o as one kind of disturbance of the current inner loop, the transfer function of compensation voltage u_o is:

$$u_o = \frac{PR_1(z)G_h(z)}{1 + PR_1(z)G_h(z)} \cdot u_i + \frac{C(z)}{1 + PR_1(z)G_h(z)} \cdot i_o \tag{18}$$

Where: the second is the effect of interference value i_o in current inner loop on the compensation voltage u_o, $C(z)$ is the transfer function of capacitance integral link in Z-domain. Because the gains of fundamental PR controller t $PR_1(z)$ end to infinity in fundamental frequency, the second item in (18) tends to zero. Simultaneously, the first item in (18) basically equals to the given compensation voltage. Therefore, the fundamental PR controller can not only achieve zero steady-state errors, also inhibit the influence of load current on output.

5 Experimental Result

In addition to compensating voltage drops, DVR-prototype still can compensate voltage harmonics of system. One single experiment only with compensating harmonics and another comprehensive experiment with contemporaneously compensating voltage sags and harmonics are respectively studied on DVR-prototype.

Therefore, DVR prototype mainly aimed at the 3rd, 5th and 7th harmonic voltage higher in load voltage to be compensated. The non linear load was added into system when the work voltage is normal, its compensation effect on harmonic voltage is shown as Fig. 2. After DVR-prototype compensation, 3rd harmonic distortion rate reduced to 0.58% from 13.25%, 5th harmonic distortion rate reduced to 0.46% from 11.37%, and 7th harmonic distortion rate reduced to 0.75% from 7.82%. Three, five and seven times harmonic content are all qualified in scope, meanwhile THD also drops from 36.28% to 4.69%.

The non-linear load also was added into system, but when the voltage sags occur, the compensation effect on harmonic voltage shown as Fig. 3. At this point, the system output voltage is stabilized at 220 ± 0.7 V, the steady-state precision is 0.32%. Additionally, 3rd times harmonic distortion rate reduced to 0.93% from 13.25%, 5th harmonic distortion rate reduced to 0.57% from

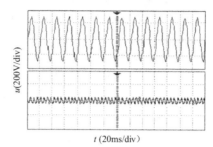

(a) System voltage and compensation voltage waveform

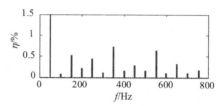

(b) System voltage frequency spectrum

Fig. 2. Compensation effect for voltage harmonics

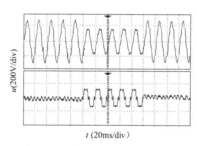

(a) System voltage and compensation voltage waveform

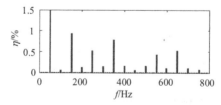

(b) System voltage frequency spectrum

Fig. 3. Compensation effect for voltage sag and harmonics

11.37%, and 7th times harmonic distortion rate reduced to 0.83% from 7.82%. As the paper can see that the 3rd, 5th and 7th harmonic content is greatly reduced. Therefore, the control strategy put forward in this article has favorable compensation effect on the voltage drop and harmonic voltage in steady-state.

6   Conclusion

The paper proposes a new-type control strategy to make DVR inverter unit to compensate the low-order harmonic waves caused by nonlinear factors, like dead time and nonlinear load, and the effect of digital control on system performance, while ensures the dynamic response characteristic of DVR inverter unit. The thesis analyzes the discretization method of RP controller in current literature, designs in a detailed way the parameters of voltage outer-loop fundamental PR controller and current inner loop third, fifth and seventh harmonic PR controller in discrete filter. In the process of analyzing the performance of control system and designing parameters, the paper introduces one time-delay, and therefore achieves the control performance and harmonic compensation effect closer to actual DVR. The harmonic PR controller can effectively restrain the influence of the specific harmonic current in load current on DVR output voltage. The fundamental PR controller can not only realize the zero steady -error, but also restrain the effect of load current on output. The experiment of DVR prototype proves the efficiency of the control strategy in this paper.

References

1. Shi, W., Tang, G., Li, J.: Double feed-forward control strategy for dynamic voltage restorer. Electric Power Automation Equipment 27(2), 11–15 (2007)
2. Li, Y., Vilathgamuwa, D.M., Lohp, C.: Design and comparison of high performance stationary-frame controllers of DVR implementation. IEEE Transactions on Power Electronics 22(7), 602–612 (2007)
3. Shen, K., Wang, J., Cai, X., et al.: Proportional-resonant control for dynamic voltage restorer. Electric Power Automation Equipment 30(5), 65–69 (2010)
4. Li, Y., Vilathgamuwa, D.M.: Investigation and improvement of transient response of DVR at medium voltage level. IEEE Transactions on Industry Applications 43(3), 1309–1319 (2007)
5. Sanchez, P.R., Acha, E., Calderon, J.E.O.: A versatile control scheme for a dynamic voltage restorer for power-quality improvement. IEEE Transactions on Power Delivery 24(1), 277–283 (2009)
6. Shi, W., Tang, G., Li, J.: Double feed-forward control strategy for dynamic voltage restorer. Electric Power Automation Equipment 27(2), 11–15 (2007)
7. Wang, T., Xue, Y., Choi, S.S.: Review of dynamic voltage restorer. Automation of Electric Power Systems 31(9), 101–107 (2007)
8. Liu, C., Ma, W., Sun, C., et al.: Design of Output LC Filter and Low Order Harmonics Suppression in High Power 400Hz Inverter. Transactions of China Electrotechnical Society 26(6), 129–136 (2011)

9. Ghosh, A., Jindal, A.K., Joshi, A.: Design of a capacitor-supported dynamic voltage restorer(DVR) for unbalanced and distorted loads. IEEE Transactions on Power Delivery 19(1), 405–413 (2004)
10. Kim, H., Lee, S.-J., Sul, S.-K.: A calculation for the compensation voltages in dynamic voltage restorers by use of PQR power theory. In: Nineteenth Annual IEEE Applied Power Electronics Conference and Exposition, APEC 2004, vol. 1(11), pp. 573–579 (2004)
11. Mihalache, L.: DSP control of 400 Hz inverters for aircraft applications. In: Industry Applications Annual Meeting, Jacksonville, Florida, USA (2002)
12. Mihalache, L.: Improved load disturbance rejection method for 400 Hz GPU inverters. In: Applied Power Electronics Conference and Exposition, Anaheim, California, USA (2004)
13. Sato, Y., Ishizuka, T., Nezu, K., et al.: A new control strategy for voltage-type PWM rectifiers to realize zero steady-state control error in input current. IEEE Transactions on Industry Applications 34(3), 480–486 (1998)
14. Li, Z., Wang, P., Li, Y., et al.: 400Hz High-Power Voltage source Inverter With Digital Control. Proceeding of the CSEE 29(6), 36–42 (2009)
15. Bojoi, R.I., Griva, G., Bostan, V., et al.: Current control strategy for power conditioners using sinusoidal signal integrators in synchronous reference frame. IEEE Transactions on Power Electronics 20(6), 1402–1412 (2005)
16. Liserre, M., Teodorescu, R., Blaabjerg: Double harmonic control for three-phase systems with the use of resonant current controllers in a rotating frame. IEEE Transactions on Power Electronics 21(3), 836–841 (2006)

Research of IMC-Fed BLDCM Servo System Based on Fuzzy Control

Hongchen Liu, Lei Liu, and Lishan Sun

School of Electrical Engineering and Automation, Harbin Institute of Technology,
Harbin, HLJ 150001, P.R. China
{fenmiao,sunlishan}@hit.edu.cn, liu515712992@yahoo.com.cn

Abstract. This paper presents a control scheme designed for Brushless DC Motor (BLDCM) drive. The proposed scheme is fed by Indirect Matrix Converter (IMC). The IMC works as an AC-AC power supply for BLDCM. The scheme includes speed servo system and position servo system. Firstly, the advantages and working principles of IMC are introduced, showing that IMC is a superior power converter. Then, the IMC-fed BLDCM servo system is designed. The command speed or command position for BLDCM could be tracked accurately and respectively. In the position servo system, a fuzzy P controller is employed as the position regulator. From the simulation results, it is illustrated that IMC-fed BLDCM servo system has satisfied performance of tracking the command speed or command position.

Keywords: IMC, BLDCM, servo, fuzzy P controller.

1 Introduction

The concept of Two Stage Matrix Converter, which is also called Indirect Matrix Converter (IMC), was first proposed in 2001 [1]. IMC contains rectifier stage and inverter stage. It has the similar configuration compared with traditional frequency converter except that the IMC's rectifier stage contains six bidirectional switches and there is no energy storage element. We can utilize the matured control strategy in IMC. Therefore, IMC is a promising matrix converter in the future.Most papers on MC-fed variable speed drive system are about IM drive. In paper [2], a vector-controlled closed-loop for MC-fed IM drive is proposed in electric vehicle applications. In paper [3], direct torque control of six-phase IM drive based on MC is proposed. Some papers focus on the improvement of MC-fed IM drive. Paper [4] presented predictive torque control for MC-fed IM drive. Papers on MC-fed PMSM drive system could also be seen. Paper [5] presented MC-fed PMSM drive system using vector control technique. Besides, more advanced control strategies were used in MC-fed PMSM drive system, such as speed anti-windup PI strategies [6], nonlinear adaptive back-stepping controller [7], sliding-mode controller [8]. Paper [9] designed a sensorless scheme for MC-fed PMSM drive system. In addition, MC could also be applied in three-phase utility power supply [10], PMSG wind turbine generation [11], unified power-flow

K. Li et al. (Eds.): ICSEE 2013, CCIS 355, pp. 19–27, 2013.

controller [12], dynamic voltage restorer [13] and so on. Paper [14] presented a new modulation scheme for three-level MC. Paper [15] presented the experimental evaluation of three-level MC. From the researches above, three-level and multi-level MCs produce less harmonic component and the output voltage waveform is much closer to standard sinusoidal wave. BLDCM is also called Trapezoidal Permanent Magnet Synchronous Motor, which has the advantages of simple structure, convenient operation, high power density and high efficiency. It is widely used in AC drive area. In this paper, a BLDCM drive system fed by IMC is proposed. The IMC is a superior AC-AC converter while BLDCM is a commonly used motor. By combination of the two, we could overlap the advantages of them.

2   IMC-FED BLDCM DRIVE

2.1   BLDCM Drive

The BLDCM is a kind of permanent magnet synchronous motor. Because of its simple structure, good performance and high reliability, BLDCM is widely used in servo system. Different from sinusoidal permanent magnet synchronous motor, the air-gap magnetic field of BLDCM is square-wave. The back EMFs of BLDCM are trapezoid-wave and the phase current is square-wave.

The mathematical equations of BLDCM are

$$V_{a,b,c} = Ri_{a,b,c} + (L - M)\mathrm{d}i_{a,b,c}/\mathrm{d}_t + E_{a,b,c}. \tag{1}$$

The mechanical equations of BLDCM are

$$E_a i_a + E_b i_b + E_c i_c$$

$$J\frac{\mathrm{d}\omega}{\mathrm{d}t} = T_e - B\omega - T_L$$

$$d\theta_r/dt = P\omega/2. \tag{2}$$

In this paper, 120 electrical degrees control method is proper for BLDCM drive. The IMC's output DC voltage applies a voltage during 120 electrical degrees to the motor inductances. This is similar to conventional converter-fed BLDCM drive system. The desired stator current is shown in Figure 1.

We know that the electromagnetic torque of BLDCM is caused by the interaction between stator winding current and rotor magnetic field [16]. Electromagnetic torque is proportional to the amplitude of stator currents. In order to produce constant electromagnetic torque, square-waves for stator currents and trapezoid-waves for back EMFs are required. Just like Figrue 1, in each half cycle, square-wave stator currents last for 120 electrical degrees and the flat top of the trapezoid-wave back EMFs are all 120 electrical degrees as well. The square-wave stator currents and the trapezoid-wave back EMFs must be synchronous. At any time, only two phases are conducted.

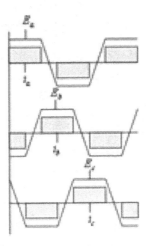

Fig. 1. The EMFs and stator currents

2.2 Modulation for IMC

MCs can be divided into two types- Direct Matrix Converter (DMC) and Indirect Matrix Converter (IMC). In this paper, IMC is used. IMC is divided into rectifier stage and inverter stage. The rectifier stage is composed of six bi directional switches and the inverter stage is the same with traditional AC-DC-AC power converter.In this paper, BLDCM is fed by IMC. As is shown in Figure 2.

$$d_{1,2} = -u_{b,c}/u_a = -\cos\theta_{b,c}/\cos\theta_a. \tag{3}$$

Under balanced input conditions , the mean value of DC voltage in a PWM period is

$$u_{\mathrm{dc}} = d_{\mathrm{ab}}u_{\mathrm{ab}} + d_{\mathrm{ac}}u_{\mathrm{ac}} = 3U_{\mathrm{im}}/2|\cos\theta_{\mathrm{a}}| \tag{4}$$

The other five sectors are similar to sector 1. The mean value of DC voltage is

$$u_{\mathrm{dc}} = 3U_{im}/2|\cos\theta_{in|} \tag{5}$$

3 Closed-Loop Control System for IMC-Fed BLDCM Drive

The rectifier stage uses PWM modulation while the inverter stage provides with 3-phase symmetry square wave current with the width of 120 electrical degrees. There are two feedback loops in the system. The inner loop is called speed loop while the outer loop is called position loop. Firstly, the actual position is compared with the command position and the error between the two is regulated

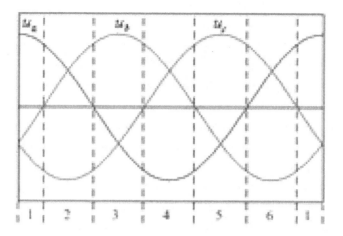

Fig. 2. The division of input voltage sectors

through a position controller. Then the command speed is produced. Secondly, the actual speed is compared with the command speed and the error between the two is regulated through a speed controller. Then the need voltage for BLDCM is produced. In position servo system, an additional controller for position loop is added. Usually, position overshoot is not allowed in position servo systems. In order to avoid overshoot phenomenon, a simple P controller is employed instead of a PI controller in position loop. However, if the P parameter is very large, overshoot phenomenon also occurs. Therefore, a appropriate P value is chosen to guarantee servo accuracy and absence of overshoot. In traditional P controller, the P parameter is fixed and can't be modified. If P is chosen too large, overshoot phenomenon would appear. Oppositely, if P is chosen too small, the response speed would be quite slow, though without overshoot phenomenon. Fuzzy P controller is one that modifies the P parameter according to the error and error-in-change based on fuzzy theory. In the start time, when the error is large, the P parameter is large to get rapid response. When the error becomes small, P parameter value is reduced to avoid overshoot phenomenon. Therefore, fuzzy P is a flexible controller. The modified P parameter is calculated as

$$K_p = K_{p0} + \Delta K_p \, (K_{p0} = 200)$$

4   Simulation Results

The simulation parameters are as follows. 3-phase input: 220V, 50Hz Stator resistance: 4.765 Stator inductances: 0.0085H Flux induced by magnets: 0.1848Wb Inertia: 0.0001051 Friction: 0.00004047 Pairs of poles: 2 Suppose the command speed was 2000r/min at 0s and changed to 3000r/min at 0.05s. The load torque was 1Nm all the time. The simulation results are as follows. From Figure 4(a), it

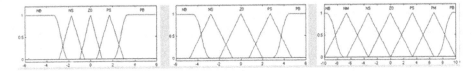

Fig. 3. The membership functions

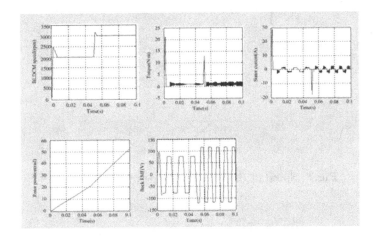

Fig. 4. Simulation results of command speed mutation

could be concluded that IMC-fed BLDCM system can track the command speed signal. Figure 4(b) is the torque waveform of BLDCM. When BLDCM operates at steady stage, the torque is about 1Nm. Figure 4(c) shows the 120 electrical degrees of stator current. Figure 4(d) is the rotor position. When BLDCM was accelerated, the rotor position changed faster. Figure 4(e) is the back EMF.To verify speed servo performance further, suppose the command speed is a liner slope signal. It declined from 2000r/min at 0s to 0r/min at 0.2s. Figure 5 illustrated that IMC-fed BLDCM speed servo system could track command speed at any time with good performance. Suppose the command speed wais 2000r/min all the time. The load torque was 1Nm at 0s and changed to 3Nm at 0.05s. The simulation results are in Figure 6.Figure 6(a) shows that when given sudden load torque, BLDCM still remained the command speed, while the load torque changed from 1Nm to 3Nm, as is shown in Figure 6(b). At the same time, the stator current became larger in order to provide larger power. Suppose the command position was 20rad at 0s and changed to 40rad at 0.2s.The simulation results are in Figure 5. Figure 7 showed that IMC-fed BLDCM position servo system could track the command position signal. When fuzzy P controller is employed in position loop, the performance is improved. IMC-fed BLDCM position servo system with fuzzy P controller could track the command position faster than that with traditional P controller. Before reach the command position,

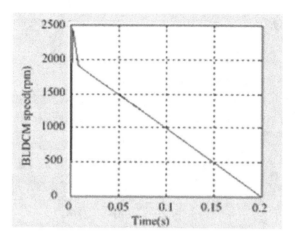

Fig. 5. Speed of BLDCM with slope command speed

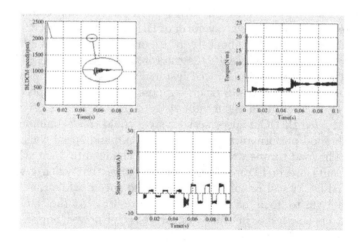

Fig. 6. Simulation results of load torque mutation

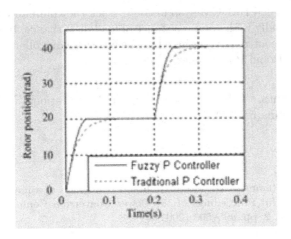

Fig. 7. Simulation results of position servo system

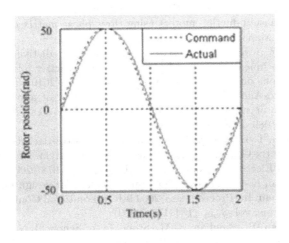

Fig. 8. Rator position of BLDCM with sinusoidal command position

BLDCM speed with fuzzy P controller rose faster than that with traditional P controller, both of them without overshoot phenomenon at the end. In order to verify the speed servo performance further, suppose the command position signal is a sinusoidal one. The simulation result is shown in Figure 8. Figure 8 shows that IMC-fed BLDCM position servo system could track the command position signal at any time with good performance.

5   Conclusions

From the simulation results, it could be concluded that IMC-fed BLDCM drive system is practical. In both speed and position servo systems, the performances are well. The IMC-fed BLDCM system could track the command speed or command position rapidly with good resistance of load disturbance. In position servo system, a fuzzy P controller is designed as the position controller which performs better than traditional P controller.

Acknowledgments. This work was supported by National Natural Science Foundation of China(51107016).

References

1. Zwimpfer, P., Stemmler, H.: Modulation and realization of a novel two-stage matrix converter. In: Proceedings of the Power Conversion Conference Electronics Conference, vol. 2, pp. 495–500 (2001)
2. Podlesak Thomas, F., Katsis Dimosthenis, C., Wheeler Patrick, W., Clare Jon, C.: Identification of Common Molecular Subsequences. IEEE Transactions on Industry Applications 40, 841–847 (2005)
3. Talaeizadeh, V., Kianinezhad, R., Seyfossadat, S.G., Shayanfar, H.A.: Direct torque control of six-phase induction motors using three-phase matrix converter. Energy Conversion and Management 51, 2482–2491 (2010)
4. René, V., Ulrich, A., Boris, H., Jose, R., Patrick, W.: Predictive torque control of an induction machine fed by a matrix converter with reactive input power control. IEEE Transactions on Power Electronics 25, 1426–1438 (2010)
5. Said, B., Gerard-Andre, C., Michel, P.: Vector control of a permanent-magnet synchronous motor using ac-ac matrix converter. IEEE Transactions on Power Electronics 13, 1089–1099 (2010)
6. Chen, D.-F., Liu, T.-H., Hung, C.-K.: Nonlinear Adaptive-Backstepping Controller Design for a Matrix-Converter Based PMSM Control System. In: IECON Proceedings (Industrial Electronics Conference), vol. 1, pp. 673–678 (2003)
7. Chen, D.-F., Yao, K.-C.: IA novel sliding-mode controller design for a matrix converter drive system. In: Proceedings - 2009 9th International Conference on Hybrid Intelligent Systems, vol. 3, pp. 134–137 (2009)
8. Snary, P., Bhangu, B., Bingham, C.M., Stone, D.A., Schofield, N.: Matrix converters for sensorless control of PMSMs and other auxiliaries on deep-sea ROVs. IEE Proceedings: Electric Power Applications 152, 382–392 (2005)
9. Dimosthenis, K., Patrick, W., Jon, C., Pericle, Z.: A three-phase utility power supply based on the matrix converter. In: Conference Record of the 2004 IEEE Industry Applications Conference, vol. 3, pp. 1447–1451 (2004)
10. Yang, G., Zhu, Y.: Application of a matrix converter for PMSG wind turbine generation system. In: 2nd International Symposium on Power Electronics for Distributed Generation Systems, vol. 4, pp. 185–189 (2010)
11. Monteiro, J., Silva, J.F., Pinto, S.F., Palma, J.: Matrix converter-based unified power-flow controllers. IEEE Transactions on Power Delivery 26, 420–430 (2010)
12. Bingsen, W., Giri, V.: Dynamic voltage restorer utilizing a matrix converter and flywheel energy storage. IEEE Transactions on Industry Applications 45, 222–231 (2009)
13. Yeong, L.M., Patrick, W., Christian, K.: A new modulation method for the three-level-output-stage matrix converter. In: Fourth Power Conversion Conference-NAGOYA, vol. 14, pp. 776–783 (2007)

14. Lee, M.Y., Klumpner, C., Wheeler, P.: Experimental evaluation of the indirect three-level sparse matrix converter. In: 4th IET International Conference on Power Electronics, Machines and Drives, vol. 14, pp. 50–54 (2008)
15. Mirtalaei, S.M.M., Moghani, J.S., Malekian, K., Abdi, B.: A novel sensorless control strategy for BLDC motor drives using a fuzzy logic-based neural network observer. In: International Symposium on Power Electronics, Electrical Drives, vol. 18, pp. 1491–1496 (2008)
16. Espina, J., Arias, A., Balcells, J., Ortega, C., Galceran, S.: Speed anti-windup PI strategies review for field oriented control of permanent magnet synchronous machines servo drives with matrix converters. In: 2009 13th European Conference on Power Electronics and Applications, pp. 279–285. EPE (2001)

Modulation Strategy to Reduce Common-Mode Voltage for Three-to-Five Phase Indirect Matrix Converter

Hongchen Liu, Hailong Yu, and Lishan Sun

School of Electrical Engineering and Automation, Harbin Institute of Technology,
Harbin, HLJ 150001, P.R. China
{fenmiao,sunlishan}@hit.edu.cn, yu_haleo@163.com

Abstract. Based on the analysis of the space vector pulse modulation (SVPWM) using four nearest space vectors, this paper put forward the improved algorithm to reduce the common-mode voltage of the five phase indirect matrix converter. The algorithm can effectively reduce the common- mode voltage by resetting the zero vectors positions. When the modulation ratio is between 0 and 1.0514, harmonic in output voltage could be restricted to zero by eliminating the voltage vectors in harmonic space. Simulation results are shown to verify the effectiveness of the proposed methods.

Keywords: multi-phase, TSMC, common-mode voltage.

1 Introduction

Recently, multi-phase drive systems have received much more attention due to their potential value in high power drives. Compared with traditional three-phase machines, the multi-phase machines have some advantages such as the reduced amplitude, the increased frequency of torque pulsations, the reduction of the stator current per phase, the improved torque per ampere, power density of the electric machine, and fault tolerance. Detail reviews on the development in the area of multi-phase motor drive are presented in [1]-[5]. Since multi-phase drive system has gained popularity, a need is felt to develop power electronic converter to supply such multi-phase systems.

Matrix converter is a direct ac-ac power converter employing bidirectional switches. In addition to its basic capacity to provide variable sinusoidal voltages to the load, a matrix converter has several attractive features: bidirectional power flow, high reliability and long life due to the absence of the capacitor, unity input power factor at the power supply side [6]-[8] .The two-stage matrix converter (TSMC) is a novel matrix converter composing of rectifier stage and inverter stage. This technology provides simplified switch commutation requirements and modulation strategy. However, the two-stage matrix converter will produce the common-mode voltage (CMV)at work. Leakage current that flows through a parasitic capacitor between stator core and stator winding can reduce the life

K. Li et al. (Eds.): ICSEE 2013, CCIS 355, pp. 28–36, 2013.

time of winding insulation, break the motor insulation easily and cause wideband electromagnetic interference. Thus, the techniques to reduce the CMV or to limit it within certain bounds are of crucial importance for adjustable speed drive system which is fed by two-level or multilevel inverter.

To make sure the TSMC get maximal DC voltage utilization and minimal output voltage harmonics, a novel space vector pulse width modulation(SVPWM) algorithm using four nearest space vectors is first discussed. A suitable SVPWM algorithm is then developed to reduce the common-mode voltage. The proposed modulation techniques are further supported by simulation.

2 Three-to-Five Phase TSMC Model and Its Inverter Stage Space Voltage Vector

Fig.1 illustrates the modified matrix converter topology presented in this paper. The equivalent transformation matrix from the five-phase variables to the synchronously rotating $d_1 - q_1 - d_3 - q_3$ axes variables can be defined as

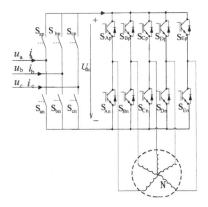

Fig. 1. Three-to-five phase TSMC model

$$T(\theta) = \begin{bmatrix} \cos(\theta_0) & \cos(\theta_1) & \cos(\theta_2) & \cos(\theta_3) & \cos(\theta_4) \\ -\sin(\theta_0) & -\sin(\theta_1) & -\sin(\theta_2) & -\sin(\theta_3) & -\sin(\theta_4) \\ \cos(3\theta_0) & \cos(3\theta_1) & \cos(3\theta_2) & \cos(3\theta_3) & \cos(3\theta_4) \\ -\sin(3\theta_0) & -\sin(3\theta_1) & -\sin(3\theta_2) & -\sin(3\theta_3) & -\sin(3\theta_4) \\ \frac{1}{2} & \frac{1}{2} & \frac{1}{2} & \frac{1}{2} & \frac{1}{2} \end{bmatrix} \quad (1)$$

where $\theta_i = \theta_r - i\alpha$; θ_r denotes the rotor angle; $\alpha = \frac{2\pi}{5}$. Since a five-phase IMC is under consideration, one deals here with a five-dimensional space. Hence two space vectors have to be defined, each of which will describe space vectors in one two-dimensional subspace (and)[9]. The third subspace is single-dimensional and it canoe be excited due to assumed star connection of the system. Space vectors of phase voltages are defined as:

$$V_{d1-q1} = \frac{2}{5}(V_a + \lambda V_b + \lambda^2 V_c + \lambda^3 V_d + \lambda^4 V_e) \tag{2}$$

$$V_{d3-q3} = \frac{2}{5}(V_a + \lambda V_c + \lambda^2 V_e + \lambda^3 V_b + \lambda^4 V_d) \tag{3}$$

In Eqs. (2) and (3): $\lambda = \exp(j2\pi/5)$; $V_a \sim V_e$ are the output voltage vectors in inverter stage.

In general, an n-phase VSI has a total of 2^n space vectors. Thus in case of a five-phase IMC, there are 32 space vectors in the inverter stage, two of which are zero vectors. The remaining 30 active vectors which are easily calculated using (2) and (3) in conjunction with inverter output phase voltages for each possible state form three decagons in both $d_1 - q_1$ and $d_3 - q_3$ planes. Fig.2 shows the space vectors in $d_1 - q_1$ and $d_3 - q_3$ plane.

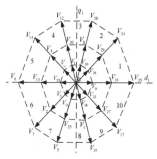

(a) Voltage space vectors in $d_1 - q_1 - 0$ plane

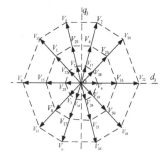

(b) Voltage space vectors in $d_3 - q_3 - 0$ plane

Fig. 2. Voltage space vectors distribution of three-to-five phase IMC

3   The Space Vector Pulse Width Modulation of IMC

It is assumed that, on the input side

$$u_a = U_{im} \cos\left(w_i t\right)$$
$$u_b = U_{im} \cos\left(w_i t - 120°\right)$$
$$u_c = U_{im} \cos\left(w_i t + 120°\right)$$

$$(4)$$

and on the output side:

$$u_A = U_{om} \cos\left(w_o t\right)$$
$$u_B = U_{om} \cos\left(w_o t - 2\pi/5\right)$$
$$u_C = U_{om} \cos\left(w_o t - 4\pi/5\right)$$
$$u_D = U_{om} \cos\left(w_o t - 6\pi/5\right)$$
$$u_E = U_{om} \cos\left(w_o t - 8\pi/5\right)$$

$$(5)$$

U_{im}, U_{om}: amplitudes of input voltage, output voltage respectively. w_i, w_o are the input and output angular frequencies.

3.1   PWM Method for the Rectifier Stage

The aim of the pulse width modulation of the rectifier is to maintain the maximum positive voltage in the dc side as well as to maintain the input power factor as unity. Since the input voltages are balanced, there are two possible conditions for the input phase voltages to separate six sectors. First, two voltages are positive, and one is negative. Second, two voltages are negative, and one is positive [10]. Finally, the average value of the dc voltage during each switch interval is

$$U_{dc} = \frac{3 \cdot U_{im}}{2 \cdot |\cos\left(\theta_{in}\right)|}$$

$$(6)$$

3.2   SVPWM Method for the Inverter Stage

The application of only large space vectors does not produce satisfactory results in terms of the harmonic content of the output phase voltage. When reference voltages are sinusoidal, the space vector reference contains only $d_1 - q_1$ components, $d_3 - q_3$ components are zero. However, if only large vectors are used to create the desired $d_1 - q_1$ voltages, $d_3 - q_3$ voltage components will inevitably be created as well. Each switching combination that gives a large vector in the $d_1 - q_1$ plane gives simultaneously a small vector in the $d_3 - q_3$ plane. Thus, the harmonic in output voltage could be restricted to zero by eliminating the voltage vectors in harmonic space when the output voltage is composed of medium and large vectors[11]. Fig.3 shows the active and zero space vectors in sector 1. Set the switch function of matrix converter for inverter side $S = [S_A, S_B, S_C, S_D, S_E]$. If $S_B = 1$, it denotes that the upper switch of leg B is ON state, **otherwise**, $S_B = 0$ indicates the lower switch is ON state. Assuming the reference output voltage space vector V_{ref} is in the sector 1, two neighboring medium and two

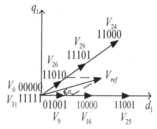

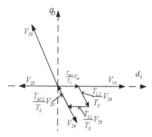

(a) Synthesis the reference output voltage vector in sector 1 for $d_1 - q_1$ plane

(b) The created output voltage vector in sector 1 for $d_3 - q_3$ plane

Fig. 3. Synthesis the reference output voltage vector in sector 1

large active vectors are used to synthesize the reference output voltage. The duty ratio of active and zero vectors is[12]

$$d_{11001} = \frac{m}{5\sin(\pi/5)} \left(\frac{\mu}{1+\mu^2}\right) \sin\left(\frac{\pi}{5} - \alpha_0\right)$$
$$d_{11000} = \frac{m}{5\sin(\pi/5)} \left(\frac{\mu}{1+\mu^2}\right) \sin\left(\alpha_0\right)$$
$$d_{10000} = \frac{m}{5\sin(\pi/5)} \left(\frac{\mu}{1+\mu^2}\right) \sin\left(\frac{\pi}{5} - \alpha_0\right) \qquad (7)$$
$$d_{11101} = \frac{m}{5\sin(\pi/5)} \left(\frac{\mu}{1+\mu^2}\right) \sin\left(\alpha_0\right)$$
$$d_{11111} = d_{00000} = \frac{1}{2}(1 - d_{11001} - d_{11000} - d_{10000} - d_{11101})$$

where $\mu = 2 \cdot (\pi/5)$. m is the modulation index of the inverter stage, $m = \frac{2V_{om}}{V_{dc}}$, the maximum modulation index in the inverter stage is

$$m_{max} = \frac{1}{\cos\left(\frac{\pi}{10}\right)} = 1.0514 \qquad (8)$$

3.3   Modulation Strategy to Reduce CMV for Three-to-Five Phase IMC

In a five phase motor driving system, there are inevitable voltage between the load and the earth. These had previously been ignored voltage is called common mode voltage u_{NO}, as shown in Fig.4. Obviously has the following relationship

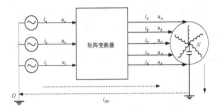

Fig. 4. The common mode voltage of system

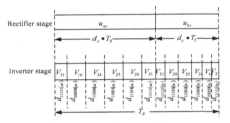

(a) SVPWM sequence before improvement

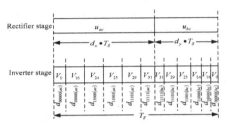

(b) The improved SVPWM sequence

Fig. 5. SVPWM sequence

$$\begin{cases} u_{AO} = u_{AN} + u_{NO} \\ u_{BO} = u_{BN} + u_{NO} \\ u_{CO} = u_{CN} + u_{NO} \\ u_{DO} = u_{DN} + u_{NO} \\ u_{EO} = u_{EN} + u_{NO} \end{cases} \tag{9}$$

From(9), the common mode voltage is

$$u_{NO} = \frac{1}{5}\left(u_{AO} + u_{BO} + u_{CO} + u_{DO} + u_{EO}\right) \tag{10}$$

Assuming $U_{im} = 100V$, the input power factor is unity. When the input reference current is in sector 2, the output reference voltage is in sector 1, from (10) it can be found that, the maximal common-mode voltage $-U_{im}$ appears in zero vector function of the moment. When the output voltage in other sectors it has the same conclusion. Hence, the maximal CMV is U_{im} when the input current sector is odd number; the maximal CMV is $-U_{im}$ when the input current sector is even number. A proper selection of zero space vectors in the inverter stage results in a reduction in the peak of CMV[13]-[14]. For example, the zero space vector V_0 is chosen when the input current sector is odd number and the zero space vector V_{31} is chosen when the input current sector is even number. The improved SVPWM pulse width modulation sequence is shown in Fig.5.

4   Simulation

Through the above analysis, simulations are carried out for a three-to-five phase R lode using Matlab software. The simulation parameters are as follows:

Input power supply: line-to-line voltage $U_{im} = 100V$, $f = 50Hz$. The switching frequency of rectifier stage: $f = 5kHz$. The desired output line-to-line voltage

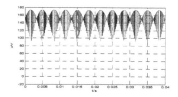

Fig. 6. The waveform of rectifier stage

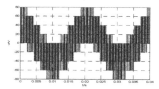

(a) The waveform of five-phase VSI

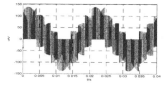

(b) The phase voltage of five-phase IMC

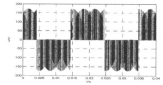

(c) The line-to-line voltage of five-phase IMC

Fig. 7. The waveform of output voltage for VSI and IMC

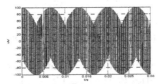

(a) The waveform of CMV before improvement

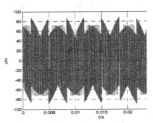

(b) The waveform of CMV after improvement

Fig. 8. The waveform of common-mode voltage

of inverter stage: 60V, $f = 50$Hz. The switching frequency of inverter stage: $f = 10$Hz. Five phase R load: $R = 10\Omega$.

Fig.6 shows the simulation results of the rectifier stage. Since using the line-to-line input voltage to synthesize the dc voltage, the output voltage waveform of three-to-five phase IMC is different from the waveform of five-phase VSI, which are shown in Fig.7.

Fig.8 shows the CMV waveforms with the conventional method and the improved method. It can be seen that the peak value of CMV before improvement is 100V. However, the peak value of CMV using improved method in inverter stage is 86.6V, the common-mode voltage reduced significantly.

5   Conclusion

In this paper, a novel space vector pulse modulation algorithm using four nearest space vectors for three-to-five phase IMC is proposed. It provides the good performance to validate the advantages of indirect matrix converter. However, the common-mode voltage is unwanted for the proposed modulation. So the relationship between the CMV and the switching states of IMC is clarified. By suitable arranging the zero space vectors of the inverter stage, it successfully reduces the CMV of IMC based on the zero dc-link current commutation. The simulations are carried out to verify the theoretical analysis.

Acknowledgments. This work was supported by National Natural Science Foundation of China(51107016).

References

1. Singh, G.K.: Multi-phase induction machine drive research-a survey. Electric Power System Research 61, 139–147 (2002)
2. Jones, M., Levi, E.: A literature survey of state-of-the-art in multiphase ac drives. In: Proc. 37th Int. Universities Power Eng. Conf. UPEC, Stafford, UK, pp. 505–510 (2002)
3. Bojoi, R., Farina, F., Profumo, F., Tenconi: Dual three induction machine drives control-A survey. IEEJ Tran. on Ind. Appl. 126(4), 420–429 (2006)
4. Levi, E., Bojoi, R., Profumo, F., Toliyat, H.A., Williamson, S.: Multi-phase induction motor drives-A technology status review. IET Elect. Power Appl. 1(4), 489–516 (2007)
5. Levi, E.: Guest editorial. IEEE Trans. Ind. Electronics 55(5), 1891–1892 (2008)
6. Alesina, A., Venturini, M.G.B.: Analysis and design of optimum-amplitude nine-switch direct AC-AC converters. IEEE Trans. Power Electron. 4(1), 101–112 (1989)
7. Wheeler, P.W., Rodriguez, J., Clare, J.C., Empringham, L., Weinstein, A.: Matrix converters: a technology review. IEEE Trans. Ind. Electron. 49(2), 276–288 (2002)
8. Klumpner, C., Nielsen, P., Boldea, I., Blaabjerg, F.: A new matrix converter motor (MCM) for industry applications. IEEE Trans. Ind. Electron. 49(2), 325–335 (2002)
9. Ryu, H.M., Kim, J.W., Sul, S.K.: Synchronous frame current control of multi-phase synchronous motor: Part I: Modeling and current control based on multiple d-q spaces concept under balanced condition. In: 39th IAS Annual Meeting, pp. 56–63. IEEE Press, United States (2004)
10. Wei, L., Lipo, T.A.: A novel matrix converter topology with simple commutation. In: Thirty-Sixth IAS Annual Meeting Conference Record of the 2001 IEEE Industry Applications Conference, September 30-October 4, vol. 3, pp. 1749–1754 (2001)
11. Iqbal, A., Levi, E.: Space vector modulation schemes for a five-phase voltage source inverter. In: European Conference on Power Electronics and Applications, pp. 1–12 (2005)
12. Nguyen, T.D., Lee, H.-H.: Carrier-based PWM Technique for Three-to-Five Phase Indirect Matrix Converter. IEEE (2011)
13. Chen, L.H., You, C.X., Yong, F.: A research on common-mode voltage for matrix converter based on two line voltage synthesis. Proceedings of the CSEE 24(12) (December 2004)
14. Chen, L.H., You, C.X.: The zero output commutation strategy for reducing common-mode voltage of matrix converter. Proceedings of the CSEE 25(3) (February 2005)

A Novel Three-Electrode Solid Electrolyte Hydrogen Gas Sensor

Min Zhu[1], Chunling Yang[1], Yan Zhang[1,*], and Zheng Jia[2]

[1] School of Computer Science and Technology, Harbin Institute of Technology,
Harbin, China
[2] School of Chemical Engineering and Technology, Harbin Institute of Technology,
Harbin, China
zyhit@hit.edu.cn

Abstract. A three-electrode solid electrolyte hydrogen gas sensor is explored in this paper. The sensor utilized phosphotungstic acid as the electrolyte material and adopted platinum, nickel and tungsten as the three-electrode materials respectively. In real applications, platinum was used as the measuring electrode, nickel was used as the adjusting electrode and tungsten was used as the reference electrode. In order to compare the performance of the new sensor with that of the traditional two-electrode sensor, the hydrogen concentrations were adjusted so as to detect the output of the two-electrode sensor and the three-electrode sensor. The dynamic range between the measuring electrode and the reference electrode is about 0.65 V and the highest detectable limit is 12% for the three-electrode solid hydrogen gas sensor. While the dynamic range is about 0.25 V and and the highest detectable limit is 1% for the two-electrode solid electrolyte gas sensor. The results demonstrate that the three-electrode solid hydrogen gas sensor has a higher resolution and detectable limit than the two-electrode sensor. *abstract* environment.

Keywords: hydrogen gas sensor, solid electrolyte, measuring electrode, reference electrode, adjusting electrode.

1 Introduction

Solid electrolyte hydrogen gas sensor is a newly developed technology. The electromotive force E between the two electrodes follows the Nernst Law theoretically when detecting the hydrogen concentrations. The solid electrolyte sensor change the traditional design method by taking new mechanism and can work in normal temperature. The solid electrolyte sensor can also reduce power consumption, avoid invalidation caused by high temperature and simultaneously possess the excellent gas selectivity. Due to the above advantages, solid electrolyte hydrogen gas sensor is of considerable interest these years.

At present, the solid electrolyte hydrogen gas sensor can mainly be divided into two types. one is the amperometric sensor which detects the hydrogen concentration by measuring the current and the other is the potentiometric sensor [1]

---

* Corresponding author.

K. Li et al. (Eds.): ICSEE 2013, CCIS 355, pp. 37–45, 2013.

which detects the hydrogen concentration by measuring the electric potential. The amperometric solid electrolyte hydrogen gas sensor need to pump the standard concentration oxygen to the reference electrode as the reference gas [2]. The amperometric sensor has the disadvantages of large volume and complex structure. In addition, the redox reaction between the two electrodes can increase the temperature of the generated substance, which in turn make the electrolyte dehydration. Therefore, the performance of this kind of sensor is not stable and the service life become shorter. The further problem is that the long recovery time of the reaction between the two electrodes will make the sensor impossible to use. The potentiometric sensor can also be divided into two kinds. one kind of potentiometric sensor adopted the same substance as its electrodes material such as platinum (Pt) and used Nafion [3] as the electrolyte. This kind of potentiometric sensor have a complex structure and require the reference gas. The other kind of potentiometric sensor employed different substance to be its electrode materials [4–7] such as using Pt for the measuring electrode and tungsten (W) for the reference electrode. This kind of sensor indicates the future development direction because it has some advantages such as working without the reference gas, simple structure, small volume, convenient use, easy miniaturization. However, this kind of sensor is only effective for low concentration hydrogen, which is below 1%. For high concentration hydrogen, It is difficult to identify because of its terrible dynamic performance.

In our study presented here we describe the properties of two-electrode solid electrolyte hydrogen gas sensor and then introduce the fabrication of the new three-electrode solid electrolyte hydrogen gas sensor. In the design of the three-electrode sensor, Pt is used as the measuring electrode, W is used as the reference electrode and nickel (Ni) is used as the adjusting electrode. The experiment were conducted to compare performance of the two-electrode solid electrolyte hydrogen gas sensor and the three-electrode solid hydrogen gas sensor.

2   Design of the Two-Electrode Solid Electrolyte Hydrogen Gas Sensor

A common approach in the solid electrolyte hydrogen gas sensor is based on the two-electrode structure. In this study, platinum is chosen to be the measuring electrode and tungsten is chosen to be the reference electrode. Phosphotungstic acid is utilized to make the electrolyte. When the hydrogen is pumped, the redox reactions will take place on the platinum electrode for hydrogen and phosphotungstic acid. The reactions can be described by equation (1) and (2).

$$H_2 \rightarrow 2H^+ + 2e^- \tag{1}$$

$$PW_{12}(VI)O_{40}^{3-} + e^- \rightarrow PW_{11}(VI)W(V)O_{40}^{3-} \tag{2}$$

$[PW_{12}(VI)O_{40}^{3-}]$ and $[PW_{11}(VI)W(V)O_{40}^{3-}]$ indicate the concentration of phosphotungstic acid in the oxidation state and in the reduced state respectively.

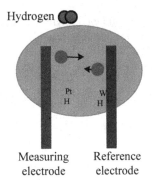

Fig. 1. The structure of the two-electrode solid electrolyte hydrogen gas sensor

The potential of the platinum electrode is determined by the dynamic relationship of the redox reaction when the concentration of the hydrogen is stable. The redox reaction is substantially mutual conjugation reaction and the potential is also called the mixed potential. The reaction rate of the hydrogen is much faster than that of the phosphotungstic acid due to the high catalytic activity of the platinum. Therefore the mixed potential established on the platinum electrode is close to the balance potential of hydrogen, which has an identified relationship with the hydrogen concentration. The redox reaction of hydrogen and phosphotungstic acid take place simultaneously on the tungsten electrode, which is used as the reference electrode. The mixed potential of the tungsten electrode is primarily determined by the balance potential of phosphotungstic acid because tungsten has a low catalytic activity. This also make the potential of tungsten to be a relatively stable reference potential.

The potential between the platinum electrode and the tungsten electrode, which is the mixed potential difference between the two electrodes, has an identified relationship with the hydrogen concentration. Although the relationship is still a half logarithmic linear relationship, it can not follow the Nernst equation entirely and the coefficient in front of the logarithmic is also uncertain. The potential varies slowly when the hydrogen concentration is higher than 1%, thus the resolution is pretty low at this condition.

3 Design of Three-Electrode Solid Electrolyte Hydrogen Gas Sensor

To extend the measuring range of the sensor, we introduce a third electrode, which is defined as the adjusting electrode. The selection of the third electrode is based on the hydrogen absorption ability of the material. By analyzing and making preliminary experiment, nickel is used to be the material of the adjusting electrode. The structure of the three-electrode solid electrolyte hydrogen gas sensor is shown in Figure 2. Pt is the measuring electrode, Ni is the adjusting

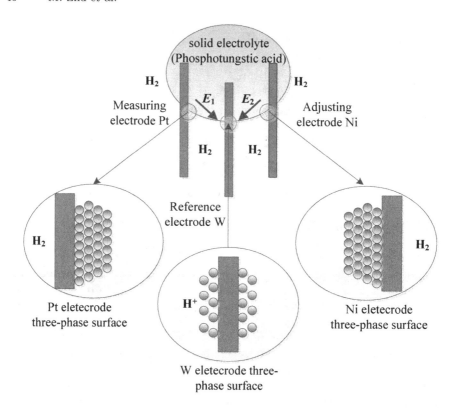

Fig. 2. The structure of the three-electrode solid electrolyte hydrogen gas sensor

electrode and W the reference electrode. The three electrodes are settled in parallel spatially. The potential difference between the measuring electrode and the reference electrode is E_1, the potential difference between the adjusting electrode and the reference electrode is E_2.

The redox reaction of the hydrogen and the phosphotungstic acid are taken place simultaneously on each electrode. the kinetics disciplines of the conjugate reaction determine the mixed potential of the electrodes. The potential is the mixed potential when the oxidization rate of hydrogen is equivalent to the reduction rate of phosphotungstic acid. The oxidation reaction of hydrogen and the reduction reaction of phosphotungstic acid are taken place in the strong polarization zone, Tafel relationship can be applied to the kinetics disciplines. The oxidation current of hydrogen and the reduction current of phosphotungstic acid can be expressed as follows:

$$i_a = i^0_{H^+/H_2} exp[\frac{2\beta_H F}{RT}(E_\mu - E_{H^+/H_2})] \tag{3}$$

$$i_c = i^0_{HPW} exp[\frac{2\alpha_{HPW} F}{RT}(E_{HPW} - E_\mu)] \tag{4}$$

$i^0_{H^+/H_2}$ is the exchange current of the redox reaction for hydrogen and i^0_{HPW} is the exchange current of the redox reaction for phosphotungstic acid. β_H is the transfer coefficient of the oxidation of hydrogen. R is the gas constant and its value is $8.314\,J^{-1}{\cdot}K{\cdot}mol^{-1}$. F is the Faraday constant and its value is $96485\,C{\cdot}mol^{-1}$. α_{HPW} is the transfer coefficient of the reduction reaction of phosphotungstic acid. T is the absolute temperature and the unit is K. E_μ is the mixed potential. E_{HPW} is the balance potential of hydrogen, which conforms to the below Nernst equation.

$$E_{H^+/H_2} = \frac{RT}{2F} \ln \frac{[H^+]^2}{pH_2} \tag{5}$$

We can get the expression of the mixed potential E_μ by solving the equation (3), (4) and (5):

$$E_\mu = \frac{\alpha_{HPW}}{2\beta_H + \alpha_{HPW}} + \frac{2.3RT}{(2\beta_H + \alpha_{HPW})H} \log \frac{i^0_{HPW}}{i^0_{H^+/H_2}}$$
$$+ \frac{\beta_H}{2\beta_H + \alpha_{HPW}} \cdot \frac{2.3RT}{F} \log \frac{[H^+]^2}{pH_2} \tag{6}$$

In equation (6), $[H^+]$ is the proton concentration of hydrogen. pH_2 is the pressure of hydrogen, whose unit is atm. E_{HPW} is the balance potential for the reduction reaction of phosphotungstic acid and can be expressed as:

$$E_{HPW} = E^0_{HPW} + \frac{RT}{F} \ln \frac{PW_{12}(VI)O^{3-}_{40}}{PW_{11}(VI)W(V)O^{3-}_{40}} \tag{7}$$

In equation (7), E^0_{HPW} is the standard electrode potential of phosphotungstic acid and it can be denoted by equation (8).

$$E^0_{HPW} \approx a + b(T - 298.15) \tag{8}$$

In equation (8), a and b are the temperature coefficients of standard electrode potential of phosphotungstic acid, which are constant. In equation (6), i^0_{HPW} is the exchange current density of the redox reaction of phosphotungstic acid and it can be expressed as:

$$i^0_{HPW} = 2Fk_{HPW}[PW_{12}(VI)O^{3-}_{40}]^{\beta_{HPW}}$$
$$\cdot [PW_{11}(VI)W(V)O^{3-}_{40}]^{\alpha_{HPW}} \tag{9}$$

k_{HPW} is the standard reaction rate constant of the redox reaction of phosphotungstic acid.

We can conclude from equation (6) that the relationship between the mixed potential and the hydrogen concentration can be described by half logarithmic function. However the coefficient in front of the logarithmic is not that of the Nernst equation (59/n mV when 25°C).

The dynamic property of hydrogen's oxidation and phosphotungstic acid's reduction varies on different electrodes, so $i^0_{H^+/H_2}$, i^0_{HPW}, β_H, α_{HPW}, E_{HPW}, the mixed potential and the discipline that it complies with the hydrogen concentration are also different.

After introducing the adjusting electrode based on the two-electrode sensor, the mixed potential and the discipline that it complies to the hydrogen concentration, of the measuring electrode and adjusting electrode who have different catalytic dynamics are different. So the potential between the measuring electrode and the reference electrode and the potential between the adjusting electrode and the reference electrode will show different trends. The responding curve of the sensor and the detecting range will extend accordingly.

4   Experiments

The sensor was assembled in a sealed container with a complete gas path. The standard concentration hydrogen was obtained by adjusting the volume ratio of the hydrogen and the insert gas. In order to avoid the external electromagnetic interference to the voltage signal, the sensor was placed in the box to shield the electromagnetic wave.

4.1   The Results of the Two-Electrode Solid Electrolyte Hydrogen Sensor in Detecting Hydrogen Concentration

To acquire the relationship between the electromotive force and the hydrogen concentration, the hydrogen concentrations were adjusted to 0, 0.1%, 0.2%, 0.3%, 0.4%, 0.5%, 0.6%, 0.7%, 0.8%, 0.9%, 1%. The results were shown in Figure 3. The results show that the output electromotive force of the sensor is sensitive to the hydrogen concentration. The sampled data were plotted in Figure 3 and the curve is similar to be logarithmic shape. However, the slope decrease with the increase of the hydrogen concentration. Especially for the concentration range higher than 1%, the two-electrode solid electrolyte hydrogen sensor has relatively lower resolution, which in turn result in the unrecognized hydrogen concentration.

4.2   The Results of the Three-Electrode Solid Electrolyte Hydrogen Sensor in Detecting Hydrogen Concentration

For the three-electrode solid electrolyte hydrogen sensor, the hydrogen concentration was adjusted to 0, 0.3%, 0.6%, 1.4%, 2.2%, 2.9%, 3.7%, 4.5%, 5.2%, 6.9%, 8.3%, 9.8%, 11.4%, 12%. The electromotive forces between Pt electrode and W electrode and between Ni electrode and W electrode were measured simultaneously. The relationship between the electromotive force of Pt-W electrodes and the hydrogen concentration is shown in Figure 4, and the relationship between electromotive force of Ni-W electrodes and the hydrogen concentration is shown in Figure 5. The measured results show that the electromotive force

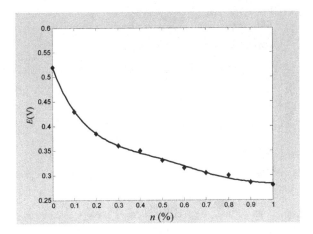

Fig. 3. The electromotive force of two-electrode solid electrolyte sensor as a function of the hydrogen concentration

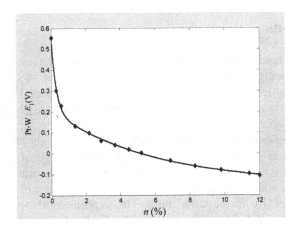

Fig. 4. The electromotive force of Pt-W electrodes E_1 of the three-electrode solid electrolyte sensor as a function of the hydrogen concentration

of Pt-W electrodes of the three-electrode solid electrolyte hydrogen sensor has better resolution when the hydrogen concentration changes in a range of 0-12%, and the dynamic range of the sensor output electromotive force can reach about 0.65 V. Comparing Figure 4 with Figure 5, it is obvious that the electromotive force of the three-electrode solid electrolyte hydrogen sensor, E_1 and E_2 present different trends. Comparing Figure 3 to Figure 4, it is clear that the detecting range of the three-electrode hydrogen sensor has been improved than that of the two-electrode sensor.

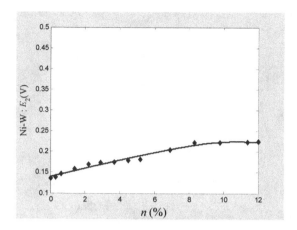

Fig. 5. The electromotive force of Ni-W electrodes E_2 of the three-electrode solid electrolyte as a function of the hydrogen concentration

5   Conclusions

A novel three-electrode solid electrolyte hydrogen sensor was investigated in this study. By introducing an adjusting electrode, the design of the three-electrode solid electrolyte hydrogen sensor was described in this paper. The experiments were conducted in a sealed container and the performance of the two-electrode solid electrolyte hydrogen sensor and the three-electrode solid electrolyte hydrogen sensor were explored. For the two-electrode sensor, the maximal detecting concentration is limited to 1% and the dynamic range of output voltage can achieve about 0.25 V. For the three-electrode sensor, the maximal detecting concentration can reach about 12% and the dynamic range of output voltage is about 0.65 V for Pt-W electrode pair. The experimental results show that the three-electrode solid electrolyte sensor can effectively improve the measurement range of the hydrogen concentration, and has a better sensitivity. The design of the new three-electrode solid electrolyte sensor can also present the feasibility for the research on the multi-electrode solid electrolyte sensor and provide the foundation for improving the sensor performance by data fusion method.

Acknowledgements. We are grateful for the support from the National Science Foundation of China (grant No. 61201017), the Fundamental Research Funds for the Central Universities (grant No. HIT.NSRIF.2013010, grant NO. HIT.NSRIF.201146).

References

1. Maffei, N., Kuriakose, A.K.: A solid-state potentiometric sensor for hydrogen detection in air. Sensors and Actuators B 98, 73–76 (2004)
2. Dong, H., Zhang, W., Hao, Y.: Research on three electrodes galvanic cell hydrogen sensor. Chinese Journal of Sensors and Actuators 20, 747–750 (2007)
3. Du, J., Bai, Y., Chu, W., Qiao, L.: Effects of temperature and humidity on the performance of hydrogen sensors based on the Nafion membrane. Journal of University of Science and Technology Beijing 32, 787–791 (2010)
4. Harada, S., Zheng, Y.S.: Hydrogen sensor. The open patent communique, Japan Patent Office, JP2009-243962.2009.10
5. Harada, S., Okada, M., Zheng, Y.S.: Hydrogen sensor. Japan Patent Office, JP2008-196903.2008.8
6. Bouchet, R., Rosini, S., Vitter, G., Siebert, E.: A Solid-State Potentiometric Sensor Based on Polybenzimidazole for Hydrogen Determination in Air. Journal of the Electrochemical Society 149, H119–H122 (2002)
7. Harada, S., Suda, T.: Characteristic Properties of New EMF-Type Hydrogen Sensor and Its Detective Method 57, 257–260 (2006)

Grid-Connected MPPT Control for MG Operating under Partially Shaded Condition

Xuemei Zheng[*], Chun Wang, and Yi Ren

School of Electrical Engineering and Automation,
Harbin Institute of Technology, Harbin, China
xmzheng@hit.edu.cn

Abstract. This paper investigates the control performance of a physical configuration of a microgrid, integrated with photo-voltaic (PV) arrays, battery energy storage systems, and variable loads. The main purpose is to achieve optimal control under grid connected for this microgrid. In order to improve the conversion efficiency of PV arrays under partially shading conditions (PSC), we use a maximum power point tracking (MPPT) algorithm developed by Ji et al [11] based on analyzing the P-V and I-V output characteristics under PSC. Then, a voltage source inverter based on swoop control is used to control the MG connection to the grid in order to get the maximum power to the grid. Both simulation and experimental results confirm that the proposed method can achieve optimal grid-connected control of the MG system.

Keywords: PV array, Partial Shading Condition, MPPT algorithm, Droop control, Microgrid.

1 Introduction

The ever-increasing demand for low-cost energy and growing concern about environmental issue has generated enormous interest in the utilization of the nonconventional energy sources such as the solar energy. The freely and abundantly available solar energy can be easily converted into electrical energy using photovoltaic (PV) cells. A PV source has the advantage of low maintenance cost, absence of moving/ rotating parts, and pollution-free energy conversion process. However, a major drawback of PV source is its ineffectiveness during the nights or low insolation periods or during partially shaded conditions (PSC). These drawbacks notwithstanding, the PV systems have emerged as one of the most popular alternatives to conventional energy because it is a environmental source.

A major challenge in use of PV is posed by its nonlinear current-voltage (I-V) characteristics, which result in a unique maximum power point (MPPT) on its power-voltage (P-V) curve. The matter is further complicated due to the dependence of these characteristic on solar insolation and temperature. As these parameters vary continuously, MPPT also varies. Considering the high initial capital cost of a PV

---

[*] The paper is supported by the Natural Science Foundation of HeiLongJiang Province of China (E200918).

K. Li et al. (Eds.): ICSEE 2013, CCIS 355, pp. 46–56, 2013.

source and its low energy conversion efficiency, it is imperative to operate the PV source at MPPT so that maximum power can be extracted.

Several tracking schemes have been proposed [2]-[12]. Among the popular tracking schemes are the perturb and observe (P& O) or hill climbing [4], [5], incremental conductance [8], short-circuit current [2], open-circuit voltage [7], and ripple correlation approaches [6]. The tracking schemes mentioned above are effective and time tested under uniform solar insolation, where P-V curve of a PV module exhibits only one MPPT for a given temperature and insolation. Under partially shaded conditions, when the entire array does not receive uniform insolation, the P-V characteristic get more complex, displaying multiple peaks, because only one of which is global peak. The presence of multiple peaks reduces the effectiveness of the existing MPPT schemes, which assume a single peak power point on the P-V characteristics. The occurrence of partially shaded conditions being quite common, e.g. clouds, trees, etc, there is need to develop special MPPT schemes that can track the global peak under these conditions.

This paper investigates the control performance of a physical configuration of a microgrid, which incorporates photo-voltaic (PV) arrays, a battery energy storage system, and variable loads. In order to improve the conversion efficiency of the PV array and the charger under PSC in MG grid-connected mode, we use a MPPT algorithm under PSC developed by Ji et al [11] based on the analysis of the P-V and I-V output characteristics. Then, the grid-connection requirements of the whole MG system are achieved using a voltage source inverter. Both simulation and experimental results confirm that the proposed algorithm can automatically track the global power point under different insolation conditions and optimal control of the microgrid is achieved.

2   Physical Configuration of the Micro-grid under Study

The paper considers the following physical configuration of a microgrid as depicted in figure 1, which consists of 2 PV units representing renewable power, and a charger unit, i.e. a battery energy storage system (BESS), all connected to an AC single-phase micro-grid (MG). The MG can operate in grid-connected mode or islanded mode. The PV power systems are subject to the atmosphere condition and thus generate variable power. An inverter consisting of a DC/DC converter, a DC-bus and an H-Bridge, can realize the interface between the PV panels and the MG. The inverter injects active power and reactive power through the point of common coupling based on different operation conditions.

The battery energy storage system ensures the power balance in the MG, acting as a load or a source according to power unbalance situation. Neglecting the power losses in the system, the active power balance is given by:

$$P_{BESS} = P_L - P_{PV} \tag{1}$$

where P_L is the load power requirement, P_{BESS} is the BESS power (positive or negative), and P_{PV} is the PV power.

The active power balance must be ensured at any time, the purpose is to maintain the frequency of the system within the required limits. The main sources of uncertainties are from the PV power systems, which inject variable active power according to the atmosphere condition, and also from the loads that vary quasi-randomly. Different strategies for maintaining the active power balance have been proposed, based on the energy storage, dump load control, or a combination of both and load shedding.

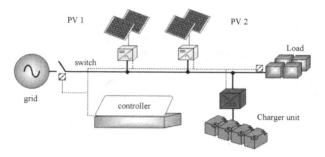

Fig. 1. The structure of microgrid under study

3   MG Control Strategy under Grid-Connected Mode

When the MG works in grid-connected mode, the frequency and the voltage of the microgrid are maintained within a tight range by the main grid. In this section, we detail the MG control design under grid-connected mode.

3.1   Normal Model of a PV Array

To control the MG with integration of PV power as shown in Fig 1, to model the PV array is necessary. A PV array is composed of several PV modules connected in series-parallel to produce desired voltage and current. Usually, more PV cells are needed to form the series-parallel PV array. The relationship of the output voltage and current of one PV cell can be represented as follows:

$$I = I_{ph} - I_s \left\{ \exp \left[\frac{q}{AKT} (U + IR_s) \right] - 1 \right\} - \frac{U + IR_s}{R_{sh}} \qquad (1)$$

where I is the output current of PV cell; U is the output voltage of PV cell; I_{ph} is the photocurrent; I_s is the reverse saturation current of diode; q is the electronic charge $(1.6 \times 10^{-19} C)$; K is Boltzmann's constant $(1.38 \times 10^{-23} J/K)$; T is Junction temperature; A is the diode ideality factor; R_s is series resister; R_{sh} is shunt resister.

Due to the large value of shunt resistance R_{sh}, the last term in (1) is often omitted, the short-circuit current and photocurrent are considered to be equal ($I_{sc} \approx I_{ph}$), and when the PV cell is on open circuit, the output current is zero, so output current of a PV cell can be approximated as :

$$I = I_{sc} \left\{ 1 - \exp \left[\frac{q}{AKT} (U + IR_s - U_{oc}) \right] \right\} \qquad (2)$$

Equation (2) is an implicit function relating the voltage and current. To help with the analysis, it can be shown that:

$$U = \frac{AKT}{q}\ln(1+\frac{I}{I_{sc}}) - IR_S + U_{oc} \tag{3}$$

Then, the output power of a PV cell is:

$$P = UI = \frac{AKTI}{q}\ln(1+\frac{I}{I_{sc}}) - I^2R_S + IU_{oc} \tag{4}$$

Equations (3) and (4) can be used to produce output characteristic curve of the PV cell. When these series-parallel PV cells constitute a $N_S \times N_P$ array (N_S is the number of PV cells in series, and N_P is the number in parallel), the corresponding PV array output voltage, current, open circuit voltage, short-circuit current and series resistance can be formulated as:

$$\begin{cases} U_A = N_SU \\ I_A = N_PI \\ U_{ocA} = N_SU_{oc} \\ I_{scA} = N_PI_{sc} \\ R_{sA} = (N_S / N_P)R_s \end{cases} \tag{5}$$

Taking an array of $N_S \times N_P = 3 \times 2$ PV as an example, Figure 2 and Figure 3 show the I-V curve and P-V curve respectively under the normal condition.

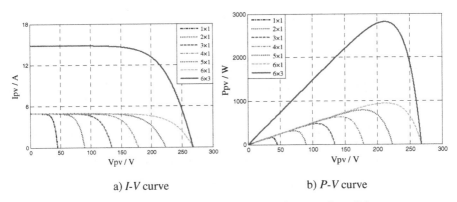

a) I-V curve   b) P-V curve

Fig. 2. I-V and P-V characteristic curve under normal condition

3.2 Output Characteristic of PV Array under Partially Shading Conditions

When the PV cells are operating in series, if the light is intense, the load current is equal to the output current of each PV cell; if shadow appears on the surface of one or more PV cells, their output currents are reduced, thus can easily cause Hot Spot phenomenon, which will accelerate the aging of PV cells and even damage the battery. So every PV cell has a bypass diode connected in parallel when PV cells work in series. When one PV cell is under shading condition, it accepts less light intensity, correspondingly produce smaller current; when the load demand current exceeds the maximum output current of the cell under shading, the bypass diode turns on. Assuming the forward voltage of the bypass diode is zero, then, the output voltage, the open circuit voltage and the equivalent series resistance of the series PV cells are given as:

$$\begin{cases} U_A = (N_S - N_D)U \\ U_{ocA} = (N_S - N_D)U_{oc} \\ R_{SA} = (N_S - N_D)R_s \end{cases} \quad (6)$$

where N_D is the number of PV cells under the shadow in series connection.

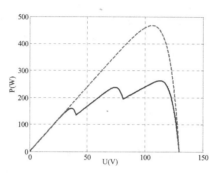

(a) The PSC ratio to the normal condition is 1,0.7,0.5

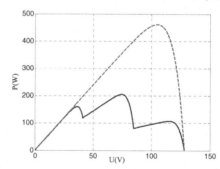

(b) The PSC ratio to the normal condition is 1,0.6,0.2

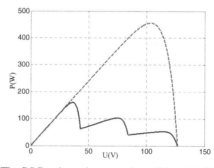

(c) The PSC ratio to the normal condition is 1,0.3,0.1

Fig. 3. P-Vcurves of PV panels under partial shading conditions

When a PV array is series-parallel structure, a diode will be added to each series branch. If all the branches are in the shadow, then the expression of the output voltage is as follows:

$$I_{scA} = \sum_{x=1}^{N_P} I_{scAx} \tag{7}$$

where I_{scAx} is the short current of the xth branch.

Therefore, the whole output current of a PV array is given as [24]:

$$I_A = \sum_{x=1}^{N_P} I_{Ax} = \sum_{x=1}^{N_P} I_{scAx} \left\{ 1 - \exp\left[\frac{q(U_A + R_S I_{Ax} - U_{ocAx})}{AKT(N_S - N_D)} \right] \right\} \tag{8}$$

where I_{Ax} is the output current of the xth branch.

Equations (6), (7) and (8) produce the I-V and P-V curves shown in Fig. 3, where non-dotted segments are the curves under PSC.

3.3   MG Unit Control under PSC

To control the MG configured in Fig. 1, we first aim to achieve the PV MPPT control and improve the efficiency of the PV array power conversion under partial shading conditions. Then, we use DC/DC converter to make the output current and the grid voltage following the same phase and frequency to realize the unit power factor control.

An important target in MPPT control is to track the global maximum power point under partial shading condition. One of the key issues is to identify the partial shading conditions, and in this paper, we adopt the following partial shading judgment criteria which was originally proposed in [11]

$$\begin{cases} \Delta U_{pv} = U_{pv}(n) - U_{pv}(n-1) < \Delta U_{set} \\ \dfrac{\Delta I_{pv}}{I_{pv}(n-1)} = \dfrac{I_{pv}(n) - I_{pv}(n-1)}{I_{pv}(n-1)} < \Delta I_{set} \approx -\dfrac{I_{pv}(n-1)}{N_p} \end{cases} \tag{9}$$

where ΔU_{pv} is the voltage change of PV array, ΔU_{set} is a preset voltage variation limit of PV array, ΔI_{pv} is the current change of PV array and ΔI_{set} is a preset current voltage variation limit of PV array.

The overall configuration of the control system for MG under PSC is shown in figure 4, and it has two separate parts that can be controlled individually. The first part incorporates the global maximal power point tracking into the inverter to connect DC power from the PV array to the MG. The back part is the inverter for controlling the integration of MG to the AC line. In this configuration, the key element is the MPPT control of the PV system with a PSC judging mechanism, which is described as follows.

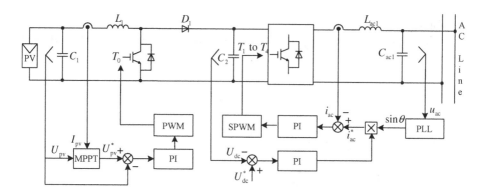

Fig. 4. The whole control system configuration of MG under grid-connected mode

In figure 4, the first converter is the Boost converter and adopts voltage loop control, which is used to achieve the MPPT control through adjusting the output voltage of the PV array. The second converter is connected with the first converter by a large capacitor C_2, and the output current of the first converter is the control target in the back converter.

DC bus line control is also necessary. Although MPPT control can make the PV array track the maximum power point, the different level of insolation will affect the power output, and the power change will cause the DC bus line drift. If the PV output energy increases sharply, the power delivery of DC bus line will be increased if there is no converter or load to consume the extra energy. On the contrary, if the PV output energy is decreased, and cannot satisfy AC line voltage peak value, the converter cannot work, therefore it is necessary to keep the DC bus line balanced.

The second part has two loops, that is the outer voltage loop and inner current loop. The function of the voltage loop is to keep the DC bus line balanced, it is controlled by a PI controller by comparing the actual value U_{dc} and the given value U^*_{dc} as the error for driving the PI controller. The output of the outer loop PI controller produces the AC given current value, which is multiplied by the sine output signal of the PLL of the AC voltage u_{ac} to produce the given current i^*_{ac}. The aim of the inner current loop is to realize the AC current control. The given current i^*_{ac} is compared with the actual i_{ac}, producing the error and is controlled by the inner loop PI controller. The output of the inner loop PI controller is compared with the triangle wave for generating the PWM signal to control T_1 to T_4, and thus produce the AC current output, which has the same frequency and phase with the grid side.

4   Simulation Results

Firstly, regarding to the influence of partial shading condition of PV, the paper studied two conditions: one is the slightly PSC, the other is the condition of heavy

PSC. For the slightly PSC, the bypassed diodes will not work and the output current of the series circuit is equal to the PV output current. While for the heavy PSC, the bypass diodes will connected to the circuit, the output current is equal to the PV output current under no PSC. Figure 5-7 are the waveforms under slightly PSC. Figure 5 is the PV output voltage waveform under PSC. From the figure, it can be seen that the output voltage and the current has dropped under the PSC. On this condition, the bypass diodes do not work, because the PV source can provide current to the load, so the output voltage is not zero. Figure 6 is the output power under PSC. Figure 7 is the *P-V* waveform to realize MPPT. From the figure, it can be seen that the Boost converter can realize MPPT (above 680W) under normal condition, that is, without PSC. If under PSC, the current and voltage will all drop under PSC point. But the MPPT can be realized also (above 500W).

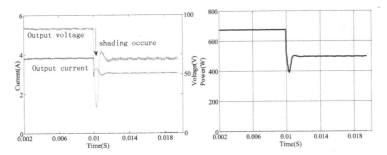

Fig. 5. V and I waveforms under PSC   **Fig. 6.** Output power under PSC

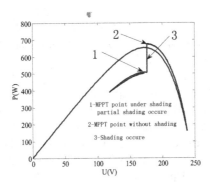

Fig. 7. P-V curve for realizing MPPT under PSC

Figure 8-10 are the waveforms under heavier PSC. Figure 8 showed that current is not zero, as for the bypass diodes are on, so the source voltage is zero. Figure 9 is the output power under PSC. From the figure, it can be seen that the output power drop quickly at the PSC point. But under the proposed method in the paper, the PV system can track the MPPT quickly. Figure 10 is the *P-V* curve of MPPT under PSC. From the figure, it can be seen that under the heavier PSC, can also get the global power maximum (above 320W).

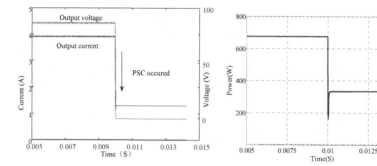

Fig. 8. V, *I* under heavier PSC   **Fig. 9.** Output power under heavier PSC

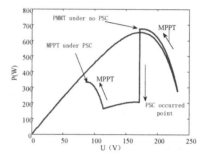

Fig. 10. P-V curve for realizing MPPT under heavier PSC

Then, we studied the proposed control in figure 4, which simulated over four different scenarios considering the variations in the produced PV power and the demand in the grid-connected mode. Figure 11 shows the simulation results adopting the control strategy in figure 4, where u_{grid}、i_{grid}、i_{load}、i_{pv1}、i_{pv2}、P_{pva1}、P_{pva2} are the grid voltage, grid current, load current, the output current of PV1 and PV2, the output power of PV1 and PV2, respectively. During 0.1s~0.2s, the PV array produced 6000W power and the total power was greater than the load demand, so excessive power was fed back to the grid and the current had the same phase with the grid voltage. During 0.2s~0.4s, as light intensity was decreased, the total PV energy was reduced to 2200W, which could not meet the demands, so the extra power was injected from the grid, and the grid side current and the grid voltage had different phase. While between 0.4s~0.6s, although the PV array produced 2200W power, the load demand was decreased, and he extra PV power was feed back to the grid again. It is evident from these figures that the quality of the output current of converter was satisfactory and the response of MPPT control strategy was swift.

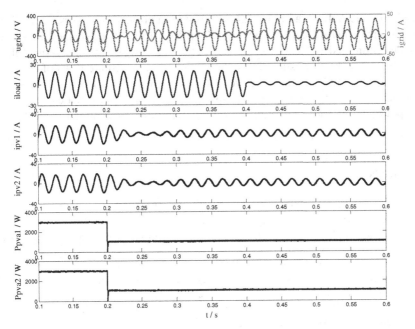

Fig. 11. The simulation results of PV array connected to the grid

The proposed control schemes in figure 4 was then applied to physical setup for the grid-connected mode, and figure 12 shows the experimental results when the PV was connected to the grid. It can be seen that the output current and the output voltage of converter has the same phase and the unit power factor has been achieved.

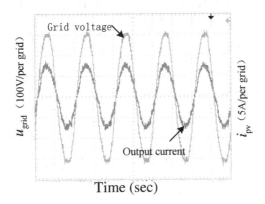

Fig. 12. The experimental results in grid-connected mode

5   Conclusions

This paper has investigated a physical configuration and performance of a microgrid integrated with PV panels, variable loads, battery energy storage systems and AC line under grid-connected operation modes. In order to improve the conversion efficiency of the PV array and the charger under PSC in MG grid-connected mode, we use a MPPT algorithm under PSC developed by Ji et al [11] based on the analysis of the P-V and I-V output characteristics. Then, the grid connection requirements of the whole MG system are achieved using a voltage source inverter. Both simulation and experimental results confirm that the proposed algorithm can automatically track the global power point under different insolation conditions and optimal control of the microgrid is achieved.

References

1. Miyatake, M., Inada, T., Hiratsuka, I., et al.: Control characteristics of a fibonacci-search-based maximum power point tracker when a photovoltaic array is partially shaded. In: The 4th International Power Electronics and Motion Control Conference, vol. 2, pp. 816–821 (2004)
2. Kobayashi, K., Takano, I., Sawada, Y.: A study on a two stage maximum power point tracking control of a photovoltaic system under partially shaded insolation conditions. In: IEEE Power Engineering Society General Meeting, vol. 4, pp. 2612–2617 (2003)
3. Al-Diab, A., Sourkounis, C.: Multi-tracking single-fed PV inverter. In: IEEE Mediterranean Electrotechnical Conference, pp. 1117–1122 (2010)
4. Patel, H., Agarwal, V.: Maximum power point tracking scheme for PV systems operating under partially shaded conditions. IEEE Transactions on Industrial Electronics 55(4), 1689–1698 (2008)
5. Cheng, Z., Zhou, H., Yang, H.: Research on MPPT control of PV system based on PSO algorithm. In: Chinese Control and Decision Conference, pp. 887–892 (2010)
6. Mishima, T., Ohnishi, T.: Power Compensation System for Partially Shaded PV Array Using Electric Double Layer Capacitors. In: IEEE Industrial Electronics Society Conference, vol. 4, pp. 3262–3267 (2002)
7. Liu, B.-Y., Liang, C.-H., Duan, S.-X.: Research on Topology of DC-module-based Building Integrated Photovoltaic System. In: Proceedings of the CSEE, vol. 28(20), pp. 99–104 (2008)
8. Karatepe, E., Hiyama, T., Boztepe, M., et al.: Power controller design for photovoltaic generation system under partially shaded insolation conditions. In: International Conference on Intelligent Systems Applications to Power Systems, pp. 1–6 (2007)
9. Nguyen, D., Lehman, B.: A reconfigurable solar photovoltaic array under shadow conditions. In: Twenty-Third Annual IEEE Applied Power Electronics Conference and Exposition, pp. 980–986 (2008)
10. Khajehoddin, S.A., Bakhshai, A., Jain, P.: A Novel topology and control strategy for maximum power point trackers and multi-string gridconnected PV inverters. In: Twenty-Third Annual IEEE Applied Power Electronics Conference and Exposition, pp. 173–178 (2008)
11. Ji, Y.-H., Jung, D.-Y., Won, C.-Y., et al.: Maximum power point tracking method for PV array under partially shaded condition. In: IEEE Energy Conversion Congress and Exposition, pp. 307–312 (2009)
12. Patel, H., Agarwal, V.: Maximum Power Point Tracking Scheme for PV Systems Operating Under Partially Shaded Conditions. IEEE Transactions on Industrial Electronics 55(4), 1689–1698 (2008)

Applications of Chaos and Fractal Theories in the Field of Electromagnetic Scattering

Chao Qi[1], Yuan Zhuang[1], Huixin Zhang[1], and Sai Qi[2]

[1] School of Electrical Engineering and Automation,Harbin Institute of Technology, Harbin, HLJ 150001, P.R. China
[2] Heilongjiang Electric Power Staff University Harbin, HLJ 150030, P.R. China
qichao@hit.edu.cn

Abstract. Chaos and fractals have gained extensive attention since they were proposed as two useful tools in studying nonlinear dynamic systems. The combination of the two theories has achieved obvious effect on practical application for their inherent relation. This paper has discoursed the applications of chaos and fractal in the field of EM (electromagnetic) scattering, mainly on rough surface modeling, analysis of sea clutters, scatters synthesis and scatter communication. It is obvious that chaos and fractal theories have great superiority over traditional ones in dealing with nonlinear problems. This paper shows that chaos and fractals have played an important part in various aspects of electromagnetic scattering.

Keywords: Chaos, Fractal, Electromagnetic Scattering.

1 Introduction

The process of which object molecules forming the electromagnetic polar moment of orientation and radiating electromagnetic waves when being irradiated, is called electromagnetic scattering. It makes significantly sense of practice to study the magnitude and distribution of the scattered electromagnetic field. For example, scattered echoes are used for searching, tracking or recognising target object in radar detection, while radar stealth techniques take the advantage of some materials absorbing a significant amount of electromagnetic waves and transferring them into heat to reduce the echoes[1]. In communication field, as a kind of beyond-the-horizon communication ,the scatter communication uses electromagnetic waves scattered by troposphere and ionosphere as the non-homogeneous of them[2]. For remote sensing, it is feasible to analyze the nature, property and state of target objects by detecting the characteristic of radiation and reaction of the electromagnetic waves from them[3]. In addition, the theory of EM scattering relates to many other fields such as subterranean exploration, electromagnetic compatibility, interference& anti-interference and so on. It's critical to model the rough surface in the analysis of the electromagnetic scattering. Generally, traditional models are Gaussian, exponential or the combinations of them, which have their limitations. The newly-born chaos and

K. Li et al. (Eds.): ICSEE 2013, CCIS 355, pp. 57–65, 2013.

fractal offer new theories to model rough surface. This paper mainly focuses on the applications of chaos and fractal in the field of electromagnetic scattering basing on acquired studies.

2   Chaos and Fractal

Put forward by the meteorologist Lorentz in the 1970s, the chaos is regarded as one of the three most important developments as well as the Relativity and Quantum theory in 20th century. Chaos refers to a reciprocating non-periodic motion which is sensitive to the initial condition in a determinate nonlinear system. The motion is deterministic in short terms but unpredictable for long. Comparing to linear systems and other nonlinear ones, chaos system has the characteristics of ergodicity, boundedness, inner randomness, sensitive to the initial value and long-term unpredictability. Withal, the chaos system always converges at some certain attractors in the phase space. The referents of Chaos are mainly the unstable divergence process in the nonlinear dynamics and the evolution rules of complex system behavior in pace with time[4].

Fractal theory was set up by the French mathematician B. B. Mandelbrot in 1975. Fractal refers to structures of self-similarity, which also means that they look the same under different magnifications. Fractal has the characteristics of self-affine, scale-free and self-similarity. The fractal theory focuses on the unsmooth and nondiferentiable geometric object generated by a nonlinear system. Studies have shown that it is more useful and suitable to use fractal theory to describe natural objects than beelines or smooth curves of Euclidean geometry, such as the boundary of clouds, the profile of the mountain peak, coastline, forked lightning and so on. It also fits for the forecasting of natural phenomenon. So fractal is referred to as the language of describing the nature. Fractal dimension is an important part of the fractal theory, as it is also a method to characterize chaos system quantitatively[5].

As important parts of nonlinear system, chaos and fractal have close connexions. The similarity in changing pattern of chaos system in time scale resembles with the self-similarity of fractal geometry in space scale.Whats more, the fact that the chaos system always converges at some certain attractors in the phase space conforms to formation process of fractal structure greatly. Therefore, a system with fractal structure usually has some characteristics of chaos, and with appearance of chaos there often follows the distinction of fractal. Although with different origins and development processes, these two disciplines interosculate closely. Chaos mainly focuses on the unsteady divergence process in time scale, while the referents of fractals relate to the irregular geometrical structure in space scale. Taking the research on Logistic mapping as a example, during the process of the system from the beginning to chaos, the phenomena of Period-doubling bifurcation appears, and the proportion between adjacent stages is a constant, which expresses the similarity of the fractal theory[6]. Thence, chaos theory and fractal theory can be applied in the field of electromagnetic scattering simultaneously. Simply speaking, chaos is the fractal in the scale of time and fractal is the chaos in the scale of space.

3 The Applications of Chaos and Fractal in EM Scattering

3.1 The Application of Chaos and Fractal in Modeling of Rough Surfaces

The multiple scattering theory for waves in mediums containing random scatterers was presented by Flody, Twersdy and Lax originally. Then others began researches on scattering of random turbulence medium. As the only Gaussian process that has self-similarity inherent, FBM (Fractional Brownian Motion) is continuous everywhere but can not be derived. So it was used to analyze wave scattering and propagation in fractals by Berry in the year 1979[7].

Numerical accuracy of the representation of a Weierstrass structure function based on FBM was examined by M.F.M Sanghaasa et al in 1994.The results showed that this representation was satisfying for a large range of parameters. With spectrum being a reliable approximation of FBM and performing well in predictability and controllability, WM(WeierstrassCMandelbrot) function was introduced to synthesize FBM[8]. As chaos and fractal theories could characterize natural surfaces with a few parameters, much attention was paid to rough surfaces modeling based on them. In the year 2008, Ren Xin-Cheng et al used a normalized two-dimensional band-limited Weierstrasss fractal function to model rough surfaces. The function shows a combination of both deterministic periodic structure and random rough structure. A general solution for the scattered field based on the KA(Kirchhoff Approximation) was given to dielectric rough surface. The influences of fractal dimension, the patch size of surface and the fundamental wavenumber on the scattering field were discussed theoretically and numerically. It was concluded that diffracted envelopes of scattering pattern can be approximated as a slope of linear equation in the near forward direction or the right of specular direction. This conclusion could be applicable for solving the inverse problem of reconstructing rough surface and remote sensing[9].

3.2 The Application of Chaos and Fractal in Studies of Sea Surfaces

The theories of chaos and fractal are widely used in the studies of electromagnetic scattering at fractal sea surface. Based on the 2D band-limited Weierstrass function, Hou Deting et al developed the model of rough sea surface. The effects of fractal dimension, frequency, amplitude factor and incidence angle were discussed in the research. The analytic expression of scattering coefficient was deduced using Helmholts integral and KA on the condition of the simulated fractal surface. The numberical simulations were carried out at the same time. Results showed that microwaves may be shaded by the rough sea surface, so the shadow function was introduced to amend the scattering coefficient at low gazing angle. Furthermore, the effects of fractal dimension, frequency, amplitude factor and incidence angle on the electromagnetic and scattering were discussed. Conclusions can be made that the peaks of scattering coefficient are more uniformly distributed with the increasing of the fractal dimension or the frequency and amplitude factor. The backscattering increases with the increase of the incidence angle while the forward scattering decreases[10].

A linear fractal Gaussian sea surface model was first presented by Berizzi and Mess in 1999. Then Longuest-Higgins and Stewart set up a nonlinear one by applying the nonlinear interaction relationship between long waves and short ones. Based on these studies, Xie Tao et al developed a new one-dimensional nonlinear fractal sea surface model and the scattering coefficient was attained, considering the nonlinearities of sea surface and the effect of wind speed. The numberical results of averaged NRCS(Normalized Radar Cross Section) of electromagnetic scattering from linear and nonlinear fractal models were compared. It suggested that the new nonlinear fractal EM backscattering model can potentially be more reliably used to numberically study, involving high sea states or rough sea[11]. Strictly considering both geometrical fractal characteristic and permittivity characteristic of sea water, improved integrated model of electromagnetic scattering for two dimensional fractal sea surface was built, especially that the effects of salinity and temperature on the electromagnetic field scattered by sea water were added. In this model, the permittivity of sea water is no longer a constant but a variable parameter. They incorporated a sea spectrum that accords with the experimental into the fractal model and the permittivity of saline water was calculated from Debye equation. This model descripts the whole electromagnetic scattering characteristics of sea surface, with the error of backscattering coefficient no more than 2 dB[12].

Haykin and Leung presented that sea clutter had a fractal dimensional attractor based on correlation dimension analysis dating back to 1992[13]. The results of a following research showed that the first Lyapunov exponent of sea clutters was indeed positive and around 0.015. The chaos inherent of sea clutter of X-band radar was proved and conclusion was that the movement of sea clutter was controlled by a low-order dynamical attractor[14]. Studies of application of sea clutter of S-band radar showed that on one hand, sequence signal can be extracted from mess radar echoes using method of chaos dynamic in ocean remote sensing . On the other hand, a more accurate model for sea clutters of S-band radar could be built based on the analysis of the reciprocity mechanism between S-band radar waves and sea waves in the field of ocean target detection, utilizing chaos dynamic and FBM. Adopting the model of FBM, Jiang Bin et al deduced the Hurst exponent based on the observed data. The fractal dimension of the S-band sea clutter was obtained primaly and it was about 1.5771.The largest Lyapunov exponent was 0.025, calculated by Rosenstein method. So the sea clutter of S-band radar was proved to be chaotic and fractal for the first time. Their research provided a new approach for target detection with the S-band sea radar[15].

3.3   The Application of Chaos and Fractal in Scatters Synthesis

Fractal theory was previously applied in many analysis problems or in graphics. In the year 1986, Y. Kim and D. L. Jaggard introduced the concept of fractal geometry into the problem of random array synthesis in a novel way. They built quasi-random antenna arrays which had fractal inherent by combining virtrues of the periodic subarray generators with those of random initiators. The current response of the fractal antenna arrays was studied and a modified fractal dimension was introduced in order to characterize the energy distribution in the radiation pattern. The research showed that the radiation pattern of the subarray became the envelope of the overall radiation pattern which forced the sidelobe pattern to be well controlled[16].

In recent researches, based on FBM, Giuseppe Ruello et al presented an innovative procedure for manufacturing fractal surfaces using Weierstrass function. A cardboard-aluminum fractal surface was built as a representation of WM fractal process. The comparison between the obtained calibrated data and the theoretical results deriving from the FBM using the KA and SPM (small perturbation method) showed matching and discrepancies between theoretical prediction and experimental results. They concluded that fractal synthesis was closer to experimental date and that the surface was efficiently described in terms of only two intrinsic parameters, the standard deviation and the correlation length[17].

It's well known that fractals can efficiently reflect(and conduct) EM waves with wavelengths much larger than fractal dimensions. In the year of 1993, Shalaev predicted that strong localization of dipole radiation in fractals results in very high local fields[18]. In order to gain a deeper understanding of the processes responsible for EM wave localization in 3-D fractals, Semouchkina and Miyamoto et al produced a second stage 3 D Menger sponge of $81 * 81 * 81mm^3$ the same way that the Cantor bar formed. In the experiment, the Menger sponge was illuminated by EM waves whose frequency range was 6 20GHz. The experiment confirmed deep attenuation of reflection and transmission characteristics of Menger sponge and an increase in 90°scattering at a previously found localization frequency. Simulations of FDTD(finite-difference time-domain) demonstrated a formation of the full wavelength resonance in the central cavity and a bandgap which could be considered as a signature of the bandgap formation of the front side. EM energy inside the fractal structure in the narrow frequency band. When it was blocked by the resonances in the front part of the sponge, equalizing and symmetrization of the EM response showed up from different parts of the structure[19].

3.4   The Application of Chaos and Fractal in Scatter Communication

In the field of scatter communication, EM waves can be scattered by troposphere, Ionosphere, meteor trail, manmade scatters and so on. EM waves could be scattered into any direction, only the ones scattered the way nearly front can reach far away. The energy dissipation in scatter communication is so much that EM waves received are usually very feeble. To solve this problem, transmitter of high-power, termination of high level sensitivity and antenna of great gain and narrow wave band are adapted traditionally[20]. Theories of chaos and fractal bring new ways to make this problem resolved. Chaos system is very propitious to detect weak signal as it is sensitive to even a small change of the initial value. Nie Chun-yan and Xu Zhen-zhong formed a chaos system for weak signal detection using Duffing function. By introducing the auto-correlation method and cross-correlation method respectively, the lowest signal to noise ratio was reduced to -77dB. The combination of chaos and auto-correlation or cross-correlation detection method has the advantages of time-domain such as simple, intuitionistic and feasible in hardware, while reducing the detection threshold of normal detection ones[21].

The surfaces of traditional High-gain antennas are usually of parabola shape with the caliber usually being circular. American scientist Nathan made the first fractal antenna in the year 1988. But it didn't gain much attention until an article on studies of fractal antenna written by him was reported[22].The newly developed fractal antenna has the characters of small size and multiband. Koch fractal antenna is studied most in antenna miniaturization, as Koch monopole performs better in space-filling than straight wire monopole at the same resonance frequency. While the number of iterations on the small fractal Koch monopole is increased, the Q of the antenna approaches the fundamental limit for small antennas[23].

Scatter communication happened mostly in troposphere, which is near to the earth surface. Being the densest layer of earth atmosphere, 75% quality of atmosphere is contained in it with almost all the vapor and aerosol. So scatter communication may be affected by clouds and rainfalls. Since 1960's,the propagation of radio waves in the earth's atmosphere involving multiple frequency bands has been widely studied[24]]. But it was difficult to forecast and calculate the absorption and scattering of electromagnetic waves in rains because of the lack of adequate statistical description of rains. To fill this vacancy in a certain degree, S. A. Zhevakin built the fractal model of spatio-temporal chaos of rain intensity permitting to calculate the fluctuations of electromagnetic wave intensity by its passage through the rain with a few parameters. It was of help in developing the physics of rains[25].

In humid atmospheric environments, soot aggregates frequently acquire a water coating which effect the radiation of electromagnetic waves. Chao Liu et al studies the influence of water coating on the scattering properties of fractal soot aggregates. Through building fractal aggregates and using effective medium approximations to the models, their studies shows that water coating of the fractal aggregates increases the forward scattering, extinction and absorption cross sections, and single-scattering albedo, but decreases the backward scattering [26]. Apart from all above, applications of chaos and fractal in scatter communication proposed involve signal encoding, encrypion or mapping, multiple access techniques, constructing reproducible broadband waveforms, image compression frequency selective channels and so on[27].

3.5   The Application of Chaos and Fractal in Other EM Scattering Fields

Chinese scientist Zhen-song Wang and Bao-wei Lu developed the general theory of multiple scattering of electromagnetic wave in fractal media by modifying the Twersky's in the way of ignoring the multiple scattered waves between the same scatterers. Statistical quantities were studied for wave propagating in a fractal medium and equation of wave field was obtained. The results of range dependence of the intensity of the backscattered radar signals showed agreement with the numerical simulation. So the developed theory and interrelated model are effective [28].

Furthermore, scattering components of targets in radar communication include creeping waves scattered by shadows of target edge, echoes reflected by mirror surface scattered waves from discontinuous area, scattered waves from derivative discontinuous area, scattered waves from accidented area and scattering waves produced by reciprocity of different parts of the surface. The reciprocity of these waves make the process of target scattering behave strongly nonlinearly. Based on it, Ming Xian and Zhao-wen Zhuang et al applied chaos and fractals into the studies of analysis and recognition of radar targets scattering signal. The Lyapunov exponents of five kinds of plane waves are calculated and the chaos inherent of radar echos was proved. The multifractal dimensions of these targets' scattering signal were further obtained, providing a reliable warrant for target recognition[29].

4 Conclusions

As efficient tools for studies of nonlinear system, chaos and fractal theories have been widely applied in the field of electromagnetic scattering. In researches of the modeling of random surface, the development of FBM and Weierstrass function have greatly improved the development of electromagnetic scattering. This paper mainly discusses the application of chaos and fractal theories on EM scattering from the random rough surface, the modeling of sea surface, the studies of the sea clutter, the scatters synthesis, the scattering communication and so on. Among these, chaos theory and the fractal theory integrate well with each other in studies of sea clutters and scattering communication.

It is obvious that the FBM is unpredictable, so do the chaos. So the process of chaos may also be used in surface modeling, especially for sea surface modeling. As attractors have fractal characteristic, it is natural for a surface formed based on the chaos dynamic to be fractal. So it is necessary to test whether the phenomenon of chaos scattering will show up in fractal scatters. Since the chaos theory mainly refers to nonlinear dynamic systems and fractal theory mainly focuses on the irregular geometric objects with infinite subdivision and self-similar structure, the combination of the these theories will have a promising future with wider applications.

Acknowledgments. This work was financially supported by the National Natural Science Foundation of China (No.61178066) and Natural Science Foundation of Heilongjiang (No.F201013).

References

1. Wang, Y., Feng, Y.J.: Study Progress of Stealthy Techniques and Stealthy Materials. Chemical Engineering 132, 43–47 (2006)
2. Ding, H., Chen, G., Xu, Z., Sadler, B.M.: Channel modelling and performance of non-line-of sight ultraviolet scattering communications. IET Communications 5, 514–516 (2012)
3. Wang, Y.F.: Regularization for inverse models in remote sensing. Progress in Physical Geography 36, 38–40 (2012)
4. Peitgen, H.O., Jürgens, H., Saupe, D.: Chaos and Fractals: New Frontiers of Science. National Defense Industry Press, Beijing (2008)
5. Zhu, H., Zhu, C.C.: Fractal Theory and Its Application. Science Press, Beijing (2011)
6. Karamanos, K., Mistakidis, I., Mistakidis, S.: The many facets of Poincare recurrence theorem of the logistic map. Kybernetes 41, 794–799 (2012)
7. Berry, M.V.: Diffractals. Phys. A: Math. Gen. 12, 789–797 (1979)
8. Sanghadasa, M.F.M., Sung, C.C.: Scattering of Electromagnetic Waves from Fractals of Large Amplitudes. Journal of Applied Physics 75, 7224–7226 (1994)
9. Ren, X.C., Guo, L.X.: Fractal Characteristics Investigation on Electromagnetic Scattering from 2-D Weierstrassfractal Dielectric Rough Surface. Chinese Physics B 17, 2956–2962 (2008)
10. Hou, D.T., Song, H., Zhou, D.F.: Model of Microwave Electromagnetic Scattering at Fractal Sea Surface. High Power Laser and Particle Beams 22, 2119–2123 (2010)
11. Xie, T., He, C., Perrie, W.: Numerical Study of Electromagnetic Scattering from One-dimensional Nonlinear Fractal Sea Surface. Chin. Phys. B 19, 024101-1–024101-6 (2010)
12. Fang, C.H., Liu, Q., Zhao, X.N.: Improved Integrated Model of Electromagnetic Scattering for Two Dimensional Fractal Sea Surface. International Journal of Modern Physics B 24, 4217–4224 (2010)
13. Haykin, S., Leung, H.: Model Reconstruction of Chaotic Dynamics: First Preliminary Radar Results. In: Proceedings of 1992 IEEE International Conference on Acoustics, Speech and Signal Processing, pp. 125–128. IEEE Press, San Francisco (1992)
14. He, N., Haykin, S.: Chaotic Modelling of Sea Clutter. Electronics Letters 28, 2076–2077 (1992)
15. Jiang, B., Wang, H.Q., Li, X., Guo, G.R.: The Analysis of Chaos and Fractal Characteristic Based on the Observed Sea Clutter of S band Radar. Journal of Electronics & Information Technology 29, 1809–1812 (2007)
16. Kim, Y., Jaggard, D.L.: The Fractal Ramdom Array. Proceedings of the IEEE 74, 1278–1280 (1996)
17. Ruello, G., Blanco-Sanchez, P., Mallorqui, J.J., Broquestas, A., Franceschetti, G.: Measurement of the Electromagnetic Field Backscattered by a Fractal Surface for the Verification of Electromagnetic Scattering Models. IEEE Transactions on Geoscience and Remote Sensing 48, 1770–1787 (2010)
18. Shalaev, V., Botet, R., Butenko, A.: Localization of Collective Dipole Excitation on Fractals. Phys. Rev. B, Condens. Matter 48, 6662–6664 (1993)
19. Semouchkina, E., Miyamoto, Y.: Analysis of Electromagnetic Response of 3-D Dielectric Fractals of Menger Sponge Type. IEEE Transactions on Microwave and Techniques 55, 1305–1312 (2007)

20. Li, M.D., Fu, G., Wu, G.L.: Antenna Technology for Meteor Scatter Communication. In: 2007 International Symposium on Microwave, Antenna, Propagation and EMC Technologies for Wireless Communications, pp. 582–585. IEEE Press, HangZhou (2007)
21. Nie, C.Y., Xu, Z.Z.: Application of Chaotic System in Detecting Weak Signal. Journal of Transducer Technology 22, 55–57 (2003)
22. Cohen, N.: Fractal Antennas: Part1. Commun. Quart. Summer, 7–22 (1995)
23. Baliarda, C.P., Romeu, J., Cardama, A.: The Koch Monopole: A Small Fractal Antenna. IEEE Trans. Antennas Propagation 48, 1773–1781 (2000)
24. Zhevaki, S.A., Naumov, A.P.: The Propagation of Centimeter, Millimeter, and Submillimeter Radio Waves in the Earth's Atmosphere. Radiophysics and Quantum Electronics 48, 678–694 (1967)
25. Zhevaki, S.A.: The Probability Distribution of Electromagnetic Wave Intensity Formed by Its Passage through Rain Fractal Model. In: MMET 1998 Proceedings, pp. 857–859. IEEE Press, Kharkov (1998)
26. Liu, C., Panetta, R.L., Yang, P.: The Influence of Water Coating on the Optical Scattering Properties of Fractal Soot Aggregates. Aerosol Science and Technology 46, 31–43 (2012)
27. El Khamy, E.: New Trends in Wireless Multimedia Communications Based on Chaos and Fractals. In: National Radio Science Conference (2004)
28. Wang, Z.S., Liu, B.W.: The Scattering of Electromagnetic Waves in Fractal Media. Taylor & Francis, London (1994)
29. Xian, M., Zhuang, Z.W., Xiao, S.P., Guo, G.R.: The Analyse and Recognition of Radar Targets Scattering Signal With Chaos Multifractal Theory. Journal of National University of Defense Technology 20, 60–64 (1998)

Comparative Study on Infrared Image De-noising Based on Matlab

Chao Qi[1], Suixing Cui[2], and Zhiyu Zhou[3]

[1] School of Electrical Engineering and Automation, Harbin Institute of Technology, Harbin, HLJ 150001, P.R. China
[2] Harbin University of Science and Technology, Harbin 150080 HLJ, China
[3] North China Electric Power University, 102206 Beijing, China
qichao@hit.edu.cn, {81299049,2472996658}@qq.com

Abstract. As the significant link of the image preprocessing, image de-noising has an effect on subsequent research work that cant be ignored. Infrared image de-noising based on the Matlab digital image processing environment is studied. The infrared image with different noises by different algorithms, such as average filtering, median filtering, multi-image average de- noising algorithm, low pass filtering and wavelet transform de-noising algorithm, are analyzed and processed. At last, via analyzing these algorithms for different noises and their owe characteristics comparatively, a suitable and effective de noising algorithm for the infrared image of electric power transmission line fault is found.

Keywords: Infrared Image, algorithm, mage Processing, Matlab.

1 Introduction

In recent years, since the progress of science and technology, infrared image instruments had been widely used in modern scientific research, military technology, electric power, medicine, meteorological, astronomy and other fields. The infrared image also called thermal imagery makes the object imaging depending on the temperature between different parts[1]. Its because the thermal imagery technology has the good features of concealment, wide detection range, accurately positioning, strong penetrating, and high recognition camouflage and so on. It enjoys the staff of different research fields. However, because of the influence of imaging equipment level, it makes the infrared images signal-to-noise ratio poor. In order to make the follow-up research more precise and effective, removing noises in infrared image becomes a difficulty in the fault diagnosis by the infrared technology[2].

Modern infrared technology began in around 1940, PbS infrared detector was developed, produced and put into application that was the sign. In 1959, the Germans infrared technology was adopted and developed by American and they announced the index of the infrared technology, which makes infrared research

K. Li et al. (Eds.): ICSEE 2013, CCIS 355, pp. 66–75, 2013.
© Springer-Verlag Berlin Heidelberg 2013

become a proven technique. And infrared image de-noising became an important subject[3]. In 1997, Johnstone with his team gave out the wavelet threshold estimator about the related wavelet de-noising method. In 2000, Chang, put forward self- adaptive wavelet threshold de-noising method in spatial domain which combines translation invariant method with selfadaptive threshold[4]. For the past few years, there were some new infrared image de-noising methods based on other theories, such as Fuzzy Mathematics[5], Neural Network[6], Curvelet Transform and Nonsubsampled Contourlet Transform and so on[7].

This paper mainly studies on the infrared image of the fault of the power transmission lines which are interfered by the various noises, where the software of Matlab is taken as the workbench.

2   Infrared Image De-noising

2.1   Neighborhood Linear Filtering

If the pixel value of processed image is $f(u, v)$, the grey value could be $g(u, v)$ after transformation, the space neighborhood mean de-noising algorithm could be given as follows

$$g(u, x) = \frac{1}{N} \sum_{(m,n) \in S} f(u - m, v - n) \tag{1}$$

Where N denotes the number of pixel value in the neighborhood, while S denotes the neighborhood. The size and shape of the neighborhood are decided by infrared image, usually rectangle is used, because the image data is matrix like $A \times B$, so the neighborhood areaS can be $3 \times 35 \times 57 \times 7$ and so on. For example, when S is 3×3, Eq.1 can be described as

$$g(u, x) = \frac{1}{9} \sum_{i=-1}^{1} \sum_{j=-1}^{1} f(u + i, v + j) \tag{2}$$

The mean template is as $1/9 * [111; 111; 111]$.

Neighborhood mean method is a simple and effective image smoothing method in spatial domain, which takes place of current pixels with the average of adjacent pixels, making the gray value in spatial domain even and playing the role of smoothing the gray value. The weighted mean method is: if the pixel value of processed image is also f(u,v), the gray value could be g(u,v) after transformation, then the weighted mean method could be defined as

$$g(u, v) = h(u, v) \times f(u, v) \tag{3}$$

Where h(u,v) denotes the matrix of weighted template. The common weighted template matrix has the kinds of form as follows

$$\frac{1}{5}\begin{bmatrix}0 & 1 & 0\\1 & 1 & 1\\0 & 1 & 0\end{bmatrix} \qquad \frac{1}{10}\begin{bmatrix}1 & 1 & 1\\1 & 2 & 1\\1 & 1 & 1\end{bmatrix} \qquad \frac{1}{16}\begin{bmatrix}1 & 2 & 1\\2 & 4 & 2\\1 & 1 & 1\end{bmatrix}$$
①   ②   ③

When the smaller the center element of $h(u,v)$ takes up, the better the smoothing resultis, but the fuzzier infrared image will be. Totally, we need to choose appropriate template practically. Fig. 1 shows the de-noised results by different mean filtering windows and different weighted templates under Gaussian noise (noise mean: 0, noise variance: 0.15).

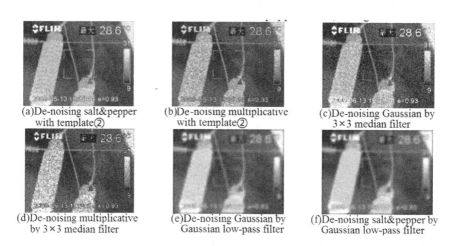

(a)De-noising salt&pepper with template②

(b)De-noising multiplicative with template②

(c)De-noising Gaussian by 3×3 median filter

(d)De-noising multiplicative by 3×3 median filter

(e)De-noising Gaussian by Gaussian low-pass filter

(f)De-noising salt&pepper by Gaussian low-pass filter

Fig. 1. The results with Gaussian noise de-noised by the mean filters

It is obvious that the neighborhood mean filtering can remove Gaussian noise well. While the bigger the filtering window is, the better the de-noising effect will be, but the fuzzier the image will be. Although the weighted mean filtering algorithm could keep the details in the image. The drawback is that while the template is going to be large, the image will be relatively poor, though the de-noising results become better. So it takes us great trouble to diagnose the fault.

2.2   Median Filtering

Median filtering is a nonlinear filtering technology, which could inhibit the random noise without making the details of the image fuzzy, and it is very effective to filtrate the impulse interference and image scanning noise. Since image signal is generally related in two dimensions, 2D sliding window is used for the median filtering, the 2D median filter could be described as

$$y_{ij} = \underset{S}{Med}\{x_{mm}\} \tag{4}$$

Where
$$m = i - M, i - M + 1, \ldots, 1, \ldots, i + M \quad n = j - N, j - N + 1, \ldots, j, \ldots, j + N$$
S denotes the 2D sliding windows, and xmn denotes the gray sequence of image pixels. So median filtering has its own filtering window as well, the window shape and size of 2D median filtering have a great effect on the filtering result. The common 2D median filtering windows are as follows

Table 1. The common window shapes of median filtering

Name	$3 \times 3\,square$	$5 \times 5\,cross$	$7 \times 7\,cross$
Window shape	○ ○ ○ ○ ○ ○ ○ ○ ○	○ ○ ○○○○○ ○ ○	○ ○ ○ ○○○○○○○ ○ ○ ○

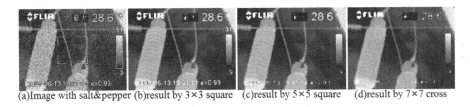

(a)Image with salt&pepper (b)result by 3×3 square (c)result by 5×5 square (d)result by 7×7 cross

Fig. 2. The results de-noised by median filters of different windows(noise intensity: 0.15)

Fig. 2 shows us that median filtering can remove salt& pepper noise in the infrared image, the bigger the median filtering window is, the clearer the de-noising effect becomes. Choosing an appropriate filtering window can reserve the useful detail information of image, on the premise that does not influence the fault diagnosis. So median filtering is desirable to remove salt& pepper noise without any details considered.

2.3   Frequency Domain Low-Pass Filtering

What stated above is spatial domain filtering, and frequency domain low-pass filtering is another effective method which is based on the image smoothing in frequency domain. Noises in the image are mostly in the high frequency part of frequency domain, the low-pass filter can retain the low-frequency information while filtering high-frequency noise. Supposing $F(u, v)$ denotes the Fourier transformation of the original image, and $G(u, v)$ represents the image through the low-pass filter, then the mathematical expression of low-pass filtering could be defined as

$$G(u, v) = H(u, v) * F(u, v) \tag{5}$$

There are so many types of low-pass filters. Here gives the common low-pass filters, such as ideal low-pass filter, Butterworth low-pass filter, Gaussian low-pass filter and soon. The transfer function models of these low-pass filters are defined as Table 2.In Table2 below, D_0 denotes the limiting frequency,$D(u,v) = \sqrt{u^2 + u^2}$ denotes the distance from point (u,v) on the frequency plane to the origin[8].

Table 2. Several transfer functions of loss-pass filters

Name	Transfer fuction
Ideal lowpass filter	$H(u,v) = \begin{cases} 1 & D(u,v) \leq D_0 \\ 0 & D(u,v) > D_0 \end{cases}$
Gaussian lowpass filter n class Butterworth	$H(u,v) = E^{-D^2(u,v)/2D_0^2}$
low-pass filter	$H(u,v) = \dfrac{1}{1 + [\frac{D(u,v)}{D_0}]^{2n}}$

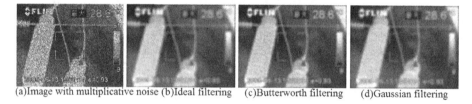

(a)Image with multiplicative noise (b)Ideal filtering (c)Butterworth filtering (d)Gaussian filtering

Fig. 3. The results de-noised by the different loss-pass filters

Fig. 3 shows the de-noised results after the filtering of the infrared image which contains multiplicative noise (noise intensity: 0.15), using different low-pass filters can all nicely remove the high frequency exponents in the image, the de-noised effects are also very desirable, but they can not avoid the image from becoming blurred. The image definition is not good. This method is not very satisfying especially for the fault diagnosis image which has a strict request to the details

2.4   Multi-image Average De-noising Algorithm

Multi-image average de-nosing algorithm is a method to eliminate noise by averaging several images of the same object. Multi-image average de-nosing algorithm is an statistical and averaging approach applied on many images of the same object to eliminate one's noise. Supposing that the original image is $f(u,v)$, $n(u,v)$ denotes additive noise, then the image with noise $g(u,v)$ could be defined as

$$g(u,v) = f(u,v) + n(u,v) \tag{6}$$

If pattern noise is the irrelative additive noise with mean index zero, then

$$f(u, v) = Eg(u, v) \qquad (7)$$

Where $Eg(u, v)$ denotes the expectation of $g(u, v)$. After averaging the M images with noise, we have

$$f(u, v) = Eg(u, v) \approx \bar{g}(u, v) = \frac{1}{M} \sum_{i=1}^{M} g_i(u, v) \qquad (8)$$

$$\sigma_{\bar{g}(u,v)}^2 = \frac{1}{M} \sigma_{n(u,v)}^2 \qquad (9)$$

Where $\sigma_{\bar{g}}^2(u, v)$ and $\sigma_g^2(u, v)$ denote the variance of $\bar{g}$ and $n(u, v)$ at the point (u, v) respectively. Eq.9 proves that averaging the M images with noise can reduce the variance of noise to $1/M$,compared with that as before. While M is magnifying, $g(u, v)$ will get closer to the original pixel values. In other words, $\bar{g}(u, v)$ will get closer to $f(u, v)$ as the number of M increasing..

Deduced from the above-mentioned model with additive noise, the noisy image with multiplicative noise n(u,v) could be defined as

$$g(u, v) = f(u, v) + n(u, v) * f(u, v) \qquad (10)$$

Where n(u,v) denotes random noise generated by the function of rand() and subjected to equidistribution(0-1). If the noise of image is uncorrelated and irrelative to the image, then

$$f(u, v) = \frac{E\{g(u, v)\}}{1 + E\{n(u, v)\}} \approx \frac{\bar{g}(u, v)}{1 + E\{n(u, v)\}} \qquad (11)$$

Since $E\{n(u, v)\} \neq 0$, letting $E\{n(u, v)\} = V$, then

$$f(u, v) \approx \frac{\bar{g}(u, v)}{1 + V} = \frac{1}{(1 + V)M} \sum_{i=1}^{M} g_i(u, v) \qquad (12)$$

Meanwhile, as Eq.8 demonstrates the noise variance after the restoration reduces to $1/M$ of the original one and the effect of de-noising is perfect to some extent. Since $V \neq 0$, the contrast may decrease. In a word, the entire pixel values of the processed infrared image diminish, compared with the original, because the number of V is not equal to zero. This paper takes five and twenty infrared images with noise respectively to average for de-nosing, so that we can take a direct judgment of its superiority with our visual ability. De-noised results by multi-image average de-nosing algorithm under the three mentioned noises are showed in Fig.4

Fig.4 indicates that de-noising effects of this method with the three above-mentioned noises are all satisfactory. But its weakness is that one has trouble in taking many images of the same object at the same time.However, this method could be deserved to be adopted if condition allows.

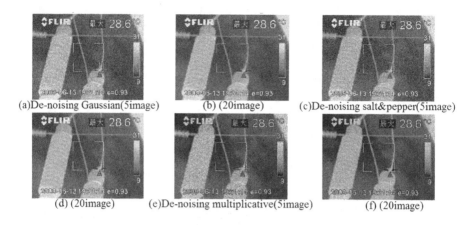

(a)De-noising Gaussian(5image)   (b) (20image)   (c)De-noising salt&pepper(5image)

(d) (20image)   (e)De-noising multiplicative(5image)   (f) (20image)

Fig. 4. The results de-noised by multi-image average de-noising algorithm under the three noises respectively

2.5 Wavelet Threshold De-noising Combined with Histogram Equalization

2D discrete WT mainly solve analysis process of the 2D multi-resolution problem. Assume that a 2D discrete image is denoted as $\{j(u,v)\}$, it can be broken down to each resolution approximate component at each level. They are approximate coefficient CA_j, horizontal component details CH_j, vertical component details CV_j, diagonal details CD_j. Two-level wavelet decomposition and reconstruction processes are shown as Fig.5. The selection of wavelet threshold must be

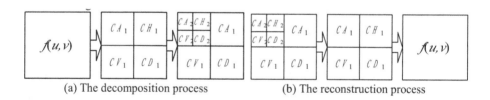

(a) The decomposition process   (b) The reconstruction process

Fig. 5. The two-level wavelet decomposition and reconstruction processes

larger than the maximum of the corresponding noise wavelet coefficients exactly. There are two coefficient estimating methods: hard threshold de-noising, and soft threshold de-noising. Since the constant error of soft threshold de-noising is little generally, we choose the soft here. We useBirge-Massart penalty function to get wavelet threshold by which process the decomposed wavelet coefficients. Fig.6 shows all the single-decomposed parts of the image with Gaussian noise.

Here, the first level wavelet decomposition is shown only, the second is that Fig.6(a) is decomposed once more. Then we eliminate Gaussian noise and multiplicative noisewhich are contained in the infrared image by this algorithm combined with histogram equalization respectively. The results as Fig.7 illustrates

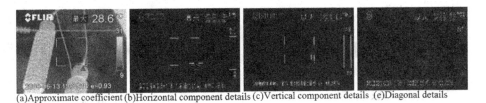

(a)Approximate coefficient (b)Horizontal component details (c)Vertical component details (e)Diagonal details

Fig. 6. The results decomposed by one-level wavelet transform

that not only the de-noised effect is perfect, but the contrast of the infrared image of transmission joints is enhanced by comparatively, which makes it easy to judge the faults. This algorithm is ideal for Gaussian and multiplicative noise, while it has little effect on salt& pepper noise.

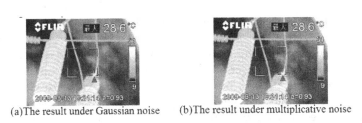

(a)The result under Gaussian noise   (b)The result under multiplicative noise

Fig. 7. The results de-noised by wavelet threshold combined with histogram equalization

3   Comparison and Analysis of Simulation Results

In this image simulation processing study, Different de- noising algorithms are taken to eliminate the Gaussian, salt& pepper and multiplicative noise respectively. Comparing the final images using different methods of de-noising, we can find out the suitable de-noising method for the specific noise.

Comparing Fig. 1 with (a), (b) of Fig. 8, we know that the de-noising effects for Gaussian noise are not well whichever method is taken. Fig. 2 and (c), (d) of Fig. 8 illustrate that the median filter is ideal for salt& pepper noise, though some details of the image which have no effect on judging the faults are removed too, so it can be for salt & pepper noise. Fig. 3 and (e), (f) of Fig. 8 shows that the de-noising results are all not ideal by the three traditional methods, but if some images need to be blurred, and then the low-pass filters would display its advantages. For example, an image of character needs to smooth the local details, and the low- pass filters can work well.Integrated analysis of Fig. 4 proves that the multi-image average de-noising algorithm could remove all the three noises and if possible, the more the average times is, the better the results will be. Fig. 7 shows wavelet threshold united with histogram equalization method not only removes the Gaussian and multiplicative noises perfectly but enhances the contrast of results which makes it much easy to find the faults. But it has poor effect on salt& pepper noise, because the high-frequency components of the noise

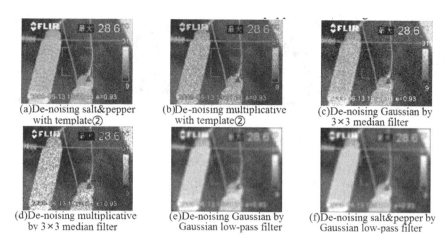

(a)De-noising salt&pepper with template②

(b)De-noising multiplicative with template②

(c)De-noising Gaussian by 3×3 median filter

(d)De-noising multiplicative by 3×3 median filter

(e)De-noising Gaussian by Gaussian low-pass filter

(f)De-noising salt&pepper by Gaussian low-pass filter

Fig. 8. Different de-noising methods for the three noises

treated as the useful component details of the original image are remained during the decomposition of the wavelet.So we can take the wavelet threshold combined with histogram equalization to process the infrared images with Gaussian and multiplicative noises, and take the multi-image average de-noising algorithm to process the infrared images with salt & pepper noise, this is the better choice.

4   Conclusions

Image de-noising is a process technology to showing advantages and disadvantages of each algorithm, contrasting of the linear theories in spatial domain and favorite image de- noising algorithms based on the image visual effects. This paper studies on the infrared application environments of different algorithms, which prompts us to explore the new image theory and method of de-noising algorithms, in order to eliminate noises in the images rapidly and effectively and make them more clear for research and visual need, and study how to eliminate the disturbing information of the image by filtering in order to satisfy one frequency domain.

Acknowledgments. This work was financially supported by the National Natural Science Foundation of China (No.61178066) and Natural Science Foundation of Heilongjiang (No.F201013).

References

1. Kang, Z.L., Xu, L.J.: An Algorithm Study on Infrared Image De-noising Based on Wavelet Transform. Computer Simulation 28, 256–367+274 (2011)
2. He, Y.J., Li, M., Lv, D., Huang, K.: Novel Infrared Image De-noising Method Based on Curvelet Transform. Computer Engineering and Applications 47, 191–193 (2011)

3. Zhou, G.A.: The Infrared Image Enhancement Algorithms. In: MS. Dissertation of Southwest Jiao-Tong University, Sichuan, pp. 11–17 (2007)
4. Shi, H.B.: The Algorithm Research of Image De-noising Based on WT. In: MS. Dissertation of Jilin University, Jilin, pp. 13–18 (2007)
5. Li, C.F., Nie, S.D.: Medical Image Processing and Analysis Based on Fuzzy Mathematic. Chinese Journal of Medical Physics 28, 2750–2753 (2011)
6. Deng, C., Wang, R., Zhang, T., Yao, Q.H.: De-noising Method of CCD Image Based on Improved Neural Network. North University China 32, 128–131 (2011)
7. Li, Y.H., Zhang, X.H., Lu, S.: Infrared Image De-noising Algorithm Using Adaptive Threshold Based on Contourlet Transform. Infrared Technology 33, 564–567+573 (2011)
8. Wang, B.: MATLAB Digital Signal Processing. China Machine Press, China (2010)

A Fast Detection Method for Bottle Caps Surface Defect Based on Sparse Representation

Wenju Zhou[1,2], Minrui Fei[1], Huiyu Zhou[3], and Zhe Li[1]

[1] School of Mechatronical Engineering and Automation,
Shanghai University, Shanghai, 200072, China
mrfei@staff.shu.edu.cn
[2] School of Information and Electronic Engineering,
Ludong University, Yantai, China
zhouwenju2004@126.com
[3] The Institute of Electronics, Communications and Information Technology,
Queen's University Belfast, United Kingdom
h.zhou@ecit.qub.ac.uk

Abstract. A machine-vision-based system is developed for detecting defects occurring on the surface of bottle caps. This system adopts a novel algorithm which uses circular region projection histogram (CRPH) as the matching feature. A fast algorithm is proposed based on sparse representation for speed-up searching. The non-zero elements of the sparse vector indicate the defect size and position. Experimental results show that the proposed method is superior to the orientation code method (OCM) and has promising results for detecting defects on the caps' surface.

Keywords: detect defection, bottle cap, circular region projection histogram(CRPH), sparse representation.

1 Introduction

Cap is a very important part of the bottling product packaging. The pattern of cap surface normally includes a company logo, but it is likely to be polluted such as surface scratch, distortion, stains, printing deviation, and other ill-defined faults during the production. Therefore, inspection of bad caps is of crucial importance for quality control. Much work has been done on the subject of defect detection, such as fabric [1], lumber [2], and bottling industries [3]. However, few studies have addressed the inspection of the surface of bottle caps. The goal of this paper focuses on the problem of defect detection of bottle caps' surface.

Most of the automated visual inspection systems for complicated textured-surfaces generally attempt to identify defects by building adequate templates of features representation using sample images. This representation is called feature dictionary. Detection accuracy is dependent on how adequate and general the dictionary is. Generally, the selection of an adequate feature set in the training process requires the help of complicated classifiers such as Bayes [4], maximum likelihood [5], and neural networks for classifying sample features and template features. The entire process is of high computational complexity and time-consuming.

K. Li et al. (Eds.): ICSEE 2013, CCIS 355, pp. 76–84, 2013.
© Springer-Verlag Berlin Heidelberg 2013

In the meantime, the production of caps is very fast. Thereby, a high-speed detection method is urgently needed. In this paper, we propose a novel circular region projection histogram (CRPH) method and a fast detection algorithm based on sparse representation for detecting defects of bottle caps' surface.

The circular region projection histogram method is inspired by the orientation code histogram method published in [6], and the fast detection algorithm is similar as sparse decomposition mothed in [7]. In literature [7], text and piecewise smooth contents in an image are separated into two different images. Different dictionaries were used for different contents, such as a dictionary of biorthogonal wavelet transforms (OWT) for piecewise smooth contents and a dictionary of discrete cosine transform (DCT) for texture contents. In our method, the image center is firstly located, and then the appropriate radius circle range is extracted as the template region of interesting (ROI). The ROI is projected as histograms along different directions. The histograms are the arrays that model the true distribution by counting the occurrences of pixel values that fall into each bin. These arrays are regarded as atoms that compose the template dictionary. Secondly, the sample cap surface image is captured whilst the sample ROI is extracted. The sample histograms at vertical and horizontal directions are computed by projecting the ROI. Lastly, the defect can be found through matching the atoms in the template dictionary to the sample histogram using a developed sparse representation method. The experimental result shows that the CRPH method and the developed sparse representation algorithm are effective for solving the rotation match.

The rest of this paper is organized as follows. Section 2 reviews the challenges and corresponding solutions based on sparse representation. Section 3 discusses the processing of extracting ROI and introduces a novel method CRPH for feature for matching, while the template dictionary and the defect dictionary are built. The fast match algorithm is proposed in section 4. Experimental results and their analysis are shown in section 5 and section 6 concludes the paper with a summary of the proposed work and discussions.

2   Sparse Representation Method

Suppose that an arbitrary bottle cap has a surface image X which contains a template pattern marks X_t and a defect component X_d. As such, the bottle cap surface can be denoted as follows,

$$X = X_t + X_d. \tag{1}$$

The defect detection is a process of separating the template image X_t and defect image X_d from the sample image X. We can use a sparse representation to solve this equation.

The template image X_t only contains original patterns which is flawless. Sparse decomposition matrix $D_t \in M_t^{N \times L}$ is written as

$$X_t = D_t \alpha_t, \tag{2}$$

where α_t is the template sparse factor. D_t is the template dictionary.

Similarly, for the defect image on the cap surface, sparse decomposition matrix $D_d \in M_d^{N \times K}$ is written as

$$X_d = D_d \alpha_d, \tag{3}$$

where α_d is the defect sparse factor. D_d is the defect dictionary. So, X can be denoted as follow,

$$X = D_t \alpha_t + D_d \alpha_d. \tag{4}$$

To seek a sparse representation over a combined dictionary containing both D_t and D_d, the generic method use the L_0 norm as a definition of sparsity. Hence, the following equations need to be solved

$$\begin{aligned} \{\alpha_t, \alpha_d\} &= \arg\min \|\alpha_t\|_0 + \|\alpha_d\|_0 \\ st. \quad X &= D_t \alpha_t + D_d \alpha_d. \end{aligned} \tag{5}$$

As well known, the problem formulated in Eq.(5) is non-convex and intractable. Its complexity grows exponentially with the number of columns in the overall dictionary. In order to obtain a tractable convex optimization solution, the basis pursuit (BP) method suggests the replacement of the L_0-norm with an L_1-norm [8]. Eq.(5) is re-written as a linear programming problem:

$$\begin{aligned} \{\alpha_t, \alpha_d\} &= \arg\min \|\alpha_t\|_1 + \|\alpha_d\|_1, \\ st. \quad \|X &- D_t \alpha_t - D_d \alpha_d\|_2 \leqslant \varepsilon . \end{aligned} \tag{6}$$

where the parameter ε stands for the residual, which is the tolerance between the sample image and the template image. Although these methods above are very effective for image separating, they are computationally complex and time-consuming. One of the reasons is that the dictionaries are redundant and over-complete, these make the dictionaries very large. The more atoms there are in a dictionary, the better the searching precision is. But everything have two sides, large dictionary increases computational complexity, and consumes large amounts of time. Therefore, we must trade off between matching precision and computational complexity. In this paper, a novel dictionary is proposed which is compactly supported and simple. A simple dictionary may reduce the accuracy of separation whilst satisfying the requirements of real-time process.

3   Feature Extract and Dictionary

The circular region of search and extraction is the key point to improve the overall speed of the detection algorithm. The projective transform are performed in the region of interest(ROI). Apparently, this way can save a lot of computing time since the ROI has a smaller size than that of the entire image. Fig. 1 shows the extracting of the region of interesting. The left is the original images. The right is the ROI. In order to carry out the rotation match between the template ROI and the sample ROI, we propose a method of using the circular region projection histogram transform as the rotation invariant feature. Such, the 2D image

Fig. 1. Region of interesting (ROI) extracting

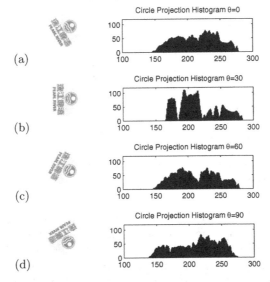

Fig. 2. Circular region (or ROI) projection histogram. (a) Original images and the projection histogram. (b) Rotated 30 degrees ROI and the projective histogram. (c) Rotated 60 degrees ROI and the projection histogram. (d) Rotated 90 degrees ROI and the projective histogram.

are projected into the 1D space. The computing complexity is greatly reduced. The projection histograms at the same angle have rotation invariant features. 360 histograms can be obtained by projecting the ROI at the 360 directions.

Fig.2 shows an example for projection histogram with the ROI and its part of rotated versions ($\theta = 0, 30, 60, 90$). Finally, projection histograms of the template ROI as the columns are grouped into a matrix, which is regarded as the sparse dictionary D_t of the template. D_t can be written as follows,

$$
D_t = \begin{bmatrix}
R_0^0 & R_0^1 & \cdots & R_0^{359} \\
R_1^0 & R_1^1 & \cdots & R_1^{359} \\
\vdots & \vdots & \ddots & \vdots \\
R_{N-1}^0 & R_{N-1}^1 & \cdots & R_{N-1}^{359}
\end{bmatrix} \tag{7}
$$

where R_i^θ denotes the value of the bin which is obtained by projecting, $i \in [0, N-1]$ is the number of the bin, which represents the position of the bin on the horizontal axis, N is the diagonal length of the template image, $\theta \in [0, 359]$ is the value of the projection angle, which denotes the number of the histogram.

Furthermore, the defect dictionary D_d will be proposed. In most cases, the defect whose width or height is more than 5 pixels cannot be tolerated. Therefore, a circular region with the 5-pixel diameter is regarded as minimum basic unit, whose projection can be got as $p^i = [3, 5, 5, 5, 3]^T$, where i is the descriptor of the central location. Motivated by the work in [9], we propose a defect dictionary D_d that can be obtained by successive change the position of p^i, where $i = 0, 1, \cdots N-1$. D_d is similar as the 'trivial templates' in literature [9], which is constituted as follows

$$D_d = \begin{pmatrix} p^0 & & 0 \\ & \ddots & \\ 0 & & p^{N-1} \end{pmatrix} \tag{8}$$

Due to five nonzero terms adjacent in the atoms of D_d, the residual less than the width of 5 pixels is ignored.

4   Fast Algorithm

When the template dictionary D_t and the defect dictionary D_d are obtained, the further task is to detect for sample bottle caps. Assume the vertical projection histogram and the horizontal projection histogram of the sample ROI are denoted by X_1 and X_2 respectively. Using expression (6), the defect sparse factors α_d^1 and α_d^2 can be obtained with respect to X_1 and X_2 respectively. The non-zero elements in α_d^1 and α_d^2 corresponds to defects, see Fig.3. As shown in Fig.3, because two defect sparse factors are orthogonal, the location and size of the defect can be determined. In order to get α_d^1 and α_d^2 quickly, a novel method which includes two steps is proposed. The detail process will be addressed as the following example of α_d^1. The way of solving α_d^2 is similar as that of α_d^1.

Firstly, according to (6), α_t^1 should be solved. α_t^1 is a template sparse factor with respect to X_1. The element in the sparse factor α_t^1, which corresponds to the best match column in the template dictionary D_t, is set to one, and the other elements of the sparse factor α_t^1 are set to zero. Thus, the following equation is obtained, $\|\alpha_t^1\|_0 = \|\alpha_t^1\|_1 = 1$. Without loss of generality, we assume that the best match column is jth. Obviously, the following resolutions is hold,

$$\alpha_t^1 = \begin{pmatrix} s_0 \cdots s_{j-1} \ s_j \ s_{j+1} \cdots\cdots s_{359} \end{pmatrix}^T = \begin{pmatrix} 0 \cdots 0 \ 1 \ 0 \cdots\cdots 0 \end{pmatrix}^T, \quad s \in R.$$

Secondly, α_d^1 is obtained. Substituting $\|\alpha_t^1\|_1 = 1$ into (6), we have

$$\alpha_d^1 = \arg\min \|\alpha_d^1\|_1,$$
$$st. \quad \|X_1 - D_t\alpha_t^1 - D_d\alpha_t^1\|_2 \leqslant \varepsilon. \tag{9}$$

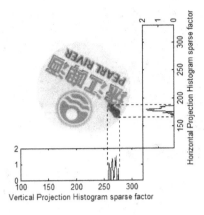

Fig. 3. Relationship for the defects with sparse factors

According to (1) and (2), we have $X_1 = X_t^1 + X_d^1$ and $X_t^1 = D_t\alpha_t^1$. Substituting the above equations, expression (9) can be re-written as

$$\alpha_d^1 = \arg\min \left\|\alpha_d^1\right\|_1,$$
$$st. \quad \left\|X_d^1 - D_d\alpha_d^1\right\|_2 \leqslant \varepsilon . \tag{10}$$

where ε is the error, which is caused by image noise and inaccurate alignment between the sample and the template. From expression (10), we get α_d^1 using the algorithm proposed in literature [10]. If we only judge the cap quality, α_d^1 that is solved is enough. According to the previous knowledge, the positioning defects in a 2D image need to know the defects sparse factors in both the vertical and horizontal directions. So, α_d^2 needs also to be further calculated. Due to the projection direction of X_2 perpendicular to the projection direction of X_1, the interval is 90 atoms between the two best matching atoms. Thus, α_t^2 is gained through α_t^1 cyclic right moving 90 units. This makes very simple and fast to getting α_t^2. Nextly, α_d^2 can be obtained easily, which uses the way same as that of α_d^1.

5 Experimental Results

The setup that is used for capturing caps surface image in this experiment is depicted in Fig.4. A Basler Aca640/100gm high-performance machine vision camera with Gigabit Ethernet interface (GigE Vision) is mounted above the stage, which supports jumbo frames and is capable of reaching a frame rate of 100Hz full frame (650 × 492). The experiments were carried out on an Intel dual-core 3.0GHz PC with 4GB RAM. All the computations were performed with MATLAB. The bottle cap images are captured by the high speed camera which is triggered by the signal come from the sensor. To obtain a high speed to deal with image, we read out only a part of the image sensor as large as

Fig. 4. Device for capturing image

320×300 pixels. Due to the monochrome images can satisfy the precision of the testing needs, we use monochrome camera in the experiment. We present the assessment on the performance of the proposed method in comparison to other similar techniques. The three main results are obtained and described below, which are rotating alignment accuracy, run time and defect detection effectiveness.

Firstly, the proposed CRPH solution in this paper is compared to the orientation code method (OCM) [6], the morphological component analysis (MCA) solution [7] in order to verify the rotation matching result. Fig.5 demonstrates the rotating alignment result using three methods above between the template image and the sample image. In OCM rotation alignment experiment, 64 orientation codes are used, that corresponding to a sector width Δ_θ of $\pi/32$ radians. We pre-built the OCM histogram set of the template image as the comparison reference, which play similar role as the sparse dictionary of our method. The motivation of creating orientation code histogram set is to reduce the time for matching, as well as to have the same environmental conditions to compare with the other two methods equivalently. A circular region with empirical radius of 36 pixels is used as the OCM base element, which is extracted orientation code histogram as a base atom of the OCM histogram set. The center of the OCM base circular region is bounded in a small field which is tolerance range for the printing deviation of caps surface image. In the case, a circular region with radius 3 pixels is used tolerance field of the image deviation. Thus, the pixels of the tolerance field are served as the center of the OCM base element. In other words, each pixel corresponds to a OCM base element. Orientation code histograms extracted from these base elements are put into the OCM histogram set. The searching and rotation alignment methods adopt the methods of literature [6]. The result is shown in Fig.5(c). In our propose method, the projection of the templates on 360 orientations (namely interval angle $\Delta\theta = 1°$) are created previously, which are regarded as the dictionary of the feature match, i.e. the number of the dictionary column is 360. Each column represents a projection unit of a certain angle. This dictionary is used both in the MCA and the proposed method. It can be found from Fig.5 that the accuracy of our method is close to that of the MCA and better than that of the OCM.

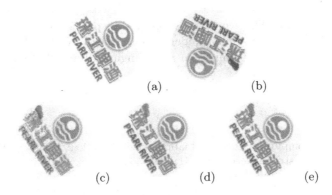

Fig. 5. Sample image of a cap matched against the template using three methods. (a) Template image. (b) Sample image with defect. (c) Rotation matching result using OCM. (d) Rotation matching result using MCA. (e) Rotation matching result using our method.

Secondly, in order to verify the running time we select three sample bottle caps surface images to run 10 times to obtain the average time. The results are presented in Table 1. Our method is 23 times faster than the OCM and nearly 2 times faster than the MCA. The reason that the OCM is slow may be caused by many invert tangent function and vast loops process. Our method is superior to the MCA because we improved algorithm of the finding α_t.

Table 1. Result of the average run-time (second) for the three methods

Sample Cap	OCM(second)	MCA(second)	Our Method(second)
PEARL RIVER 1	0.448263	0.038351	0.019446
PEARL RIVER 2	0.446221	0.038062	0.019001
PEARL RIVER 3	0.450014	0.040433	0.021135

6 Conclusion

A new method base on circular region projection histogram and sparse representation for detecting the defect of caps surface is presented. The method employs the projection of the template on different orientation as the template dictionary and the circular point projection structure the diagonal matrix as the defect dictionary. The closest match position can be found by searching the template dictionary. Experiments on real world images of bottle caps surface show that the proposed method is the effectiveness in searching the defect of caps rotated by arbitrary angles. However, our method is more sensitive to changes in light and in printing density. Our future work is to improve the algorithm so that it is robust to changes in brightness and printing density.

Acknowledgments. This work was supported in part by the National Natural Science Foundation of China under Grant No.61074032 and No.61104089, Science and Technology Commission of Shanghai Municipality under Grant No.10JC1405000.

References

1. Kumar, A.: Computer-vision-based fabric defect detection: A survey. IEEE Transactions on Industrial Electronics 55(1), 348–363 (2008)
2. Todoroki, C.L., Lowell, E.C., Dykstra, D.: Automated knot detection with visual post-processing of Douglas-fir veneer images. Computers and Electronics in Agriculture 70(1), 163–171 (2010)
3. Yazdi, L., Prabuwono, A.S., Golkar, E.: Feature extraction algorithm for fill level and cap inspection in bottling machine. IEEE (2011)
4. Friedman, N., Geiger, D., Goldszmidt, M.: Bayesian network classifiers. Machine Learning 29(2), 131–163 (1997)
5. Kleinbaum, D.G., Klein, M.: Maximum likelihood techniques: An overview. Logistic Regression, 103–127 (2010)
6. Ullah, F., Kaneko, S.: Using orientation codes for rotation-invariant template matching. Pattern Recognition 37(2), 201–209 (2004)
7. Starck, J.L., Elad, M., Donoho, D.L.: Image decomposition via the combination of sparse representations and a variational approach. IEEE Transactions on Image Processing 14(10), 1570–1582 (2005)
8. Donoho, D.L.: For most large underdetermined systems of linear equations the minimal-norm solution is also the sparsest solution. Communications on Pure and Applied Mathematics 59(6), 797–829 (2006)
9. Mei, X., Ling, H.B.: Robust Visual Tracking and Vehicle Classification via Sparse Representation. IEEE Transactions on Pattern Analysis and Machine Intelligence 33(11), 2259–2272 (2011)
10. Huang, J.B., Yang, M.H.: Fast sparse representation with prototypes. In: Proceedings of the IEEE Computer Society Conference on Computer Vision and Pattern Recognition, San Francisco, CA, United States, pp. 3618–3625 (2010)

A Parameter Identification Scheme for Second-Order Highway Traffic Model Based on Differential Algebraic Methodology

Nan Li[1,2] and Guangzhou Zhao[1]

[1] College of Electrical Engineering, Zhejiang University, Hangzhou 310027, China
[2] Faculty of Electronic and Control Engineering, Liaoning Technical University, Huludao 125105, China
{happyapple,zhaoz}@zju.edu.cn

Abstract. In this paper, we develop a fast on-line and parameter identification scheme for the second-order macroscopic traffic flow model. The proposed parameter identification scheme is devised in the framework of the algebraic identification, with differential algebra and operational calculus as major mathematical tools. Compared to conventional methods, the new identification scheme allows the parameters of second-order macroscopic traffic model, namely free speed and critical density, be estimated in an on-line and computationally efficient fashion are identified by means of differential algebra and operational calculus. The simulation example of a hypothetical scenario demonstrates these advantages numerically.

Keywords: parameter identification, highway model, differential algebra.

1 Introduction

Traffic congestion on highways is becoming an increasingly severe problem in many countries all over the world. An effective and practically feasible approach to tackle the congestion problem is through the application of various traffic control measures. Many traffic control strategies are employed extensively, e.g., ramp metering, variable speed limit and route guidance. Design and evaluation of these strategies usually involve macroscopic traffic flow models in different ways [1][2].

Due to the high non-linearity and time-varying features of highway traffic flow, which are coupled with the uncertainties of measurements, the parameters of macroscopic models are usually hard to know exactly and influenced by exogenous conditions, such as climate conditions (snow, rain or frog) or traffic incident, infrastructure downgrade and so on [3].

Therefore, for the purpose of effective management and control of transportation system, parameter identification (calibration) of traffic model calls for on-line and fast algorithms, which can deal with the real-time variations of traffic dynamics and randomness of observations.

K. Li et al. (Eds.): ICSEE 2013, CCIS 355, pp. 85–93, 2013.
© Springer-Verlag Berlin Heidelberg 2013

The traffic parameter identification methods can be classified into two categories, according to their data requirements: the first is off-line identification method [4], which mainly depends on the historical traffic flow data, and parameters are obtained by nonlinear programming or other intelligent learning methods. The drawback of off-line schemes lies in that the parameters may change due to variation of environmental factors. This method thus can not reflect the variation of the traffic dynamics promptly. In addition, the demanding computational loads usually renders such estimation methods useless in the context of real-time traffic control. Therefore, in view of the serious disadvantages of off-line approach for traffic estimation, the other method, i.e. the on-line parameter identification method is more attractive and desirable. This is also the research focus of the traffic control and automation. The classical on-line identification techniques include the well-known recursive least square algorithms (RLS) [5], optimal filtering [6], (e.g. Kalman filter, EKF, UKF and PF), and asymptotic observer [7]. In spite of many advantages of these methods, the existing identification techniques are usually complicated and have some common limitations for practical use. For instance, these methods are usually sensitive to random perturbations, precise priori information must be acquired beforehand, and the computational load is usually heavy, which results in unsatisfactory identification speed. These shortcomings pose the primary obstacles their employment for the real-time traffic control.

Algebraic identification is a novel kind of on-line and non-asymptotic identification method proposed by M.FLESS et al [8][9]. Compared with the common probability meaning methods such as observers and filtering method, the merits of algebraic parameter identification are its fast speed and robustness. Due to these two important features, algebraic identification method has been applied in various fields, for instance, continuous-time system identification, fault diagnosis, and signal processing [10]. In intelligent traffic system (ITS) field, the algebraic identification has been applied in the traffic state estimation [11], description of the freeway network [12], vehicle stop-and-go control [13]. In particular, a kind of algebraic parametric estimation scheme of LWR model is proposed by [14]. Nonetheless, it is well-known that in LWR model the speed of vehicles is assumed to be the equilibrium speed. This assumption cannot describe the non-equilibrium characteristics which are pertinent to safety and environmental concerns, for example, the stop-and-go wave and phantom congestion. To be more realistic and address the non-equilibrium features, in most highway traffic control schemes, the second-order traffic model is employed [15]. In summary, the parameter identification for second-order macroscopic traffic flow model plays a very essential role in the context of highway real-time traffic control and management.

Based on differential algebraic framework in [14], a parameter estimation scheme for non-equilibrium macroscopic traffic flow is proposed in this paper. The rest of the paper is organized as follows: Section 2 presents the highway second-order macroscopic traffic flow model METANET which shall be used as the prototype model in parameter estimation. Section 3 describes the basic framework of the algebraic identification theory, which utilizes the differential algebraic and operator calculus as main mathematical tools. Section 4 gives the key parameters identification scheme for second-order highway traffic model based on differential algebraic methodology. In section 5, a number of simulations are conducted to verify the effectiveness of the proposed the algebraic identification scheme. Finally, the main conclusions are summarized and a few open problems are discussed in section 6.

2   Dynamic Model of Highway Stretch

Macroscopic traffic flow model are employed to describe the dynamic behavior of traffic flow of a highway stretch using aggregate variables, including space-mean speed, density (or occupancy), and flow rate. The second-order macroscopic traffic flow model METANET is employed in this paper. It is also the foundation of a large number of traffic control studies, due to the ability to realistically reproduce traffic phenomena [16].

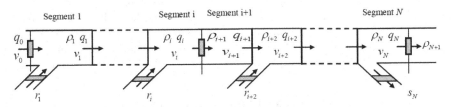

Fig. 1. Highway section divided into N sections

METANET model is discrete both in time and space. To be specific, one highway stretch is subdivided into segments of equal length (usually about 500m), as is shown in figure 1. The METANET models are as follows:

$$\rho_i(k+1) = \rho_i(k) + \frac{T}{\Delta_i \lambda_i}[q_{i-1}(k) - q_i(k) + r_i(k) - s_i(k)] \tag{1}$$

$$s_i(k) = \beta_i(k) \cdot q_{i-1}(k) \tag{2}$$

$$v_i(k+1) = v_i(k) + \frac{T}{\tau}[V(\rho_i(k)) - v_i(k)] + \frac{T}{\Delta_i}v_i(k)[v_{i-1}(k) - v_i(k)]$$
$$\underbrace{\qquad\qquad\qquad}_{\text{relaxation term}} \qquad \underbrace{\qquad\qquad\qquad}_{\text{convection term}}$$
$$- \frac{vT}{\tau\Delta_i}\frac{[\rho_{i+1}(k) - \rho_i(k)]}{\rho_i(k) + \kappa} - \frac{\delta T}{\Delta_i \lambda_i}\frac{r_i(k)v_i(k)}{\rho_i(k) + \kappa} \tag{3}$$
$$\underbrace{\qquad\qquad\qquad}_{\text{anticipation term}} \qquad \underbrace{\qquad\qquad}_{\text{ramp influence term}}$$

$$V(\rho) = v_f \exp[-\frac{1}{\alpha}(\frac{\rho}{\rho_{cr}})^\alpha] \tag{4}$$

$$q_i(k) = \rho_i(k) \cdot v_i(k) \cdot \lambda_i \tag{5}$$

where T denotes the sampling time, $k = 0, \ldots, K$ denotes the k-th time interval, $i = 1, \ldots, N$ denotes the i-th section of the highway, and N is the number of sections. The parameter meanings are shown in Table 1. According to previous studies, the parameters τ, δ, κ and α are relatively deterministic and have a small variation for a specific sections of highway, and the traffic state estimation and control results are known to be most sensitive to variations of the free speed v_f and critical density ρ_{cr}. Therefore, this paper only treats the free speed and critical density as the unknown model parameters that need to be identified. In practice, other parameters can be determined by off-line model calibration and regarded as known parameters in an on-line algorithm. The physical meanings of the free speed and critical density are clearly illustrated through a fundamental diagram, e.g. in the form of function (4). Fundamental diagram is a basic tool in understanding the behavior of traffic system through establishing the functional relationship of the traffic flow and the traffic density of a given highway stretch. One example of fundamental diagram is shown in Fig.2. Based on the fundamental diagram and traffic measurements, it is indicated that highway segment has a maximal flow rate, i.e. capacity. Once the traffic flow reaches the regime behind ρ_{cr}, it becomes congested and throughput starts to decrease with the backward propagation of shock waves. One of the major aims of the traffic control is to avoid the onset of congestions through changing the flow input at boundaries.

Table 1. Model parameter meanings

Parameter	Unit	Physical meaning
τ	hour	Driver's reaction time
β	%	Turning rate of vehicles leaving off-ramps
κ	veh/km	Additional tuning parameter
δ		Ramp effect parameter
ρ_{cr}	veh/km	Critical density
v_f	km/h	free flow speed
$r(k)$	veh/h	On-ramp flow rate
$s(k)$	veh/h	Off-ramp flow rate
α		Additional tuning parameter

3   Philosophy of Algebraic Identification Theory

Algebraic identification is based on elementary algebraic 4manipulations of the following mathematical tools: differential algebra, operational calculus, and module theory for a linear system [13][17][18][19].

Let us denote s as the differentiation operator and d/ds corresponds in the time domain to the multiplication by t. $\mathbb{R}(s)$ is the field of rational functions in the variable s with real coefficients, and $\mathbb{R}(s)[\frac{d}{ds}]$ is the set of linear differential

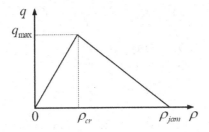

Fig. 2. The flow-density fundamental diagram of highway traffic

operators of the form $\sum_{\text{finite}} a^\alpha \frac{d^\alpha}{ds^\alpha}$ where $\alpha \in \mathbb{R}(s)$. This is a non-commutative ring according to Weyl algebra, since $\frac{d}{ds} s - s \frac{d}{ds} = [\frac{d}{ds} s, s \frac{d}{ds}] = 1$.

The finite set $\theta = \{\theta_1, \dots, \theta_r\}$ of constant parameters is said to be linearly identifiable with respect to a finite set $\mathbf{X} = \{x_1, \dots, x_k\}$ of signals, the input and output variables of a linear system for instance, if and only if it leads in the operational domain:

$$P\theta = Q \tag{6}$$

where P and Q are respectively $r \times r$ and $r \times 1$ matrices, and the entries of P and Q belong to span $_{\mathbb{R}(s)[\frac{d}{ds}]}(1, x_1, \dots, x_k)$. Moreover, $det(P) \neq 0$. Consider the additive perturbation, that is $y_i = x_i + \tilde{\omega}_i$, then equation (6) become (7): (7):

$$P\theta = Q + Q' \tag{7}$$

where, Q' is a $r \times 1$ matrix with entries depending now on $\tilde{\omega}_i$. If $\tilde{\omega}$ is structured, it means that $\tilde{\omega}_i$, $i = 1, 2, \dots, k$ satisfies a linear differential equation with polynomial coefficients, and there would exist $\Delta \in \mathbb{R}(s)[\frac{d}{ds}]$, such that by multiplying both sides of equation (7) by Δ annihilates the structured perturbations:

$$\Delta P\theta = \Delta Q \tag{8}$$

Multiplying both sides of equation (8) by suitable proper rational functions in $\mathbb{R}(s)$ yields proper rational functions in all the coefficients.

The unstructured perturbations are modeled as highly fluctuating noises, which can be attenuated by invariant low-pass filters, such as $F(s) = 1/s^\nu$, where $\nu \geq 1$ is a large enough real number.

4   The Parameter Identification Scheme

In the algebraic identification framework, a novel key parameters identification method is proposed, and the non-linearity characteristic of traffic dynamic is locally transformed into a linearization model approximately. The identification procedure consists of five steps.

Step 1: Convert the equilibrium speed-density function into a linearization form so that the linear parameter identification procedure can be acquired. The logarithmic style of equation (4) is shown as follows:

$$\ln V = \ln v_f + [-\frac{\rho^\alpha}{\alpha}(\frac{1}{\rho_{cr}})^\alpha] \tag{9}$$

Let $\ln V = V'$, $\rho^\alpha = \rho'$, $\theta_1 = \ln v_f$, $\theta_2 = -\frac{1}{\alpha}(\frac{1}{\rho_{cr}})^\alpha$, then the static speed equation would be the linear equation of the form:

$$V'(t) = \theta_1 + \theta_2 \rho'(t) \tag{10}$$

Step 2: Transform function (10) from time-domain to complex frequency domain by means of Laplace transformation. The operational calculus of (10) yields:

$$V'(s) = \frac{\theta_1}{s} + \theta_2 \rho'(s) \tag{11}$$

Step 3: In most parameter identification problem, the initial conditions are usually difficult to get and regarded as unknown disturbances, so the unknown initial conditions and the constant disturbance can be treated as structured terms and be annihilated by differentiating both sides with respect to s. That is, select the derivation operator $\Delta_1 = \frac{d}{ds}$ to multiply both sides of equation (11), we obtain

$$V'(s) + s\frac{d}{ds}V'(s) = \theta_2\rho'(s) + \theta_2 s\frac{d}{ds}\rho'(s) \tag{12}$$

Step 4: It is well known that the differential of noisy signals would amplify the noise. Moreover, a (strict) proper estimator can ensure the causality property of the system. Therefore, we select the derivation operator $\Delta_2 = s^{-\nu}$ $(\nu \geq 1)$ to multiply both sides of equation (12), in order to avoid the derivatives of the input and output signal in the constructed estimator. We obtain the algebraic expressions that only contain the integral of noisy signal.

$$\frac{V'(s)}{s^2} + \frac{1}{s}\frac{d}{ds}V'(s) = \theta_2[\frac{\rho'(s)}{s^2} + \frac{1}{s}\frac{d}{ds}\rho'(s)] \tag{13}$$

Furthermore, according to the identifiable definition aforementioned, there are two parameters to be identified, so another equation is required to construct a square matrix system. According to (11), it is readily obtained as follows:

$$\frac{V'(s)}{s} = \frac{\theta_1}{s^2} + \theta_2\frac{\rho'(s)}{s} \tag{14}$$

Step 5: Transfer equation (13) and (14) simultaneously to time-domain from frequency domain by inverse-Laplace transformation. Then make use of Cauchy formulation $(\int \cdots \int y(\tau_1)d\tau_1 \cdots \tau_1 = \int_0^t \frac{(t-\tau)^{l-1}y(\tau)}{(l-1)!}d\tau)$ to simplify the results. Finally, the parameter estimator in time-domain would be:

$$\theta_{1e} = \frac{1}{T}[\int_0^T v(\tau)d\tau - \theta_{2e}\int_0^T \rho(\tau)d\tau], \theta_{2e} = \frac{\int_0^T (T-2\tau)v(\tau)d\tau}{\int_0^T (T-2\tau)\rho(\tau)d\tau}$$

where θ_{1e} and θ_{2e} are the identification results of θ_1 and θ_2 respectively.

In the end, we can obtain the identification results:

$$\begin{cases} \hat{v}_f = e^{\theta_{1e}} \\ \hat{\rho}_{cr} = \frac{1}{\sqrt{-\alpha\alpha}} \end{cases} \tag{15}$$

5 Numerical Example

The simulation is validated using synthetic traffic data from a section of simulated highway and the inputs signals are the speed measurements and flow measurements from loop detectors. We assume $\xi_i^v(k)$ and $\xi_i^q(k)$ are zero-mean Gaussian white measurement noises respectively to reflect the measuring inaccuracies. That means the noisy speed and traffic flow measurements are [20]: $v_{mi}(k) = v_i(k) + \xi_i^v(k)$, $q_{mi}(k) = q_i(k) + \xi_i^q(k)$, and $\xi_i^v(k) \sim N(0, 0.01)$, $\xi_i^q(k) \sim N(0, 0.01)$. The identification result following the proposed procedure in noisy conditions is shown in figure 3. The initial parameter estimated are selected: $v_f(0) = 100$ km/h, $\rho_{cr}(0) = 20$ veh/km/lane, $\alpha = 1.5$.

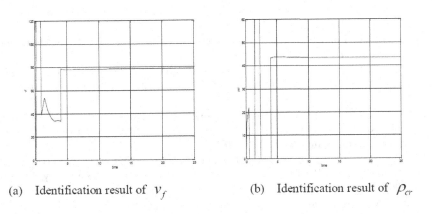

(a) Identification result of v_f (b) Identification result of ρ_{cr}

Fig. 3. Identification results

It can be seen that after a very short "warm up" period, with the evolution of the highway traffic dynamic, the key parameters of traffic model can be identified rapidly and the identification results are consistent with the pre-setting values. Therefore, we can apply the on-line identification result to subsequent traffic control scheme, such as ramp metering, and design the control algorithm, including model-free control, adaptive control, PI control and so on. Ultimately, this allows people to maneuver the traffic in desired fashion.

6 Conclusion

In this study, we have presented an on-line algebraic parameters identification scheme by means of differential algebra and operational calculus. The free speed and critical density of second macroscopic traffic model METANET are identi-fied form the linear style equation of static speed equation. Its effectiveness is demonstrated through numerical examples.

In an ongoing work, we perform further analysis and simulations based on real life highway data, moreover, in future we will integrate the on-line identification results with some important traffic control strategies, such as model predictive control (MPC), adaptive control and model-free control, so as to further improve the real-time highway traffic automation feasibility.

References

1. Luspay, T., Kulcs, B., Peni, T., et al.: Freeway ramp metering: an LPV set the-oretical analysis. In: American Control Conference, USA, CA, San Francisco, pp. 733–738 (2011)
2. Papageorgiou, M., Kostialos, A.: Freeway ramp metering: An overview. IEEE Transactions on Intelligent Transportation Systems 3, 271–281 (2002)
3. Yan, J.: Parameter identification of freeway traffic flow model and adaptive ramp metering. In: International Symposium on Electronic Commerce and Security, China, Nan Chang, pp. 235–238 (2005)
4. Cremer, M., Papageorgiou, M.: Parameter identification for a traffic flow model. Automatica 17, 837–843 (1981)
5. Kohan, R.R., Bortoff, S.A.: An observer for highway traffic systems. In: Proceeding of the 37th IEEE Conference on Decision and Control, USA, Florida, pp. 1012–1017 (1998)
6. Wang, Y., Ioannou, P.A.: Real-time parameter estimators for a second-order macro-scopic traffic flow model. In: IEEE Intelligent Transportation Systems Conference, Canada, Toronto, pp. 1466–1470 (2006)
7. Wang, Y.B., Papageorgiou, M., Messmer, A.: A real-time freeway network traffic surveillance tool. IEEE Transactions on Control Systems Technology 14, 18–32 (2006)
8. Fliess, M., Sira Ramirez, H.: An algebraic framework for linear identification. Con-trol Optimisation and Calculus of Variations 9, 151–168 (2003)
9. Fliess, M., Cedric, J., Mamadou, M.: Algebraic change-point detection. Applicable Algebra in Engineering Communication and Computing 21, 131–143 (2010)
10. Fliess, M., Sira Ramirez, H.: Closed-loop parametric identification for continuous-time linear systems via new algebraic techniques. In: Identification of Continuous-time Models from Sampled Data, pp. 362–391. Springer (2008)
11. Hassane, A.: An algebraic framework for traffic state estimation. International Journal of Innovative Computing, Information and Control 5, 2499–2510 (2009)
12. Hassane, A., Fliess, M., et al.: Towards a nonlinear characterization of a freeway network. In: 3èmes Journées Identification et Modélisation Expérimentale, JIME 2011, Douai, France (2011)
13. Villagra, J., Choi, S., et al.: Robust stop-and-go control strategy: an algebraic approach for nonlinear estimation and control. International Journal of Vehicle Autonomous Systems, 270–291 (2009)
14. Hassane, A., Fliess, M., et al.: Fast parametric estimation for macroscopic traffic flow model. In: 17th IFAC World Congress (2008)

15. Darbha, S., Rajagopal, K.R.: A review of mathematical models for the flow of traffic and some recent results. Nonlinear Analysis: Theory, Methods & Applications 69, 950–970 (2008)
16. Wang, Y.B., Papageorgiou, M.: Real-time freeway traffic state estimation based on extended Kalman filter: a general approach. Transportation Research (Part B) 39, 141–167 (2008)
17. Rudolph, J., Woittennek, F.: An algebraic approach to parameter identification in linear infinite dimensional systems. In: 16th Mediterranean Conference on Control and Automation Congress Centre, Ajaccio, France, pp. 332–337 (2008)
18. Sira-Ramirez, H.: Some Applications of Differential Algebra in Systems Identification. In: 3rd IFAC Symposium on System Structure and Control, Plenary Lecture
19. Mboup, M.: Parameter estimation via differential algebra and operational calculus. Tech. rep. Submitted to Signal Processing (2007)
20. Wang, Y.B., Papageorgiou, M., Messmer, A., et al.: An adaptive freeway traffic state estimator. Automatica 45, 10–24 (2009)

Influence of Coil Parameters on Rayleigh Waves Excited by Meander-Line Coil EMATs

Shujuan Wang, Zhichao Li, Lei Kang, and Guofu Zhai

School of Electrical Engineering and Automation, Harbin Institute of Technology,
150001, Harbin, P.R. China
hitclaire@163.com, victorkang11@126.com

Abstract. A 3-D model for Rayleigh wave EMATs operating on the Lorentz force mechanism has been established. Rayleigh waves generated by two Lorentz forces are calculated respectively. The ratio of wire width to spacing interval between neighboring wires (RWWSI) is chosen to study the coil parameter influence of Lorentz forces. The vibration amplitude of Rayleigh waves due to the dynamic magnetic field is almost proportional to the reciprocal of the RWWSI, whereas that due to the static magnetic field decreases slowly with the increase of the RWWSI. The divergence angles of Rayleigh waves due to dynamic magnetic field keep invariable when the RWWSI is less than 0.5, and that due to static magnetic field reach the minimum values and have better detectability when the RWWSI is 0.5. The critical excitation current, at which Rayleigh waves due to static and dynamic magnetic fields are equal, changes sharply when the RWWSI differs.

Keywords: Rayleigh wave, EMATs, Lorentz forces, RWWSI.

1 Introduction

In the field of non-destructive detection and non-destructive evaluation, Rayleigh waves are widely applied in detection of surface and sub-surface slots and defects [1-4]. There are various non-contact technologies for generating and detecting Rayleigh waves and the frequently used are lasers and electromagnetic acoustic transducers (EMATs) [1-4]. Independence of couplant opens up a wide variety of applications for EMATs [5], such as inspections at high speed or at high temperatures, which are unavailable to conventional piezoelectric transducer (PT) techniques.

The main disadvantage of EMATs is their low efficiency of generation and detection. Comparing with PT techniques, electromagnetic acoustic signals are extremely weak. In [1, 2], EMATs are designed as detectors of ultrasound because of their better detection performance than generation. Several theoretical models and experimental work have been done to study the coupling mechanism of EMATs [6-13]. Some work has been published on optimizing the design of EMATs for Rayleigh wave generation [11-13].

K. Li et al. (Eds.): ICSEE 2013, CCIS 355, pp. 94–103, 2013.
© Springer-Verlag Berlin Heidelberg 2013

In early researches, the influence of the dynamic magnetic field generated by the excitation current was ignored. With the development of electronics, the excitation current becomes larger and larger, and the dynamic magnetic field plays an unignorable role in generating ultrasounds. Some researchers have established Rayleigh wave EMAT models considering the influence of the dynamic magnetic field [11-13]. Two Lorentz forces are calculated, including the Lorentz force due to the static magnetic field and the Lorentz force due to the dynamic magnetic field. The relationship between Rayleigh wave displacements generated by these two Lorentz forces individually and the excitation current are calculated for spiral coil EMATs [11, 12] and meander-line coil EMATs [13] operating on the Lorentz force mechanism.

In industrial application, EMATs consisting of a meander-line coil and a static magnetic field are widely applied to generate and detect Rayleigh waves. Meander-line coils are designed with different lengths, widths and spacing intervals. These parameters will affect Rayleigh waves generated by these two Lorentz forces.

In this paper, finite element method and analytical solutions are combined to calculate Rayleigh waves generated by meander-line coil EMATs working on the Lorentz force mechanism. Based on this method, the influence of coil parameters on generating Rayleigh waves are compared and analyzed.

2   Modeling Method

Electromagnetic acoustic transducers mainly consist of three parts: a coil, a static magnetic field and a sample under test. The working process of an EMAT operating on Lorentz force mechanism is illustrated in Fig.1.

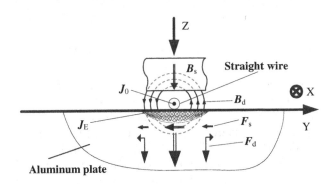

Fig. 1. Lorentz force mechanism in aluminum plate

When an alternating current I_0 passes through the coil, the current density J_0 will generate a dynamic magnetic field B_d in the surrounding air and the aluminum plate, and the dynamic magnetic field will induce an eddy current J_E

in the aluminum plate. This eddy current experiences Lorentz forces under the interaction with the dynamic magnetic field B_d from the coil itself and also with the external static magnetic field B_s. The Lorentz force F_d resulted from the dynamic magnetic field always acts like a repulsive force between the aluminum plate and the coil, whereas the Lorentz force F_s due to the static magnetic field is equal to the cross product of the eddy current vector and the external static magnetic field. These two forces work together to generate vibrations under the coil. The Cartesian coordinate system is shown in Fig. 1, and the origin locates on the surface of the aluminum plate. Finite element method (FEM) is an effective way to calculate Lorentz forces generated by EMATs [6-9,11-13].

In isotropic homogeneous elastic material, the propagation of elastic wave is

$$\mu \bigtriangledown \times (\bigtriangledown \times \boldsymbol{u}) - (\lambda + 2\mu) \bigtriangledown (\bigtriangledown \boldsymbol{.} \boldsymbol{u}) + \rho \partial^2 \boldsymbol{u}/\partial t^2 = \boldsymbol{F} \ . \tag{1}$$

where, $\boldsymbol{u}$ is the displacement vector of elastic wave; λ and μ are Lamé constants; $\boldsymbol{F}$ is the body forces.

Driven by the Lorentz forces, Rayleigh waves will be generated in the aluminum plate. Rayleigh waves' displacements consist of two components: the in-plane displacement and the out-of-plane displacement. Rayleigh waves generated by surface point forces have been documented in early researches [10, 11-16].

Lorentz forces mainly exist under the meander-line coil within the skin depth. Assuming Lorentz forces have been decomposed into N point forces and the Rayleigh wave displacement at (X, Y, Z) generated by the i_{th} point Lorentz force component F_{Li} is $\boldsymbol{u}_i$, the total displacement $\boldsymbol{u}$ for the whole force components of a meander-line coil EMAT is given by

$$\boldsymbol{u}(t; X, Y, Z) = \sum_{i=1}^{N} \boldsymbol{u}_i(t; X, Y, Z; \boldsymbol{F}_{Li}) \ . \tag{2}$$

3   Modeling Process of Rayleigh Wave EMATs

Rayleigh wave displacements generated by Lorentz force due to dynamic magnetic field $\boldsymbol{f}_d$ and that due to static magnetic field $\boldsymbol{f}_s$ should be calculated separately [13, 17]. In this paper, two EMAT models are built: 1) a whole model which consists of a coil, a magnet and an aluminum plate; and 2) a dynamic magnetic field model (DMF model) which consists of a coil and an aluminum plate. The whole model describes a complete working process of an EMAT which contains the influence of both $\boldsymbol{f}_d$ and $\boldsymbol{f}_s$, and the Lorentz force $\boldsymbol{f}_L$ in the whole model is the composite force of $\boldsymbol{f}_d$ and $\boldsymbol{f}_s$. The DMF model describes a working process of an EMAT without the static magnetic field, and the Lorentz force is completely generated by the dynamic magnetic field. As a result, the Lorentz force in DMF model is $\boldsymbol{f}_d$. Thus, $\boldsymbol{f}_s$ can be acquired by (7). The 3-D modeling process of Rayleigh wave EMATs is illustrated in Fig. 2.

This modeling method is valid and effective to calculate Rayleigh waves generated by EMATs working on Lorentz force mechanism [13].

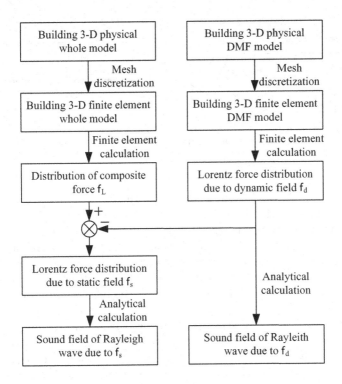

Fig. 2. 3-D modeling process of Rayleigh wave EMATs

The Lorentz force distribution depends on the EMAT configuration, including the coil and the magnet configuration. Some parameters of the coil affects the Lorentz force distribution, such as wire number N_C, width W_C, length L_C and thickness T_C, spacing interval between wires a, lift-off distance between the coil and the aluminum plate G, and excitation current amplitude A_I. For meander-line coil EMATs, cuboid magnets are mainly adopted to generate Rayleigh waves, so the magnet width W_M, length L_M and thickness T_M should be considered. Structural parameters of meander-line coil EMATs are illustrated in Fig. 3.

In early research [18], we found that the wire width is the key parameter affecting Rayleigh waves generated by meander-line coil EMATs. The parameter values of meander-line coil EMAT are chosen according to the practical application, and the working frequency is 500 kHz. In aluminum, the Rayleigh wave velocity is about 2930 m/s, so the spacing intervals between neighboring wires are 2.93 mm. The coil has 12 wires and each wire has a length of 35 mm, and a thickness of 0.035 mm. The lift-off distance between the coil and the aluminum plate is 0.1 mm. Only the part of aluminum plate where the electrical-acoustic energy conversion takes place is modeled and its size is 50 mm×50 mm×1.58 mm. The resistivity of the aluminum is 2.6×10^{-8} Ω·m. The mostly used Nd-Fe-B is taken as permanent magnet, whose remnant magnetism is 1.21 T.

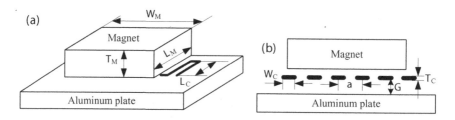

Fig. 3. Structural parameters of meander-line coil EMATs. (a) is the 3-D view and (b) is the X-direction view.

A meander-line coil EMAT with a certain parameter combination is taken as an example to illustrate the modeling process. First, 3-D physical whole model and DMF model of the meander-line coil EMAT are established. Tetrahedron elements are adopted to divide the physical models into finite element models. The region under the magnet is further subdivided. The refinement significantly improves the calculation accuracy, whereas the calculation time increases greatly. In practice, when the calculation results hardly vary with subdivision, further refinement is unnecessary.

The Lorentz force density distribution can be calculated. To make sure the obtained Lorentz forces approximate to point forces, the force action region should be subdivided into tiny hexahedron elements. Simulation results show that Lorentz forces mainly concentrated with an area about 1.5 times of the coil's outline on the plate surface within the skin depth. Beyond this region, Lorentz forces are less than 1% of that in the region. So this region is exported to point forces by a 3-D grid.

To clearly describe the Rayleigh waves, the aluminum plate surface is also divided into many square elements. After dividing Lorentz forces into point forces, Rayleigh wave displacements at each element on the aluminum plate surface can be calculated, and displacements at other points can be acquired by interpolation method.

4   Simulation and Analysis

In these models, tone-burst current signals with different amplitudes are sent into meander-line coils. The energy of Rayleigh waves is proportional to the square of their vibration amplitudes, and the divergence angle is defined as sound beam at -3dB value below the peak response, which is related to the detection area. The detectability of Rayleigh wave EMATs depends on the vibration amplitude and the divergence angle. In this paper, these two parameters are chosen as main targets to study Rayleigh wave distributions.

Fig. 4 illustrates the Rayleigh wave sound distribution on aluminum plate generated by meander-line coil EMATs, which shows that Rayleigh waves have good directivity. The Rayleigh wave vibration amplitude and divergence angle can be obtained through the sound field distribution.

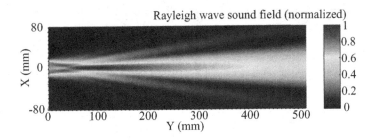

Fig. 4. Calculated Rayleigh wave sound field distribution on aluminum plate surface. The maximum Rayleigh wave vibration amplitude is normalized.

It was reported that for a spiral coil EMAT, the Lorentz force due to the dynamic magnetic field F_d generates Rayleigh waves more efficiently than that due to the static magnetic field when the p-p amplitude of the excitation current is about 300 A [11, 12]. However, for a meander-line coil EMAT, this excitation current is 528.9 A [13]. This difference shows that the EMAT configuration affects the Lorentz force distribution and also the generated Rayleigh wave distribution. It is interesting to study the influence of Lorentz forces due to static and dynamic magnetic fields for commonly used meander-line coil EMATs with different parameters.

First, the current amplitude is set to be constant. The current amplitude is 100 A. The relationships between the divergence angle and vibration amplitude of Rayleigh waves and the ratio of wire width to spacing interval between neighboring wires (RWWSI) is shown in Fig. 5 and Fig. 6.

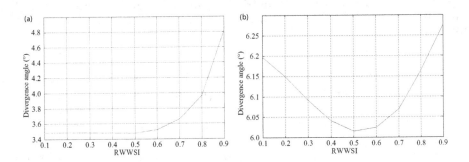

Fig. 5. Relationship between the Rayleigh wave divergence angle and the RWWSI. Rayleigh waves are generated by Lorentz forces (a) due to the dynamic magnetic field and (b) due to the static magnetic field respectively.

Fig. 5 indicates that the divergence angle of Rayleigh wave generated by Lorentz force due to the dynamic magnetic field keeps invariable when the RWWSI is less than 0.5, whereas that due to the static magnetic field reaches

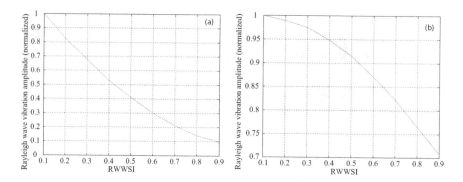

Fig. 6. Relationship between the Rayleigh wave vibration amplitude and the RWWSI. Rayleigh waves are generated by Lorentz forces (a) due to the dynamic magnetic field and (b) due to the static magnetic field respectively. The maximum vibration amplitudes are normalized.

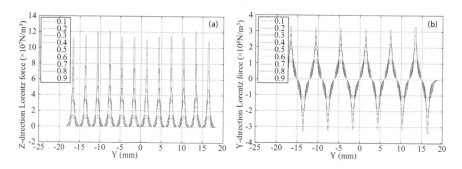

Fig. 7. (a) Z-direction Lorentz force distributions due to dynamic magnetic field and (b) Y-direction Lorentz force distributions due to static magnetic field with different RWWSI

the minimum value when the RWWSI is 0.5. Fig. 6 illustrates that Rayleigh waves generated by Lorentz force due to the dynamic magnetic field decreases sharply with the increase of the RWWSI. Whereas Rayleigh waves generated by Lorentz force due to the static magnetic field decreases slowly when the RWWSI increases.

The excitation current density decreases almost linearly with the increase of the RWWSI. The Lorentz force distribution due to the dynamic magnetic field has close relationship with the excitation current amplitude, and the force amplitude is proportional to the inverse square of the RWWSI, whereas the Lorentz force due to the static magnetic field is proportional to the reciprocal of the RWWSI, as shown in Fig. 7. Z-direction force is the main component of the Lorentz force due to the dynamic magnetic field, whereas Y-direction force is the main component of the Lorentz force due to the static magnetic field.

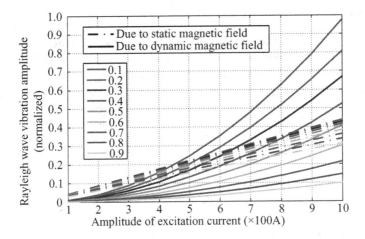

Fig. 8. Relationship of Rayleigh wave vibration amplitude with excitation current amplitude for different RWWSI. The maximum Rayleigh wave vibration amplitude is normalized.

The Lorentz force acting region is decomposed into many point forces, and each point force has contribution to Rayleigh waves generated on the aluminum plate. With the increase of the RWWSI, not only the Lorentz force amplitude changes but also the force distribution changes. The interference between each point force components differs. This may cause the Rayleigh wave vibration amplitude is proportional to the reciprocal but not the inverse square of the RWWSI. When the RWWSI is less than 0.5, the acting area of the Lorentz force due to the dynamic magnetic field hardly changes, so the divergence angles keep invariable. However, the acting area becomes larger when the RWWSI is larger than 0.5, which causes the divergence angle becomes larger. In contrast, the acting areas of the Lorentz force due to the static magnetic field changes linearly with the increase of the RWWSI. When the RWWSI is 0.5, the interference of each point Lorentz force component is most obvious and generates Rayleigh waves with the minimum divergence angle.

Rayleigh waves generated by the Lorentz force due to the dynamic magnetic field have larger energy and better directivity when the RWWSI is smaller. Whereas for Rayleigh waves generated by the Lorentz force due to the static magnetic field, the RWWSI is recommended to be 0.5.

As has mentioned, the RWWSI has different influences on the Rayleigh waves generated by Lorentz forces due to dynamic and static magnetic fields respectively. So it is assumed the critical current mentioned in [13] will also be affected by the RWWSI. The relationship between the Rayleigh wave vibration amplitude due to static and dynamic magnetic fields respectively with the excitation current amplitude are calculated as shown in Fig. 8. The critical current varies with the RWWSI and changes sharply when the RWWSI increases. It is proved that ignoring the contribution of the dynamic magnetic field is problematic [11-13]. The divergence angles of Rayleigh waves keep invariable when the RWWSI changes.

5   Conclusion

To study the coil parameter influence on Rayleigh waves generated by meander-line coil EMATs, 3-D models have been established by combination of FEM and analytical solutions. Lorentz force distributions due to static and dynamic magnetic fields can be calculated respectively by FEM. Rayleigh waves generated by the Lorentz forces are obtained by analytical solutions.

The Rayleigh wave vibration amplitude and divergence angle have close relationship with the RWWSI. The vibration amplitude of Rayleigh waves generated by Lorentz forces due to the dynamic magnetic field is almost proportional to the reciprocal of the RWWSI, whereas that due to the static magnetic field decreases slowly when the RWWSI increases. The divergence angle of Rayleigh waves generated by Lorentz forces due to the dynamic magnetic field keeps invariable when the RWWSI is less than 0.5, and that due to the static magnetic field reach a minimum value when the RWWSI is 0.5.

Rayleigh waves generated by Lorentz force due to the dynamic magnetic field have larger energy and better directivity when the RWWSI is smaller. Whereas for Rayleigh waves generated by Lorentz force due to the static magnetic field, the RWWSI is recommended to be 0.5.

The critical excitation current, at which Rayleigh wave displacements generated by Lorentz forces due to static and dynamic magnetic fields respectively are equal, changes sharply with the increase of the RWWSI. When the RWWSI is small, Lorentz forces due to the dynamic magnetic field will be more efficient in generating Rayleigh waves at a low excitation current.

References

1. Boonsang, S., Dewhurst, R.J.: Enhancement of Laser-Ultrasound/Electromagnetic Acoustic Transducer Signals From Rayleigh Wave Interaction at Surface Features. Applied Physics Letters 82(19), 3348–3350 (2003)
2. Boonsang, S., Dewhurst, R.J.: Signal Enhancement in Rayleigh Wave Interactions Using a Laser-Ultrasound/EMAT Imaging System. Ultrasonics 43, 512–523 (2005)
3. Edwards, R.S., Dixon, S., Jian, X.: Characterisation of Defects in the Railhead Using Ultrasonic Surface Waves. NDT&E International 39, 468–475 (2006)
4. Jian, X., Dixon, S., Guo, N., Edwards, R.S., Potter, M.: Pulsed Rayleigh Wave Scattered at a Surface Crack. Ultrasonics 44, e1131–e1134 (2006)
5. Hirao, M., Ogi, H.: EMATs for Science and Industry: Noncontacting Ultrasonic Measurements. Kluwer Academic Publishers (2003)
6. Thompson, R.B.: A Model for the Electromagnetic Generation and Detection of Rayleigh and Lamb Waves. IEEE Trans. Sonics and Ultrasonics SU-20(4), 340–346 (1973)
7. Ludwig, R.: Numerical Implementation and Model Predictions of a Unified Conservation Law Description of the Electromagnetic Acoustic Transduction Process. IEEE Transactions on Ulrtasonics, Ferroelectrics, and Frequency Control 39(4), 481–488 (1992)
8. Shapoorabadi, R.J., Sinclair, A.N., Konrad, A.: Improved Finite Element Method for EMAT Analysis and Design. IEEE Transactions on Magnetics 37(4), 2821–2823 (2001)

9. Shapoorabadi, R.J., Sinclair, A.N., Konrad, A.: The Governing Electrodynamic Equations of Electromagnetic Acoustic Transducers. J. Appl. Phys. 10E102, 1–3 (2005)
10. Kawashima, K.: Electromagnetic Acoustic Wave Source and Measurement and Calculation of Vertical and Horizontal Displacements of Surface Waves. IEEE Transactions on Sonics and Ultrasonics SU-32(4), 514–522 (1985)
11. Jian, X., Dixon, S., Edwards, R.S.: Ultrasonic Field Modeling for Arbitrary Non-Contact Transducer Source. In: Third Intl. Conf. on Experimental Mechanics and Third Conf. of the Asian Committee on Experimental Mechanics, vol. 5852, pp. 515–519 (2005)
12. Jian, X., Dixon, S., Grattan, K.T.V., Edwards, R.S.: A Model for Pulsed Rayleigh Wave and Optimal EMAT Design. Sensors and Actuators A 128, 296–304 (2006)
13. Wang, S.J., Kang, L., Li, Z.C., Zhai, G.F., Zhang, L.: 3-D Modeling and Analysis of Meander-Line-Coil Surface Wave EMATs. Mechatronics 22(6), 653–660 (2012)
14. Lamb, H.: On the Propagation of Tremors Over the Surface of an Elastic Solid. Phil. Trans. R. Soc. Ser. 203, 1–42 (1904)
15. Nakano, H.: Some Problems Concerning the Propagations of the Disturbance in and on Semi-Infinite Elastic Solid. Geophys. Mag. 2, 189–348 (1930)
16. Bresse, L.F., Hutchins, D.A.: Transient Generation of Elastic Waves in a Solid by a Disk-Shaped Normal Force Source. J. Acoust. Soc. Am. 86(2), 810–816 (1989)
17. Wang, S.J., Li, Z.C., Kang, L., Hu, X.G., Zhan, X.: Modeling and Comparison of Three Bulk Wave EMATs. In: Proc. 37th Annual Conference of the IEEE Industrial Electronics Society, pp. 2658–2673 (2011)
18. Wang, S.J., Kang, L., Li, Z.C., Zhai, G.F.: 3-D Finite Element Analysis and Optimum Design of Electromagnetic Acoustic Transducers. Proceedings of the CSEE 29(30), 12–128 (2009) (in Chinese)

Recursive Model Predictive Control
for Fast Varying Dynamic Systems

Da Lu, Guangzhou Zhao, and Donglian Qi

College of Electrical Engineering, Zhejiang University,
310027 Hangzhou, China
http://www.springer.com/lncs

Abstract. A well known drawback of model predictive control (MPC) is that it can only be adopted in slow dynamics, where the sample time is measured in seconds or minutes. The main reason leads to the problem is that the optimization problem included in MPC has to be computed online, and its iterative computational procedure requires long computational time. To shorten computational time, a recursive approach based on Iterative Learning Control (ILC) and Recursive Levenberg Marquardt Algorithm (RLMA) is proposed to solve the optimization problem in MPC. Then, recursive model predictive control (RMPC) is proposed to realize MPC for fast varying dynamic systems. Simulation results show the effectiveness of RMPC compared with conventional MPC.

Keywords: Model Predictive Control, Recursive Levenberg Marquardt Algorithm, Iterative Learning Control.

1 Introduction

The basic idea of Model Predictive Control (MPC) is to improve the future plant behavior by computing a sequence of future manipulated variable adjustments based on a dynamic model of plant. Only the first element in the optimal sequence is applied to the system. This process is repeated at every sampling interval to update information [1]. It is well-known that an optimization problem needs to be solved online to obtain future manipulated variable in MPC. The heavy computational burden has limited the application of MPC to slow dynamic systems for a long time. However, in recent years, some cheering improvements have been made, which lead to the possibility of applying MPC to fast varying dynamic systems [2–4].

Generally speaking, the emerging MPC methods for fast varying dynamic systems can be divided into two groups. The first is to reduce MPC optimization problem to the selection of system behavior from finite input sequences[2, 3]. However, since the set of possible input sequences is required to be finite, modulation methods, such as space vector modulation, are difficult to be enrolled in the scheme. In the second approach, MPC optimization problem is solved offline, and the input sequence can be obtained through simple online calculation,

K. Li et al. (Eds.): ICSEE 2013, CCIS 355, pp. 104–112, 2013.
© Springer-Verlag Berlin Heidelberg 2013

as presented in[4]. The limitation of the approach is that the mismatch between the plant and the model can cause control performance degrading.

The main purpose of this paper is to further expand the application of MPC in fast dynamic systems. The possible contribution is that a recursive MPC (RMPC) algorithm is proposed. In traditional MPC, several passes are made through the data to iteratively improve the optimal result, which is major part of the computational burden[5]. It is a natural idea to replace the iteration by recursive method. However, recursive methods can not work here effectively. The reason is that only the information in the first predictive step can be utilized to obtain the first term which is actual input in MPC. To solve the problem, Iterative Learning Control (ILC)[10] is adopted in this paper to use the limited information sufficient to improve the control performance of the first term. Moreover, the first term obtained from ILC can provide a satisfied start point for recursive algorithms. Then the recursive Levenberg Marquardt Algorithm (RLMA) [6] is adopted to obtain other inputs in control horizon.

As the structure of ILC is chosen to be very simple, the amount of computation in ILC is quite small. Meanwhile, RLMA, which derives input one by one, only needs short computational time. Therefore, RMPC can reduce the computational burden of conventional MPC significantly.

2   RMPC Method

In MPC, the controller selects the next input sequence based on the prediction of the future system state behavior. Precisely speaking, the sequences that optimize a given cost function is chosen. The controller in RMPC also utilizes the above strategy.

2.1   Problem Formulation

For convenience to compare, MPC and RMPC use the same plant model:

$$
\begin{aligned}
x(k+1) &= Ax(k) + Bu(k) \\
y(k) &= Cx(k)
\end{aligned}
\tag{1}
$$

A quadratic cost function has been preferred as follows:

$$
J = \sum_{i=1}^{P} (e(k+i)Qe(k+i))
\tag{2}
$$
$$
e(k+i) = y(k+i) - y^r(k+i)
$$
$$
y \in \Psi \in \Re^P, u \in \Upsilon \in \Re^M
$$

where Q is positive semidefinite matrix to weight the output vector, y^r is reference trajectory. The time interval over which process inputs are computed to optimize the future plant behavior is known as prediction horizon. P denotes the length of the prediction horizon. Meanwhile, the time interval for adjusting inputs is named control horizon. M denotes the length of the control horizon. The process outputs are referred.

Instant punishing in the cost function (2), a predefined output reference trajectory is used here to avoid aggressive MV moves:

$$y^r(k+i) = \alpha^i y(k) + (c - y(k))(1 - \alpha^i)$$
$$\alpha \in [0, 1]$$

(3)

where c is the setpoint, α is a tunable parameter.

2.2   Solving Optimization Problem in MPC

For controllers use a quadratic cost function as (2), the dynamic optimization takes the form of a quadratic program (QP)[7]. There are a variety of methods that are commonly used to solve QP. Most of them replace the inequality constraints in the QP with linear equality constraints, or replace the constrained optimization problem with unconstrained optimization problem. Then, the replacing optimization problem can be solved by iterative algorithms such as Levenberg Marquardt Algorithm (LMA) or Gauss-Newton algorithm (GNA).

During the operation of LMA or GNA, several passes are made through the data to iteratively improve the optimal result[9]. However, for very fast processes, there may not be sufficient time available to complete the iteration. As the iteration part of solving QP is the main cause of computational burden in MPC, it is a nature idea to use a recursive method to replace the iterative method.

2.3   Solving Optimization Problem in RMPC

RLMA [6] can solve an optimization problem effectively. However, RLMA cannot be directly adopted to minimize the cost function in MPC. The reason is that RLMA operates based on information in the present step and steps before. So RLMA can only improve the result of optimization problem gradually. As a consequence, the first control input obtained from RLMA which is chosen as actual input in MPC often lead to a poor result.

Considering the first MV needs to make the predictive trajectory somewhat close to the final solution and the available information is just the predict result of the first step, we should use the limited information sufficiently. As a result, the ILC [8] is adopted here to obtain the first MV. Comparing with RLMA which uses the information of the first step only once, the ILC utilize the information of the first step several times to obtain a better result.

As the target of the proposed algorithm is to control fast varying dynamic systems, a P-type ILC is chosen to obtain the first MV as follows[10]:

$$u_{j+1}(k) = u_j(k) + Le_j(k+1) \tag{4}$$

where subscript j represent the iteration index. is a parameter matrix in P-type ILC.

It should be noted that, (4) works under unconstrained condition. Although constrained ILC [11, 12] can be used here conveniently, it does not need to do so for three reasons. Firstly, it has been proved that model (1) using input derived from (4)can converge to the setpoint [10], so it can satisfy the output constraint. Secondly, choosing a suitable α in reference trajectory (3) can avoid too aggressive MV moves. Thirdly, unconstrained ILC need less computational time.

As the first MV has made the predictive trajectory close to the final solution, it is relatively simple to obtain other MVs $u(k+i)i = 1...M-1$ using RLMA [6]:

$$u(k+i+1) = u(k+i) + N(k+i)\Psi(k+i)e(k+i)$$
$$N(k+i) = N(k+i-1) - N(k+i-1)\Psi^*(k+i)S(k+i)^{-1}\Psi^*(k+i)^T N(k+i-1)$$
$$S(k+i) = \Psi^*(k+i)^T N(k+i-1)^{(} - 1)\Psi^*(k+i) + \Lambda^*(k+i) \tag{5}$$
$$\Psi(k) = \frac{\partial J}{\partial u(k)}, \Psi^*(k+i) = \left(\frac{\Psi(k+i)^T}{0\cdots1\cdots0}\right)^T, \Lambda^*(k+i)^{-1} = \begin{pmatrix} 1 & 0 \\ 0 & \eta \end{pmatrix}$$

$\Psi^*(k+i)$ is a M by 2 matrix, the second column of $\Psi^*(k+i)$ is designed to deal with the damped term in LMA. The 1 in the second column of $\Psi^*(k+i)$is placed at $(tmod(M-1))+1$. η is the damping factor. After an elapse of $M-1$ time units, (5) is virtually the same as LMA.

The flowchart of RMPC in each predictive horizon is shown in Fig.1.

Suppose the dimension of $u(k+i)$ is d. It can be observed from (4) and (5) that RMPC requires the computation complexity of $\varnothing(d^2)$ to solve an optimization problem recursively. In contrast, the computation complexity of conventional

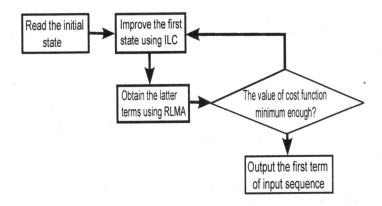

Fig. 1. Flowchart of RMPC in one predictive horizon

MPC to solve an optimization problem is $\emptyset((dP)^3)$ [6]. As the order of magnitude of P is usually 1, the computational burden of RMPC is much smaller than that of MPC.

3   Convergence Analysis of RMPC

Set the weights matrix Q in (2) to I for convenience. The RMPC converges if the cost function satisfies.

$$J \leq \varepsilon \tag{6}$$

where ε is a sufficient small real number given as required tolerance. Then (6) can be achieved if all the steps in a predictive horizon satisfy:

$$J_i = e(k+i)e(k+i) < \frac{\varepsilon}{P} \qquad i = 1 \cdots P \tag{7}$$

The convergence of the P-type algorithm for LTI plant (1) has been well established in the literature [10]. Some important analysis results are described below. System (1) is equivalent to:

$$y_i(k) = C(qI - A)^{-1}Bu_j(k) + CAx_0 \tag{8}$$

where $x_j(k) = x_0$, q is the forward time-shift operator $qx(k) \equiv x(k+1)$.

Let $H = C(qI - A)^{-1}B$, $\rho(A) = max_i |\lambda_i(A)|$ be the spectral radius of the matrix A , and $\lambda_i(A)$ be the i th eigenvalue of A ranked in descending (ascending) order. Then system (1), (4) is convergent if

$$\rho(I - LH) < 1 \tag{9}$$

Hence, if ILC runs enough number of circles, the first MV in input sequences can achieve (7), i.e.:

$$J_1 \leq \varepsilon \tag{10}$$

As RLMA is virtually the same as LMA after an elapse of $M-1$ time sample, its convergence properties are like LMA. As is known to all, the damping factor h in LMA can be adjusted to guarantee local convergence of the algorithm. However, LMA may not converge nicely if the initial guess is far from the solution [13]. Fortunately, the first MV derived from ILC can provide good initial value(s) for LMA. So the local convergence of LMA can be guaranteed. As (2) is a convex function, if the problem has a feasible solution, the global optimum is unique. So LMA can guarantee global convergence in this problem. As RLMA is virtually same as LMA after an elapse of $M-1$ time sample, a nice convergence result can be expected from RLMA. Hence

$$\frac{\varepsilon}{P} \geq J_1 \geq J_2 \geq \cdots \geq J_p \tag{11}$$

According to (6), RMPC is convergent.

4   Case Study

The RMPC described above is applied to the Permanent Magnet Synchronous Motor (PMSM) current control to verify its effectiveness. The dynamics of the PMSM are modeled in the dq reference frame [14]:

$$\begin{cases} i_d(k+1) = i_d(k) + \frac{T}{L_d}(-Ri_d(k) + n_p\omega L_q i_q(k)) + \frac{T}{L_d}u_d(k) \\ i_q(k+1) = i_q(k) + \frac{T}{L_q}(-Ri_q(k) - n_p\omega L_d i_d(k) - n_p\omega\phi) + \frac{T}{L_q}u_q(k) \end{cases} \tag{12}$$

where i_d, i_q are d and q components of the stator current, u_d,u_q are d and q components of the stator voltage. The parameters of (12) are listed in Table.1. It should be noted that T in (12) is the sample interval, which is required to be around 0.2 ms to obtain good performance for PMSM[15].

Table 1. Specifications of the PMSM

Symbols	Values	Units
L_d	0.000334	H
L_q	0.000334	H
R	0.4578	Ω
n_p	4	
ϕ	0.171	Wb
ω	20	rad/s
T	0.2	ms

The RMPC controller is designed with cost function as shown in (2), where $y(k+i) = [i_d(k+i) \quad i_q(k+i)]^T$, initial value $y_0(k+i) = [3 \quad 0]^T$,$y^r(k+i) = [10 \quad 5]^T$. An input sequence is acceptable if $J < 2$. Design parameter Q is set as diag[1 1], learning filter$L$ in ILC is diag[10 10]. The initial value of η is 1, but η is updated in every predict step to guarantee that every step is effective in decreasing J. αin (3) is 0.6.The simulation is based on Matlab/Simulink.

When P is 20 and M is 10, the values of the cost function in a prediction horizon of RMPC are shown in Fig.2. It can be seen from Fig.2 that, all the CVs are within limits. However, J converged before 10, but diverged after that. It is because the RMPC only utilize the information of present step. All the

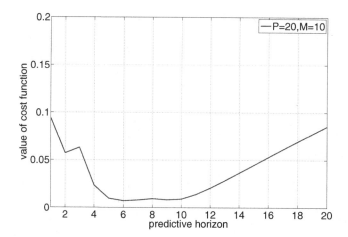

Fig. 2. Values of the cost function in a prediction horizon of RMPC when P is 20 and M is 10

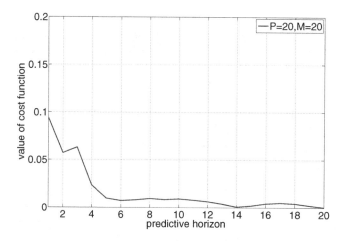

Fig. 3. Values of the cost function in a prediction horizon of RMPC when P is 20 and M is 20

information beyond present step is unknown for RMPC. If P is larger than M, the performance of CVs out of manipulated horizon can not guaranteed.

When RMPC consider all CVs, it is P is 20 and M is 20 in this example, the values of the cost function in a prediction horizon of RMPC is shown in Fig.3. When M is equal to P, the algorithm can achieve satisfied result.

The computational time of one prediction horizon under different conditions for RMPC and MPC is shown in Table.2. The computational time of RMPC is much less than that of MPC. Furthermore, the computational time of RMPC shows small increases with the increase on M and P. So the advantage of RMPC in computational time is more apparent as M and P increase. It should be mentioned that the computational time of RMPC is longer than $0.2ms$.

Table 2. Specifications of the PMSM

	RMPC			MPC	
M	P	Time(s)	M	P	Time(s)
5	20	0.00071	5	20	0.074
10	20	0.00075	10	20	0.19
20	20	0.00092	20	20	0. 097
10	35	0.0015	10	5	> 5
10	50	0.0022	10	50	> 5

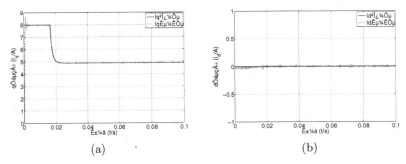

(a)   (b)

Fig. 4. The performance of RMPC controller. (a) q-axis current, (b) d-axis current.

The reason is that the approach is completed in Matlab. In practice, C language is used and controller is often DSP. So it is expectable that the approach can operate faster in practice.

The full picture of the performance of RMPC is shown in Fig.4. The set point of i_d is 0. To provide setpoints for i_q , a PI controller is used in PMSM speed control. The set point of PMSM speed is $100 rad/s$. Fig.4 shows that actual values can tract setpoints quickly and precisely.

5   Conclusion

In this paper, the RMPC suitable for fast varying dynamic systems has been proposed. While maintaining the general structure of a conventional MPC, the proposed approach reduces the computational time significantly. Simulation results have confirmed the effectiveness of the proposed RMPC with a comparison with the conventional MPC.

References

1. Chen, W.H., Ballance, D.J., Gawthrop, P.J.: Optimal control of nonlinear systems: a predictive control approach. Automatica 39, 633–641 (2003)
2. Geyer, T., Papafotiou, G., Morari, M.: Model predictive direct torque control Part I: Concept, algorithm, and analysis. IEEE Transactions on Industrial Electronics 56, 1894–1905 (2009)

3. Kouro, S., Cortes, P., Vargas, R.: Model predictive control: simple and powerful method to control power converters. IEEE Transactions on Industrial Electronics 56, 1826–1838 (2009)
4. Mattavelli, P.: An improved deadbeat control for UPS using disturbance observers. IEEE Transactions on Industrial Electronics 52, 206–212 (2005)
5. Cao, Y.: A formulation of nonlinear model predictive control using automatic differentiation. Journal of Process Control 15, 851–858 (2005)
6. Ngia, L.S.H., Sjoberg, J.: Efficient training of neural nets for nonlinear adaptive filtering using a recursive Levenberg-Marquardt algorithm. IEEE Transactions on Signal Processing 48, 1915–1927 (2000)
7. Mishra, S., Topcu, U., Tomizuka, M.: Optimization-Based Constrained Iterative Learning Control. IEEE Transactions on Control Systems Technology 19, 1–9 (2011)
8. Lee, J.H., Lee, K.S., Kim, W.C.: Model-based iterative learning control with a quadratic criterion for time-varying linear systems. Automatica 36, 641–657 (2000)
9. Nelles, O.: Nonlinear system identification: from classical approaches to neural networks and fuzzy models. Springer, Berlin (2001)
10. Moore, K.L.: An observation about monotonic convergence in discrete-time, P-type iterative learning control. In: 2001 IEEE International Symposium on Intelligent Control, pp. 45–49. IEEE Press, New York (2001)
11. Freeman, C., Tan, Y.: Point-to-point iterative learning control with mixed constraints. In: 2011 American Control Coference, pp. 3657–3662. Conference Publications, San Francisco (2011)
12. Mishra, S., Topcu, U., Tomizuka, M.: Iterative learning control with saturation constraints. In: 2009 American Control Coference, pp. 943–948. Conference Publications, St. Louis (2009)
13. Nocedal, J., Wright, S.J.: Numerical optimization. Springer, Berlin (1999)
14. Qian, W., Panda, S.K., Xu, J.X.: Torque ripple minimization in PM synchronous motors using iterative learning control. IEEE Transactions on Power Electronics 19, 272–279 (2004)
15. Mohamed, Y.A.R.I.: Design and implementation of a robust current-control scheme for a PMSM vector drive with a simple adaptive disturbance observer. IEEE Transactions on Industrial Electronics 54, 1981–1988 (2007)

Distributed Optimization and State Based Ordinal Potential Games

Jianliang Zhang, Guangzhou Zhao, and Donglian Qi

Department of Electrical Engineering, Zhejiang Uinversity, Hangzhou 310027, China
{jlzhang,zhaogz,qidl}@zju.edu.cn

Abstract. The focus of this paper is to develop a theoretical framework to analyze and address distributed optimization problem in multi-agent systems based on the cooperative control methodology and game theory. First the sensing/communication matrix is introduced and the minimal communication requirement among the agents is provided. Based on the matrix communication model, the state based ordinal potential game is designed to capture the optimal solution. It is worth noting that the proposed methodology can guarantee the distributed optimization problem converge to desired system level objective, even though the corresponding communication topologies may be local, time-varying and intermittent. Simulations on a multi-agent consensus problem are provided to verify the validness of the proposed methodology.

Keywords: Distributed optimization, multi-agent system, potential games, consensus problem.

1 Introduction

Distributed coordination of dynamic agents in a group plays an important role in many practical applications ranging from unmanned vehicles, automated highway systems, weapon target assignment and wireless sensor networks communication, etc. As a result, the central problem for multi-agent system is to design local control laws such that the group of agents can reach consensus on the shared information in the presence of limited and unreliable information exchange as well as dynamically changing interaction topologies [1]. In the past decades, numerous studies have been conducted on the consensus problems [1–7]. However, designing local control laws with real-time adaption and robustness to dynamic uncertainties would come with several underlying challenges[7][14].

Recently, the appeal of applying game theoretic methodology to multi-agent systems is receiving significant attention [4–9]. The most advantage of the game theoretic approach is that it provides a hierarchical decomposition between the game design and the distributed learning algorithm design. Marden [4] established a relationship between cooperative control and potential game by using the potential function to capture the global objective in game model. Based on but different from the trial and error learning procedure of Young [11], Pradelski [12] propose a variant of log linear learning in order to simply compute the states

K. Li et al. (Eds.): ICSEE 2013, CCIS 355, pp. 113–121, 2013.

of the process. In Li [13], an extension of potential games, termed as state based potential game, was presented to cope with the design challenges by introducing an additional state variable into the game environment.

The main contribution of this paper is to extend the results of Li [13] to a more general game framework based on matrix theory. The matrix theory [3] are used in this paper to develop a new framework for analyzing the interaction behavior and the minimal information requirement among agents are provided. Using the canonical form of matrix theory, the results in Li [13] are extended to the general form of state based ordinal potential games. We will show that the game designed in Li [13]is a special case of the general game model proposed in this paper. Furthermore, this general game model provides us with much more degree of freedom to design local control laws and both cases with connected and time-invariant interaction topology in Li [13]are extended to the practical conditions including time-variant and not always connected communication topology.

2   Problem Setup

2.1   System Model

We are interested in optimization algorithm that can be distributed across the decision-makers. Suppose there is a multi-agent system consisting of $n \geq 2$ agents denoted by the set $N = \{1, 2, \ldots, n\}$. Each agent $i \subseteq N$ is endowed with a set of possible decisions (or values) denoted by A_i which is a nonempty convex subset of $\mathbb{R}$. We denote a specific joint decision profile by the vector $v \triangleq \{v_1, v_2, \ldots, v_n\}$ and $v \subseteq V \triangleq \prod_{i \in N} V_i$, where V is the closed, convex and non-empty set consisting of all possible joint decisions. Suppose the global objective $\phi : V \to R$ that system designer seeks to minimize is differentiable and convex. More specifically, the distributed optimization problem takes on the general form:

$$\min_{v_i} \phi(v_1, v_2, \ldots, v_n) \qquad \text{s.t.} v_i \in V_i, \forall i \in N \tag{1}$$

2.2   Problems to Be Solved

The sensing and communication among agents is described mathematically by a time-varying and piecewise-constant matrix whose dimension is equal to the number of dynamical agents and the elements assume binary values. The matrix can be defined without loss of any generality [3]:

$$\mathbf{S}(t) = \begin{pmatrix} s_{11} & s_{12} & \cdots & s_{1n} \\ s_{21} & s_{22} & \cdots & s_{2n} \\ \cdots\cdots\cdots\cdots\cdots \\ s_{n1} & s_{n2} & \cdots & s_{nn} \end{pmatrix}$$

where $s_{ii} = 1$ because agent can always acquire its own information. In general, $s_{ij} = 1$ if the agent i can get the information of agent j for any $j \neq i$ at time

t; $s_{ij} = 0$ if otherwise. Over time, binary changes of $S(t)$ occur at an infinite sequence of time instants, denoted by $\{t_k : k \in \Omega\}$, where $\Omega \triangleq \{1, 2, \ldots, \infty\}$,and $S(t)$ is piecewise constant as $S(t) = S(t_k)$ for all $t \in [t_k, t_{k+1})$.

Problem 1: Design the local control law for agent i to designate how the agent processes available information at time $t - 1$ in order to formulate a decision v_i at time t as $v_i(t) = U_i\big(s_{i1}(t)v_1(t-1), s_{i2}(t)v_2(t-1), \ldots, s_{in}(t)v_n(t-1)\big)$, where $i = \{1, 2, \ldots, n\}$, $U_i(\cdot)$ is the local control law for agent i at time t.

Intuitively, it would be sufficient for all the agents to be controlled properly if each of them can receive enough information from its neighboring units. However, it is not practical. So the minimum requirement on communication needs to be specified properly to ensure the system level objective is satisfied. So it will be the problem as below.

Problem 2: Determine the sensing/communication matrix in order to ensure the global objective is desirable while minimizing the communication costs.

Through the design of local communication topology, the candidate sequences of $\{S(t_0), S(t_1, \ldots))\}$ should be chosen appropriately to ensure model developed by using game theory converge to the desired equilibrium.

3   Rule of Communication Topology Design

In order to ensure the global objective can be achieved, local information needs to be shared among the agents. Heuristically, the more information channels there are, the faster the convergence to the desired global behavior. In order to guarantee the validness of the proposed control strategy with the minimal information requirement, we will give the rule of the communication topology in this section.

Rule: The sequence of sensing/communication matrices $S_{\infty:0} = \{S(t_0), S(t_1), \ldots\}$ should be sequentially complete [15].

The sequentially completeness condition is a very precise method to schedule local communication among agents. Especially, it gives the cumulated effects in an interval of time and shows that the cumulated communication network can be connected even if the network may not be connected at some time instants.

4   State Based Ordinal Potential Game Design

4.1   State Based Ordinal Potential Game

Different from but complementary to the result in Marden [6] and Li [13], we focus on a more general potential game, termed as state based ordinal potential games, to get the desirable solutions for the optimization problem in 1.

Definition 1 (state based ordinal potential game): A state based ordinal potential game denoted by $G = \{N, \{A_i\}_{i \in N}, \{U_i\}_{i \in N}, X, f, \varphi\}$, consists of a player set N and an underlying finite state space X. Each agent i has a state invariant action set A_i, and a state dependent payoff function $U_i : X \times A \to \mathbf{R}$, also a both state and action dependent state transition function $f : X \times A \to X$. For every $i \in N$, $a_{-i} \in \prod_{j \neq i} A_j$, $a_i \in A_i$, $a'_i \in A_i$,given the old state action pair (x, a_i, a_{-i}) and the new state action pair (x, a'_i, a_{-i}) ,if there exists a differential and convex potential function $\varphi : X \times A \to \mathbf{R}$ that satisfy the following property

$$sgn\big(U_i(x, a_i, a_{-i}) - U_i(x, a'_i, a_{-i})\big) = sgn\big(\varphi(x, a_i, a_{-i}) - \varphi(x, a'_i, a_{-i})\big) \quad (2)$$

Then G is called state based ordinal potential game with potential funciton φ.

4.2   State Based Ordinal Potential Game Design

The design process of state based ordinal potential game is based on the work of Li [13]. In contrary to Li [13], the desirable global objective can be acquired even if the interaction topology is not connected at some time intervals based on the rule of communication.

Given $v = (v_1, v_2, \ldots, v_n)$ as the tuple of value profile for n players. In Marden [14], the average of local objective functions of agent i's neighboring agents is chosen to act as a new kind of local objective function for agent i, termed as equally shared utility, which is $U_i(v) = \sum_{j=1}^n s_{ij} U_j(v) / \sum_{j=1}^n s_{ij}$.

Because an agent may not have the complete knowledge about the true value of the local agents' action as quickly as possible, especially in systems with intermittent communication or time delays. So we make use of the estimation term $e = (e_1, e_2, \ldots, e_n)$ in the state space to estimate the true value of actions. Accordingly, the equally shared utility above can be rewritten as $U_i(e_j|_{s_{ij}=1}) = \sum_{j=1}^n s_{ij} \phi(e_j^1, e_j^2, \ldots, e_j^n) / \sum_{j=1}^n s_{ij}$.

Next, we will introduce the equation above as one of the components in the following design for local objective function in the framework of state based ordinal potential game. Meanwhile, considering the error caused by the introduction of the estimation items, thus we will have to minimize the errors so that the global objective can be achieved. Accordingly we define the local objective function as $U_i(x, a) = U_i^\phi(x, a) + \alpha U_i^e(x, a)$, where $U_i^e(x, a) = \sum_{j=1}^n \sum_k s_{ij}(e_i^k - e_j^k)^2 / \sum_{j=1}^n s_{ij}$, which is the component in order to minimize the error between the true value and the estimation items in our game model, and α is a positive tradeoff parameter. By inspired from the notion of equally shared utility, we define $U_i^\phi(x, a) = U_i(e_j|_{s_{ij}=1})$.

4.3   Analytical Properties of State Based Ordinal Potential Games

Next, we need to further analyze the analytical properties of the model and verify whether the model we designed meets the desired goals or not. First and foremost, a very important issue is to verify whether the designed model results in the framework of potential game or not.

Theorem 1. *Model the distributed optimization problem in (1) as game model in Section 4.2 with any positive constant α. Given the potential function as $\varphi(x, a) = \varphi^\phi(x, a) + \alpha \varphi^e(x, a)$ where $\varphi^\phi(x, a) = \sum_{j=1}^n \phi(e_j^1, \ldots, e_j^n)/n$ and $\varphi^e(x, a) = \sum_i \sum_{j=1}^n \sum_k s_{ij}(e_i^k - e_j^k)^2/2n$. Then the game model in Section 4.2 is potential game.*

Proof: It is straightforward to verify the state based ordinal game designed in Section 4.2 is a potential game with the potential function $\varphi(x,a)$.

The potential function $\varphi(x,a)$ captures the global objective of distributed optimization problem in multi-agent system. As we know in [4], the potential game will guarantee the existence of an equilibrium while at the same time allow the existing learning algorithms to be directly used in our game model. However, whether the equilibriums of our designed game are solutions to the optimization problem in (1) becomes our main concern in the following theorems.

Theorem 2. *Model the optimization problem in (1) as the state based ordinal potential game proposed in section 4.2 with any positive constant α. Suppose the interaction topology is undirected, time-varying, and the sequence of sensing/communication matrixes is sequentially complete, then $\forall i, k \in N, e_i^k = v_k$.*

Proof: Suppose state action pair $(x,a) = ((v,e),(\hat{v},\hat{e}))$ is the equilibrium of the game we design. $\forall i \in N$, and for any action profile $a' = ((\hat{v}_i', \hat{v}_{-i})(\hat{e}_i', \hat{e}_{-i}))$, we can get $U_i(x,a) \le U_i(x,a')$ according to the notion of Nash equilibrium.

$\forall i, k \in N$, the adjacent agents of the agent i is denoted by $L_i = \{l : s_{il}(t) = 1\}$, where s_{il} is the element of the sensing/communication matrix in ith row and lth column at time instant t. $\forall j_1, j_2 \in L_i$, the new action for the agent i, denoted by $a_i' = (\hat{v}_i', \hat{e}_i')$, is defined as $\hat{v}_i' = \hat{v}_i$ and

$$\hat{e}_{i \to j}'^k = \begin{cases} \hat{e}_{i \to j}^k + \delta, & j = j_1 \\ \hat{e}_{i \to j}^k - \delta, & j = j_2 \\ \hat{e}_{i \to j}^k, & j \in L_i \setminus \{j_1, j_2\} \end{cases} \tag{3}$$

where $\forall \delta \in \mathbf{R}$ and $\hat{e}_{i \to j}^k$ is the estimation that agent i passes to agent j regarding the value of agent k. Accordingly the change in the local objective function for agent i can be expressed as $\sum_{j=1}^n s_{ij} \Delta U_i = \sum_{j=1}^n s_{ij} U_i(x,a') - \sum_{j=1}^n s_{ij} U_i(x,a)$. It is noticed that the estimate items regarding to the value of player k haven't changed for all the agents except the agent j_1 and j_2, therefore, the change in the local objective function for agent i simplifies to

$$\sum_{j=1}^n s_{ij} \Delta U_i = \phi(e_{j_1}^1, \ldots, e_{j_1}^k + \delta, \ldots, e_{j_1}^n) + \phi(e_{j_2}^1, \ldots, e_{j_2}^k - \delta, \ldots, e_{j_2}^n)$$
$$- \phi(e_{j_1}^1, \ldots, e_{j_1}^k, \ldots, e_{j_1}^n) - \phi(e_{j_2}^1, \ldots, e_{j_2}^k, \ldots, e_{j_2}^n) \tag{4}$$
$$+ \alpha \sum_{k \in N} (2\delta e_{j_1}^k - 2\delta e_{j_2}^k + \delta^2)$$

When $\delta \to 0$, we can express the equation in (4) as

$$\sum_{j=1}^n s_{ij} \Delta U_i = \left(\frac{\partial \phi}{\partial e_{j_1}^k} - \frac{\partial \phi}{\partial e_{j_2}^k} + \alpha \sum_{k \in N} (2e_{j_1}^k - 2e_{j_2}^k) \right) \delta + o(\delta^2) \tag{5}$$

As we suppose the state action pair $(x,a) = ((v,e),(\hat{v},\hat{e}))$ is the equilibrium of the game model, we know that $\forall \delta \in \mathbf{R}, \Delta U_i \ge 0$. Furthermore, as δ can be positive or negative, so $\forall i, k \in N, \forall j_1, j_2 \in L_i$, the equation in (5) can be translated to $\frac{\partial \phi}{\partial e_{j_1}^k} - \frac{\partial \phi}{\partial e_{j_2}^k} + \alpha \sum_{k \in N}(2e_{j_1}^k - 2e_{j_2}^k) = 0$.

As the global objective function in 1 is assumed to be convex over the set $V \subset \mathbf{R}$, by applying the mean value theorem for the convex function, we have

$$\frac{\partial \phi}{\partial e_{j_1}^k} - \frac{\partial \phi}{\partial e_{j_2}^k} = H(\phi)|_{\xi e_{j_1^k} + (1-\xi)e_{j_2}^k}(e_{j_1}^k - e_{j_2}^k) \tag{6}$$

where $\xi \in (0,1)$, $H(\phi)$ is the Hessian matrix of function $\phi(\cdot)$. Multiply $(e_{j_1}^k - e_{j_2}^k)$ left to both side of equation (6), we have

$$(e_{j_1}^k - e_{j_2}^k)(\frac{\partial \phi}{\partial e_{j_1}^k} - \frac{\partial \phi}{\partial e_{j_2}^k}) = H(\phi)|_{\xi e_{j_1^k} + (1-\xi)e_{j_2}^k}(e_{j_1}^k - e_{j_2}^k)^2 \tag{7}$$

Substituting the equation (7) with the equation (6), we have

$$0 \geq -\alpha \sum_{k \in N}(2e_{j_1}^k - 2e_{j_2}^k)^2 = H(\phi)|_{\xi e_{j_1^k} + (1-\xi)e_{j_2}^k}(e_{j_1}^k - e_{j_2}^k)^2 \tag{8}$$

According to the nature of the convex function $\phi(\cdot)$, we know that its Hessian matrix will be positive semi-definite, that is $H(\phi)|_{\xi e_{j_1^k} + (1-\xi)e_{j_2}^k} \geq 0$.

So the equation (8) can be simplified to $0 \geq -\alpha \sum_{k \in N}(2e_{j_1}^k - 2e_{j_2}^k)^2 \geq 0$, which implies $\forall i, k \in N, \forall j_1, j_2 \in L_i$, we have $e_{j_1}^k = e_{j_2}^k$.

In the process of state space design, it is noticed that the sum of the estimation from all the agents regarding any specific agent $k's$ value is equal to the n times the agent $k's$ value, that is $\sum_{i \in N} e_i^k(t) = n v_k(t)$. Coupled with $\forall i, k \in N$, $\forall j_1, j_2 \in L_i$, $e_{j_1}^k = e_{j_2}^k$, we can have $\forall i, k \in N$, $e_i^k = v_k$. This completes the proof.

Next, we will need to examine the relationship between the Nash equilibrium and the optimal solution of the distributed optimization problem.

Theorem 3. : *Model the optimization problem in (1) as the state based ordinal potential game proposed in section (4.2) with any positive constant α. Suppose the interaction topology is undirected, time-varying, and the sequence of sensing/communication matrixes is sequentially complete, then the resulting Nash equilibrium $(x, a) = ((v, e)(\hat{v}, \hat{e}))$ is optimal solution of the distributed optimization problem in (1).*

Proof: According to the *theorem 2*, we know that all the estimations from any agent $i \in N$ regarding the value of any specific agent k is equal to the true value of agent k. Therefore, consider the following class of change in the value instead of the change in the estimation. That is, a new action profile $a' = (a_i', a_{-i}) = ((\hat{v}_i', \hat{v}_{-i})(\hat{e}_i', \hat{e}_{-i}))$ which can be specifically expressed as $\hat{v}_i' = \hat{v}_i + \delta$ and $\hat{e}_i' = \hat{e}_i$, where $\forall \delta \in \mathbf{R}, v_i + \hat{v}_i + \delta \in V_i$.

Accordingly, the change in the local objective function for agent i can be expressed as follows.

$$\sum_{j=1}^{n} s_{ij} \Delta U_i = \sum_{j=1}^{n} s_{ij} \phi(v_1, \ldots, v_i + \delta, \ldots, v_n) - \sum_{j=1}^{n} s_{ij} \phi(v_1, \ldots, v_i, \ldots, v_n) \tag{9}$$

When $\delta \to 0$, we can express the equation in(9)as

$$\triangle U_i = \frac{\partial \phi}{\partial v_i} \delta \tag{10}$$

As we suppose the state action pair $(x, a) = \big((v, e)(\hat{v}, \hat{e})\big)$ is the Nash equilibrium for the game model, we have $\forall \delta \in \mathbf{R}, \triangle U_i \geq 0$. Furthermore, as δ can be positive or negative, so the equation (10) can be translated to $\frac{\partial \phi}{\partial v_i} = 0$, which implies that supposing the state action pair $(x, a) = \big((v, e)(\hat{v}, \hat{e})\big)$ is the Nash equilibrium, then the derivative of the convex global objective function $\phi(\cdot)$ at the point of value v_i equals zero. According to the definition of convex function, the minimum for the convex function can be acquired at the point where the derivative of the function equals zero. That is to say, at the value of v_i , which also is Nash equilibrium of our game model, the minimum of the distributed problem in (1) can be acquired. Therefore, the resulting Nash equilibrium $(x, a) = \big((v, e)(\hat{v}, \hat{e})\big)$ is the optimal solution of the distributed optimization problem. This completes the proof.

5   Simulation Results

In this section we will illustrate the applicability of the theoretical results on the consensus problem. Consider a set of agents $N = \{1, 2, \ldots, n\}$, each agent $i \in N$ has an initial value $v_i(0) \in \mathbf{R}$ which represents physical location of the agent. The goal of the consensus problem is to establish a set of local control laws $\{U_i(\cdot)\}_{i \in N}$ such that all the agents seek to an agreement upon a common scalar value by repeatedly interacting with one another. By reaching consensus, we mean converging to the agreement space characterized by $\lim_{t \to \infty} = v^* = \sum_{i \in N} v_i(0)$. Given the number of agents $n = 12$, the initial interaction topology and the corresponding sensing/communication matrix $S(0)$ is shown in Fig. 1.

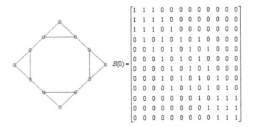

Fig. 1. Initial interaction topology of 12 agents and the corresponding sensing/communication matrix $S(0)$

It is possible that some of the entries in matrix $S(t)$ may switch from 1 to 0 intermittently as the system evolves, but the communication rule should still be observed unless many of the communication channels stop working at the same time. This shows that the local communication network can be designed to be robust to the time varying and intermittent conditions according to the rule in section (3).

Consequently, the consensus problem can be formalized as the following optimization problem: $\min_{v \in V} \phi(v) = \Sigma_{i \in N, s_{ij}=1}(v_i - v_j)^2$.

We simulated this average consensus problem by applying the state based ordinal potential games with the parameter $\alpha = 0.2$ and the gradient play learning process. The results are presented in Fig. 2. In the top of Fig. 2, it illustrates the dynamics of the values of the agents and shows that all the values converge to the common scalar value after about $1,000$ iterations by applying the model we design and the gradient play learning algorithm for potential games. In the bottom of Fig. 2, it demonstrates the evolution of the potential function $\varphi(x, a) = \varphi^{\phi}(x, a) + \alpha \varphi^e(x, a)$. And it is obvious to notice that at the beginning, the value of the potential game will be very large since the agents are initially dispersed. After about 500 iterations, the value of potential game will rapidly converge to zero. This plot demonstrates the potential function designed in our game model will tolerate the error caused by the introduction of estimate items and achieve the desired global objective rapidly.

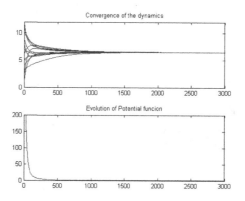

Fig. 2. Top figure: convergence of the dynamics of the 12 agents. Bottom figure: evolution of the potential function.

6   Conclusions

In this paper, a new theoretical framework for analysis and design of distributed optimization problem is developed based on the cooperative control methodology and game theory. The matrix theoretical approach provides the basis for state based ordinal potential game design, which gives the system designer additional freedom to design local control laws with the locality of information and the efficiency of the resulting equilibrium. Also the minimal requirement on the interaction topology among the agents is provided. Further direction includes two problems (i) developing systematic procedures for designing the local objective functions in the framework of state based ordinal potential games and (ii) extending the matrix theory analysis to the directed and time-delay information networks.

References

1. Ren, W., Beard, R.W.: Consensus Seeking in Multiagent Systems under Dynamically Changing Interaction Topologies. IEEE Transactions on Automatic Control 50, 655–661 (2005)
2. Jadbabaie, A., et al.: Coordination of Groups of Mobile Autonomous Agents using Nearest Neighbor Rules. IEEE Transactions on Automatic Control 48, 988–1001 (2003)
3. Qu, Z.H., et al.: Cooperative Control of Dynamical Systems with Application to Autonomous Vehicles. IEEE Transactions on Automatic Control 53, 894–911 (2008)
4. Marden, J.R., et al.: Cooperative Control and Potential Games. IEEE Transactions on Systems Man and Cybernetics Part B-Cybernetics 39, 1393–1407 (2009)
5. Lin, W., et al.: A Design of Distributed Nonzero-sum Nash Strategies. In: Proceedings of the 49th IEEE Conference on Decision and Control, CDC 2010, Atlanta, Georgia, USA, December 15-17, pp. 6305–6310 (2010)
6. Arslan, G., et al.: Autonomous Vehicle Target Assignment: A Game Theoretical Formulation. Journal of Dynamic Systems, Measurement and Control-Transactions of the ASME 129, 584–596 (2007)
7. Li, N., et al.: Designing Games to Handle Coupled Constraints. In: Proceedings of the 49th IEEE Conference on Decision and Control, CDC 2010, Atlanta, Georgia, USA, December 15-17, pp. 250–255 (2010)
8. Marden, J.R., et al.: Joint strategy Fictitious Play with Inertia for Potential Games. IEEE Transactions on Automatic Control 54, 208–220 (2009)
9. Marden, J.R., et al.: Payoff based Dynamics for Multi-player Weakly Acyclic Games. SIAM Journal on Control and Optimization 48, 373–396 (2009)
10. Monderer, D., Shapley, L.S.: Potential Games. Games and Economic Behavior 14, 124–143 (1996)
11. Young, H.P.: Learning by Trial and Error. Games and Economic Behavior 65, 626–643 (2009)
12. Pradelski, B.S.R., Young, H.P.: Learning Efficient Nash Equilibria in Distributed Systems, University of Oxford, Discussion Paper Series (2010)
13. Li, N., et al.: Designing Games for Distributed Optimization. In: Proceedings of the 50th IEEE Conference on Decision and Control and European Control Conference, Orlando, FL, USA, December 12-15, pp. 2434–2440 (2011)
14. Marden, J.R., Wierman, A.: Distributed Welfare Games with Applications to Sensor Coverage. In: Proceedings of the 47th IEEE Conference on Decision and Control, Cancún, México, pp. 1708–1713 (2008)
15. Qu, Z.: Cooperative Control of Dynamical Systems: Applications to Autonomous Vehicles. Springer, London (2009)

Enhanced Two-Step Satisfactory Method for Multi-Objective Optimization with Fuzzy Parameters

Chaofang Hu, Qizhi Liu, and Yanwen Liu

School of Electrical Engineering and Automation, Tianjin University,
Tianjin, China
cfhu@tju.edu.cn, lqz59@126.com, wen654321@qq.com

Abstract. An enhanced two-step satisfactory method for multi-objective optimization problem with fuzzy parameters is proposed in this paper. By means of the α-level sets of the fuzzy numbers, all the objectives with fuzzy parameters are modeled as the fuzzy goals. The order of satisfactory degrees about different denoting that the higher priority achieves the higher satisfactory degree is applied to preemptive priority requirement. The strict order constraints are relaxed by decreasing the maximum overall satisfactory degree. The original optimization problem is divided into two models to be solved iteratively. The satisfactory solution can be acquired by changing parameter or regulating α. The numerical example demonstrates the power of the proposed method.

Keywords: Multi-objective optimization, fuzzy parameter, priority.

1 Introduction

Recently, Multi-Objective Optimization (MOO) problem has become more and more obvious and important in production, economy, and everyday life, where multiple objectives are conflicting, non-commensurable and imprecise. Its study and development have attracted many researchers [1][2]. In the real world, MOO problem takes place in a vague environment in which the goals or parameters of the objectives and constraints are not known precisely. This problem is called Fuzzy Multi-Objective Optimization (FMOO) [3][4][5]. In FMOO, the preemptive priority requirement, as a common and practical preference, is often given by Decision Maker (DM), which means all the objectives being classified into the different levels in terms of their importance. Traditionally, the lexicographic optimization is interesting [6], where the multiple subproblems including different objectives are solved in lexicographic order. However, this maybe results in the complex computation or the degenerative optimization. Chen et al. [7] propose the higher priority having higher satisfactory degree for the preemptive priority. Nevertheless, the satisfactory even feasible solution for the strict comparison doesn't possibly exist. The generalized varying-domain optimization method is presented by Hu et al. [8]. But it strengthens the nonlinearity of the original

K. Li et al. (Eds.): ICSEE 2013, CCIS 355, pp. 122–129, 2013.
© Springer-Verlag Berlin Heidelberg 2013

problem. Although Hu et al. [9] propose the two-step satisfactory method to handle preemptive priority, it is only concerned with fuzzy goals.

In this paper, MOO problem with fuzzy parameters having possibilistic distributions and being treated as fuzzy numbers is studied. By the corresponding α-level sets, fuzzy parameters are regarded as variables. Accordingly, the objectives are treated as fuzzy goals under the α-level sets. Then the MOO problem with fuzzy parameters is called an α-FMOO. For preemptive priority requirement, the two-step satisfactory method [9] is introduced and enhanced here. Firstly, the order of satisfactory degrees denoting that the higher priority achieves the higher satisfactory degree is applied to the priority structure. The original problem is reformulated into two models about α-level sets. Then the maximum overall satisfactory degree of the first model is obtained, and the second model is solved through relaxing maximum overall satisfactory degree. The two models are solved iteratively by changing the result of the first model or regulating α-level sets. With this method, DM can easily get the satisfactory solution.

In this paper, Section 2 describes MOO problem with fuzzy parameters and preemptive priority requirement. The enhanced two-step satisfactory method is presented in Section 3. Section 4 summarizes the algorithm. The numerical example demonstrates its power in Section 5. Section 6 draws the conclusions.

2 MOO with Fuzzy Parameters and Priority

2.1 MOO Problem with Fuzzy Parameters

Generally, there are multiple objectives to be optimized in MOO. In practice, it would be appropriate to consider that the possible values about the parameters of the objectives or constraints involve the ambiguity of DM's understanding of the real system. These parameters are called fuzzy parameters. Then the MOO problem with fuzzy parameters is described as

$$\left. \begin{array}{ll} \min & (f_1(x, \tilde{a}_i), \cdots, f_k(x, \tilde{a}_k)) \\ s.t. & x \in G(\tilde{b}) = \left\{ x | g_j(x, \tilde{b}_j) \leq 0, j = 1, \cdots, m \right\} \end{array} \right\} \quad (1)$$

where $f_i(x, \tilde{a}_i)$, $(i = 1, \cdots, k)$ are multiple objectives to be minimized; $G(\tilde{b}) \subset R^n$ is system constraints, and $\tilde{a}_i = (\tilde{a}_{i1}, \tilde{a}_{i2}, \cdots, \tilde{a}_{ir_i})$ and $\tilde{b}_j = \left(\tilde{b}_{j1}, \tilde{b}_{j2}, \cdots, \tilde{b}_{js_j} \right)$ are respectively vectors including fuzzy parameters.

The fuzzy parameters are characterized as the fuzzy numbers [5][10]. It is proper to take a real fuzzy number as a convex continuous fuzzy subset. There are various kinds of membership functions for fuzzy number. All of them are continuously mapping, monotonously increasing or decreasing, and they lie in the interval [0,1]. For example, fuzzy number $\tilde{c}$ is shown in the Fig.1.

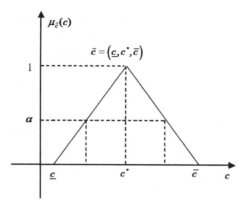

Fig. 1. Fuzzy number $\tilde{c}$

$$\mu_{\tilde{c}}(c) = \begin{cases} 0 & c \geq \bar{c} \\ 1 - \dfrac{c - c^*}{\bar{c} - c^*} & c^* \leq c \leq \bar{c} \\ 1 & c = c^* \\ 1 - \dfrac{c^* - c}{c^* - \underline{c}} & \underline{c} \leq c \leq c^* \\ 0 & otherwise \end{cases} \tag{2}$$

Then α-level set of $\left(\tilde{a}, \tilde{b}\right)$ is defined as

$$\left(\tilde{a}, \tilde{b}\right)_\alpha = \left\{ (a,b) \;\middle|\; \begin{array}{l} \mu_{\tilde{a}_{id}}(a_{id}) \geq \alpha, \mu_{\tilde{b}_{je}}(b_{je}) \geq \alpha \\ i = 1, 2, \cdots, k; d = 1, 2, \cdots, r_i \\ j = 1, 2, \cdots, m; e = 1, 2, \cdots, s_j \end{array} \right\} \tag{3}$$

Then α-MOO problem with fuzzy parameters is presented as follows

$$\left. \begin{array}{ll} \min & (f_1(x,a), \cdots, f_k(x,a)) \\ s.t. & x \in G(b) = \{x | g_j(x,b) \leq 0, j = 1, \cdots, m\} \\ & (a,b) \in \left(\tilde{a}, \tilde{b}\right)_\alpha \end{array} \right\} \tag{4}$$

where the parameters a and b are treated as the decision variables about α.

In a fuzzy environment, DM usually gives all objectives the implicit targets. For minimization problem, DM permits the objective value $f_i(x, \tilde{a})$, ($i = 1, \cdots, k$) are more than aspiration level up f_i^* to stated tolerant limit $f_i^{\max}$. The triangle-like membership function under α-level set is defined for the fuzzy objective.

$$\mu_{f_i}(x,a) = \begin{cases} 1 & f_i(x,a) \leq f_i^* \\ 1 - \dfrac{f_i(x,a) - f_i^*}{f_i^{\max} - f_i^*} & f_i^* \leq f_i(x,a) \leq f_i^{\max} \\ 0 & f_i(x,a) \geq f_i^{\max} \end{cases} \tag{5}$$

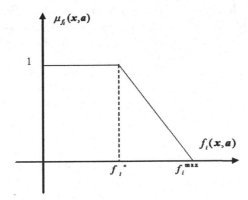

Fig. 2. Membership function $\mu_{f_i}(x, a)$

The Fig. 2 shows the shape of this membership function.

The value of the membership function about one solution is also called the satisfactory degree of the objective.

2.2 Preemptive Priority Requirement

Preemptive priority requires all the objectives be optimized in an order. Usually, there are one or several objectives in one level, which means all the objectives are grouped according to priority. Then α-FMOO problem with preemptive priority is formulated as

$$
\left.
\begin{aligned}
\max \quad & \left[P_1\left(\mu_{f_1^1}(x, a), \cdots, \mu_{f_{l_1}^1}(x, a)\right), \cdots, P_L\left(\mu_{f_1^L}(x, a), \cdots, \mu_{f_{l_L}^L}(x, a)\right)\right] \\
s.t. \quad & x \in G(b) = \{x | g_j(x, b) \le 0, j = 1, \cdots, m\} \\
& (a, b) \in \left(\tilde{a}, \tilde{b}\right)_\alpha
\end{aligned}
\right\} (6)
$$

where P_j is the priority factor. $f_1^j(x, \tilde{a}), \cdots, f_{l_j}^j(x, \tilde{a})$ represent some objectives in jth the priority level.

For example, suppose the priority of $f_s(x, \tilde{a})$ is higher than that of $f_{s'}(x, \tilde{a})$, $(s, s' \in \{1, \cdots, k\}, s \ne s')$. Then $f_s(x, \tilde{a})$ should be optimized before $f_{s'}(x, \tilde{a})$.

3 Enhanced Two-Step Satisfactory Method

3.1 Order of Satisfactory Degrees

According to the assumption in Section 2, $f_s(x, \tilde{a})$ has the higher priority than $f_{s'}(x, \tilde{a})$. That means the former objective has the higher satisfactory degree than the latter. Then the preemptive priority requirement can transformed into the order of the satisfactory degrees, i.e.

$$
\mu_{f_{s'}}(x, \tilde{a}) \le \mu_{f_s}(x, \tilde{a}), s, s' \in \{1, \cdots, k\}, s \ne s' \tag{7}
$$

However, it is seen that the comparison is too strict. If (7) is taken as the constraint, the feasible or satisfactory solution maybe can not be obtained. In addition, the bigger difference between the objectives is not reflected from (7). Thus the priority variable γ is used to release the order of satisfactory degrees. Then the released preemptive priority requirement is reformulated as

$$\mu_{f_{s'}}(x, \tilde{a}) - \mu_{f_s}(x, \tilde{a}) \leq \gamma, s, s' \in \{1, \cdots, k\}, s \neq s' \tag{8}$$

3.2   The First Step

For preemptive priority requirement, α-FMOO problem (6) is divided into two models. They include the preliminary optimization model and the priority model. The first step is to solve the former to get the maximum overall satisfactory degree of all objectives by max-min decisions regardless of priority. Thus the preliminary optimization model is equivalent to

$$\left. \begin{array}{ll} \max & \lambda \\ s.t. & \mu_{f_i}(x, a) \geq \lambda, i = 1, \cdots, k \\ & \mu_{f_i}(x, a) \leq 1 \\ & x \in G(b) = \{x | g_j(x, b_j) \leq 0, j = 1, \cdots, m\} \\ & (a, b) \in \left(\tilde{a}, \tilde{b} \right)_\alpha \end{array} \right\} \tag{9}$$

By means of (9), all objectives can be optimized simultaneously as much as possible. The optimization result, i.e. maximum overall satisfactory degree λ^* under certain α-level set will be treated as the given condition of the next step optimization.

3.3   The Second Step

After optimization of the first step, preemptive priority requirement needs to be considered. On basis of the first step, the second model is utilized to balance optimization and the priority order. The maximum overall satisfactory degree λ^* about α-level set is relaxed by means of the releasing parameter $\Delta\delta(\Delta\delta \geq 0)$, which is determined by the interaction between DM and the analyzer. And the comparing inequality is incorporated as the constraint. Then the second optimization model is constructed in the following expression

$$\left. \begin{array}{ll} \min & \gamma \\ s.t. & \mu_{f_i}(x, a) \geq \lambda^* - \Delta\delta, \ i = 1, \cdots, k \\ & \mu_{f_{s'}}(x, a) - \mu_{f_s}(x, a) \leq \gamma, \ s, s' = 1, \cdots, k, s \neq s' \\ & -1 \leq \gamma \leq 1 \\ & x \in G(b) = \{x | g_j(x, b_j) \leq 0, j = 1, \cdots, m\} \\ & (a, b) \in \left(\tilde{a}, \tilde{b} \right)_\alpha \end{array} \right\} \tag{10}$$

By the releasing parameter $\Delta\delta$, the feasibility of (10) can be ensured. And the preemptive priority requirement is realized by minimizing the priority variable γ

as far as possible until $\gamma < 0$ and DM is satisfactory. If $\gamma > 0$, the solution does not satisfy priority. Then the maximum overall satisfactory degree λ^* needs to be relaxed continuously through changing the parameter $\Delta\delta$ and solve the second model again, or α is regulated.

4   Algorithms

The corresponding algorithms of the proposed method is summarized as follows

Step 1. Initialization: Calculate the individual minimum $f_i^{\min}$ and maximum $f_i^{\max}$ of the objective function $f_i(x, \tilde{a})$, $(i = 1, \cdots, k)$, under the given constraints for $\alpha = 0$ and $\alpha = 1$.

Step 2. Determine the desirable target and the tolerance, construct the membership functions of the objectives in α-FMOO problem, and ask DM to select the initial value of α.

Step 3. Solve (9), and get the maximum overall satisfactory degree λ^*.

Step 4. Let the initial releasing parameter $\Delta\delta = 0$, and formulate the second model (10) according to the priority order.

Step 5. Solve (10). If there is no feasible solution, then go to step 7. On the contrary, continue.

Step 6. Judge: if $\gamma > 0$, go to next step. If $\gamma \leq 0$ but not satisfactory, go to step 7. Otherwise stop optimization, and the satisfactory solution is acquired.

Step 7. Relax the maximum overall satisfactory degree λ^* by increasing the releasing parameter $\Delta\delta$ and back to step 5; or decrease α and back to step 3.

5   Numerical Example

We demonstrate for the effectiveness of the proposed optimization method by the following numerical example

$$\left. \begin{array}{ll} min & f_1(x, \tilde{a}) = \tilde{a}_1 x_1 + 3x_2 + \tilde{a}_2 x_3 \\ min & f_2(x, \tilde{a}) = (x_1 - 1)^2 + 2(x_2 - 3)^2 + \tilde{a}_3(x_3 - 2)^2 \\ s.t. & x_1 + x_2 + x_3 \leq 10 \\ & 0 \leq x_1, x_2, x_2 \leq 10 \end{array} \right\}$$

$\tilde{a} = (\tilde{a}_1, \tilde{a}_2, \tilde{a}_3)$ are fuzzy parameters, whose distributions are given in Table 1.

Table 1. Fuzzy Parameters

$\tilde{c}$	$(\underline{c}, c^*, \bar{c})$
$\tilde{a}_1$	(-2, -1, 0)
$\tilde{a}_2$	(2, 3, 4)
$\tilde{a}_3$	(-1, 0, 1)

The preemptive priority requirement is that $f_2(x, \tilde{a})$ is higher than $f_1(x, \tilde{a})$.

Then the membership functions about $\tilde{a}$ are formulated by (2) and the corresponding constraints about α-level set are denotes as (3). The four individual optimum values of each objective function for $\alpha = 0$ and $\alpha = 1$ are computed according to Table 1. Then the aspiration values and the tolerant limits of the two objectives are respectively (-20, 40) and (-45, 83). The corresponding membership functions under certain α are presented as

$$\left.\begin{array}{l} \mu_{f_1}(x,a) = (40 - a_1 x_1 - 3x_2 - a_2 x_3)/60 \\ \mu_{f_2}(x,a) = (83 - (x_1 - 1)^2 - 2(x_2 - 3)^2 - a_3(x_3 - 2)^2)/128 \end{array}\right\}$$

Therefore, the reformulated first model is written in the following expression

$$\left.\begin{array}{ll} \max & \lambda \\ s.t. & \mu_{f_i}(x,a) \geq \lambda, \mu_{f_i}(x,a) \leq 1, i = 1,2 \\ & \mu_{f_1}(x,a) = (40 - a_1 x_1 - 3x_2 - a_2 x_3)/60 \\ & \mu_{f_2}(x,a) = (83 - (x_1 - 1)^2 - 2(x_2 - 3)^2 - a_3(x_3 - 2)^2)/128 \\ & x_1 + x_2 + x_3 \leq 10 \\ & 0 \leq x_1, x_2, x_2 \leq 10 \\ & a \in (\tilde{a})_\alpha \end{array}\right\}$$

The second model is

$$\left.\begin{array}{ll} \min & \gamma \\ s.t. & \mu_{f_i}(x,a) \geq \lambda^* - \Delta\delta, \ i = 1,2 \\ & \mu_{f_1}(x,a) - \mu_{f_2}(x,a) \leq \gamma \\ & \mu_{f_1}(x,a) = (40 - a_1 x_1 - 3x_2 - a_2 x_3)/60 \\ & \mu_{f_2}(x,a) = (83 - (x_1 - 1)^2 - 2(x_2 - 3)^2 - a_3(x_3 - 2)^2)/128 \\ & x_1 + x_2 + x_3 \leq 10 \\ & 0 \leq x_1, x_2, x_2 \leq 10 \\ & -1 \leq \gamma \leq 1 \\ & a \in (\tilde{a})_\alpha \end{array}\right\}$$

According to the algorithms, the above modes are solved iteratively. The corresponding optimization results are given in Table 2.

Table 2. Optimization Results

α	$\Delta\delta$	γ	$f_1(x,\tilde{a})$	$f_2(x,\tilde{a})$	$\mu_{f_1}(x,a)$	$\mu_{f_2}(x,a)$
	0.15	-0.1788	11.74	-0.1773	0.4710	0.6498
0.8	0.25	-0.2819	17.74	-0.5663	0.3710	0.6529
	0.35	-0.4023	23.74	-3.1832	0.2710	0.6733
	0.15	-0.1757	11.344	-06185	0.4776	0.6533
0.6	0.25	-0.2863	17.344	-1.9778	0.3776	0.6639
	0.35	-0.4378	23.344	-8.5726	0.2776	0.7154

From the various results listed in Table 2, it is known that the alternation of the priority variable γ conforms to that of the releasing parameter $\Delta\delta$, and the values of γ are always less than 0. This means that all the results satisfy the preemptive priority requirement. Then DM can choose one from them as the preferred solution according to his requirement.

6   Conclusions

In this paper, the enhanced two-step satisfactory method is proposed for MOO problem with fuzzy parameters. The strict preemptive priority structure is transformed by the relaxed order of satisfactory degrees. The original problem is divided into two optimization models about α-level sets. The result of the first step is relaxed by the releasing parameter. Moreover, the priority variable is minimized to achieve the preemptive priority requirement in the second step. By the numerical example, the power of our approach is demonstrated.

Acknowledgments. The authors would like to thank the anonymous reviewers for their helpful comments and constructive suggestions with regard to this paper. This work was supported in part by National Natural Science Foundation of China under Grant (61074064 & 91016018) and Natural Science Foundation of Tianjin (12JCZDJC30300 & 11JCZDJC25100).

References

1. Bornstein, C.T., Macuian, N., Pascoal, M., Pinto, L.L.: Multiobjective combinatiorial optimization problems with a cost and several bottleneck objective functions: an algorithm with reoptimization. Computers and Operations Research 39(9), 1969–1976 (2012)
2. Wei, X., Zhang, W., Weng, W., Fujimura, S.: Multi-objective local search combined with NSGA-II for Bi-criteria permutation flow shop scheduling problem. IEEJ Transactions on Electronics, Information and Systems 132(1), 32–41 (2012)
3. Bellman, R.E., Zadeh, L.A.: Decision-making in a fuzzy environment. Management Science 17, 141–164 (1970)
4. Zimmermann, H.J.: Fuzzy programming and linear programming with several objective functions. Fuzzy Sets and Systems 1, 45–55 (1978)
5. Li, S.Y., Hu, C.F.: An interactive satisfying method based on alternative tolerance for multiple objective optimization with fuzzy parameters. IEEE Transactions on Fuzzy Systems 16(5), 1151–1160 (2008)
6. Tiwari, R.N., Dharmar, S., Rao, J.R.: Fuzzy goal programming - an additive model. Fuzzy Sets and Systems 24, 27–34 (1987)
7. Chen, L.H., Tsai, F.C.: Fuzzy goal programming with different importance and priorities. European Journal of Operational Research 133, 548–556 (2001)
8. Hu, C.F., Teng, C.J., Li, S.Y.: A fuzzy goal programming approach to multiobjective optimization problem with priorities. European Journal of Operational Research 175, 1319–1333 (2007)
9. Li, S.Y., Hu, C.F.: Two-step interactive satisfactory method for fuzzy multiple objective optimization with preemptive priorities. IEEE Transactions on Fuzzy Systems 15(3), 417–425 (2007)
10. Sakawa, M., Yano, H.: An interactive fuzzy satisficing method for multiobjective nonlinear-programming problems with fuzzy parameters. Fuzzy Sets and Systems 17, 654–661 (1989)

Analysis on Data-Based Integrated Learning Control for Batch Processes

Li Jia[1], Luming Cao[1], and Minsen Chiu[2]

[1] Shanghai Key Laboratory of Power Station Automation Technology,
Department of Automation, College of Mechatronics Engineering and Automation,
Shanghai University, Shanghai 200072, China
[2] Department of Chemical and Biomolecular Engineering,
National University of Singapore, Singapore

Abstract. A novel integrated learning control system is presented in this paper. It systematically integrates discrete-time (batch-axis) information and continuous-time (time-axis) information into one uniform frame. More specifically, the iterative learning controller is designed in the domain of batch-axis, while an adaptive single neuron predictive controller (SNPC) in the domain of time-axis. In addition, the convergence and tracking performance of the proposed integrated learning control system are firstly given rigorous description and proof. Lastly, to verify the effectiveness of the proposed integrated control system, it is applied to a benchmark batch process, in comparison with ILC recently developed.

Keywords: batch processes, integrated learning control system, single neuron predictive controller (SNPC).

1 Introduction

SINCE batch process satisfies the requirements of the modern market, it have been widely used in the production of low volume and high value added products, such as special polymers, special chemicals, pharmaceuticals, and heat treatment processes for metallic or ceramic products[1]. For the purpose of deriving the maximum benefit from batch process, it is important to optimize the operation policy of batch process. Therefore, optimal control of batch process is very significant. However, with strong nonlinearity and dynamic characteristics, the optimal control of batch process is more complex than that of continuous process and thus it needs new non-traditional techniques.

Iterative learning control (ILC) has been used in the optimization control of batch process because of its repeatability [1], [2]. However, in ILC system, only the batch-to-batch performance of the batch process is taken for consideration but not the real-time feedback performance. Thus, ILC is actually an open-loop control from the view of a separate batch because the feedback-like control just plays role between different batches. Thus it is difficult to guarantee the performance of the batch process when

K. Li et al. (Eds.): ICSEE 2013, CCIS 355, pp. 130–138, 2013.

uncertainty and disturbance exist. Therefore, the integrated optimization control technology is required in order to derive the maximum benefit from batch process, in which the performance of time-axis and batch-axis are both analyzed synchronously, such as the works done by Amann, Gao, Lee, Xiong, Rogers, Kurek and et.al [3-17].

Motivated by the previous works, an integrated learning control system based on input-output data is proposed in our previous work. Based on that paper, the convergence and tracking performance of the proposed integrated learning control system are firstly given rigorous description and proof in this paper.

The paper is structured as follows. Section 2 presents the proposed data-driven based integrated learning control system. Section 3 presents performance analysis. Simulation example is given in Section 4, followed by the concluding remarks given in Section 5.

2   Data- Based Integrated Learning Control System Design for Batch Processes

The proposed integrated learning control system consists of: the iterative learning control (ILC) working as feedforward controller and adaptive single neuron predictive controller (SNPC) playing as feedback controller. For the convenience of discussion, the number of batch and batch length are respectively defined as k and t_f which is divided into T equal intervals. $u_{ILC}(k,t)$ and $u_{SNPC}(k,t)$ are ILC control variable and SNPC control variable of time t in k-th batch, $u(k,t) = u_{ILC}(k,t) + u_{SNPC}(k,t)$, y_d is the targeted end-product quality, $y(k,t)$ is the corresponding product quality of two control actions, $\hat{y}(k,t)$ is the predicted output of data-based model. Since batch process is repetitive in nature, the model prediction at the end of the k-th batch, $\hat{y}(k,t_f)$ can be corrected by $\tilde{y}(k,t_f)$. During k-th batch, the control policy of $U_{ILC,k}$ obtained from ILC optimization controller and the control policy of $U_{SNPC,k}$ computed from SNPC controller are summed as U_k and is sent into batch process to improve the performance, Y_k and $\hat{Y}_k$ are respectively product quality variables and predicted product quality variables. As discussed above, the proposed integrated learning optimization control action can be described as

$$U_k = U_{ILC,k} + U_{SNPC,k} \tag{1}$$

$$u(k,t) = u_{ILC}(k,t) + u_{SNPC}(k,t) \tag{2}$$

$$u^{low} \leq u(k,t) \leq u^{up}$$
$$y^{low} \leq y(k,t) \leq y^{up}$$

where u^{low}, u^{up} are the lower and upper bound of control input sequence respectively. y^{low}, y^{up} are the lower and upper bound of end-product quality.

The predicted output of the data-based model [18] is

$$\hat{y}(k,t+j) = \text{Model}\big(y(k,t+j-1), y(k,t+j-2),\ldots, y(k,t+1), u(k,t+j-1), \\ (k,t+j-2),\ldots,u(k,t+1)\big) \tag{3}$$

The model prediction error of end quality is written as

$$\hat{e}(k,t_f) = y(k,t_f) - \hat{y}(k,t_f) \tag{4}$$

where

$$\tilde{y}(k+1,t_f) = \hat{y}(k+1,t_f) + \alpha \bar{\hat{e}}(k,t_f) \tag{6}$$

$$\bar{\hat{e}}(k,t_f) = \frac{1}{k}\sum_{i=1}^{k}\hat{e}(i,t_f) = \frac{1}{k}\sum_{i=1}^{k}\big(y(i,t_f) - \hat{y}(i,t_f)\big) \tag{5}$$

where α is error correction term parameter.

The batch-axis iterative learning control optimization problem can be formulated as

$$\min J\big(U_{ILC,k+1}\big) = \Big\|y_d(t_f) - \tilde{y}\big(U_{ILC,k+1}, t_f\big)\Big\|_Q^2 + \big\|U_{ILC,k+1} - U_k\big\|_R^2 \tag{7}$$

$$\Delta U_{ILC,k} = U_{ILC,k+1} - U_k \tag{8}$$

where $\mathbf{Q}$ is selected as constant matrix here, $\mathbf{Q} = q \times \mathbf{I}_T$, $\mathbf{R}$ is dynamic matrix, $\mathbf{R} = r_k \times \mathbf{I}_T$, where r_k is bounded and its upper bound is M_r. $\mathbf{I}_T$ is T-dimensional matrix.

The proposed SNPC is described as

$$u_{SNPC}(k,t) = u_{SNPC}(k-1,t) + \Delta u_{SNPC}(k,t) \tag{9}$$

$$\Delta u_{SNPC}(k,t) = w_1(k,t)e(k,t) + w_2(k,t)\Delta e(k,t) \\ + w_3(k,t)\delta e(k,t) \tag{10}$$

$$\Delta e(k,t) = e(k,t) - e(k,t-1) \tag{11}$$

$$\delta e(k,t) = \Delta e(k,t) - \Delta e(k,t-1) \tag{12}$$

where $w_i(k,t)(i=1,2,3)$ is adjustable parameter. In order to apply it to practical batch process, the following transformation is taken

$$w_i(k,t) = \begin{cases} e^{\varsigma_i(k,t)}, & \left(w_i(k,t) \geq 0\right) \\ -e^{\varsigma_i(k,t)} & \left(w_i(k,t) < 0\right) \end{cases} \quad i = 1,2,3 \tag{13}$$

where $\varsigma_i(k,t)$ is a real number.

By using Lyapunov method [19, 20], the adjustment algorithm of parameters can be obtained .

3   Performance Analysis

3.1   Convergence Analysis

For the convenience of discussion, it defines that C_1 is an optimization controller (ILC) of the batch-axis, C_2 is a feedback controller (SNPC) of the time-axis, and G denotes the batch process.

Theorem 1: If the feedback controller satisfies the condition of (14), the proposed integrated learning optimization control policy converges with respect to the batch number k, namely $\Delta U_k \to 0$ as $k \to \infty$.

$$\left|1 + G(j\omega)C_2(j\omega)\right| \geq 1 \tag{14}$$

Proof: By using the condition of (14), we have

$$\frac{\left|E_2(j\omega)\right|}{\left|E_1(j\omega)\right|} = \left|\frac{1}{1 + G(j\omega)C_2(j\omega)}\right| \leq 1$$

where $E_1 = y_d - C_1 G y_d = (1 - C_1 G) y_d$ and $E_2 = y_d - \dfrac{G(C_1 + C_2) y_d}{1 + GC_2} = \dfrac{(1 - C_1 G) y_d}{1 + GC_2}$

The conclusion of $\left|E_2(j\omega)\right| \leq \left|E_1(j\omega)\right|$ means that the tracking error of the integrated optimization control system is less than or equals to that of the system without real-time feedback control.

Therefore, the time-axis feedback controller satisfies the following inequality

$$\left\|e(U_k)\right\|_Q \leq \left\|e(U_{ILC,k})\right\|_Q \tag{15}$$

Similar to most new controller design methods developed in the literature, perfect model assumption is assumed in this work in order to develop the first of its kind that guarantees the convergence of control policy with the proposed integrated control scheme derived from a rigorous proof. As a result, (7) can be simplified as

$$\min J\left(U_{ILC,k+1}, k+1\right) = \left\|e\left(U_{ILC,k+1}\right)\right\|_Q^2 + \left\|U_{ILC,k+1} - U_k\right\|_R^2 \tag{16}$$

where $e\left(U_{ILC,k+1}\right) = y_d - y_{ILC}\left(k+1, t_f\right)$.

Therefore, we obtain

$$\lim_{k \to \infty} \|\Delta U_{k+1}\|_R^2$$

$$\leq \lim_{k \to \infty} \left(\left(\|e(U_k)\|_Q + \|e(U_{ILC,k+1})\|_Q \right) \cdot \left(\|e(U_k)\|_Q - \|e(U_{ILC,k+1})\|_Q \right) \right) \quad (17)$$

$$\leq M \cdot \lim_{k \to \infty} \left(\|e(U_k)\|_Q - \|e(U_{ILC,k+1})\|_Q \right)$$

where M is the upper bound of $\|e(U_k)\|_Q + \|e(U_{ILC,k+1})\|_Q$, namely

$$\|e(U_k)\|_Q + \|e(U_{ILC,k+1})\|_Q \leq M.$$

Thus, we have the conclusion that $\|\Delta U_k\|_R$ convergences to 0, namely ΔU_k convergences to 0. This completes the proof. Q.E.D.

3.2   Tracking Performance Analysis

M^* is defined as the minimum of $\|e(\cdot)\|_Q^2$ (the ideal value of M^* is zero). Motivated by our previous work [21], the definition of bounded tracking and zero tracking of the integrated learning optimization control system are defined as

Definition 1: Bounded-tracking. If there exists a $\delta = \delta(\varepsilon) > 0$ for every $\varepsilon > 0$ and $U_k = U_{SNPC,k} + U_{ILC,k}$ such that the inequality $\left| \|e(U_k)\|_Q^2 - M^* \right| < \varepsilon$ holds when $r_{k_0+1} < \delta$ for every $k > k_0$.

Definition 2: Zero-tracking. If it is bounded-tracking and there exists $\delta > 0$ and $U_{k+1} = U_{SNPC,k+1} + U_{ILC,k+1}$ such that the equality $\lim_{k \to \infty} \left| \|e(U_{k+1})\|_Q^2 - M^* \right| = 0$ holds when $r_{k_0+1} < \delta$.

Theorem 2 The tracking error $e(U_k)$ of the proposed integrated optimization problem is bounded-tracking for arbitrary initial control profiles $U_{k_0} = U_{SNPC,k_0} + U_{ILC,k_0}$. Moreover, if the function of $\|e(U_k)\|_Q^2$ ($U_k = U_{SNPC,k} + U_{ILC,k}$) is derivable and the optimization solution is not in the boundary, it is zero-tracking for arbitrary initial control profiles $U_{k_0} = U_{SNPC,k_0} + U_{ILC,k_0}$.

Proof: It is easy to know that for every $\varepsilon > 0$, there exists $\delta = \delta(\varepsilon) > 0$ such that the optimal solution in $k_0 + 1$-th batch satisfies $\left\| e(U_{ILC,k_0+1}) \right\|_Q^2 - M^* < \varepsilon$ when $r_{k_0+1} < \delta$.

Therefore, for any $\mathbf{U}_{k_0}$ at time k_0, the proposed integrated optimization problem is bounded-tracking.

Set the optimal point of $\left\| e(U_k) \right\|_Q^2$ $(U_k = U_{SNPC,k} + U_{ILC,k})$ as $(\mathbf{U}^*, M^*)$. Considering a small neighborhood Θ_{U^*}. (if there exists not only one optimal solution in function $\left\| e(U_k) \right\|_Q^2$ $(U_k = U_{SNPC,k} + U_{ILC,k})$, then we consider Θ_{U^*} as the union of the optimal solutions), in which there are no other extremal solutions except global optimal solution only. Because the optimal solution of $\left\| e(U_k) \right\|_Q^2$ $(U_k = U_{SNPC,k} + U_{ILC,k}$) is not in the boundary, then we can find a set Θ_{U^*} uncovering the boundary of Θ_U.

Namely, we have $\Theta_{U^*} = \left\{ \mathbf{U} \Big| \left\| e(\mathbf{U}_{ILC,k}) \right\|_Q^2 - M^* \Big| < \varepsilon \right\}$ and $U \in \Theta_{U^*}$ is not in the boundary of Θ_U. We make conclusion that $\left\| e(\mathbf{U}_{ILC,k}) \right\|_Q^2$ converges to global optimal solution, namely

$$\lim_{k\to\infty} \left\| e(\mathbf{U}_{ILC,k+1}) \right\|_Q^2 - M^* \right| = \lim_{k\to\infty} \left\| e(\mathbf{U}_{ILC,k+1}) \right\|_Q^2 - M^* = 0 \qquad (18)$$

Thus, we get

$$\lim_{k\to\infty} \left\| e(\mathbf{U}_{ILC,k+1}) \right\|_Q = \lim_{k\to\infty} \left\| e(U_k) \right\|_Q$$

Therefore

$$\lim_{k\to\infty} \left\| e(\mathbf{U}_{k+1}) \right\|_Q^2 - M^* = 0$$

This completes the proof. Q.E.D.

4   Example

To demonstrate the effectiveness of the proposed scheme, this example considers the following batch process, in which a first-order irreversible exothermic reaction takes place [22]

$$\dot{x}_1 = -4000\exp(-2500/T)x_1^2$$
$$\dot{x}_2 = 4000\exp(-2500/T)x_1^2 - 6.2\times10^5\exp(-5000/T)x_2$$

where x_1 and x_2 are respectively the reactant concentration of component A and B, T is the reactor temperature.

Firstly, T is normalized using $T_d = (T - T_{min})/(T_{max} - T_{min})$, in which T_{min} and T_{max} are 298（K）and 398（K）respectively. T_d is the control variable bounded between $[0,1]$, and $x_2(t)$ is the output variable. The nominal operating conditions are: $x_1(0) = 1$, $x_2(0) = 0$.

The control objective is to maximize the endpoint concentration of B , $x_2\left(t_f\right)$ by manipulating the reactor temperature, T_d . To proceed with the proposed method, 30 batches of independent random signal with uniform distribution between [0, 1] are used to obtain input-output data for training purpose. Applying the identification procedure in [15] results in a neuro-fuzzy model with 6 fuzzy rules.

The robustness of the proposed integrated control system is evaluated by introducing 5% Gaussian white noise to the measured batch process variables at fifth batch. As illustrated in Fig. 1 and 2, the proposed integrated control system has reasonable robustness to stochastic noise.

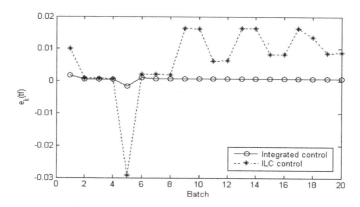

Fig. 1. 5-th batch error curves of two methods under the same disturbance

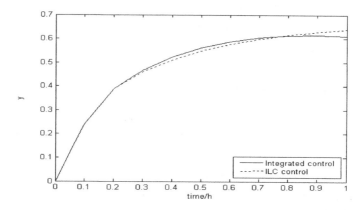

Fig. 2. 5-th batch production quality curves of two methods under the same disturbance

5   Conclusion

The proposed system integrates discrete-time (batch-axis) information and continuous-time (time-axis) information into one uniform frame. More specifically, the iterative learning controller is designed in the domain of batch-axis, while an adaptive single neuron predictive controller (SNPC) in the domain of time-axis. The convergence and tracking performance of the proposed integrated learning control system are firstly given rigorous description and proof. It showed that the integrated scheme not only enhanced the control performance of the batch processes but also guaranteed the convergence and robustness of batch processes.

Acknowledgement. Supported by National Natural Science Foundation of China (61004019), Research Fund for the Doctoral Program of Higher Education of China (20093108120013), Shanghai University, "11th Five-Year Plan" 211 Construction Project.

References

1. Xiong, Z.H., Zhang, J., Wang, X., Xu, Y.M.: Integrated tracking control strategy for batch processes using a batch-wise linear time-varying perturbation model. Control Theory & Applications 1(1), 178–188 (2007)
2. Lu, N., Gao, F.: Stage-based process analysis and quality prediction for batch processes. Industrial and Engineering Chemistry Research 44(10), 3547–3555 (2005)
3. Amann, N., Owens, D.H., Rogers, E.: Iterative learning control for discrete-time system with exponential rate of convergence. IEE Proc., Control Theory Appl. 143, 217–224 (1996)
4. Liu, T., Gao, F.-R.: IMC-based iterative learning control for batch processes with uncertain time delay. Journal of Process Control 20, 173–180 (2010)
5. Lee, K.S., Lee, J.H.: Convergence of constrained model-based predictive control for batch processes. IEEE Trans. Autom. Control 45, 1928–1932 (2000)
6. Lee, K.S., Chin, I.S., Lee, H.J., Lee, J.H.: Model predictive control technique combined with iterative learning control for batch processes. AIChE J. 45, 2175–2187 (1999)
7. Lee, J.H., Lee, K.S., Kim, W.C.: Model-based iterative learning control with a quadratic criterion for time-varying linear systems. Automatica 36, 641–657 (2000)
8. Lee, K.S., Chin, I.-S., Lee, J.H.: Model Predictive Control Technique Combined with Iterative Learning for Batch Processes. AIChE Journal 45(10), 2175–2187 (1999)
9. Xiong, Z.H., Zhang, J.: Tracking control for batch processes through integrating batch-to-batch interative learning control and within-batch on-line control
10. Xiong, Z.H., Zhang, J.: Product quality trajectory tracking of batch processes using iterative learning control based on linear perturbation model. Ind. Eng. Chem. Res. 42, 6802–6814 (2003); American Chemrican Society 44, 3988–3992 (2005)
11. Xiong, Z.H., Zhang, J.: Batch-to-batch optimal control of nonlinear batch processes based on incrementally updated models. IEE Proc., Control Theory Appl. 151, 158–165 (2004)
12. Rogers, E., Owens, D.H.: Stability analysis for linear repetitive processes. Springer, Heidelberg (1992)
13. Kurek, J.E., Zaremba, M.B.: Iterative learning control synthesis based on 2-D system theory. IEEE Transactions on Automatic Control 38(1), 121–125 (1993)
14. Fang, Y., Chow, T.W.S.: 2-D analysis for iterative learning controller for discrete-time systems with variable initial conditions. IEEE Transactions on Circuits and Systems-I: Fundamental Theory and Applications 50(5), 722–727 (2003)

15. Shi, J., Gao, F., Wu, T.: 2D Model Predictive Iterative Learning Control Schemes for Batch Processes. In: IFAC International Symposium on Advanced Control of Chemical Processes 2006, Gramado, Brazil, pp. 215–220 (2006)
16. Shi, J., Gao, F., Wu, T.: From Two-dimensional Linear Quadratic Optimal Control to Iterative Learning Control Paper 1. Two-dimensional Linear Quadratic Optimal Controls and System Analysis. Industrial & Engineering Chemistry Research 45(13), 4603–4616 (2006)
17. Shi, J., Gao, F., Wu, T.: From Two-dimensional Linear Quadratic Optimal Control to Iterative Learning Control Paper 2. Iterative Learning Controls for Batch Processes. Industrial & Engineering Chemistry Research 45(13), 4617–4628 (2006)
18. Jia, L., Shi, J.-P., Chiu, M.-S.: Fuzzy neural model with global convergence for batch process. Information and Control 38(6), 1–7 (2009)
19. Jia, L., Cao, L.-M., Chiu, M.-S.: Data Driven-Adaptive Single Neuron Predictive Controller Based on Lyapunov Approach. In: WCICA 2011 - 2011 World Congress on Intelligent Control and Automation, Conference Digest, pp. 7–12 (2011)
20. Jia, L., Cao, L., Chiu, M.-S.: Data-driven Based Integrated Learning Control for Batch Processes. In: IFAC 2012 International Symposium on Advanced Control of Chemical Processes, Singapore, pp. 234–238 (2012)
21. Jia, L., Shi, J.-P.: R-adjustable learning control for batch process. Control Theory & Applications 28(9), 1159–1162 (2011)
22. Lu, N., Gao, F.: Stage-based process analysis and quality prediction for batch processes. Industrial and Engineering Chemistry Research 44(10), 3547–3555 (2005)

Passive Control of Sampling System

Wenyuan Li

Department of Computer, Harbin Financial Institute
Harbin, 150080, China
wxl_jsj@163.com

Abstract. The passive control problem of a class of uncertain state-delay sampling system is discussed. Applying Lyapunov method, and combining the properties of matrix inequality, the sufficient condition of passive stability is given, and passive controller is designed. Finally, a numerical example illustrates the effectiveness and availability for the design.

Keywords: Uncertain state-delay system, Sampling system, Passive control, Linear Matrix Inequality.

1 Introduction

Researches on passive stability of the sampling system were very active in the past decades[1, 2]. Several approaches have been proposed to solve passive stability of the sampling system. Literature [3] discusses passive stability of sampled system using Riccati equation approach. Literature [4] discusses passive stability of sampled system for a class of non-structural perturation models using time-domain approach. Literature addresses passive control of a class of sampled systems with structured uncertainty using LMI approach. Literature addresses passive state feedback control of uncertain sampled system with pole constrain. But it isn't perfect, they only consider uncertainty of the system parameters without taking into account of uncertainty of controller gain. when there is perturbation in controller parameters (for example, which often occurs that the system initially runs, the controller finely tunes and the controller gain parameters change when controller performance deterioration) , the traditional passive control methods shows a highly fragile [5–8], which results in that closed-loop system performance decreased and stability damage. Therefore, we must take into account uncertainties in the controller to guarantee the stable and high-performance operation of the system.

Based on this, this article considers passive control problem of the uncertain time-delay a sampled system with perturbations of controller gain, in the controller gain of two forms with the addition and multiplication, we discusses with the design method of the controller gain and gives passive controller's existence sufficient condition of the uncertain time-delay sampled system , which depends on the LMIs feasible solution, and is verified by simulation examples.

K. Li et al. (Eds.): ICSEE 2013, CCIS 355, pp. 139–144, 2013.

2   Preliminaries and Problem Statement

Consider continuous control system corresponding to sampling system

$$
\begin{cases}
\dot{x}(t) = (A_0 + \Delta A_0)x(t) + (A_1 + \Delta A_1)x(t - \tau) + (B_0 + \Delta B_0)u(t) + \\
\quad f(x, u, t) + B_1\omega(t) \\
z(t) = C_1 x(t) + C_2 x(t - \tau) + H_2\omega(t) \\
x(t) = x_0, \ t \in [-\tau, \ 0]
\end{cases}
\tag{1}
$$

where $x(t) \in R^n$ is the state vector, $u(t) \in R^n$ is control input vector, A_0, A_1 and B_0 are system matrix and control matrix with appropriate dimension respectively, $\Delta A_0, \Delta A_1$ are uncertain system matrix with the appropriate dimension respectively. Constant time-delay $d > 0$.

to discretization equation (1)

$$
\begin{cases}
x(k + 1) = (G_0 + \Delta G_0)x(k) + (G_1 + \Delta G_1)x(k - d) + (H_0 + \Delta H_0)u(k) + \\
\quad \bar{f}(x, u, t) + H_1\omega(k) \\
z(k) = C_1 x(k) + C_2 x(k - d) + H_2\omega(k) \\
x(k) = x_0, \ k \in [-d, \ 0]
\end{cases}
\tag{2}
$$

where

$$
G_0 = e^{A_0 h}, \ G_1 = \int_0^h e^{A_0(h-w)} dw A_1
$$

$$
H_0 = \int_0^h e^{A_0(h-w)} dw B_0, \ H_1 = \int_0^h e^{A_0(h-w)} dw B_1
$$

$\Delta G_0, \Delta G_1, \Delta H_0$ are uncertain matrix, and satisfy the following form

$$
\begin{bmatrix} \Delta G_0 & \Delta G_1 & \Delta H_0 \end{bmatrix} = MF(k) \begin{bmatrix} E_0 & E_1 & E_2 \end{bmatrix}
$$

Lemma 1. [9] For a given symmetric matrix $S = S^{\mathrm{T}} = \begin{bmatrix} S_{11} & S_{12} \\ S_{12}^{\mathrm{T}} & S_{22} \end{bmatrix}$ with $S_{11} \in \mathbf{R}^{r \times r}$, the following conditions are equivalent:

(1) $S < 0$

(2) $S_{11} < 0, \quad S_{22} - S_{12}^{\mathrm{T}} S_{11}^{-1} S_{12} < 0$

(3) $S_{22} < 0, \quad S_{11} - S_{12} S_{22}^{-1} S_{12}^{\mathrm{T}} < 0$

Lemma 2. [10] For given matrices $Q = Q^{\mathrm{T}}$, H, E with appropriate dimensions,

$$
Q + HF(t)E + E^{\mathrm{T}} F^{\mathrm{T}}(t) H^{\mathrm{T}} < 0
$$

holds for all $F(k)$ satisfying $F^{\mathrm{T}}(t)F(t) \le I$ if and only if there exists $\varepsilon > 0$, such that

$$
Q + \varepsilon^{-1} H H^{\mathrm{T}} + \varepsilon E^{\mathrm{T}} E < 0
$$

3   Main Results

Theorem 1. *Consider the time-delay discrete-time systems (2) is asymptotically stable, if there is an appropriate dimension of the positive definite symmetric matrix $P > 0$ and $S > 0$, at the same time satisfy the following LMI*

$$\begin{bmatrix} -P_1 + P_2 & * & * \\ 0 & -P_2 & * \\ P_1 G_0 & P_1 G_1 & -P_1 \end{bmatrix} < 0 \qquad (3)$$

Proof. Define the Lyapunov function as

$$V(x(k)) = x^{\mathrm{T}}(k) P x(k) + \sum_{t=k-h}^{k-1} x^{\mathrm{T}}(i) S x(i)$$

The full differential form of $\Delta V(x(k))$, along the trajectories of $V(k)$ is given by

$$\begin{aligned} \Delta V(x(k)) &= V(x(k+1)) - V(x(k)) \\ &= x^{\mathrm{T}}(k+1) P_1 x(k+1) - x^{\mathrm{T}}(k) P_1 x(k) \\ &\quad + \sum_{i=k+1-h}^{k} x^{\mathrm{T}}(i) P_2 x(i) - \sum_{i=k-h}^{k-1} x^{\mathrm{T}}(i) P_2 x(i) \\ &= \begin{bmatrix} x^{\mathrm{T}}(k) & x^{\mathrm{T}}(k-h) \end{bmatrix} \Omega \begin{bmatrix} x(k) \\ x(k-h) \end{bmatrix} \end{aligned}$$

where

$$\Omega = \begin{bmatrix} G_0^{\mathrm{T}} P G_0 - P_1 + P_2 & G_0^{\mathrm{T}} P_1 G_1 \\ G_1^{\mathrm{T}} P_1 G_0 & G_1^{\mathrm{T}} P_1 G_1 - P_2 \end{bmatrix}$$

Hence when $\Delta V < 0$, we can obtain $\Omega < 0$, It can be proved.

Theorem 2. *Consider the uncertain time-delay discrete-time systems (2) with additive controller gain perturbations (4), if there is an appropriate dimension of the positive definite symmetric matrix $X > 0$ and $W > 0$ and constant value $\varepsilon_i > 0$, $(i = 0, 1, 2)$, at the same time satisfy the following LMI*

$$\begin{bmatrix} M & G_0 X + H_0 Y & G_1 X & 0 & 0 & 0 \\ * & -X + W & 0 & X^{\mathrm{T}} E_0^{\mathrm{T}} & X^{\mathrm{T}} E_2^{\mathrm{T}} & 0 \\ * & * & -W & 0 & 0 & X E_1 \\ * & * & * & -\varepsilon_0 I & 0 & 0 \\ * & * & * & * & -\varepsilon_2 I & 0 \\ * & * & * & * & * & -\varepsilon_1 I \end{bmatrix} < 0 \qquad (4)$$

then time-delay closed-loop system is asymptotically stable, at this time, there is passive controller in the uncertain time-delay system $K = Y X^{-1}$.

Proof. By passive controller, the closed-loop time-delay system

$$x(k+1) = G_0 x(k) + G_1 x(k-h) + H_0 K x(k)$$
$$= (G_0 + H_0 K)x(k) + G_1 x(k-h) \tag{5}$$

Let $P > 0$, $S > 0$ are symmetric positive matrix, define the Lyapunov function as

$$V(x(k)) = x^{\mathrm{T}}(k)Px(k) + \sum_{t=k-h}^{k-1} x^{\mathrm{T}}(i)Sx(i)$$

The full differential form of $\Delta V(x(k))$, along the trajectories is given by

$$
\begin{aligned}
\Delta V(x(k)) &= V(x(k+1)) - V(x(k)) \\
&= x^{\mathrm{T}}(k+1)Px(k+1) - x^{\mathrm{T}}(k)Px(k) \\
&\quad + \sum_{i=k+1-h}^{k} x^{\mathrm{T}}(i)Sx(i) - \sum_{i=k-h}^{k-1} x^{\mathrm{T}}(i)Sx(i) \\
&= \begin{bmatrix} x^{\mathrm{T}}(k) & x^{\mathrm{T}}(k-h) \end{bmatrix} \Psi \begin{bmatrix} x(k) \\ x(k-h) \end{bmatrix}
\end{aligned}
$$

where

$$
\Psi = \begin{bmatrix} (G_0 + H_0 K_1)^{\mathrm{T}} P(G_0 + H_0 K_1) & (G_0 + H_0 K_1)^{\mathrm{T}} P G_1 \\ * & G_1^T P G_1 \end{bmatrix}
$$
$$
+ \begin{bmatrix} -P + S & 0 \\ * & -S \end{bmatrix}
$$

Thereupon $\Delta V < 0$ is equivalent with that

$$
\begin{bmatrix} (G_0 + H_0 K_1)^{\mathrm{T}} P (G_0 + H_0 K_1) & (G_0 + H_0 K_1)^{\mathrm{T}} P G_1 \\ * & G_1^T P G_1 \end{bmatrix}
$$
$$
+ \begin{bmatrix} -P + S & 0 \\ * & -S \end{bmatrix} < 0
$$
$$
= \begin{bmatrix} (\bar{G}_0 + H_0 K_1)^{\mathrm{T}} \\ \bar{G}_1^{\mathrm{T}} \end{bmatrix} P \begin{bmatrix} \bar{G}_0 + H_0 K_1 & \bar{G}_1 \end{bmatrix} + \begin{bmatrix} -P + S & 0 \\ * & -S \end{bmatrix} < 0
$$

Introducing passivity,

$$\Delta V - 2z^{\mathrm{T}}(k)\omega(k) < 0$$

That is to say,

$$
\begin{bmatrix} -P + S & * & * \\ 0 & -S & * \\ -C_1 & -C_2 & -H_2 - H_2^T \end{bmatrix} + \begin{bmatrix} G_0^T \\ G_1^T \\ H_1^T \end{bmatrix} P_1 \begin{bmatrix} G_0 & G_1 & H_1 \end{bmatrix} < 0
$$

By lemma1,

$$
\begin{bmatrix}
-P+S & * & * & * \\
0 & -S & * & * \\
-C_1 & -C_2 & -H_2 - H_2^T & * \\
G_0 & G_1 & H_1 & -P^{-1}
\end{bmatrix} < 0
\tag{6}
$$

Pre-and post-multiplying the matrix (16) by diag $(I,\ I,\ I,\ P^{-1})$, let $X = P^{-1}$, $W = P^{-1}SP^{-1}$, $Y = KX$, we have inequality (7), It can be proved.

4 Numerical Example

Consider the uncertain time-delay system (1), and as is known

$$
A_0 = \begin{bmatrix} -1 & 4 \\ 0 & -1 \end{bmatrix},\ A_1 = \begin{bmatrix} -1 & 0.5 \\ 1 & 0.1 \end{bmatrix}
$$

$$
B_0 = B_1 = \begin{bmatrix} 0 \\ 1 \end{bmatrix},\ D_0^T = \begin{bmatrix} 0.1 \\ 0.1 \end{bmatrix},\ D_1^T = \begin{bmatrix} 0.1 \\ 0.1 \end{bmatrix},\ D_2^T = \begin{bmatrix} 0.1 \\ 0.1 \end{bmatrix}
$$

$$
D_1^T = \begin{bmatrix} 0.1 \\ 0.1 \end{bmatrix},\ D_2^T = \begin{bmatrix} 0.1 \\ 0.1 \end{bmatrix}
$$

$$
E_0 = \begin{bmatrix} 0.2 & 0.3 \\ 0.1 & 0.4 \end{bmatrix},\ E_1 = \begin{bmatrix} 0.2 & 0.4 \\ 0.1 & 0.3 \end{bmatrix},\ E_2 = \begin{bmatrix} 0.2 & 0.1 \\ 0.3 & 0.4 \end{bmatrix}
$$

we are known from theorem 2 using Matlab to solve

$$
X = \begin{bmatrix} 0.1291 & -0.2015 \\ -0.2015 & 0.9243 \end{bmatrix},\ Y = \begin{bmatrix} -0.1291 & -0.4304 \end{bmatrix}
$$

$$
K = YX^{-1} = \begin{bmatrix} -2.6174 & -1.0362 \end{bmatrix}
$$

thus passive controller of the system can be designed

$$
u(k) = YX^{-1}x(k) = -2.6174x_1(k) - 1.0362x_2(k)
$$

References

1. Dugard, L., Verriest, E.I.: Stability and Control of Time-delay Systems. Springer, Berlin (1998)
2. Chen, T., Francis, B.: Optimal Sampled-Data Control System. Springer, New York (1995)
3. Wu, J.F., Wang, Q., Chen, S.B.: Robust stability for sampled-data systems. Control Theory & Applications 18(8), 99–102 (2001)
4. Cui, B.T., Hua, M.G.: Robust passive control for uncertain discrete-time systems with time-varying delays. Chaos Solitons & Fractials 29(2), 331–341 (2006)

5. Keel, L.H., Bhattacharyya, S.P.: Robust, fragile, or optimal. IEEE Transactions Automatic Control 42(8), 2678–2683 (1997)
6. Wang, S.G.: Non-fragile $\mathcal{H}_\infty$ control with pole constraints for a class of nonlinear sampled-data system. Lecture Notes in Electrical Engineering 87(2), 587–594 (2011)
7. Wang, S.G., Wu, J.F.: Observer-based non-fragile $\mathcal{H}_\infty$ control for a class of uncertain time-delay sampled-data systems. Systems Engineering and Electronics 33(6), 1358–1361 (2011)
8. Wang, S.G., Wu, J.F.: Non-fragile passive filtering for a class of sampled-data system with long time-delay. Research Journal of Applied Sciences, Engineering and Technology 4(17), 2861–2865 (2012)
9. Albert: Conditions for positive and non-negative definiteness in terms of pseudoinverses. SIAM Journal on Appiled Mathematics 17(2), 434–440 (1969)
10. Bamieh, B., Pearson Jr., J.B.: A general framework for linear periodic system with applications to $\mathcal{H}_\infty$ sampled-data control. IEEE Transactions Automatic Control 37(4), 418–435 (1992)

Endocrine-Immune Network and Its Application for Optimization

Hao Jiang, Tundong Liu*, Jing Chen, and Jiping Tao

College of Information Science and Technology, Xiamen University,
Xiamen 361005, China
jianghao_1017@163.com, {ltd,taojiping}@xmu.edu.cn, chenjxmu@126.com

Abstract. A novel artificial immune network model (EINET) based on the regulation of endocrine system is proposed. In this EINET for optimization, several operators are employed or revised which aim at faster convergence speed and better optimal solution. Further speaking, a new operator, hormonal regulation, exerts a bidirectional regulatory mechanism inspired from endocrine system, which undergoes elimination and mutation according to hormone updating function, to increase the diversity of antibody population. And antibody learning is an evolution of individuals through learning from memory antibody in immune network. Then, a local search procedure called enzymatic reaction is utilized to facilitate the exploitation of the search space and speed up the convergence. To evaluate whether the proposed model can be directly extended to an effective algorithm for solving combinatorial optimization problem, EINET-TSP algorithm is designed. Comparative experiments are conducted using some benchmark instances from the TSPLIB, and the results compared with the existing immune network applied to combinatorial optimization problem shows that the EINET-TSP algorithm is capable of improving search performance significantly in solution quality.

Keywords: Artificial Immune Network, Hormonal regulation, Enzymatic reaction, Traveling Salesman Problem.

1 Introduction

Over the past few years, based on principles of the immune system, a new paradigm, called artificial immune system (AIS), has been employed for developing interesting algorithms in many fields such as pattern recognition, computer defense, optimization, and others. Artificial Immune Network (AIN) is inspired from immune network theory originally proposed by Jerne [1], which has been one of the most important immune theories. In recent years, s a large number of AINs have been developed, two popular approaches are RLAIS model [2],which is modified from a earlier version named AINE, and aiNet model presented by de Castro [3,4,5]. The aiNet enhanced the clonal selection algorithm (CLONALG) [6] by combining it with immune network theory. In a subsequent work,

---

* Corresponding author.

K. Li et al. (Eds.): ICSEE 2013, CCIS 355, pp. 145–159, 2013.

de Castro and Timmis [5] developed a modified version of aiNet to solve multi-modal optimization problems, called opt-aiNet. The AIN-based metaheuristics for optimization are receiving increasing attention in recent research. The Copt-aiNet algorithm[7,8] is an improved version of opt-aiNet to solve combinatorial optimization problems. Artificial Immune Network for Dynamic Optimization [9] (dopt-aiNet) is applied to optimize time-varied functions . Another algorithm, Concentration-based Artificial Immune Network (cob-aiNet) devoted to real-parameter optimization [10,11,12]. Given its complexity, only a small part of the immune mechanism model has been used in the studies mentioned above.

The immune network is a regulated network of cells and molecules which maintain interactions between not only an antibody and an antigen, but also antibodies themselves. Regulatory mechanisms play a crucial role in maintaining the immune network in a given dynamic steady state. Therefore, in this work, we propose a novel AIN model-Endocrine-Immune Network (EINET) for optimization, which combining artificial endocrine system (AES) with the immune network. Specifically, two characteristics of hormone in the artificial endocrine system, hormonal regulation mechanisms and highly effective enzymatic reaction, are taken advantage of to achieve a faster convergence and better diversity for immune network algorithm, respectively. The main difference between this model and current models is that the elimination and mutation probability in the process of hormonal regulation is according to the hormone updating function.

Based on the framework of EINET, we present EINET-TSP algorithm, which is applied to solve Traveling Salesman Problem, a classical NP-complete problem in discrete or combinatorial optimization, and result in high-performance solution. Furthermore, some comparative experiments are conducted for demonstrating the effectiveness and high-performance of the proposed EINET for optimization.

The remainder of the paper is organized as follows. In Section 2, the new Endocrine-Immune Network for optimization is described in details and the framework of the model is given. The design of EINET-TSP algorithm, which is applied to solve traveling salesman problem, is provided in Section 3. In Section 4, Experimental results are presented and discussed. This paper is concluded in Section 5.

2   Endocrine-Immune Network for Optimization

In this section the framework of a novel AIN model, Endocrine-Immune Network (EINET) for Optimization, is outlined. Inspired from the hormonal mechanisms in AES, two specific operators, Hormonal Regulation and Enzymatic Reaction, are given in EINET.

2.1   Description of Endocrine-Immune Network

The immune network hypothesizes that antibody not only capable of recognizing antigens but also each other, which forms a regulated network. In Endocrine-Immune Network, the objective problem which needs to be solved is regarded as antigen, and the feasible solutions are regarded as candidate antibodies (Ab). Affinity $aff(.)$ between the antibody and the antigen is used to evaluate the combination of each candidate antibody with the particular antigen. There will be found one best antibody with a relatively high affinity of each generation that can be viewed as memory antibody Ab_m. In particular, the memory antibody of each generation will be learned by other general antibodies, and the difference $diff(.)$ between the general antibody and memory antibody evaluate the antibody is good or bad. That is to say, an antibody similar to the memory antibody is a relatively good antibody, and, conversely, a relatively bad antibody is quite different from memory. Meanwhile, the antibody cluster around the memory antibody is a set of immune networks . $Net = \{Ab_1, Ab_2, Ab_3...Ab_n\}$Therefore, topology structure of EINET is a similar star network structure which center on the memory antibody and consider the antibody difference as linkers connected into a network, as shown in Fig.1. Finally, the highest affinity memory antibody(Ab_{opt})will be the optimal solution.

On the other hand, inspired from the hormonal mechanisms in artificial endocrine system[13,14], the process of Hormonal Regulation and Enzymatic Reaction is provided to improve the immune network. In fact, the biological immune system and the endocrine system are integrated into one single system of information communication, and they interact and cooperate with each other to organize an intelligent regulatory network. The model integrated the advantages of two functional characteristics of hormone, including the cooperation and antagonism among hormone and the enzymatic reaction. The former mechanism has a bidirectional regulation effects on immune system, which will result in an activation or suppression on antibodies, that is, help increase diversity of immune network with whole optimization. And the enzymatic reaction is a local search algorithm that can speed up the evolution of antibody and then speed up the convergence of the solution. When combined, these two mechanisms would lead to significant possibilities for improvements of the model.

2.2   Details of the General Framework

In the proposed EINET for optimization, five operators, including affinity evaluation, difference evaluation, hormonal regulation, antibody learning and enzymatic reaction, are designed to improve and enhance the adaptability of immune network and the extreme research target is to upgrade the performance of the proposed model in complex optimization problem. The framework of EINET is shown in Fig.2. These operators are repeated until the termination conditions are satisfied.

The main five operators of the new model is explained in detail as follows:

- Affinity evaluation: Calculate affinity aff between antigen and antibodies including memory antibodies. The affinity function is given based on the objective function of an actual optimization problem. In addition, the antibody with the highest affinity is set as memory cells.

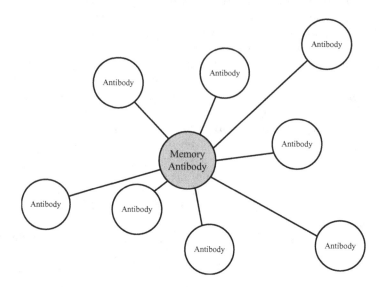

Fig. 1. EINET network structure

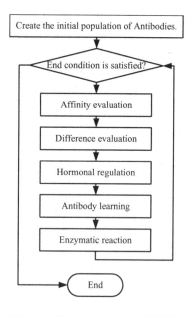

Fig. 2. The framework of EINET

- Difference evaluation: The difference *diff* between one antibody and the memory antibody is used to measure the quality of the antibody. In general, we can define *diff* by using Euclidian distance, Haming distance, etc.
- Hormonal regulation:Hormones are efficient bio-active substances, secreted by endocrine cells and endocrine glands. They have an important role in effecting physiological function and adjusting the metabolism of tissue cells in our body [15]. Simulating the behavior of hormone in endocrine system,H, exert a bidirectional regulation of immunity to increase the diversity of antibody population in EINET. The function of hormones may take two forms, suppression and activation on antibodies. According to the *diff* of antibody, hormone suppression generates low-level elimination probability and mutation probability to avoid the good antibodies being affected; by means of the high-level elimination probability and mutation probability, hormone activation can improve the bad antibody to be better.

Here, hormone updating function which determines the amounts of the hormone can be expressed as follow:

$$H = Fun(diff) \tag{1}$$

where,$H \epsilon [0,1]$. In general, H updating function is an increasing function. Then the H updating function can be defined by

$$Fun(diff) \epsilon [0,1] \tag{2}$$

$$\begin{cases} Fun(diff) < 0.5, diff < 0.5 & (3a) \\ Fun(diff) > 0.5, diff > 0.5 & (3b) \end{cases}$$

where, the constraints ensure that there is a balance point at 0.5 between the H suppression and H activation operator.

- Antibody learning: For any antibody Ab, learning from the segment or characteristic of memory antibody Ab_m is a process of immune network's evolution, in which evolution itself is a type of learning. Once an antibody whose affinity is higher than the affinity of memory antibody is found, that antibody will take the place of the original memory antibody as the new memory antibody after the learning process.
- Enzymatic reaction: Enzymatic reaction is a local search algorithm that is adopted in order to quickly improve the quality of the population of EINET. Equivalently, enzymatic reaction is a greedy algorithm implemented to optimize the antibodies to relatively good. Since that there is no randomness in the calculate process of the greedy algorithm, the convergence can be reached by iteration.

As seen from the model main framework of Fig. 2 and the above description, there are some differences between our proposed model and the existing AINs. Unlike opt-aiNet, there is no clonal selection process in the EINET. Instead, we mainly utilize the hormonal regulation related to the difference between the general antibody to memory antibody as a diversity approach. Furthermore, in our proposed model, the reactions between antibodies and memory antibody promote the evolution individually, which is suitable for parallel computing in further research. However, the reactions among antibodies (crossover operator) are not considered here. In particular, through combing the advantages of the hormonal regulation and enzymatic reaction, the EINET provides a good trade-off between the diversity and convergence.

3   EINET-TSP Algorithm Design

In order to evaluate whether the proposed model, EINET for optimization, can be directly extended to an effective algorithm capable of solving combinatorial optimization problems, an approach based on EINET called EINET-TSP are presented to solve TSP.

The TSP is a typical example of a very hard combinatorial optimization problemand can be simplified as: a search for the shortest closed tour that visits each city at lease and only once. Considering a traveling salesman problem with k cities to visit, the traveling salesman problem can be formulated as:

$$\min\{\sum_{j=1}^{k-1} d_{j,j+1}(.) + d_{k,1}(.)\} \tag{4}$$

where $d_{j,j+1}(.)$ is the distance between city j to city $j+1$.

Here, the EINET-TSP is incorporated with the necessary design to make it applicable to TSP based on the given framework of EINET for optimization. Details of the five operators of the EINET-TSP algorithm are discussed in the following subsections.

3.1   Affinity Evaluation

In the TSP, one feasible route through all the cities can be regarded as a candidate antibody. Then, the affinity function defined by the objective function of TSP (Eq.1) can be written as:

$$aff(Ab_i) = \frac{1}{\sum_{j=1}^{k-1} d_{j,j+1}(Ab_i) + d_{k,1}(Ab_i)} \tag{5}$$

As stated above, the antibody with max $aff(Ab_i)$ is selected as the memory antibody.

3.2   Difference Evaluation

Evaluate the difference value of each antibody is equal to calculate a distance between the antibody route and the memory antibody route. Denote the route of Ab_i as

$$\{c_1(Ab_i), c_2(Ab_i), c_3(Ab_i)...c_j(Ab_i)...c_k(Ab_i)\}$$

where $c_j(Ab_i)$ is the serial number of jth city in the route. Here, the formulation of the difference function is given as follows:

$$f_{diff}(Ab_i) = (\sum_{j=1}^{k} \mid c_j(Ab_m) - c_j(Ab_i) \mid).(\mid aff(Ab_m) - aff(Ab_i) \mid) \qquad (6)$$

In this formulation, $\sum_{j=1}^{k} \mid c_j(Ab_m) - c_j(Ab_i) \mid$ is the cumulative difference, and $\mid aff(Ab_m) - aff(Ab_i) \mid$ is the affinity difference. In the proposed algorithm, the product of these two difference values is considered as a measurement of the difference of each two antibody.

Then, the normalized antibody difference ($[0,1]$) can be computed as

$$diff(Ab_i) = \frac{f_{diff}(Ab_i) - \min(f_{diff}(.))}{\max(f_{diff}(.)) - \min(f_{diff}(.))} \qquad (7)$$

Considering the difference of memory antibody, we have

$$\min(f_{diff}) = f_{diff}(Ab_m) = 0 \qquad (8)$$

So that,

$$diff(Ab_i) = \frac{f_{diff}(Ab_i)}{\max(f_{diff}(.))} \qquad (9)$$

3.3   Hormonal Regulation

To implement regulation of hormone, there are two operators for the bidirectional hormonal regulation: Elimination and Mutation. And then the H updating function is defined by

$$H(Ab_i) = \frac{1}{90}(atan(2 * diff(Ab_i) - 1) + 45) \qquad (10)$$

The elimination probability of any antibody Ab_i is defined by $H(Ab_i)$. Randomly generate a number $r \epsilon [0,1)$, if, $r < H(Ab_i)$ antibody Ab_i would be eliminated. In this way, bad individuals with higher difference value will be eliminated.

The mutation probability of remaining antibody Ab_i after elimination operator is defined by $H(Ab_i)$. Randomly generate a number $r \epsilon [0,1)$, if $r < H(Ab_i)$, antibody Ab_i would mutate to a random position. Here, we adopt a mutation method of multipoint mutation, in which points or positions of mutation operators change randomly.

Note that the main feature of this the hormonal regulation is that it implement a two-way regulation intended to maintain the network in a dynamic steady state. That is, antibodies with low difference are suppressed to mutate; while antibodies with high difference are promoted. Meanwhile, new antibodies will be generated after elimination and mutation such that the scale of antibody population remains the same size n.

3.4   Antibody Learning

In the TSP, the process of antibody learning is performed in the following way: Select a segment route from memory antibody Ab_m randomly, and then add the segment to the end of the learning antibody Ab_i. Besides, remove the duplicate city number of antibody Ab_i. As shown in Fig.3.

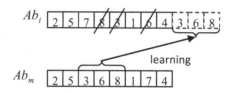

Fig. 3. antibody learning operator

3.5   Enzymatic Reaction

Here, we apply a greedy algorithm, the Interpolation-based Optimization Method for Antibody algorithm, to implement the enzymatic reaction of endocrine system for immune network, whose aim is to improve the convergence. According to the characteristics of TSP, the interpolation-based optimization method is designed as follows: Select the first visited city $c_j(Ab_i)$ in Ab_i, and try to insert it into next position. If it can shorten the route, the action will be performed; if not, undo it. The operator is repeated until the route can't be shortened. The pseudocode of the interpolation-based optimization method is given below.

Algorithm:the Interpolation-based
                  Optimization Method for Antibody
do while Path can shorten
    for n=1 to k **do**
    for j=2 to k-1 **do**
      if
$[d_{k,1}(Ab_i) + d_{1,2}(Ab_i) + d_{j,j+1}(Ab_i)] - [d_{j,1}(Ab_i) + d_{1,j+1}(Ab_i) + d_{k,2}(Ab_i)] > 0$
      then
         remove $c_1(Ab_i)$
         insert $c_1(Ab_i)$ between $c_j(Ab_i)$ and $c_{j+1}(Ab_i)$
         break
      end if
    end for
    if Path has not been modified **then**
        move $c_1(Ab_i)$ to the end of the Ab_i
    end if
    end for
end do

To summary, the whole pseudocode of the proposed algorithm EINET-TSP is presented in the following.

Algorithm Framework: EINET-TSP Algorithm
Initialization
Create the initial population of *Net*.
Main Phase
do while (End Condition)
    Affinity evaluation:
        for Ab_i **in** Net **do**

$$aff(Ab_i) = \frac{1}{\sum_{j=1}^{k-1} d_{j,j+1}(Ab_i) + d_{k,1}(Ab_i)}$$

        end for
        Find Ab_m
    Difference evaluation:
        for Ab_i **in** Net **do**

$$diff(Ab_i) = \frac{f_{diff}(Ab_i)}{\max(f_{diff})(.)}$$

        end for
    Hormonal regulation:
        for Ab_i **in** Net **do**

$$H(Ab_i) = \frac{1}{90}(atan(2 * diff(Ab_i) - 1) + 45)$$

        if rand(0,1)$<$ $H(Ab_i)$***then***
          Remove Ab_i
        else
          if rand(0,1)$<$ $diff(Ab_i)$ ***then***
            Swap some city values $(c_j|Ab_i$) randomly.
          end if
        end if
        end for
    Create new antibodies randomly to keep the population size of Net .
Antibody learning:
    for Ab_i **in** Net **do**
        Copy a segment in random position and add it to the end of Ab_i.
        Delete city values $(c_j|Ab_i)$, which are same as the values in the segment,
from Ab_i.
    end for
    Enzyme reaction:
        for Ab_i **in** Net **do**
          " Insert Optimal Method for Ab_i " **used**
        end for
end do

4   Experiments

In this section, some comparative experiments are arranged to examine the performance of the proposed EINET-TSP algorithm, where some TSP benchmark instances from TSPLIB [16], a standard library of sample instances for the TSP (and related problems) from various sources and of various types, are chosen. Firstly, a diversity analysis of EINET is discussed.

4.1   Diversity Analysis

Population diversity is an important evaluation parameter for evolution algorithms. The study of Ref. introduced Phenotypical Diversity (PDM) and Genotypical Diversity (GDM) evaluation parameters of the population in the search process [17, 18]. PDM and GDM are defined as:

$$PDM = \frac{aff_{avg}}{aff_{max}} \tag{11}$$

$$GDM = \frac{\overline{E} - E_{min}}{E_{max} - E_{min}} = \frac{diff_{avg} - diff_{min}}{diff_{max} - diff_{min}} \tag{12}$$

where aff_{avg} and aff_{max} represent the average and maximum fitness values of the population in the current generation, respectively. is the average Euclidean distance between all individuals and the best individuals of the population in the current generation, which defined by the average antibody difference; E_{max} and E_{min} represent the maximum and minimum difference of the antibody, respectively. Both PDM and GDM belong to the interval [0,1]. Usually, if $PDM > 0.9$ and $GDM < 0.1$, the algorithm tends to converge. If $0 < PDM \leq 0.9$ and $GDM \geq 0.1$, then the algorithm is in the normal search process.

To analyze the diversity of EINET, here, EINEI-TSP was applied to the two benchmark instances from TSPLIB, eil51and kroA100. From Fig.4 and Fig.5, the PDM curve before the enzymatic reaction operator and the PDM curve after the enzymatic reaction operator, we can see that EINET can better keep diversity in the search process and converge to the optimal point in a stable manner. Note that, before enzymatic reaction operator, there is a good diversity with the PDM below 0.6, which avoid the algorithm running into the local optimization solution. This verifies that hormonal regulation operator can recruit new antibodies to keep high population diversity during the run. While, after enzymatic reaction, each antibody was one local optimal solution, the PDM is always below 0.9. More analytically, the optimal solution curves of each generation show that when the algorithm is close to the local optimization solution, the diversity can help it jump out local optimization and tend to converge.

4.2   Comparing with Cob-aiNet[C] and Copt-aiNet

In this subsection, we compared the EINET-TSP with the two recent competitive immune network algorithms for combinatorial optimization such as cob-aiNet [C] [12] and copt-aiNet [8]. These four instances, namely att48, eil75, kroC100 and ch150, were extracted from the TSPLIB.

The quality of the best route found by EINET-TSP for each problem was compared with the best results obtained by cob-aiNet[C] and copt-aiNet from the literature [8, 12]. From Table 1 it is possible to see that both cob-aiNet[C] and copt-aiNet were able to find the optimal route for all the problems studied here, except ch150. However, even for ch150 the best solutions found by both algorithms are very close to the optimal one. When comparing, EINET-TSP were able to solve all the four problems studied here, which indicates that the technique is effectively capable of dealing with such problem.

Table 1. Best Results Obtained by Each Algorithm

	Opt	cob-aiNet[C]	cob-aiNet[C]	EINET-TSP
att48	10628	10628	10628	10628
eil76	538	538	538	538
kroC100	20749	20749	20749	20749
ch150	6528	6529	6531	6528

Table 2 shows the mean and standard deviation of results obtained by cob-aiNet[C] and our EINET-TSP, evaluated over the 10 runs for each problem, together with their difference (in percentage) from the global optima. As it is possible to see, EINET-TSP is capable of obtaining a set of solutions for each problem with costs very close to the optimal routes. Compared with cob-aiNet[C], our algorithm EINET-TSP has lower average percent difference or better accuracy solutions than cob-aiNet[C].

Table 2. Mean and Standard Deviation, Over 10 Runs, and Difference (in %) of These Mean Values From the OTIMA

	cob-aiNet		EINET	
	Average (std,dev.)	Percentage fromOpt	Average (std,dev.)	Percentage fromOpt
att48	1074593.6	1.1%	1066327.3	0.3%
eil76	555.146.4	3.2%	5402.0	0.3%
kroC100	21418498.0	3.2%	20856106.2	0.5%
ch150	6747.896.2	3.4%	670786.8	2.7%

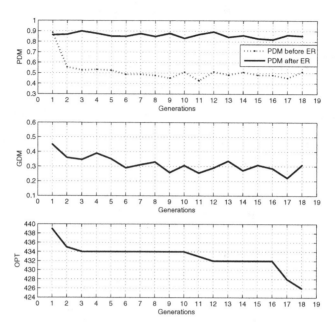

Fig. 4. PDM, GDM and Optimal solution curves of the EINET-TSP during the run for eil51

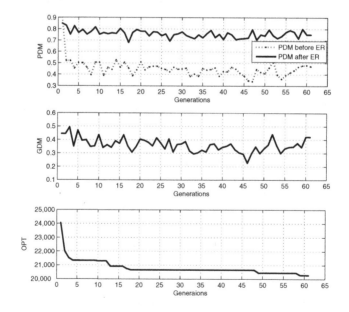

Fig. 5. PDM, GDM and Optimal solution curves of the EINET-TSP during the run for kroA100

4.3   Comparing with Neuro-Immune Network Algorithms

In order to compare the algorithm with another AIS model, Neuro-immune network, which combing immune system with the neural network, we have implemented the proposed method with 25 datasets from TSPLIB. Table 3 makes a comparison of the experimental results of the proposed EINET with the two neuro-immune network algorithms, Pasti and Castro's method [19] and Masutti and Castro's method [20], with respect to the best solutions and the average solutions for 30 independent runs. In Table 3, the best results are emphasized in bold. These result indicate that the proposed algorithm has a good capability to search global optima. For the data sets eil51,eil76,berlin52,bier127,ch130,ch150,rd -100,lin105, kroA100,kroA150,kroB200,kroC100,kroD100,kroE100 we also can see that EINET-TSP show better performance than other methods. However, for some cities scale more than 100, our algorithm EINET-TSP can find best solution while the average level is not consistently better than the two algorithms. Considering the population size, the EINET may improved by the parallelization technique with the good suitability.

Table 3. A Comparison of the Experimental Results

Instance	OPT	Pastiand Castro's method[19]		Masuttiand Castro's method[20]		EINET	
		Average	Best	Average	Best	Average	Best
eil51	426	438.70	429	437.47	427	426.45	426
eil76	538	556.10	542	556.33	541	540.08	538
eil101	629	654.83	641	648.63	638	633.32	629
berlin52	7542	8073.97	7716	7932.50	7542	7612.00	7542
bier127	118282	121780.33	118760	120886.33	118970	119701.38	118282
ch130	6110	6291.77	6142	6282.40	6145	6262.75	6110
ch150	6528	6753.20	6629	6738.37	6602	6707.24	6528
lin105	14379	14702.23	14379	14400.17	14379	14551.56	14379
lin318	42029	43704.97	42975	43696.87	42834	43698.27	42537
kroA100	21282	21868.47	21369	21522.73	21333	21296.70	21,282
kroA150	26524	27346.43	26932	27355.97	26678	26934.67	26524
kroA200	29368	30257.53	29594	30190.27	29600	30202.23	29580
kroB100	22141	22853.60	22596	22661.47	22343	22246.21	22,141
kroB150	26130	26752.13	26395	26631.87	26264	26545.36	26133
kroB200	29437	30415.60	29831	30135.00	29637	30325.17	29562
kroC100	20749	21231.60	20915	20971.23	20915	20856.15	20,749
kroD100	21294	22027.87	21457	21697.37	21374	21491.00	21,294
kroE100	22068	22815.50	22427	22715.63	22395	22172.53	22,068
rd100	7910	8253.93	7947	8199.77	7982	8147.30	7910
rat575	6773	7125.07	7039	7115.67	7047	7132.23	6982
rat783	8806	9326.30	9185	9343.77	9246	9325.73	9104
rl1323	270199	300286.00	295060	305314.33	300770	303480.25	291044
fl1400	20127	21070.57	20745	21110.00	20851	21092.63	20665
d1655	62,128	71431.70	70323	72113.17	70918	72203.46	68324

5 Conclusions

In our study, we proposed a new artificial immune network model, Endocrine-Immune Network (EINET) for optimization. Inspired by high-level regulation principle of endocrine system, we utilized two mechanisms of endocrine, hormonal regulation and enzymatic reaction mechanisms, into the evolution process of AIN model for EINET. In order to evaluate the performance of EINET, a version of EINET, EINET-TSP algorithm, is designed and implemented successfully to solving Traveling Salesman Problem. Experiments based on TSP benchmark instances from TSPLIB shows that EINET has a promising performance for combinatorial optimization problem. We also compare our algorithm with the nine recent algorithms such as copt-aiNet, cob-aiNet[C], neuro-immune network, etc. The computational results indicate that EINET-TSP was able to obtain the global optimum for most of the problems studied, while simultaneously providing a set of high quality and diverse solutions. As future steps, our future research will be mainly focused on the application of EINET in other optimization problems. In particular, a parallel version of proposed model will be an important objective in future work, which allows us to apply EINET to more real problems.

References

1. Jerne, N.: Towards a Network Theory of the Immune System. Ann. Immunol. 125C, 373–389 (1984)
2. Timmis, J., Neal, M.: A Resource Limited Artificial Immune System for Data Analysis. Knowledge-Based Syst. 14, 121–130 (2001)
3. de Castro, L.N., Von Zuben, F.J.: An Evolutionary Immune Network for Data Clustering. In: Proceedings of the IEEE SBRN 2000, Brazil (2000)
4. de Castro, L.N., Von Zuben, F.J.: Ainet: an Artificial Immune Network for Data Analysis. Int. J. Comput. Intell. Appl. 1(3) (2001)
5. de Castro, L.N., Timmis, J.: An Artificial Immune Network for Multimodal Optimization. In: Congress on Evolutionary Computation Part of the 2002 IEEE World Congress on Computational Intelligence, Honolulu, Hawaii, USA (2002)
6. de Castro, L.N., Von Zuben, F.J.: The Clonal Selection Algorithm With Engineering Applications. In: Proceedings of GECCO 2000 Workshop on Artificial Immune Systems and their Applications, pp. 36–37 (2000)
7. de C.T. Gomes, L., de Sousa, J.S., Bezerra, G.B., de Castro, L.N., Von Zuben, F.J.: Copt-aiNet and the Gene Ordering Problem. In: Proceedings of WOB, Rio de Janeiro, Brazil, pp. 28–37 (2003)
8. de Franca, F.O., de C.T. Gomes, L., de Castro, L.N., Von Zuben, F.J.: Handling Time-varying TSP Instances. In: Proc. of the 2006 IEEE Congress on Evolutionary Computation (CEC), pp. 2830–2837 (2006)
9. de Franca, F.O., Von Zuben, F.J., de Castro, L.N.: An Artificial Immune Network for Multimodal Function Optimization on Dynamic Environments. In: Proceedings of the 2005 Conference on Genetic and Evolutionary Computation, Washington, USA, pp. 289–296 (2005)

10. Coelho, G.P., Von Zuben, F.J.: A Concentration-based Artificial Immune Network for Continuous Optimization. In: Proc. of the 2010 IEEE Congress on Evolutionary Computation (CEC), pp. 108–115 (2010)

11. Coelho, G.P., Von Zuben, F.J.: A Concentration-Based Artificial Immune Network for Multi-objective Optimization. In: Takahashi, R.H.C., Deb, K., Wanner, E.F., Greco, S. (eds.) EMO 2011. LNCS, vol. 6576, pp. 343–357. Springer, Heidelberg (2011).

12. Coelho, G.P., De Franca, F.O., Von Zuben, F.J.: A Concentration-based Artificial Immune Network for Combinatorial Optimization. In: 2011 IEEE Congress of Evolutionary Computation, CEC 2011, pp. 1242–1249 (2011)

13. Chen, D.-B., et al.: A Multi-objective Endocrine PSO Algorithm and Application. Applied Soft Computing 11, 4508–4520 (2011)

14. Xu, Q.-Z., Wang, L.: Recent advances in the artificial endocrine system. Journal of Zhejiang University Science C 12, 171–183 (2011)

15. Xu, Q.-Z., Wang, L.: Lattice-based Artificial Endocrine System Model and Its Application in Robotic Swarms. Science China Information Sciences 54, 795–811 (2011)

16. TSPLIB-A Traveling Salesman Problem Library, http://www.iwr.uni-heidelberg.de/groups/comopt/software/TSPLIB95

17. Herrera, F., Lozano, M.: Adaptation of Genetic Algorithm Parameters Based on Fuzzy Logic Controllers. In: Herrera, F., Verdegay, J.L. (eds.) Genetic Algorithms and Soft Computing, pp. 95–125 (1996)

18. Shi, X., Qian, F.: A Multi-Agent Immune Network Algorithm and Its Application to Murphree Efficiency Determination for the Distillation Column. Journal of Bionic Engineering 8, 181–190 (2011)

19. Pasti, R., de Castro, L.N.: A Neuro-immune Network for Solving the Traveling Salesman Problem. In: Proceedings of 2006 International Joint Conference on Neural Networks, pp. 3760–3766 (2006)

20. Masutti, T.A.S., de Castro, L.N.: A Self-organizing Neural Network Using Ideas From the Immune System to Solve the Traveling Salesman Problem. Information Sciences 179(10), 1454–1468 (2009)

Hybrid NSGA-II Algorithm on Multi-objective Inventory Management Problem

Lin Lin and Shiji Song

Dept. of Automation, TNList, Tsinghua University, Beijing 100084, P.R. China
hattielin@126.com, shijis@mail.tsinghua.edu.cn

Abstract. Inventory management is a key issue in supply chain management. Under the circumstances that there are plenty of risks, it is more usable and appropriate if the risk problem is also taken into consideration when addressing the issue of inventory management. In this paper, we firstly introduces the classifications of inventory model, introduces two parameters, VaR and CVaR to measure risks. Also, we established a bi-objective model considering inventory cost and CVaR at the same time. Heuristic method to solve the problem is addressed then. We examined the application of Genetic Algorithm on multi-objective problems, i.e. the NSGA-II algorithm. We proposed an analytic method to simplify the solution of the problem. Besides, we examined the local search method based on the problem and proposed a Hybrid Genetic Algorithm. Simulation verifies the usability of our model and the efficiency of our algorithm.

1 Introduction

Supply chain is a network that integrates the suppliers, manufacturers, retails and the end customers. And to manage a supply chain is to manage the flow of raw materials, products and information flow that involves. It is shown by research that inventory system occupies $20\%-60\%$ value of all the manufacturing industry, making inventory management a focus in supply chain management. Inventory management is to control the behavior of when to order and how much to order so that the manufacturing process can be sustained and products can meet the needs of retailers or customers.

As to the supply chain that consists of many layers, there is much risk to be considered. Risk can be from exterior or interior. The delay in supply and less demand from customers are both interior risks [1]. Terrorism and SARS are on the other hand the exterior ones. These risks exert great influence on the company or the entire supply chain. Therefore, it is meaningful to add risk dimension to inventory problem. This passage in organized like this. Section 2 introduces the literature about inventory models and risk measurements. Section 3 builds a bi-objective model which can show the inventory cost and risk at the same time. At last, section 4 proposes a problem-based hybrid algorithm to solve this problem. Through simulation, we verified the feasibility and efficiency of this algorithm.

K. Li et al. (Eds.): ICSEE 2013, CCIS 355, pp. 160–168, 2013.

2   Literature Review

2.1   Inventory Models

Single-Period and Multi-period Models. Single-period model has another name as newsvendor model. In this model, only one period is considered and it is assumed that at the end of one period, there is always the same amount of products left so that every period would start the same. This means that managers would have to discard, sell at a lower price to get rid of extra products, and they also have to bear the shortage cost if products are not enough at the end [2]. On the contrary, multi-period model is defined on a longer time line. This time line is segmented into many time cycles. The cycles are not the same. At the beginning of a cycle, there will be an order so that certain inventory would exist in the beginning. The difference in demand in every time cycle leads to the time cycles difference. With larger demand, the inventory can only last for a shorter period [3].

Static and Dynamic Models. In static model, the manager would order the same amount of raw material or products at every beginning of time cycle. While in dynamic models, (r, Q), (r, nQ) or (s, S) are strategies often accepted. Inventory order position (IOP) is the concept we use often later in this passage. IOP = inventory on hand + outstanding orders in transit + outstanding orders backlogged − backorders [4]. (r, Q) says that if IOP falls lower than r ,then the manager would order Q units of products. (r, nQ) is a extended form of (r, Q) . As for (s, S) , it means that if IOP falls below s, then order enough units of products so that the total units can reach S .

2.2   Risk Measurements

Two kinds of risk measurements are widely accepted and used: Value at Risk(VaR) and Conditional Value at Risk(CVaR).

VaR. VaR answers the maximum lost a company will suffer under certain confidence level. JP Morgan firstly applied VaR in their Risk Metrics System [5]. Later, many assessing departments began to use VaR to measure risks. For financial institutions, they have to be put away so that the probability of its inability to survive an adverse market would be small. Similarly, in inventory management, some products should be put away to deal with risks generated by demand, price or other elements [6]. VaR is defined as following, $VaR_\alpha(x) = \min\{u|P\{f(x,y) \leq u\} \geq \alpha\}$. $f(x,y)$ is loss. And this equation means the probability of loss to be lower or equal to VaR is bigger or equal to a given parameter α.

CVaR. VaR has been widely used. However, according to Atzner, VaR is not a consistent measurement and this may cause problems in calculating. CVaR was brought up as Optimization of Conditional Value-at-Risk [7] as a better substitute for VaR. CVaR is defined as $CVaR = E[f(x, y) > VaR]$. VaR has the disadvantage of not being convex, thus conceptually and numerically being problematic [8]. CVaR, on the other hand, can be written in the form of a convex optimizing problem.

$$CVaR_\alpha(x) = \min\{\eta_\alpha(x, u)|u \in R\}$$

where $\eta_\alpha(x, u) := u + \frac{1}{1-\alpha}E[f(x, y) - u]^+$ and $[t]^+ := \max\{0, t\}$. With better attributes, we use CVaR for our model.

Kee H. Chung(1990)[9] analyzed single-stage inventory model with risk. Charles S. Tapiero(2003)[10] examined the use of VaR, which is not consistent, in supply chain management. Xin Chen, Melvyn Sim(2007)[11] studied multi-stage inventory model with risk using expected utility function, which cannot show the risk a company is facing directly. So in this passage, we use CVaR as the risk measurement to build a bi-objective model combing risk and inventory.

Genetic Algorithm and NSGA-II. Genetic Algorithm (GA) was first brought about by Holland and his colleagues [12]. The basic idea originates from the natures evolution. In the evolution process, the more adaptable individuals tend to have a bigger chance of surviving, thus the next generation is with less weak genes. Applying this to GA, it means that individuals with lower cost tend to be combined and produce offspring more often.

In multi-objective problems, parents cannot be chose just based on one single value. Therefore, the concept of Pareto front is used to compare the individuals with several objectives. Multi-objective problems differ from single-objective problems in comparison and selection parts. In comparison, we use the concept of Pareto dominance to measure two solutions. In selection, we need to choose individuals to be parents according to Pareto rule. Abdullah Konak[13] summarized the genetic methods used in multi-objective problem, among which are NSGA[14] and more effective NSGA-II(non-dominated-sort-algorithm)[15]. We here use NSGA-II for our problem.

3   Model Formulation

3.1   Problem Description

There are K suppliers and in every single stage, they each have independent demand. And in every stage they make decisions as following

1) To meet the demand according to inventory at hand and instantaneous demand.

2) If demand can be fully satisfied, then order using policy after the demand is fulfilled.

3) If demand cannot be fully satisfied, then first recourse to other suppliers. After fulfillment, if other suppliers still have inventory left, then they offer help. If other suppliers dont have enough inventories, either, shortage penalty occurs.

3.2 Mathematical Formulation

With the demand being stochastic, we first make the model discrete, then build one-objective and bi-objective models, respectively.

Monte Carlo Generating Stochastic Environment. D stands for demand and stochastic demand function is written as $f(D)$. In this problem, decision variables are r, Q, so ultimate cost are decided upon these three variables D, r, Q. For every (r, D) , the expected cost has such form

$$E(C) = \int p(r, Q, D) f(D) dD$$

However, it is hard to implement a continuous function. We here use Monte Carlo sampling method [16] to reach a precise estimation of this expectation. Monte Carlo is to obtain many stochastic realizations under the condition that D follows its distribution function $f(D)$. And then, function value will be calculated under every realization. Through weighted combining, we get the estimation we wanted. In this problem, we use M realizations, so that each has a probability of $1/M$. Eventually, the estimation can be written as

$$E(C) = \int p(r, Q, D) f(D) dD \approx \frac{1}{M} \sum_{m=1}^{M} Mp(r, Q, D_m).$$

Multi-objective Model. The parameters and decision variables are listed below.

IOP	:	Inventory order position
$Shortage$	:	Shortage quantity
$Holdinv$	:	Holding inventory
$Recourse$	:	Borrowed quantity
$Install$	:	Fixed cost
$Profit$	:	Profit
$Demand$	:	Demanded quantity
$Sellout$	:	Shortage quantity
S	:	Sold quantity
H	:	Cost for holding
R	:	Cost for borrowing
P	:	Price for selling
N	:	Stage number
K	:	Supply number
r	:	Supply number
Q	:	Ordering quantity

The total cost is

$$cost = \sum_{m=1} M \sum_{i=1} N \sum_{k=1} K \left(C_{hkim} + C_{pkim} + C_{rkim} + C_{akim} - P_{kim}\right)$$

where $C_{hkim} = holdinv \times h$, $C_{pkim} = shortage \times s$, $C_{rkim} = resource \times r$,

$$\eta_{kim} = \begin{cases} 1, \text{ if } IOP < r \\ 0, \text{ otherwise} \end{cases}$$

$C_{akim} = \eta_{kim} \times install$, $sellout_{kim} = \min(\sum demand_{kim}, \sum IOP_{kim})$, $P_{kim} = sellout_{kim} \times p$.

Then $cvar = u + \frac{1}{1-\alpha}E[C - u]^+$, $u \in R$. In bi-objective model, the objectives are $\min(cost, cvar)$ Solving this bi-objective model, we can get several combinations of (cost, cvar) for the decision makers to choose rather than a single solution as in the one-objective model.

3.3   Optimization and Algorithm

3.4   Problem-Based NSGA-II

Analytic Property of Problem. When using CVaR as the risk measurement, we use scenario analysis method to simulate risk element u. u and (r, Q) in combination will become the gene in GA and NSGA-II. We here use the property of this problem to simplify algorithms through linear programming.

Suppose that limited samples of u, which are $u_1, u_2, \cdots, u_n$ are known. We use the law of large number to generate normal distribution. According to central limit theorem for independent and identically distribution, we have

$$\left(\sum_{k=1} nX_k - n\mu\right) / \left(\sqrt{n}\sigma\right) \sim N(0, 1)$$

It is proven that satisfying approximation effect can be gained when $n=12$. Since this is standard normal distribution that we get, we should use linear transformation for non-standard normal distribution $N(a, b)$. This transformation is $x = x_0 \times b + a$.

In testing process, through changing the μ and σ namely, the mean and variance to have different demand functions. And thus we can see whether the result would be different in relatively stable and unstable market environments.

We have $cvar = u + \frac{1}{1-\alpha}E[C-u]^+$, $u \in R$. In this, u is also a decision variable. One way to solve u is to combine it with (r, Q) in gene and get them through crossover, mutation and selection. However, we find that when stochastic demand and sample have been fixed, we can use linear programming to calculate u.

The mathematical process is as follows. Assume that we conduct M times sampling, then we have $cvar = u + \frac{1}{1-\alpha} \times \frac{1}{M} \sum_{m=1}^{M} [C_m - u]^+$.

Suppose $t_m = [C_m - u]^+$, Then the problem can be transformed into linear form. When $\lambda = \frac{1}{1-\alpha} \times \frac{1}{M}$, we have

$$\begin{aligned} \min \quad & cvar = u + \lambda \sum t_m \\ \text{s.t.} \quad & t_m \geq 0, \\ & t_m + u \geq C_m \end{aligned}$$

We now discuss when varies λ, how to decide u so that $cvar$ can reach its minimum.

1. $\lambda = 1$
   (a) when $u = \max C_m$, $cvar = u = \max C_m$
   (b) when $u = \frac{1}{M} \sum C_m = \bar{C}$, $cvar = \bar{C} + \sum [C_m - \bar{C}]^+$, if we take $c_m = \max C_m$, then $cvar > \bar{C} + \max C_m - \bar{C} = \max C_m$, so the circumstance a is better.
   (c) when $u = \min C_m = C_0$, $cvar = C_0 + \sum [C_m - C]^+$, if we take $C_m = \max C_m$, then we have $cvar > C_0 + \max C_m - C_0 = \max C_m$, in this case, circumstance a is still better.
2. $\lambda > 1$
   This is similar to the situation when $\lambda = 1$, we can get that the best u is $\max C_m$ through similar analyzing method.
3. $\lambda < 1$
   Different λ will result in different best u. In this case, we will have to calculate the best u through linear programming. And the original problem can be presented,

$$\min \quad cvar = u + \lambda \sum_{m=1}^{M} t_m$$
$$\text{s.t.} \quad \begin{pmatrix} -1 & -1 & 0 & \cdots & 0 \\ -1 & 0 & -1 & & \\ \vdots & \vdots & & \vdots \\ -1 & 0 & & & -1 \end{pmatrix} \begin{pmatrix} u \\ t_1 \\ \vdots \\ t_M \end{pmatrix} \leq \begin{pmatrix} -C_1 \\ -C_2 \\ \vdots \\ -C_M \end{pmatrix}$$
$$t_m \geq 0, \ m = 1, \cdots, M$$

Local Search. Local search is usually conducted in this way. First, find an initial solution x. Make use of a simple disturbing method to generate several adjacent solutions to get a neighborhood. If solution in neighborhood is better than x, then the initial solution for next local search is this better one, otherwise, is still x.

In this problem, we generate the neighborhood using the problems property. When the variance of demand function stays unchanged, based on the results of simulation, we can know the impact of several intermediate variables to the decision variables. And the impact is listed below.

Table 1. Diections for local search

	r	Q
Shortage cost high	Should increase	
Install cost high	Should decrease	Should decrease

After the NSGA-II, we examine every solution for its shortage cost and install cost, if shortage cost is high (higher than $\alpha \times mean$), then r increases to see if new solutions are better under Pareto rule. Also, we check for solutions to see if the install cost is too high and respond accordingly.

Serial Algorithm Combing NSGA-II and Local Search. GA is a solution for global search so that it has worse local search ability. To combine it with local search is good for problem solving. The combination has often been used in single-objective problem. And some results have been made in multi-objective problems. We here use a serial method to combine NSGA-II and local search. In this algorithm, the property of this problem we analyzed before is also adopted to simplify all this process. The algorithm process contains .

4   Simulation and Remarks

4.1   Different Demand Situations

It can be seen that when the market, or the demand is more unstable (distribution variance is larger), the value of cost become larger. And we can see the degree of risk from the value of *cvar* more obviously. This verifies the feasibility of our model.

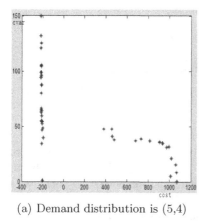
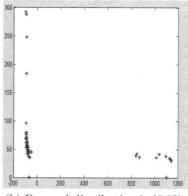

(a) Demand distribution is (5,4)   (b) Demand distribution is (5,16)

Fig. 1. Result of cost and cvar under different demand distribution

4.2 Result of Different Algorithms

We can see from the results that using hybrid NSGA-II, we can have more centralized result (comparing Fig.3 to Fig.2), especially when the cost is high. This verifies that Hybrid NSGA-II based on the problem can have more satisfactory results than the ordinary NSGA-II.

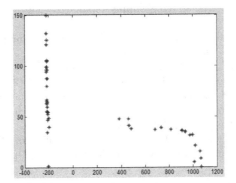

Fig. 2. Result of cost and cvar using NSGA-II algorithm

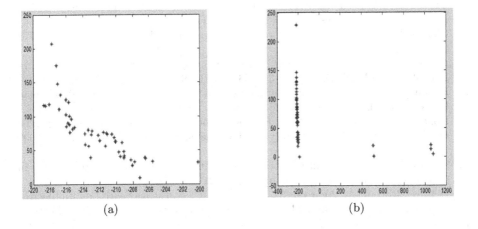

(a) (b)

Fig. 3. Different result using Hybrid NSGA-II

5 Conclusion

This paper focuses on mainly inventory and the risk in it. We built a bi-objective model, and examined several approach to solving it. Particularly, this paper introduces a serial hybrid NSGA-II algorithm which is based on problem property. Through simulation, we verified the feasibility and efficiency of our model and algorithm.

Acknowledgment. This work was supported by Natural Science Foundation of China (No.61273233) and Research Fund for the Doctoral Program of Higher Education (No. 20090002 110035).

References

1. Goh, M., Meng, F.W.: Managing supply chain risk and vulnerability. Springer (2009)
2. Jun-ya Gotoh, J., Takano, Y.: Newsvendor solutions via conditional value-at-risk minimization. Eur. J. Oper. Res. 179, 80–96 (2007)
3. Luciano, E., Peccati, L., Cifarelli, D.M.: VaR as a risk measure for multiperiod static inventory models. Int. J. Prod. Res. 81-82, 375–384 (2003)
4. Zhang, Y.L., Song, S.J., Zhang, H.M., Wu, C., Yin, W.J.: A hybrid genetic algorithm for two-stage multi-item inventory system with stochastic demand. Neural Comput. Applic. 21(6), 1087–1098 (2012)
5. Axsater, S.: Inventory control. Kluwer Academic Publishers (2000)
6. Charles, S.T.: Value at risk and inventory control. Eur. J. Oper. Res. 163, 769–775 (2005)
7. Rockafellar, R.T., Uryasev, S.: Optimization of conditional Value-at-Risk. Journal of Risk 2, 21–41 (2000)
8. Acerbi, A., Tasche, D.: On the coherent of expected shortfall. Journal of Banking & Finance 26, 1487–1503 (2002)
9. Chung, K.H.: Risk in inventory models: the case of the newsboy problem-optimality conditions. J. Oper. Res. Soc. 41(2), 173–176 (1990)
10. Charles, S.T.: Value at risk and inventory control. Eur. J. Oper. Res. 163, 769–775 (2005)
11. Axsater, S., Zhang, W.F.: A joint replenishment policy for multi-echelon inventory control. Int. J. Prod. Econ. 59(1-3), 243–250 (1999)
12. Holland, J.H.: Adaptation in natural and artificial systems. University of Michigan Press, Ann Arbor (1975)
13. Konaka, A., David, W.C., Alice, E.S.: Multi-objective optimization using genetic algorithms: a tutorial. Reliability Engineering and System Safety 91, 992–1007 (2006)
14. Srinivas, N., Deb, K.: Multiobjective optimization using nondominated sorting in genetic algorithms. J. Evol. Comput. 2(3), 221–248 (1994)
15. Deb, K., Pratap, A., Agarwal, S., Meyarivan, T.: A fast and elitist multiobjective genetic algorithm: NSGA-II. IEEE Trans. Evol. Comput. 6(2), 182–197 (2002)
16. Khouja, M., Mehrez, A., Rabinowitz, G.: A two-item news-boy problem with substitutability. Int. J. Prod. Econ. 44, 267–275 (1996)

An Evolutionary Objective Cluster Analysis-Based Interpretable Fuzzy Identification Method

Na Wang[1], Chaofang Hu[2], and Wuxi Shi[1]

School of Electrical Engineering and Automation, Tianjin Polytechnic University,
399 Binshuixi Road, Tainjin 300387, China
wangna@tjpu.edu.cn
School of Electrical and Automation Engineering, Tianjin University,
92 Weijin Road, Tianjin 300372, China
cfhu@tju.edu.cn

Abstract. In this paper, an Evolutionary Objective Cluster Analysis-based Interpretable Fuzzy Identification Method (EOCA-IFIM) is proposed for constructing Mamdani fuzzy model. Firstly, the Enhanced Objective Cluster Analysis (EOCA) algorithm is presented to obtain the robust and the moderate compact initial fuzzy partition. Following, the (1+1) Evolutionary Strategy (ES) is introduced to improve the semantics of the initial parameters. Based on that, a complexity-accuracy trade-off is well realized. The simulation results of the Box-Jenkins and the electrical application example show the superiority of the presented method.

Keywords: Fuzzy modeling, Mamdani model, Interpretable, Fuzzy identification, Objective Cluster Analysis.

1 Introduction

Maintain the interpretability of the models is one of the main objectives in fuzzy modeling for complex systems [1]. Compared with the T-S model [1-2], the Mamdani model is more approved than the T-S model for its full fuzzy sets in structure. Thus it gets wide focus in interpretable fuzzy modeling [3-4].

The identification of the Mamdani model relies on two factors: the structure identification and the estimation of parameters [5]. However, in the data-driven methology, due to the effect of noise, the redundant, inconsistent rules, or the undistinguishable fuzzy sets are usually generated. Thus the interpretability of the model is decreased.

As for that, the Ad-Hoc methods, such as WM, WCA and IRL are presented [6]. Even though, the identification of single rule still relies on the samples and the initial fuzzy partition. As a result, it is difficult to fit the local dynamics. Additionally, it possibly leads to redundancy and inconsistency. In this paper, an Evolutionary Objective Cluster Analysis-based Interpretable Fuzzy Identification Method (EOCA-IFIM) is proposed to construct the Mamdani model.

K. Li et al. (Eds.): ICSEE 2013, CCIS 355, pp. 169–177, 2013.

At first, the original Objective Cluster Analysis (OCA) algorithm [7] is improved, incorporated with the Fuzzy c-Means (FCM) algorithm [8] and the Least Square Estimation (LSE) approach [2]. Thus the redundant rules are simplified effectively and the concise initial rule base is gained. Besides, in the proposed Enhanced Objective Cluster Analysis (EOCA) method, the Dipole Partition (DP), the introducing of the Relative Dissimilarity Measure (RDM) [9] and the presented Enhanced Consistency Criterion (ECC) are used to increase the robustness of OCA algorithm. Thus the proper accuracy of the initial Mamdani model is guaranteed. Following, the (1+1) Evolutionary Strategy (ES) [10] is adopted to optimize the initial parameters. During the evolutionary learning process, the fitness function is designed by the combination of two constraints, the Covering Criteria (CC) and the Genetic Niching Principle (GNP). So the compatibility among the rules and the appropriate over-lapping between the fuzzy sets could be considered simultaneously. The example of Box-Jenkins [11] and electrical application [6] demonstrate the compactness, distinguishability and the moderate accuracy of the presented model.

2   Initial Fuzzy Partitioning via EOCA

By means of the result of EOCA, the initial clustering result is gained and afforded for Fuzzy c-Means (FCM) clustering [8] to form the initial fuzzy partition [1-4].

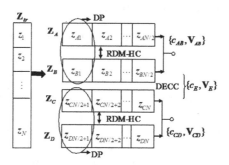

Fig. 1. EOCA principle

The principle of EOCA is shown in Fig.1 and described as follows.

*Step 1:*Partition the sample set Z into subset Z_A and Z_B by dipoles.

*Step 2:*Execute hierarchical clustering on Z_A and Z_B, respectively, then obtain c_{AB} and V_{AB} by minimum enhanced consistency index η_{AB}; In the similar way, obtain c_{CD} and V_{CD} by η_{CD} from Z_C and Z_D.

*Step 3:*Determine the final clustering number c_E and the vector of clusters centers V_E from $\{c_{AB}, V_{AB}\}$ and $\{c_{CD}, V_{CD}\}$ by selecting the minimum between η_{AB} and η_{CD}. Then the algorithm is completed.

In EOCA, Dipole Partition (DP) reduces the effect from noisy clusters with low similarity. Then through the Relative Dissimilarity Measure-based Hierarchical Clustering (RDM-HC), the effect of the unmerged clusters is considered [9]. Finally, in the Discriminating via Enhanced Consistent Criterion (DECC) process, the effect of nearest neighbor in OCA is reduced. DP, RDM and ECC criteria are referred to [4].

Then with c and V_0 from EOCA, execute the FCM clustering, the clustering centers and the fuzzy partition matrix V and the fuzzy partition matrix U are gained. Followingly, by LSE, the muti-dimension fuzzy sets of clustering result are fitted into the triangle membership functions. Thus the initial parameters of Mamdani rules are formed.

3 Semantic Parameters Optimizing via (1+1) ES

In (1+1) Evolution Strategy (ES) [10], each rule is expressed as the gene and denoted as $C = (a_1, b_1, c_1)...(a_p, b_p, c_p)(a_{p+1}, b_{p+1}, c_{p+1})$, where a_i, b_i, c_i denotes the left boundary point, the center and the right margin, respectively, $i = 1, 2, ..., p+1$ is the dimension of samples. Additionally, the fitness function consists of two parts: the Covering Criterion (CC) and the Genetic Niching Principle (GNP).

CC judge the satisfying degree of samples and includes three criteria: Average Activation (AA), Average Covering Ratio of Satisfying Samples (ACRSS) and Covering Ratio of Dissatisfying Samples (CRDS).

[Definition] *Satisfying (Dissatisfying) samples*
Given the rule R_i, z_k is called as the satisfying sample iff it is satisfies:

$$\begin{cases} A_i(X_k) = *(A_i(x_1), ..., A_i(x_p)) > 0 \\ B_i(y_k) > 0 \end{cases} \tag{1}$$

where A_i is the membership of premises; $*$ is the fuzzy inference operator and set product or minimum; B_i is the fuzzy sets of consequences; $X_k = (x_1, ..., x_p)$ is the input of z_k; y_k is the output of z_k.

In the same way, z_k is called the dissatisfying sample if $B_i(y_k) = 0$ is satisfied in the formula (4).

Average Activation (AA)
The Average Activation (AA) is defined as follows:

$$\Psi_{Z_N}(R_i) = \frac{\sum\limits_{l=1}^{N} R_i(z_l)}{N} \tag{2}$$

where $R_i(z_l)$ is the activating degree of z_l by R_i.

Average Covering Ratio of Satisfying Samples (ACRSS)
The Average Covering Ratio of Satisfying Samples (ACRSS) is defined as follows:

$$G_\omega(R_i) = \sum_{z_l \in Z_\omega^+(R_i)} \frac{R_i(z_l)}{n_\omega^+(R_i)} \tag{3}$$

where the satisfying sample set is defined as $Z_\omega^+(R_i) = \{z_l \in Z_N, \text{s.t.} R_i(z_l) \geq \omega\}$.

Covering Ratio of Dissatisfying Samples (CRDS)

The CRDS is defined as follows:

$$
g_n(R_i) = \begin{cases} 1 & \text{if} n_{R_i}^- \leq k \cdot n_\omega^+(R_i) \\ \dfrac{1}{n_{R_i}^- - k \cdot n_\omega^+(R_i) + \exp(1)}, & \text{otherwise} \end{cases} \tag{4}
$$

where the dissatisfying sample set is described as $Z^-(R_i)\{z_l \in Z_N, \text{s.t.} R_i(z_l) = 0 \text{AND} A_i(x_l) > 0\}$, $A_i(\cdot)$ is the covering degree of x_l for the premises of R_i, $n_{R_i}^- = |Z^-(R_i)|$ is the number of dissatisfying samples, $n_\omega^+(R_i)$ is the number of satisfying samples, k is an satisfying degree of the users.

As is seen in the formulas (2)-(4), the proper covering degree for the ordinary and the good samples is guaranteed by the indexes (2)and(3), respectively. Additionally, the suitable covering degree for the bad examples is also considered in the formula (4). Therefore the consistency factors among the rules are included effectively.

GNP is denoted $LNIR(\cdot)$ and described as follows:

$$
LNIR(R_i) = 1 - NIR(R_i) \tag{5}
$$

$$
NIR(R_i) = Max_i\{h_i\} \tag{6}
$$

$$
h_i = *(A(N_i x), B(N_i y)) \tag{7}
$$

$$
A(N_i x) = *(A_1(N_i x_1^i), ..., A_n(N_i x_p^i)) \tag{8}
$$

where R_i: IF x_1 is A_1^i AND $\cdots$ AND x_p is A_p^i THEN y_i is B_i is the adjusting rule in each iteration; x_j is the jth input; A_j^i is the jth fuzzy set in R_i; y_i is ith output; B_i is the membership function in the consequences of R_i ; $i = 1, ..., n$ is the number of rules; $j = 1, ..., p$ is the dimension of the input; $Max\{\cdot\}$ is the maximum operator; $*$ is the minimum operator; $N_i = (N_i x, N_i y)$ is the membership center of R_i.

Since the GNP index ranges from 0 to 1. So when there is no superposition between R_i and the current rule set $\sum_{k=1}^{i-1} R_k$, $LNIR(R_i)$ is 1. Or else, $LNIR(R_i)$ is 0. By the search for classic centers of rules, the overlapping of fuzzy sets is reduced.

Summarily, the fitness function is designed as follows:

$$
F(R_i) = \Psi_{Z_N}(R_i) \cdot G_w(R_i) \cdot g_n(R_i) \cdot LNIR(R_i) \tag{9}
$$

4   Description of EOCA-IFIM Algorithm

In a summary, the EOCA-IFIM algorithm is described as follows:

Step 1. Determine the number of clustering c and the initial centers of clusters V_0 by EOCA;

Step 2. Initialize FCM with c and V_0, execute FCM clustering and decide the final result of clustering: the centers of clusters V and the fuzzy partition matrix U.

Step 3. Obtain the parameters (a_{ij}, b_{ij}, c_{ij}) and the candidate rule set $GP = \{R_i\}$ by LSE. Then let RB^i be an empty set, $i = 1, ..., c$, $j = 1, ..., p + 1$.

Step 4. Select the rule R_i from GP to be adjusted which is the most consistent with the rules in RB^i.

Step 5. Optimize the parameters of R_i by (1+1) ES.

Step 6. Remove R_i from GP.

Step 7. Judge: if GP is not an empty set, let $i = i + 1$ and $RB^{i+1} = RB^i \cup R_i$, then return *step 4*; or else, let $RB = RB^i$ and end the algorithm.

5   Simulation Examples

In this section, the examples of Box-Jenkins gas furnace [11] and low voltage electrical application [6] are applied to verify the effectiveness of EOCA-IFIM.

Case I. Box-Jenkins example

The Box-Jenkins gas furnace system is a SISO dynamic nonlinear process with 296 samples. At each sampling time k, the input $x(k)$ is the gas flow rate, and the output $y(k)$ is the CO_2 concentration. For verifying the robustness of model, the sample set Z is added on a white gauss noise with 5dB signal-noise ratio. Then the input is $X = x(k-1), x(k-2), y(k-1), y(k-2)$, and the output is $y(k)$.

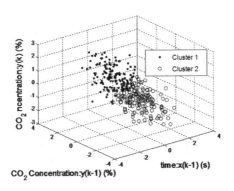

Fig. 2. Distribution of clusters under the input $y(k-1), x(k-1)$

By EOCA, the minimum enhanced consistency index are $ita_{AB} = 0.8248$ and $c_{AB} = 9$. Similarily, $ita_{CD} = 0.0211$ and $c_{CD} = 2$. Thus the initial number of clusters c_0 is 2. The results of clustering by EOCA and FCM are shown in Fig.2.

As is shown, the inner of each cluster distributes uniformly and the margin is obvious and easily to be distinguished. Thus the result of clustering is effective. Fig. 3 shows the consistency relation on Z_C and Z_D.

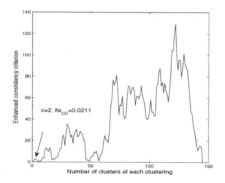

Fig. 3. Enhanced consistency criterion-cluster number at each clustering in $\mathbf{Z}_C$ and $\mathbf{Z}_D$

As is seen in Fig. 3., the clustering results are directly obtained without any presetting parameters or human judgment. Therefore they have stronger robustness than other Ad-Hoc algorithms.

The error between the actual output and the proposed model is shown in Fig. 4.

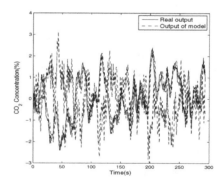

Fig. 4. Comparison between the real output and the output of model

Case II. Low voltage electrical application

In this problem, the number of users and the area of villages is the input (x_1, x_2), and the output y is the length of low voltage line. Fig. 5. shows the unadjusted membership functions. As is shown, in each variable domain, the fuzzy sets have more overlapping and are difficult to be distinguished.

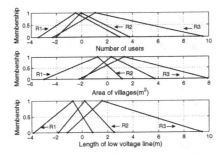

Fig. 5. Membership functions based on clustering projection and least square fitting in each variable

The parameters of the fitness function in (1+1) ES is designed as follows: $\omega = 0.05$, $k = 0.1$; the initial coefficient of step is $\sigma_0 = 1$; if the rate of success is lower than 20%, the new coefficient of step is got by $\sigma_{i+1} = \sigma_i * 0.65$ in the $(i+1)$ generation; Or else $\sigma_{i+1} = \sigma_i/0.65$; The maximum generation of evolution is $g = 100$.

The generated Mamdani rule is described as follows:

$$R_i : \text{IF} x_1 \text{is} A_1^i \text{and} x_2 \text{is} A_2^i, \text{THEN} y_i \text{is} B_i, i = 1, ..., 3 \qquad (10)$$

The tuned membership functions are shown in Fig.6.

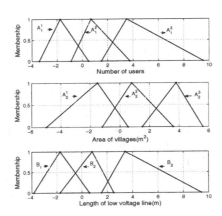

Fig. 6. Tuned membership functions in each variable by (1+1) ES

As is shown in Fig.6., on the one hand, the overlapping area located in the two adjacent membership functions in the domain of each variable is shortened; on the other hand, the centers of the membership functions are easily to be distinguished. As a result, the distinguishability of the original fuzzy partition is improved.

Table 1. Comparison of different Methods in Electrical Applications

Method		Identification			Rules reduction			Parameters tuning	
		Rn	Tre	Tee	Rn	Tre	Tee	Tre	Tee
FC[8]	FCM	49	508426	464130		-			-
	SC	37	401998	444724					
Ad-Hoc[6]	WM-TUN	13	298446	464130			-	175337	180102
	WCA-TUN	20	356434	282058			-	175887	180211
	MOG-UL	25	707773	311195	12	101071	908329	455210	599616
	COR-WN	22	180995	497910			-		-
	IRL	31	179345	220320	26	175480	165046	134020	147758
EOCA-IFIM		3	347606	278778			-		

Table 1. shows the performance comparison among EOCA-IFIM, Fuzzy Clustering (FC) [8] and Ad-Hoc. Where the performance index is 1/2 Mean Square Error (MSE), Rn is the number of rules, Tre is the training error, and Tee is the test error. By EOCA-IFIM, the minimum but compact rule base, i.e., 3 rules are got. In addition, even though the errors of training and test in our model are not optimal, it is still in the same range of magnitude compared with the other methods. Furthermore, the procedures of rules reduction and parameters tuning in Ad-Hoc are avoided. Thus the computation efficiency is also increased. Generally, our model is not of well approximation and generalization ability, but also easily to be implemented.

6   Conclusions

In this paper, based on the combination of EOCA and ES, an EOCA-IFIM approach is proposed to construct an interpretable Mamdani fuzzy model. Firstly, an EOCA-based initial fuzzy partition strategy is presented to decrease the effect of noise and reduce the redundancy of rules. Then the semantic parameters of rules are optimized by a (1+1) ES. By means of the CC and GNP principles, the consistency of rules and the suitable distinguishability between fuzzy sets are considered simultaneously by one-pass quick learning. Thus the interpretability of the Mamdani model is enhanced properly with certain accuracy. The simulation of Box-Jenkins and the electrical applications demonstrates the robustness, the compactness and the appropriate accuracy of the presented model.

Acknowledgments. The authors would like to thank the anonymous reviewers for their helpful comments and constructive suggestions with regard to this paper. This work was supported in part by National Natural Science Foundation of China under Grant (60974054 & 61074064), Natural Science Foundation of Tianjin (11JCYBJC07000) and Key Laboratory of Advanced Electrical Engineering and Energy Technology, Tianjin Polytechnic University.

References

1. Casillas, J., Cordon, O., Herrera, F., et al.: Interpretability improvements to find the balance interpretability-accuracy in fuzzy modeling: an overview. In: Cassilas, J., Cordon, O., Herrera, F., et al. (eds.) Interpretability Issues in Fuzzy Modeling, pp. 3–22. Springer, Herdelberg (2003)
2. Takagi, T., Sugeno, M.: Fuzzy identification of systems and its application to modeling and control. IEEE Transactions on Systems, Man and Cybernetics 15, 116–132 (1985)
3. Guerra, T.M., Kerkeni, H., Lauber, et al.: An Efficient Lyapunov Function for Discrete T-S Models: Observer Design. IEEE Transactions on Fuzzy Systems 20, 187–192 (2012)
4. Wang, N., Yang, Y.P.: A fuzzy modeling method via Enhanced Objective Cluster Analysis for designing TSK model. Expert Systems with Applications 36, 12375–12382 (2009)
5. Meda-Campana, J.A., Gomez-Mancilla, J.C., Castillo-Toledo, B.: Exact Output Regulation for Nonlinear Systems Described by Takagi-Sugeno Fuzzy Models. IEEE Transactions on Fuzzy Systems 20, 235–247 (2012)
6. Alcala, R., Fdez, J.A., Casillas, J.: Local identification of prototypes for genetic learning of accurate TSK fuzzy rule-based systems. International Journal of Intelligent Systems 22, 909–941 (2007)
7. Madala, H.R., Ivakhnenko, A.G.: Inductive learning algorithms for complex systems modeling. CRC Press Inc., Boca Raton (1994)
8. Bezdek, J.C.: Pattern Recognition with Fuzzy Objective Function Algorithms. Plenum Press, New York (1981)
9. Mollineda, R.A., Enrique, V.: A relative approach to hierarchical clustering. Pattern Recognition and Applications 56, 19–28 (2000)
10. Kim, J.H., Han, J.H., Kim, Y.H., et al.: Preference-Based Solution Selection Algorithm for Evolutionary Multiobjective Optimization. IEEE Transactions on Evolutionary Computation 16, 20–34 (2012)
11. Box, G.E.P., Jekins, G.M., Reinsel, G.: Times Series Analysis, Forecasting, and Control. Holden Day, San Francisco (1970)
12. Kim, D.-W., Park, G.-T.: Using Interval Singleton Type 2 Fuzzy Logic System in Corrupted Time Series Modelling. In: Khosla, R., Howlett, R.J., Jain, L.C. (eds.) KES 2005. LNCS (LNAI), vol. 3684, pp. 566–572. Springer, Heidelberg (2005)

Free Weight Exercises Recognition Based on Dynamic Time Warping of Acceleration Data

Chuanjiang Li[1,2], Minrui Fei[1], Huosheng Hu[3], and Ziming Qi[4]

[1] School of Mechanical Engineering & Automation, Shanghai Key Laboratory of Power Station Automation Technology, Shanghai University, 200072, China
[2] College of Information, Mechanical & Electrical Engineering, Shanghai Normal University, Shanghai, 200234, China
[3] Department of Computer Science, University of ESSEX, Colchester, UK, CO4 3SQ
[4] School of Architecture Building & Eng., Otago Polytechnic, Dunedin, New Zealand, 9054
licj@shnu.edu.cn, mrfei@staff.shu.edu.cn, hhu@essex.ac.uk,
Tom.Qi@op.ac.nz

Abstract. To maximize training effects in free weight exercises, people need to remember repetitions of each type of exercises, which is tedious and difficult. Recognizing exercises type and counting automatically can overcome this problem, and multiple accelerometers were used in the existing exercises recognition. This paper presents a new recognition method based on one tri-axial accelerometer, in which a filtered acceleration data stream is divided into time series with unequal length for peak analysis instead of conventional fixed length window. Based on this time series, Dynamic Time Warping (DTW) is deployed to recognize weight exercise types. 3D Euclidean distance and Itakura parallelogram constraint region are used to improve recognition performance. A reference template is set up for each class based on many examples instead of one in the conventional way. The proposed procedures are compared with other popular methods with both the user-dependent protocol and the user-independent protocol. Results show that proposed approach is feasible and can achieve good performance.

Keywords: Free weight exercise, Dynamic time warping, Acceleration data.

1   Introduction

A recent research report reveals that 35.7% of American adults and 17% of American children were obese [1]. To reduce obesity and other chronic diseases, such as heart/cardiovascular diseases, diabetes, hypertension, etc., proper weight exercise plays an important role [2-4]. However, no machine is currently available for tracking the weight exercises, apart from specially designed treadmills and cycle machines. Because the weight training involves many types of exercises, and each type should be done in many repetitions, people may forget their progress or miscount the number of repetition.

In [4], free-weight exercises are tracked based on two tri-axis accelerometers, in which one is placed on the back of the hand and the other on the waist. The overall

K. Li et al. (Eds.): ICSEE 2013, CCIS 355, pp. 178–185, 2012.
© Springer-Verlag Berlin Heidelberg 2012

recognition accuracy of 9 types of exercises is around 90% using Hidden Markov Models classifiers. A RFID based system for monitoring free weight exercises was proposed in [5]. Gaussian model-based classifiers and two tri-axis accelerometers were deployed in [6] to recognize 16 gym exercises, which included 11 weight lifting exercises and the average precision is 92%. However, two accelerometers are not only high-cost, but also inconvenient to wear.

Dynamic time warping (DTW) is one of the commonly used methods for similarity measurement in the time series classification [7], which is widely used in speech recognition, medical analysis, and moving object identification, e.g. multi-motion recognition based on accelerometer [8][13]. Recently, much work has been done to improve DTW accuracy [9-11] and to reduce computation time [13].

In this paper, a single tri-axis accelerometer is attached to the lifting hand glove to recognize nine weight lifting exercises. Improved DTW is adopted to classify exercises types by calculating the similarity between reference template series and test series. Acceleration series data in each repetition is separated by peak analysis and used as the input to 3D DTW classifiers. Experimental results revealed that the proposed method can be applied to various weight exercises. Exercise recognition accuracy is 98% for the user dependent case, and the accuracy is around 95% for user-independent cases.

The remainder of this paper is organized as follows. Section 2 describes a recognition method for weight lifting exercises. Section 3 presents the development of the recognition algorithms. Experimental results and analysis are given in Section 4 to show the feasibility and performance of the proposed algorithm. Finally, a brief conclusion and future work are presented in Section 5.

2   Method of Weight Exercises Recognition

2.1   Response of Every Type of Exercises

Nine common weight lifting exercises are adopted: Biceps curl(BP), Triceps curl(TC), Bench press(BP), Fly(FL), Bent-over row(BR), Literal raise(LR), Overhead dumbbell press(OP), Deadlift(DL), Stand calf raises(SCR). Posture of each type can be found in [4].

A MTx 3-DOF orientation tracker (produced by Xsens) is used in the experiment, as in Fig. 1. Each MTx unit has a tri-axial accelerometer, a tri-axial gyroscope, and a tri-axial magnetometer. But only the accelerometer is used in this experiment. The tracker is installed on back of the right hand glove. As in Fig. 1, with the palm down, the fingers pointing the x-axis, and y-axis at the left, and the z-axis is up. Measurement range of acceleration is ±2g. Fig. 2 shows the curve of acceleration data filtered by 5th order Butterworth low pass filter with sampling rate of 100Hz. Most types of exercises are quite distinct from each other, but BP and OP are very similar as they have the same motion path. The only difference is that body posture of BP is lying, and OP is standing.

Fig. 1. MTx sensor unit

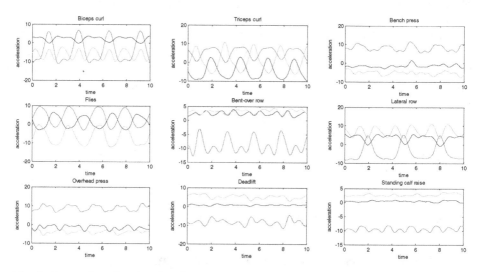

Fig. 2. Curve of acceleration data filtered by 5th Order Butterworth Filter in each of the nine exercises. X-axis is shown with red line. Y-axis is shown with blue line. Z-axis is shown with green line.

2.2   Recognition Method

The process of recognition is listed below:

(1) Get acceleration series of each repetition from data stream using peak analysis.

(2) Set up reference series library of all types of weight exercises.

(3) Calculate warping distance between test series and every reference series.

(4) Recognize class of exercises based on the minimum distance based on DTW.

How to get time sequence data of each repetition from acquisition data stream is an important step of this method. As shown in Fig. 2, the x-axial acceleration fluctuates violently, which is chosen as the major axis. Then time sequence can be split from data stream by two adjacent peaks.

Acceleration wave of some exercises have more than one peak in one repetition. For examples, two proximity peaks occur in one period of exercise TC, BP and OP. Considering that each motion period is usually between 1.5 to 4.5 seconds, we take the algorithm as the following:

(1) Search the peak point in the data stream in length of 2.5 seconds.

(2) If the interval between two adjacent peak points is less than 1 second, the later peak point is canceled.

(3) If the interval between two adjacent peaks is between 1 to 1.5 seconds, and one peak value is less than 70% of the other one, the minor peak is canceled.

The average accuracy rate of this method of splitting sequence is up to 98%. And for some exercises such as BC, TC, LR and FL, it can be up to 100%.

3   Recognition Algorithm

Traditional DTW Algorithms can be seen in [8] In this paper, the data captured by a 3-axial accelerometer is a 3-dimension (3D) time series. Though acceleration of x, y and z-axis can be done respectively using standard DTW, 3D-DTW usually gets good performance [12]. Given 3-axial acceleration reference series

$$\mathbf{R} = \{Rx_1, Rx_2, \cdots Rx_M; Ry_1, Ry_2, \cdots Ry_M; Rz_1, Rz_2, \cdots Rz_M\}^T$$

and test series ,

$$\mathbf{T} = \{Tx_1, Tx_2, \cdots Tx_N; Ty_1, Ty_2, \cdots Ty_N; Tz_1, Tz_2, \cdots Tz_N\}^T ,$$

the distance function is defined as

$$Dist(i, j) = \sqrt{(Rx_i - Tx_j)^2 + (Ry_i - Ty_j)^2 + (Rz_i - Tz_j)^2}. \tag{1}$$

The warping path then gives the minimum normalized total warping distance between $\mathbf{R}$ and $\mathbf{T}$

$$DTW(\mathbf{R}, \mathbf{T}) = \min \frac{1}{|\mathbf{w}|} Dist(\mathbf{w}_{k_i}, \mathbf{w}_{k_j}). \tag{2}$$

where $\mathbf{w}_k$ is the warping path, $|\mathbf{w}|$ is the length of $\mathbf{w}_k$.

The global constraint conditions on the admissible warping paths can not only speed up the DTW computation but also prevent the pathological alignments by globally controlling the route of a warping path. Itakura parallelogram [14] is adopted in this paper, which describes a region that constrains the slope of a warping path.

Reference template building is a main factor to the recognition rate of DTW. Usually selecting only one example for each class as reference template cannot obtain a high recognition rate. The method of multi-template average is adopted in this paper. The process of template building is listed below[15]:

(1)Calculate average length of each type and choose the series nearest to the average length as the initial reference template.

(2) The other training series are time aligned by the DTW process such that their lengths will be equal to the chosen initial template.

(3) The final reference template will be created by averaging the time-aligned series.

4   Experimental Results

4.1   Experiment Setup

The experiments are conducted by three male and a female subjects, whose heights are from 160 to 185cm. Every subject did each exercise in three sets using dumbbell of different weight. They were told to try their best to 15 repetitions in each set. Finally the acceleration data of 1610 repetitions were collected with a sampling rate of 100Hz. The data were filtered by the 5th order Butterworth low-pass filter with a cut-off frequency of 5 Hz.

4.2   User-Dependent Results

The user-dependent protocol checks the algorithmic robustness for individual users with different weights. The reference templates were built using one of three sets for each subject and the other two sets were used to test. The confusion matrices are shown as Table 1. The algorithm proposed in this paper is named as the improved DTW (IDTW) in order to distinguish it from the standard DTW (STDW). The same dataset is tested by Artificial Neural Network (ANN), Support Vector Machine (SVM) and SDTW in order to compare their performance.

Table 1. Confusion matrices of IDTW by user-dependent protocol

	BC	TC	BP	FL	BR	LR	OP	DL	SCR
BC	120	0	0	0	0	0	0	0	0
TC	0	115	0	0	0	0	0	0	0
BP	0	0	117	0	0	0	1	0	0
FL	0	0	0	120	0	0	0	0	0
BR	0	0	0	0	108	0	0	9	0
LR	0	0	0	0	0	120	0	0	0
OP	0	0	6	0	0	0	114	0	0
DL	0	0	0	0	0	0	0	120	0
SCR	0	0	0	0	0	0	0	0	120

Table 2 lists the recognition accuracy of these classifiers. In the tests of ANN and SVM, 15 features are selected, which are three correlation (x-y, y-z, x-z), acceleration average of each axis, and the first three FFT coefficients of each of three axis.

From the table we can see that ANN can hardly discriminate BP because its acceleration signals are similar to OP's. SVM has better performance because it has higher capability in approximation and generalization than ANN. SDTW has much lower performance than IDTW proposed in this paper because it doesn't adopt constraint path, reference template training and 3D Euclidean distance.

Table 2. Recognition accuracy comparison of classifiers by user-dependent protocol

	ANN	SVM	SDTW	IDTW
BC	1.0	0.98	1.0	1.0
TC	1.0	1.0	0.82	1.0
BP	0.1	1.0	0.67	0.99
FL	1.0	1.0	0.96	1.0
BR	0.95	0.95	0.5	0.92
LR	1.0	1.0	1.0	1.0
OP	0.95	0.95	0.92	0.95
DL	1.0	1.0	1.0	1.0
SCR	0.95	0.95	1.0	1.0

Table 3. Recognition accuracy comparison of classifiers by user-independent protocol

	SVM	SDTW	IDTW
BC	0.64	0.96	1.00
TC	0.72	0.80	0.98
BP	0.58	0.61	0.94
FL	0.91	0.90	0.99
BR	0.85	0.45	0.91
LR	0.90	0.93	0.98
OP	0.72	0.83	0.92
DL	0.75	0.88	0.98
SCR	0.86	0.92	0.98

4.3   User-Independent Results

In this experiment, data of four subjects is divided into two parts. One is used as training data, the other as test data. And recognition process is repeated after exchange training dataset and test dataset. The recognition rate of three classifiers is listed in Table 3. The result of the ANN classifier is not given because it has the performance that is bad and not acceptable. The recognition accuracy of IDTW is around 96%, which is much better than 81% for SDTW and 77% for SVM. The confusion matrices of IDTW are listed in Table 4. It is clear that the confusion between BP and OP is getting worse.

Table 4. Confusion matrices of IDTW by user-independent protocol

	BC	TC	BP	FL	BR	LR	OP	DL	SCR
BC	180	0	0	0	0	0	0	0	0
TC	0	172	0	0	0	0	0	3	0
BP	0	0	167	0	0	0	11	0	0
FL	0	0	0	178	0	0	2	0	0
BR	0	0	0	0	161	0	0	16	0
LR	0	0	0	0	0	177	0	3	0
OP	0	0	14	0	0	0	166	0	0
DL	0	0	0	0	0	0	0	176	4
SCR	0	0	0	0	2	0	0	2	176

4.4   Discussion

Some exercises would impose similar acceleration responses on the hardware setting, which is the main reason for having recognition errors. The main characteristics that one exercise different to others is the dynamics of gravity effect and the acceleration changing mode. Therefore, it is important for the user to keep the specified starting status of each exercise as the motion process to improve recognition accuracy.

The BP and the OP almost have the similar starting status and motion process. Though we can distinguish them in high accuracy by user-dependent protocol, the performance becomes low obviously in user-independent situation, especially to the SVM and the SDTW. The reason is that the difference between the BP and the OP to specific user is almost uniform that it can be distinguish easily. So adding the data of specific user to the training dataset can improve recognition performance.

The separation of acceleration data stream is one of the main factors to recognition accuracy. When using very heavy dumbbell, the motion is very slow and shaking, which results in very close peaks with small value. In this situation, accuracy of separation of each repetition decreases, which may reduce recognition performance.

Nine types of dumbbell exercises are given in this paper. If more types need to be recognized, we only need to get the reference template of new type of exercise and add to the reference library, and don't need to train examples of old types. It is an advantage of DTW, compared to other classifier such as SVM, HMM and ANN.

5   Conclusion and Future Work

A novel recognition method of dumbbell exercises is proposed in this paper, which is based on the improved DTW. Compared with the conventional weight exercise recognition method, it uses only one tri-axis accelerometer instead of two, and has achieved better performance. A separation of acceleration data stream is proposed to obtain signal sequences for each repetition of the exercise. The separated data stream is deployed as the input of the DTW classifier. Experimental results show that the recognition accuracy of the proposed DTW algorithm is 98.4%.

Our future work will focus on the recognition of movements that don't belong to any of the target exercise through setting a threshold to each reference of exercises. In addition, the detailed template of each exercise will be built to help users to correct their movement form.

References

1. Ogden, C.L., Carroll, M.D., Kit, B.K., et al.: Prevalence of Obesity in the United States, 2009-2010. NCHS Data Brief 82, 1–7 (2012)
2. Sato, Y., Nagasaki, M., Kubota, M., et al.: Clinical aspects of physical exercise for diabetes/metabolic syndrome. Diabetes Research and Clinical Practice 77S, 87–91 (2007)
3. Galvao, D.A., Newton, R.U.: Review of exercise intervention studies in cancer patients. J. Clin. Oncol. 23, 899–909 (2005)
4. Chang, K.-H., Chen, M.Y., Canny, J.: Tracking Free-Weight Exercises. In: Krumm, J., Abowd, G.D., Seneviratne, A., Strang, T. (eds.) UbiComp 2007. LNCS, vol. 4717, pp. 19–37. Springer, Heidelberg (2007)
5. Chaudhri, R., Lester, J., Borriello, G.: An RFID based system for monitoring free weight exercises. In: Proceedings of the 6th ACM Conference on Embedded Network Sensor Systems, pp. 431–432 (2008)
6. Seeger, C., Buchmann, A., Van Laerhoven, K.: myHealth Assistant: A Phone-based Body Sensor Network that Captures the Wearer's Exercises throughout the Day. In: The 6th International Conference on Body Area Networks, Beijing, China (2011)
7. Yu, D., Yu, X., Hu, Q., et al.: Dynamic time warping constraint learning for large margin nearest neighbor classification. Information Sciences 181, 2787–2796 (2011)
8. Akl, A., Feng, C., Valaee, S.: A Novel Accelerometer-Based Gesture Recognition System. IEEE Transactions on Signal Processing 59, 6197–6205 (2011)
9. Jeong, Y.-S., Jeong, M.K., Omitaomu, O.A.: Weighted dynamic time warping for time series classification. Pattern Recognition 44, 2231–2240 (2011)
10. Yu, D., Yu, X., Hu, Q., et al.: Dynamic time warping constraint learning for large margin nearest neighbor classification. Information Sciences 181, 2787–2796 (2011)
11. Adwan, S., Arof, H.: On improving Dynamic Time Warping for pattern matching. Measurement 45, 1609–1620 (2012)
12. Wollmer, M., Al-Hames, M., Eyben, F., et al.: A Multidimensional Dynamic Time Warping Algorithm for Effcient Multimodal Fusion of Asynchronous Data Streams. Neurocomputing 73, 366–380 (2009)
13. Muscillo, R., Schmid, M., Conforto, S., et al.: Early recognition of upper limb motor tasks through accelerometers: real-time implementation of a DTW-based algorithm. Computers in Biology and Medicine 41, 164–172 (2011)
14. Itakura, F.: Minimum prediction residual principle applied to speech recognition. IEEE Transactions on Acoustics, Speech and Signal Processing 23, 67–72 (1975)
15. Abdulla, W.H., Chow, D., Sin, G.: Cross-words Reference Template for DTW-based Speech Recognition Systems. In: Proceeding of Conference on Convergent Technologies for Asia-Pacific Region, vol. 4, pp. 1576–1579 (2003)

Study on Offshore Wind Farm Integration Mode and Reactive Power Compensation

Xiaoyan Bian, Lijun Hong, and Yang Fu

Power and Automation Engineering Department, Shanghai University of Electrical Power, 200090, Shanghai, China
{kuliz,honglijunv}@163.com

Abstract. Two typical offshore wind farm grid-connected modes are introduced and dynamic characteristics under their modes are compared from the simulation by PSS/E. The result shows that offshore wind farm with VSC-HVDC has better dynamic characteristics on fault isolation, reactive power compensation, and fault ride through ability. In addition, STATCOM has been applied to the offshore wind farm, the simulation results indicates that it can improve the bus voltage stability in fault and maintain the voltage level under a small perturbation.

Keywords: Wind Farm, Integration Mode, VSC-HVDC, STATCOM.

1   Introduction

With the rapid development of wind power technology and expansion of scale, land-based large-scale wind farms cannot meet the requirements of wind power industry development. Abundant resources of offshore wind attracted international community to give high attention on the development of offshore wind farms. Currently, offshore wind farm is limited to the intertidal zone and near the shore, but the construction of wind farm far from shore will be realized in sooner. Compared with land-based wind farms, the main features of the offshore wind as follows: (1) Abundant offshore wind resources, high and stable wind speed as well as a large continuous area available and less development space constraints; (2)The smaller environmental impact (3) Difficulty in power transmission and grid connection. Offshore wind farm located in sparsely populated areas and away from land with the character of large energy production made ordinary low-voltage cable difficult to transfer energy. Mainly used three methods of grid integration are HVAC, PCC-HVDC and VSC-HVDC. (4) Difficulty in construction and maintenance technology with high cost. The initial investment of offshore wind power projects is higher than onshore wind power projects owing to the higher cost of grid connection and support structure. Therefore, the choice of wind farm grid connection has a major impact on the offshore wind power projects, such as dynamic stability, power quality.

Concept ensuring wind farm and grid security capacity is proposed, using different measures to improve transient stability of power grid and improve safety

K. Li et al. (Eds.): ICSEE 2013, CCIS 355, pp. 186–193, 2013.

capacity of wind farm in [1]. Advantages and disadvantages of HVDC and AC transmission system access to wind farm are compared in [2]. The static and dynamic performance influence of wind farm connecting into power grid are analyzed in [3-6]. STATCOM is introduced to improve the dynamic performance of wind farm. Two different integration methods HVAC and VSC-HVDC based on PSS/E software are compared, on this basis STATCOM has been applied to dynamic performance of wind farm.

2 Method of Wind Farm Integration

2.1 HVAC Integration Mode

Offshore wind turbine generated AC power and then transmitted to collector bus, after the step-up transformer, AC cable integrated the offshore wind farm into the main onshore grid, shown in Figure 1. AC cable has large charging capacitance, generated reactive power caused the voltage of wind farm exports rose, so reactive power compensation device need to be installed in front of access grid. When the offshore wind farm in normal operation, reactive power compensation device adsorb reactive power, while in fault, the reactive power compensation need to provide reactive power to improve capability of low voltage ride through. Static Var Compensator (SVC) can be used in small capacity while static synchronous compensator (STATCOM) can be used in lager capacity.

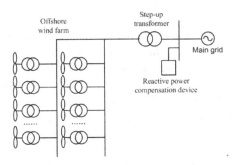

Fig. 1. Offshore wind farm integration method of HVAC

The biggest advantage of offshore wind farm AC transmission is simple system structure and low cost. However, if the offshore wind farm capacity and transmission distance increased, the line loss will seriously increase. As a consequence, transmission voltage level and devices (such as transformers, reactive power compensation devices, etc.) capacity need to increase, as well as submarine cable cross sectional area, all these will increase dramatically in costs. In addition, the grid connected must keep pace with the output of offshore wind farm, which will lead to the mutual influence between two systems. So this integration method is used in offshore wind farm system of small capacity and short transmission distance.

2.2 VSC-HVDC Integration Mode

VSC-HVDC technology is a new type of transmission technology in recent years, its core was the use of insulated gate bipolar transistor (IGBT) of VSC. Converter stations based on the power supply type DC transmission system structure shown in Figure 2, voltage source inverter used in both sides of VSC-HVDC, capacitance paralleled on DC side can provide voltage support, buffer cliffs impact of the shutdown current, reducing the role of the DC side harmonics.

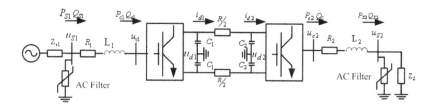

Fig. 2. VSC-HVDC transmission system structure

Advantages of offshore wind farms using VSC-HVDC transmission:

(1) VSC can both absorb reactive power generated by wind farm and can provide reactive power to the wind turbine to provide dynamic reactive power compensation. Wind turbine without load generally supplies reactive power directly by the power factor correction capacitors in parallel with the wind turbine side. On the opposite, in the case of wind turbine with load, corresponding reactive power required in accordance with the size of the active power output of wind turbines, this dynamic reactive power can be provided by the port to the bus side of VSC.

(2) Frequency control to achieve maximum wind energy capture. By the VSC control of the bus side when the wind turbine with variable speed control, in according with the wind speed to adjust to the wind turbine speed, it can achieve maximum wind energy capture.

(3) Good flexible power characteristics. VSC-HVDC can control the power exchange of sending end and receiving end, any change of the wind energy can be detected through the control system, so that the input and output power of the generator would soon reach equilibrium.

(4) Capability of black start. VSC technology of HVDC lines with induction motor reverse so that the wind turbine start-up and connected with a black network.

2.3   Comparison of Two Integration Modes

Two integration methods HVAC and VSC-HVDC have their own characteristics, as shown in Table 1.

Table 1. Comparison of Offshore Wind Farm Integration Modes

Function	HVAC	VSC-HVDC
Maximum transmission capacity	800MW(400 kV) 380MW(220kV) 220MW(132kV)	Has been established up to 350MW; Has announced up to 500MW; proposed the design of 1080MW
Voltage level	132kV has completed and 220kV and 400kV in development	Can reach ±150kV Proposed ±300kV
Engineering construction	Many small projects	Experimental project in Norway
Black start capability	Have	Have
Support grid capacity	Limited and need SVC to inductive reactive power	VSC can absorb and generate the inductive reactive power
Offshore substation	Have	On an oil plantform
Decoupling the connected grid	No	Yes
Cable model	Resistors, capacitors and inductors	Resistors
Require auxiliary equipment	No need	Needed when low wind speed
Total system loss	Depend on distance	4% ~ 6%
Fault level	Higher than HVDC	Lower than HVAC
Construction cost	High cost on power plant (transformers, the thyristor valves, filters, capacitor banks), lower cost on cable	More expensive than the cost of the PCC power station 30% ~ 40% (IGBT is more expensive than thyristor), cable expensive than PCC

3   STATCOM Theory

FACTS technology developed in response of wind farm booming development. Wind turbine is application of FACTS technology. The intermittent nature of wind speed led to the uncertainty of the wind farm voltage and output power. STATCOM in FACTS technology can solve the voltage and reactive power problem properly.

STATCOM and the system connection diagram were shown in Figure 3. The fundamental voltage amplitude of inverter output was V_{ASVG}, system voltage is V_S, compensation current is I.

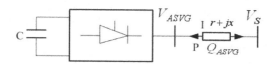

Fig. 3. STATCOM and the system connection diagram

STATCOM adopted voltage bridge circuit, therefore it must be incorporated into the system by connecting the reactor or transformer. The function of reactor first is to connect the inverter and AC bus of different voltage; second inhibition of the high harmonic currents. X in figure is equivalent for leakage inductance or connection reactance, r is equivalent for copper consumption and active power loss of STATCOM. STATCOM is represented as an ideal synchronous condenser, the steady-state operation the vector diagram shown in Figure 4.

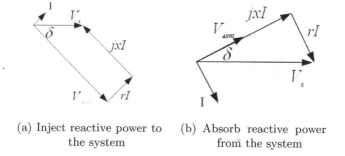

(a) Inject reactive power to   (b) Absorb reactive power
    the system   from the system

Fig. 4. the phasor diagram of STATCOM steady state operation

Known from the phasor diagram, STATCOM injected reactive power into system:

$$Q_{ASVG} = \pm I V_S \cos \delta \tag{1}$$

In formula, δ is the angle of phasor V_{ASVG} lag in vector V_S, $\delta > 0$ corresponds to a positive sign, $\delta < 0$ corresponds to the negative sign. The amplitude of the compensation current:

$$I = \mp \frac{V_S}{r} \sin \delta \tag{2}$$

4   Simulation System and Anylysis

In this paper a simulation example using improved IEEE9 node system, the wind farm access system topology shown in Figure 5. Topology of the wind farm integrating system was shown in Figure 6. Wind farm capacity is 84MW, the master device fan using GE1.5MW from the United States, rated terminal voltage 0.69kV, rated power of 1.5MW and rated frequency of 50Hz, the reactive power regulation range -0.73-0.49Mvar.Each wind turbine with a step-up transformer connected to the collector bus, than through VSC-HVDC or HVAC, offshore wind farm can access to the step-up substation on land, that is the main power grid.

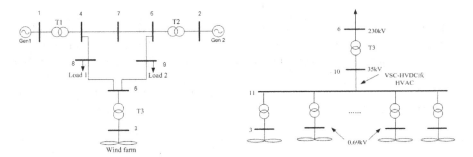

Fig. 5. topology of wind farm integrate to system

Fig. 6. Wind Farm tie line diagram

Simulate the improved IEEE9 node system in PSS / E software, the wind farm model using PSS / E fan package of GE doubly-fed wind turbine model. It uses the rotor AC excitation to the rotor excitation current by the PWM controller, this generator can be in a certain wind speed range. This article did not simulate wind speed, so the rated wind speed 14m/s, the output power was the rated power. The entire system has reached steady state at t = 0. No.8 bus occurred three-Phase short-circuit fault at t=1s, fault continued for 0.1s, and cleared at t=1.1s. Observing the voltage of bus 11 high-pressure side of offshore wind farm step-up substation and bus 8 fault occurred shown in Figure 7, Figure 8.

STATCOM can be applied to improve dynamic recovery characteristics of wind farm under the integration method of HVAC. STATCOM installed in bus 10, the capacity is ±30Mvar. Bus 8 occurred three-Phase short-circuit fault at t=1s, fault continued for 0.1s, and cleared at t=1.1s. The simulation results shown in Figure 9, under normal operating conditions, STATCOM can appropriately improve the voltage as well as the recovery speed, so it can improve fault ride through capability.

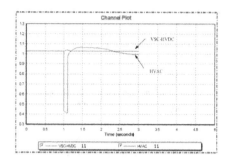

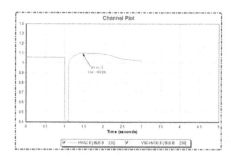

Fig. 7. bus11 voltage under different integrate method

Fig. 8. bus 8 voltage under different integrate method

Figure 10 is the voltage of bus 10 in the case of small perturbations. Load on bus 8 increased 30% at t=1s. The simulation indicated that whether installed STACOM or not, the voltage of bus 10 appeared small perturbations. In the case of small perturbations, voltage of bus installed STATCOM drop smaller amplitude, which improved point of common coupling (PCC) voltage and anti-disturbance capacity.

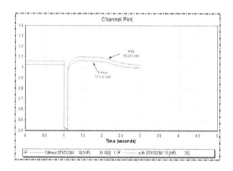

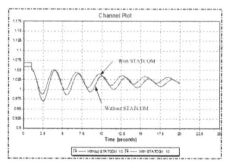

Fig. 9. voltage of bus 10 in the case of fault

Fig. 10. bus 10 voltage under small disturbance

5   Simulation System and Anylysis

This paper analyzed the characteristics of two typical offshore wind farm grid integration methods HVAC and of VSC-HVDC. On this basis, the IEEE 9 bus system was improved by adding a large-scale offshore wind farm, comparison is made under the simulation of HVAC and VSC-HVDC integration method performance. VSC-HVDC integration method has the capacity of fault isolation, and can provide reactive power to system, which improved fault ride through capability. STATCOM applied to the wind farm improved the dynamic nature of the wind farm voltage recovery of fault, also enhanced ability of anti-small perturbations and the fault ride through capability of wind farms. This paper analyzes the rated wind speed for each fan using the same approach, but actually situation is different. If using more specific and accurate models, better response characteristics can be achieved.

Acknowledgement. This work was financially supported by the Natural Science Foundation (51007054), Innovation Program of Shanghai Municipal Education Commission (12ZZ172), National High Technology Research and Development Program of China (2012AA051707).

References

1. Zhang, H.-G., Zhang, L.-Z., Chen, S.-Y., et al.: Studies on the transient behavior and dispatching strategy of power system integrated with large scale wind farms. Proceedings of the CSEE2007273147-53
2. Chang, Y., Xu, Z., Yu, P.: A comparison of the integration types of large wind farm. Automation of Electric Power Systems 31(14), 70–75 (2007)
3. Zhao, Q.-S., Wang, Z.-X.: Simulation study on grid integration and steady operation of doubly-fed wind turbine generators. Power System Technology 31(22), 69–74 (2007)
4. Qian, S.-F., Lin, L., Shen, H., et al.: Dynamic characteristic analysis of transmission grid including large wind farm by PSS/E wind package. Power System Protection and Control 37(6), 11–16 (2009)
5. Tian, C.-Z., Li, Q.-L., Song, X.-K.: Modeling and analysis of the stability for the power system considering the integration of the wind farms. Power System Protection and Control 37(19), 45–50 (2009)
6. Chi, Y.-N., Wang, W.-S., Liu, Y.-H., et al.: Impact of large scale wind farm integration on power system transient stability. Automation of Electric Power Systems 30(15), 10–14 (2006)
7. Li, G., Li, G., et al.: Development and prospect for HVDC light. Automation of Electric Power System 27(4), 77281 (2003)
8. Power Technologies Inc.: PSS/E application manual and operation manual. PSS/E brochure. Power Technologies Inc., New York (2001)
9. Ni, L., Yuan, R., Zhang, Z., Liu, C., et al.: Research on control method and dynamic characteristic of large wind farm integration. Power System Protection and Control 39(8), 75–85 (2011)
10. Zhang, S., Zhu, L., Du, L., Qi, Y., Qu, L.-K., et al.: Wind Farm Synchronization Modes Effect on Regional Power Voltage. Central China Power 25(2), 65–70 (2012)
11. Sun, T., Wang, W., Dai, H., et al.: Voltage fluctuation and flicker caused by wind power generation. Power System Technology 27(12), 62–66 (2003)
12. Chang, Y., Xu, Z.: VSC-HVDC Models of PSS/E and Their Applicability in Simulations. Power System Technology 31(8), 37–41 (2007)
13. Liu, G., Shi, X., et al.: Simulation study of affecting dynamic fault recovery performance of HVDC system with STATCOM. Journal of North China Electric Power University 37(2), 20–28 (2012)
14. Xu, Q., Xu, Z.: Dynamic Reactive Compensation Device Models for Electromechanical Transient Simulation in PSS/E. Southern Power System Technology 2(5), 39–41 (2008)

Static Voltage Stability Analysis Based on PSS/E

Xiaoyan Bian, Fangqi Yuan, and Yang Fu

Power and Automation Engineering Department, Shanghai University of Electrical
Power, 200090, Shanghai, China
{kuliz,yuanfangqi2}@163.com

Abstract. In this paper, the static voltage stability is studied by PSS/E.
Firstly, the author gives the mechanism of branch contingency ranking,PV,QV curve in the analysis of voltage stability of PSS/E.PV curve
is to find the maximum transfer power limit,while comprehensive sensitivity indicator is used to analyze static voltage stability in the method
of QV curve. Then,the author takes advantage of the above-mentioned
methods to analyze the static voltage stability of a standard test sys-
tem.In the process of analysis, simulation results are demonstrated and
analyzed.IEEE 30 Bus Test Case is analyzed and explained as an
example.

Keywords: Contingency Ranking, Transfer Power Limit, Voltage Stability, PV Curve, QV Curve.

1 Introduction

Voltage stability refers to the ability of a power system to maintain steady voltages at all buses in the system after being subjected to a disturbance from a
given initial operating condition. In recent years, voltage instability or collapse
continue to emerge. Great importance has been attached to the study of voltage
stability problems. At the same time, the voltage stability studies, especially in
the field of static voltage stability, have make great progress in recent years. At
present, the method of the static voltage stability analyze includes sensitivity
analysis and the maximum power method, the maximum power method including the PV curve method, QV curve and bifurcation analysis [1]. The objective
of a PV and QV curves is to determine the ability of a power system to maintain voltage stability at all the buses in the system under normal and abnormal
steady-state operating conditions. The PV and QV curves are obtained through
a series of ac power flow solutions. The PV curve is a representation of voltage
change as a result of increased power transfer between two systems, and the QV
curve is a representation of reactive power demand by a bus or buses as voltage
level changes [2].

However, the P-V curve has neglected the system load characteristic influence,
and its inflection point reflects the electric power network transmission power

K. Li et al. (Eds.): ICSEE 2013, CCIS 355, pp. 194–202, 2013.

limit and the corresponding running voltage. While the voltage instability or voltage collapse is a dynamic phenomenon under large or small disturbance, easily influenced by load restore characteristic and generator system, the maximum power transfer point is not necessarily the voltage instability point[3].

The Siemens PTI Power System Simulator (PSS/E) is a package of programs for studies of power system transmission network and generation performance in both steady-state and dynamic conditions. Many scholars have researched and explored the software application in depth [4,5] .

2   Model Description

A. Contingency Ranking
In large systems with many possible contingencies, especially single branch outages, it is often useful to minimize the computational and subsequent analyses by identifying the most severe contingencies prior to performing the contingency analysis. The method of contingency ranking has been discussed and studied in many references[6-8].This paper uses voltage depression criteria to rank branch contingencies.

In the voltage ranker, the performance index is defined as:

$$PI = \sum_{i=1}^{L} X_i P_i^2 \tag{1}$$

where:

X_i - is the reactance of branch i.

P_i - is the active power flow on branch i.

L - is the set of monitored branches contributing to PI.

PI gives an indication of reactive power losses under different system conditions. As line loadings increase, their $I^2 X$ losses also increase. This increase in reactive demand generally results in a depression of system voltages.

B. PV Curve
PV curves are parametric study involving a series of ac power flows that monitor the changes in one set of power flow variables with respect to another in a systematic fashion. This approach is a powerful method for determining transfer limits that account for voltage and reactive flow effects. As power transfer is increased, voltage decreases at some buses on or near the transfer path. The transfer capacity where voltage reaches the low voltage criterion is the low voltage transfer limit. Transfer capacity can continue to increase until the solution identifies a condition of voltage collapse; this is the voltage collapse transfer limit. It can be explained as Fig.1.

$$P_R = V_R I \cos \phi = \frac{Z_{LD}}{F} \left(\frac{E_s}{Z_{LN}} \right) \cos \phi \tag{2}$$

$$V_R = \frac{1}{\sqrt{F}} \frac{Z_{LD}}{Z_{LN}} E_s \tag{3}$$

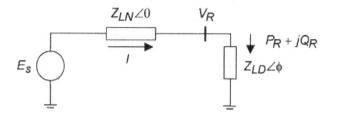

Fig. 1. Two terminals simple network

where:

$$F = 1 + \left(\frac{Z_{LD}}{Z_{LN}}\right)^2 + 2\left(\frac{Z_{LD}}{Z_{LN}}\right)\cos(\theta - \phi) \tag{4}$$

From (2), the load power, P_R, increases as Z_{LD} is decreased gradually, hence the power transmitted will increase. As the value of Z_{LD} approaches Z_{LN} the value of PR starts to decrease gradually due to F. However, from (3), the receiving voltage V_R decreases gradually as Z_{LD} decreases.

C. QV Curve

From (2) and (3), it can be seen that the power factor ϕ of the load has a significant impact on the overall equations. This is to be expected because the voltage drop in the line is a function of both active and reactive power transfer. Hence, the QV curves may also be used to assess voltage stability of the system. The bottom of the QV curves, in addition to identifying the stability limit, defines the minimum reactive power requirement for the stable operation.

This paper takes account the sensitivity and the sensitivity change rate of comprehensive sensitivity indicators L_i.

$$L_i = LM_i/\overline{LM} + \Delta LM_i/\overline{\Delta LM} \tag{5}$$

Where:

$$LM_i = dV_i/dQ_i \tag{6}$$

$$\Delta LM_i = LM_{i1} - LM_{i2} \tag{7}$$

In the above equations,

LM_i - the sensitivity of the voltage of node i on the load;
LM_{i1} - the sensitivity of the voltage of node i on the load at the first state;
LM_{i2} - the sensitivity of the voltage of node i on the load at the second state;
ΔLM_i - the change rate of the sensitivity of node i;
$\overline{LM}$ -the average value of the sensitivity of each node;
$\overline{\Delta LM}$ -the average value of the change rate of the sensitivity of each node.

3   Example

The IEEE 30 Bus Test Case represents a portion of the American Electric Power System (in the Midwestern US) in December, 1961. The test case has buses at either 132 KV or 33 KV. This paper studies steady state voltage stability base on IEEE 30 BUS Test Case. IEEE 30 Bus Test Case diagram is illustrated as Fig.2.

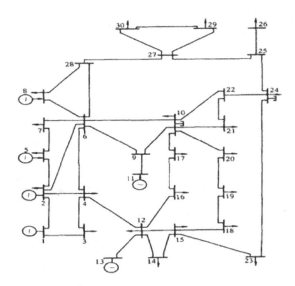

Fig. 2. IEEE 30 Bus Test Case diagram

A. Contingency Ranking

The IEEE 30 Bus Test Case includes 5 generators and 35 non-transformer branches. Among these lines, 13 branches are at 132 KV voltage level, and 22 branches are at 33 KV voltage level. Table 1 shows the result of five severe contingency ranking.

In table 1, the severe branch contingencies mainly occur at high voltage level. That is, the branch contingencies at high voltage level have a large impact on system voltage stability.

B. Base Case

Fig.3 gives the PV curve and QV curve of bus 26 at base case. From Fig.3.a, it can be seen that the bus voltage continues to decrease as the transmission power increase. From Fig.3.b, an increase in Q is accompanied by an increase in V, the operation on the right side of the QV curve is stable, whereas the operation on the left side is unstable. Thus Fig.3 verifies the model mentioned in section 2.

Table 2 shows the former eight buses' reactive power margin in order in IEEE 30 Bus Test Case. It shows that weak grid areas mainly occurs on the 33 KV

Table 1. Branch Contingency Ranking in Order of Severity

order	branch number	PI value
1	Bus 1 to Bus 2	0.5756
2	Bus 2 to Bus 5	0.4604
3	Bus 1 to Bus 3	0.3922
4	Bus 3 to Bus 4	0.3807
5	Bus 6 to Bus 7	0.3721

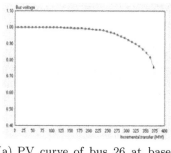   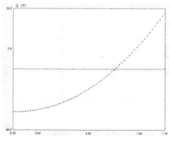

(a) PV curve of bus 26 at base   (b) QV curve of bus 26 at base
             case   case

Fig. 3. Base case

bus. Among them, bus 26,30,29 reactive power margin are small, which easily leads to system voltage instability. Therefore, the strengthening of reactive power compensation of these nodes is of great significance to system voltage stability.

Table 2. Reactive Power Margin in Order at Base Case

order	node number	reactive power margin (MVAR)
1	26	31.17
2	30	33.14
3	29	36.83
4	25	66.75
5	27	72.52
6	18	81.58

Table 3 shows the former eight buses' voltage sensitivity LM_i and the comprehensive sensitivity indicator Li in order. Compared to Table 2, we can see that the voltage sensitivity has no necessary relationship with the reactive power margin, as they are two kinds of different evaluation indicators. In addition, from the order of LM_i and Li in Table 3, although some changes occurred in the node sort, but the bus of large sensitivity still maintained at the previous position. From the order of bus 14,26, it shows that the change rate of the sensitivity has an significant impact on the voltage sensitivity.

Table 3. The Sensitivity and Comprehensive Sensitivity of Former Eight Buses

order	node number	LM_i	order	node number	Li
1	10	2.075	1	12	4.4253
2	12	2.065	2	15	4.4125
3	29	2.061	3	20	4.3598
4	15	2.058	4	10	4.3560
5	20	2.056	5	29	4.2681
6	14	1.041	6	19	2.1591
7	26	1.036	7	21	2.1035
8	19	1.030	8	14	2.0159

C. N-1 Contingency Analysis
a. Branch Contingency
Fig.4 shows the transfer power limit at base case and the first five contingencies. It can be seen, due to the change of network structure, the transfer power limit of the system changes, and then the system load margin has been greatly reduced, namely, the ability of the system with a large load diminishes greatly.

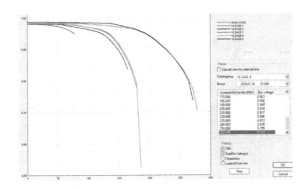

Fig. 4. PV curves at base case and former five contingencies

In Table 4, it can be seen that at the three kinds of contingencies, the order of node has not changed, and the transfer reactive power limits have little change. This is due to that the contingencies occur at the 132 KV voltage level. However, the weak buses always appear on the 33 KV voltage level.

From Fig.5, it's easily seen that the transfer reactive power limit of node 4 has a big change at base case and the first three contingencies, which is greatly different from Table 4.

b. Generator Contingency
Fig.6 shows the PV curves of bus 4 at base and five single generator contingencies. It can be seen that the single generator outage has a significant impact on the transfer power limit.

Table 4. The Bus Order of Reactive Power Margin of Former Three Contingencies

	VLTAGE 1			VLTAGE 2			VLTAGE 3	
order	node number	reactive power margin (MVAR)	order	node number	reactive power margin (MVAR)	order	node number	reactive power margin (MVAR)
1	26	30.64	1	26	30.78	1	26	30.55
2	30	32.68	2	30	32.76	2	30	32.53
3	29	36.24	3	29	36.35	3	29	36.04
4	25	62.62	4	25	64.04	4	25	62.91
5	27	67.4	5	27	69.34	5	27	68.14
6	19	73.99	6	19	78.04	6	18	74.96
7	23	74.20	7	18	78.10	7	19	74.98
8	18	74.23	8	23	78.52	8	23	75.24

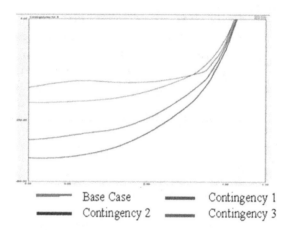

━━━━━ Base Case   ━━━━━ Contingency 1
━━━━━ Contingency 2   ━━━━━ Contingency 3

Fig. 5. QV curves of bus 4

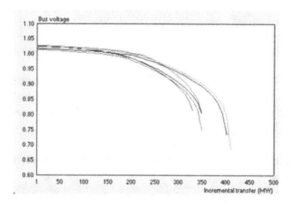

Fig. 6. PV curves of bus 4

4 Conclusion

This paper studies the static voltage stability of IEEE 30 Bus Test Case, and determines the voltage weak region of the test case. The following conclusions can be drawn. ·

(1) Contingency ranking is very effective in the static voltage stability analysis for a large power system.

(2) In a large power system, the weak grid area always occurs at the low voltage level. In this paper, the weak area, buses 26,30,29, are at the low voltage level.

(3) Branch contingency has a significant impact on the buses nearby, yet little impact on the buses far away.

(4) Generator outage changes the transfer power limit in a certain extent, further affects the voltage stability of the power system.

Besides, the strong function about the analysis of voltage stability of the PSS/E software is reflected in the paper.

Acknowledgment. This work was financially supported by the Natural Science Foundation (51007054), Innovation Program of Shanghai Municipal Education Commission (12ZZ172), National High Technology Research and Development Program of China (2012AA051707).

References

1. Lu, Y.: Research on Voltage Safety and Stability of Beijing Electric Network Based on BPA, Thesis. North China Electric Power University, Beijing (2005)
2. PSS®E 32.0 Program Operation Manual (2009)
3. Taylor, C.W.: Power System Voltage Stability. McGraw-Hill, Companies, Inc. (1994)
4. Gao, H., Wang, C., Pan, W.: A Detailed Nuclear Power Plant Model for Power System Analysis Based on PSS/E. In: Power Systems Conference and Exposition, pp. 1582–1586 (2006)
5. Leong, K.K.: Modeling and Co-simulation of AC Generator Excitation and Governor Systems Using Simulink Interfaced to PSS/E. In: Power Systems Conference and Exposition, vol. 2, pp. 1095–1100 (2004)
6. Mikolinnas, T.A., Wollenberg, B.F.: An Advanced Contingency Selection Algorithm. IEEE Transactions on Power Apparatus and Systems PAS-100(2), 608–617 (1981)
7. Lauby, M.G., Mikolinnas, T.A., Reppen, N.D.: Contingency Selection of Branch Outages Causing Voltage Problems. IEEE Transactions on Power Apparatus and Systems PAS-102(12), 3899–3904 (1983)
8. Flueck, A.J., Gonella, R., Dondeti, J.R.: A new power sensitivity method of ranking branch outage contingencies for voltage collapse. IEEE Transactions on Power Systems 17, 265–270 (2002)
9. Lu, J., Sun, H., Yuan, X., He, W.: Static voltage stability of Ningxia Power Grid based on PSS/E. Relay 35(18) (2007)

10. Zhang, Y., Zhang, J., Yang, J., Huang, W.: Research on the Critical Point of Steady State Voltage Stability. In: The 7th International Power Engineering Conference, IPEC 2005 (2005)
11. Yuan, J., Duan, X., He, Y., Shi, D.: Summarization of the Sensitivity Analysis Method of Voltage Stability in Power Systems. Power System Technology 21(9) (1997)
12. Zhou, S., et al.: Power System Voltage Stability and Control, pp. 21–25. Electric Power Press (2004) (Chinese)

The Loss-Averse Retailer's Ordering Policy under Yield and Demand Uncertainty

Wei Liu[1,2], Shiji Song[1,*], and Cheng Wu[1]

[1] Department of Automation, Tsinghua University, Beijing 100084, China
[2] Department of Basic Science, Military Economics Academy, Wuhan 430035, China
liuwei09@mails.tsinghua.edu.cn,
shijis@mail.tsinghua.edu.cn,
wuc@tsinghua.edu.cn

Abstract. This paper studies a single-period inventory problem with random yield and demand. In general, most of the previous works are based on the assumption of risk neutrality. We incorporate loss-averse preferences into this problem and the retailer's objective is to maximize the expected utility. We obtain the retailer's optimal ordering policy and then investigate the impact of loss aversion on it. Especially, if the shortage cost is small enough, the loss-averse retailer will always order less than the risk-neutral one. Moreover, the impacts of price and cost parameters on the loss-averse retailer's optimal order quantity are analyzed. Then numerical experiments are conducted to illustrate our results.

Keywords: Inventory, Loss aversion, Random yield.

1 Introduction

Inventory control under risk is one of the main subjects in supply chain management and interest in it remains unabated in recent decades. One of the assumptions of the traditional models is that the quality of the product is perfect and the lot produced or ordered does not contain any defective unit. However, this is usually not the case in reality. Most production processes do not make parts that are completely defect-free, e.g., the yield of chips in semiconductor manufacturing is usually less than 50% ([1]). On the other hand, the lot the retailer receives often contains some units that are spoiled or broken in transit and then the quantity of good units is less than the quantity ordered. Such yield uncertainty will cause the output or quantity received to be highly variable and thus affect the production or ordering policy.

This phenomenon has received considerable attention and the production or inventory problem with random yield has been studied in various contexts. The review of the literature on this problem can be found in [2] and [3]. Gerchak et al. [4] investigate both single-period and multi-period production models with random yield. They obtain the optimal policy and show that it is not an order-up-to policy. Wang and Gerchak [5] further extend the above model by considering

---

* Corresponding author.

K. Li et al. (Eds.): ICSEE 2013, CCIS 355, pp. 203–212, 2013.

the stochastic production capacity and characterize the structure of the optimal policy. Keren [6] studies a single-period problem with deterministic demand and random yield. The cases of multiplicative production risk and additive production risk are considered, respectively. Inderfurth [7] considers a single-period production-inventory problem where random yield and demand are uniformly distributed. They indicate that the optimal policy may be non-linear. Note that all these studies are based on the assumption of risk neutrality.

However, there is growing evidence that managers' decisions often challenge the risk-neutral assumption (see, e.g., [8] and [9]). Recently, incorporating loss-averse preferences into inventory model has become an important and growing area of research. Loss aversion is originated from prospect theory ([10]) and means for equivalent losses and gains, people have different perceived values and they are more averse to losses. Since it can better describe the individual decision-making behavior under risk, the inventory model based on loss aversion has been studied by some researchers over the past few years. Schweitzer and Cachon [11] show that the optimal order quantity of a loss-averse newsvendor is always less than a risk-neutral newsvendor. Wang and Webster [12] further extend Schweitzer and Cachon's model with consideration of shortage cost and find that under certain conditions, the loss-averse newsvendor may order more than the risk-neutral one. Wang [13] investigates a loss-averse newsvendor game and shows that there exists a unique Nash equilibrium with respect to order quantity. Geng et al. [14] study a single-period inventory problem in which demand is exponentially distributed. They demonstrate that a state-dependent order-up-to policy is optimal. Other papers considering the loss-averse preferences include [15], [16] and [17], etc.

In this paper, we jointly consider these two factors and investigate a single-period inventory problem with random yield and demand, where the retailer is loss-averse. To the best of our knowledge, this model has not been considered in the literature. In Section 2, we obtain the loss-averse retailer's ordering policy and then carry out the analysis. In Section 3, we conduct numerical experiments to illustrate our results. In Section 4, we conclude our paper.

2   Model Analysis

Consider a single-period inventory problem with random yield and demand. We adopt the commonly used stochastically proportional yield model (see e.g., [4] and [7]). That is, the fraction of good units is a random variable and independent of the batch size. At the beginning of the period, the retailer makes an order quantity decision and then orders product from a supplier. The lead time is zero and the order arrives instantaneously. The retailer performs 100% inspection, then pays for the good units and returns the defective units to the supplier. The inspection time and cost are not considered.

The following notations will be used throughout this paper:

c: purchasing cost per unit.
s: selling price per unit.

v: salvage value per unit.

h: shortage cost per unit.

Q: order quantity.

X: random demand. Its probability density function is $f(x)$ and cumulative distribution function is $F(x)$.

Y: the fraction of good units, i.e., the amount of good units received is YQ. It is independent of the demand, and its probability density function is $g(y)$ and mean is μ.

It is reasonable to assume that $s \geq c \geq v \geq 0$ and $h \geq 0$. Then for any $X = x$ and $Y = y$, the retailer's realized profit is

$$\pi(Q, x, y) = \begin{cases} sx - cyQ + v(yQ - x), & x \leq yQ, \\ syQ - cyQ - h(x - yQ), & x > yQ. \end{cases} \tag{1}$$

Suppose the retailer is loss-averse and we use the following piecewise-linear loss aversion utility function:

$$U(\pi) = \begin{cases} \pi, & \pi \geq 0, \\ \lambda\pi, & \pi < 0, \end{cases} \tag{2}$$

where $\lambda \geq 1$ is the retailer's loss aversion coefficient. Note that if $\lambda = 1$, the retailer is risk-neutral. The above function has been used in the literature on the inventory management because of its simplicity (see e.g., [12], [15]). Then the retailer's expected utility is

$$E[U(\pi(Q, X, Y))] = \int_0^1 \int_0^\infty U(\pi(Q, x, y))f(x)g(y)\,dx\,dy. \tag{3}$$

To calculate (3), we divide the region of integration into four subregions, as shown in Fig. 1. It is shown that the retailer's profit $\pi(Q, x, y)$ is negative in S_1 and S_4 while positive in S_2 and S_3. Thus the precise expression for (3) is as follows:

$$E[U(\pi(Q, X, Y))] = \lambda \iint_{S_1 \cup S_4} \pi(Q, x, y)f(x)g(y)\,dx\,dy + \iint_{S_2 \cup S_3} \pi(Q, x, y)f(x)g(y)\,dx\,dy$$

$$= (\lambda - 1) \int_0^1 \int_0^{\frac{(c-v)yQ}{s-v}} [(s - v)x - (c - v)yQ]f(x)g(y)\,dx\,dy$$

$$+ (\lambda - 1) \int_0^1 \int_{(1+\frac{s-c}{h})yQ}^\infty [(s - c + h)yQ - hx]f(x)g(y)\,dx\,dy$$

$$+ \int_0^1 \int_0^{yQ} [(s - v)x - (c - v)yQ]f(x)g(y)\,dx\,dy$$

$$+ \int_0^1 \int_{yQ}^\infty [(s - c + h)yQ - hx]f(x)g(y)\,dx\,dy.$$

$$\tag{4}$$

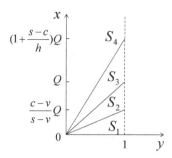

Fig. 1. A graphical presentation of demand and the fraction of good units

The retailer's objective is to choose an order quantity Q to maximize his expected utility.

The following theorem characterizes the retailer's optimal ordering policy.

Theorem 1. *The retailer's expected utility function $E[U(\pi(Q, X, Y))]$ is concave in Q. Thus there exists a unique optimal order quantity Q^* that satisfies the following first-order optimality condition:*

$$(\lambda - 1)(c - v) \int_0^1 yg(y)F[\frac{(c-v)yQ^*}{s-v}]dy$$

$$+(\lambda - 1)(s - c + h) \int_0^1 yg(y)F[(1 + \frac{s-c}{h})yQ^*]dy \qquad (5)$$

$$+(s + h - v) \int_0^1 yg(y)F(yQ^*)dy = \lambda\mu(s - c + h).$$

Proof. Taking the first-order and second-order derivatives of $E[U(\pi(Q, X, Y))]$ with respect to Q, we have

$$\frac{dE[U(\pi(Q, X, Y))]}{dQ} = -(\lambda - 1)(c - v) \int_0^1 yg(y)F[\frac{(c-v)yQ}{s-v}]dy$$

$$-(\lambda - 1)(s - c + h) \int_0^1 yg(y)F[(1 + \frac{s-c}{h})yQ]dy \qquad (6)$$

$$-(s + h - v) \int_0^1 yg(y)F(yQ)dy + \lambda\mu(s - c + h),$$

and

$$\frac{d^2 E[U(\pi(Q, X, Y))]}{dQ^2} = -\frac{(\lambda - 1)(c - v)^2}{s - v} \int_0^1 y^2 g(y)f[\frac{(c-v)yQ}{s-v}]dy$$

$$-\frac{(\lambda - 1)(s - c + h)^2}{h} \int_0^1 y^2 g(y)f[(1 + \frac{s-c}{h})yQ]dy$$

$$-(s + h - v) \int_0^1 y^2 g(y)f(yQ)dy < 0.$$

$$(7)$$

Then $E[U(\pi(Q, X, Y))]$ is concave in Q. We also have $\frac{dE[U(\pi(Q,X,Y))]}{dQ}\Big|_{Q=0} = \lambda\mu(s-c+h) > 0$ and $\lim_{Q\to\infty} \frac{dE[U(\pi(Q,X,Y))]}{dQ} = -\lambda\mu(c-v) < 0$. Thus there exists a unique Q^* that satisfies $\frac{dE[U(\pi(Q,X,Y))]}{dQ}\Big|_{Q=Q^*} = 0$, i.e., expression (5).   □

We have mentioned above that the retailer is risk-neutral when $\lambda = 1$, then from (5) the risk-neutral retailer's optimal order quantity Q_0 satisfies

$$(s + h - v) \int_0^1 yg(y)F(yQ_0)dy = \mu(s - c + h). \tag{8}$$

Next we will investigate the impact of loss aversion on the retailer's optimal order quantity. Let

$$M(h, Q) = \frac{c - v}{s - c + h} \int_0^1 yg(y)F[\frac{(c - v)yQ}{s - v}]dy$$
$$+ \int_0^1 yg(y)F[(1 + \frac{s - c}{h})yQ]dy - \mu, \tag{9}$$

and

$$N(Q^*, \lambda, c, s, v, h) = \frac{dE[U(\pi(Q, X, Y))]}{dQ}\Big|_{Q=Q^*}. \tag{10}$$

Note that $N(Q^*, \lambda, c, s, v, h) = 0$ and $\frac{\partial N}{\partial Q^*} < 0$. Then we have the following theorem:

Theorem 2. *For any $\lambda > 1$, if $M(h, Q_0) < 0$, then $Q^* > Q_0$ and $\frac{\partial Q^*}{\partial \lambda} > 0$; if $M(h, Q_0) = 0$, then $Q^* = Q_0$ and $\frac{\partial Q^*}{\partial \lambda} = 0$; otherwise, $Q^* < Q_0$ and $\frac{\partial Q^*}{\partial \lambda} < 0$.*

Proof. Plugging Q_0 into (6) and combining (8), then

$$\frac{dE[U(\pi(Q_0, X, Y))]}{dQ}$$
$$= -(\lambda - 1)(s - c + h)\left\{\frac{c - v}{s - c + h} \int_0^1 yg(y)F[\frac{(c - v)yQ_0}{s - v}]dy\right.$$
$$\left. + \int_0^1 yg(y)F[(1 + \frac{s - c}{h})yQ_0]dy - \mu\right\} \tag{11}$$
$$= -(\lambda - 1)(s - c + h)M(h, Q_0).$$

If $M(h, Q_0) < 0$, then $\frac{dE[U(\pi(Q_0,X,Y))]}{dQ} > 0$, which implies that $Q^* > Q_0$. Furthermore, since $N(Q^*, \lambda, c, s, v, h) = 0$, using the implicit function theorem we can obtain $\frac{\partial Q^*}{\partial \lambda} = -\frac{\partial N}{\partial \lambda} \Big/ \frac{\partial N}{\partial Q^*}$, where

$$\frac{\partial N}{\partial \lambda} = -(c - v) \int_0^1 yg(y)F[\frac{(c - v)yQ^*}{s - v}]dy$$
$$- (s - c + h) \int_0^1 yg(y)F[(1 + \frac{s - c}{h})yQ^*]dy + \mu(s - c + h). \tag{12}$$

Since $Q^* > Q_0$, from (8) we have $\mu(s-c+h)-(s+h-v)\int_0^1 yg(y)F(yQ^*)dy < 0$. Thus combining (5) we can obtain that (12) is positive, and then $\frac{\partial Q^*}{\partial \lambda} > 0$.

The other two results can be proved in a similar way.   □

This theorem establishes the conditions under which the optimal order quantity of the loss-averse retailer will be larger than, equal to or less than the risk-neutral retailer. Note that the impact of loss aversion on the order quantity is similar to the results found in [12].

Corollary 1. *For any $\lambda > 1$, there exists a shortage cost h'. If $h < h'$, then $Q^* < Q_0$ and $\frac{\partial Q^*}{\partial \lambda} < 0$.*

Proof. From Theorem 2 it is sufficient to prove that $M(h, Q_0) > 0$ for any $h < h'$. Let $\hat{Q}_0$ satisfy

$$(s-v)\int_0^1 yg(y)F(y\hat{Q}_0)dy = \mu(s-c), \tag{13}$$

then from (8) we can have $\hat{Q}_0 < Q_0$. Since $M(h, \hat{Q}_0)$ is decreasing in h, and $M(0, \hat{Q}_0) > 0$ and $\lim_{h \to \infty} M(h, \hat{Q}_0) < 0$, there exists a shortage cost h' such that $M(h', \hat{Q}_0) = 0$. If $h < h'$, then $M(h, Q_0) > M(h, \hat{Q}_0) > 0$.   □

This corollary shows that the loss-averse retailer will always order less than the risk-neutral retailer if the shortage cost is small enough. Moreover, the more loss-averse the retailer, the less his order quantity. Note that this corollary holds under the common assumption that the unsatisfied demand is lost, i.e., $h = 0$.

From (8) it is easy to see that the risk-neutral retailer's optimal order quantity is always increasing in shortage cost, selling price and salvage value, while decreasing in purchasing cost. These observations motivate us to further analyze the impacts of these parameters on the loss-averse retailer's order quantity.

Theorem 3. *When price or cost parameter changes, the change in the optimal order quantity Q^* is as follows:*

(i) Q^* is increasing in v;

(ii) if $\frac{(\lambda-1)(c-v)^2 Q^*}{(s-v)^2}\int_0^1 y^2 g(y)f[\frac{(c-v)yQ^*}{s-v}]dy - (\lambda-1)\int_0^1 yg(y)F[(1+\frac{s-c}{h})yQ^*]dy$ $- \frac{(\lambda-1)(s-c+h)Q^*}{h}\int_0^1 y^2 g(y)f[(1+\frac{s-c}{h})yQ^*]dy - \int_0^1 yg(y)F(yQ^*)dy + \lambda\mu < 0$, then Q^* is decreasing in s, otherwise, Q^* is increasing in s;

(iii) if $(\lambda-1)\int_0^1 yg(y)F[\frac{(c-v)yQ^*}{s-v}]dy + \frac{(\lambda-1)(c-v)Q^*}{s-v}\int_0^1 y^2 g(y)f[\frac{(c-v)yQ^*}{s-v}]dy -$ $(\lambda-1)\int_0^1 yg(y)F[(1+\frac{s-c}{h})yQ^*]dy - \frac{(\lambda-1)(s-c+h)Q^*}{h}\int_0^1 y^2 g(y)f[(1+\frac{s-c}{h})yQ^*]dy +$ $\lambda\mu > 0$, then Q^* is decreasing in c, otherwise, Q^* is increasing in c;

(iv) if $(\lambda-1)\int_0^1 yg(y)F[(1+\frac{s-c}{h})yQ^*]dy - \frac{(\lambda-1)(s-c+h)(s-c)Q^*}{h^2}\int_0^1 y^2 g(y)f[(1+$ $\frac{s-c}{h})yQ^*]dy + \int_0^1 yg(y)F(yQ^*)dy - \lambda\mu > 0$, then Q^* is decreasing in h, otherwise, Q^* is increasing in h.

Proof. (i) Using the implicit function theorem we have $\frac{\partial Q^*}{\partial v} = -\frac{\partial N}{\partial v} \Big/ \frac{\partial N}{\partial Q^*}$, where

$$
\begin{aligned}
\frac{\partial N}{\partial v} =& (\lambda - 1) \int_0^1 yg(y)F[\frac{(c-v)yQ^*}{s-v}]dy \\
&+ \frac{(\lambda - 1)(c-v)(s-c)Q^*}{(s-v)^2} \int_0^1 y^2 g(y)f[\frac{(c-v)yQ^*}{s-v}]dy \\
&+ \int_0^1 yg(y)F(yQ^*)dy > 0.
\end{aligned}
\tag{14}
$$

Since $\frac{\partial N}{\partial Q^*} < 0$, then Q^* is increasing in v.

(ii)-(iv) It is easy to calculate that

$$
\begin{aligned}
\frac{\partial N}{\partial s} =& \frac{(\lambda - 1)(c-v)^2 Q^*}{(s-v)^2} \int_0^1 y^2 g(y)f[\frac{(c-v)yQ^*}{s-v}]dy \\
&- (\lambda - 1) \int_0^1 yg(y)F[(1 + \frac{s-c}{h})yQ^*]dy \\
&- \frac{(\lambda - 1)(s-c+h)Q^*}{h} \int_0^1 y^2 g(y)f[(1 + \frac{s-c}{h})yQ^*]dy \\
&- \int_0^1 yg(y)F(yQ^*)dy + \lambda\mu,
\end{aligned}
\tag{15}
$$

$$
\begin{aligned}
\frac{\partial N}{\partial c} =& - (\lambda - 1) \int_0^1 yg(y)F[\frac{(c-v)yQ^*}{s-v}]dy \\
&- \frac{(\lambda - 1)(c-v)Q^*}{s-v} \int_0^1 y^2 g(y)f[\frac{(c-v)yQ^*}{s-v}]dy \\
&+ (\lambda - 1) \int_0^1 yg(y)F[(1 + \frac{s-c}{h})yQ^*]dy \\
&+ \frac{(\lambda - 1)(s-c+h)Q^*}{h} \int_0^1 y^2 g(y)f[(1 + \frac{s-c}{h})yQ^*]dy - \lambda\mu,
\end{aligned}
\tag{16}
$$

and

$$
\begin{aligned}
\frac{\partial N}{\partial h} =& - (\lambda - 1) \int_0^1 yg(y)F[(1 + \frac{s-c}{h})yQ^*]dy \\
&+ \frac{(\lambda - 1)(s-c+h)(s-c)Q^*}{h^2} \int_0^1 y^2 g(y) \\
&\times f[(1 + \frac{s-c}{h})yQ^*]dy - \int_0^1 yg(y)F(yQ^*)dy + \lambda\mu.
\end{aligned}
\tag{17}
$$

Then we can prove these three results in a similar way.   □

It follows from this theorem that the loss-averse retailer's optimal order quantity may be decreasing in shortage cost and selling price, and increasing in purchasing cost. These will never occur in the risk-neutral case.

3 Numerical Experiments

In this section, we conduct the numerical experiments to illustrate our results. The parameters are set as follows: $s = 20$, $c = 10$, $v = 5$. Consider two different demand distributions: one is the exponential distribution with mean $\mu = 100$, the other is the truncated normal distribution with mean $\mu = 100$ and standard deviation $\sigma \in \{25, 50, 100\}$. Note that the truncated normal distribution is defined as $F(x) = \frac{I(x) - I(0)}{1 - I(0)}$, where $I(x) = \frac{1}{\sqrt{2\pi}\sigma} \int_{-\infty}^{x} e^{-(t-\mu)^2/2\sigma^2} dt$. Suppose that Y is uniformly distributed and given by $g(y) = 1, 0 \leq y \leq 1$. We first analyze the optimal ordering policy by fixing $h = 5$ and varying λ from 1 to 5 in steps of 0.1. Then we test Corollary 1 by varying h from 0 to 50 in steps of 1.

Figures 2 and 3 illustrate the retailer's optimal order quantity with respect to the loss aversion coefficient when demand follows the exponential distribution and truncated normal distribution, respectively. As shown in Fig. 3, the optimal order quantity may be increasing or decreasing in λ under different levels of

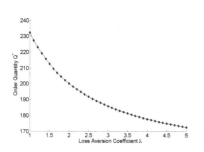

Fig. 2. Optimal policies with respect to different loss aversion coefficients. Demand follows an exponential distribution.

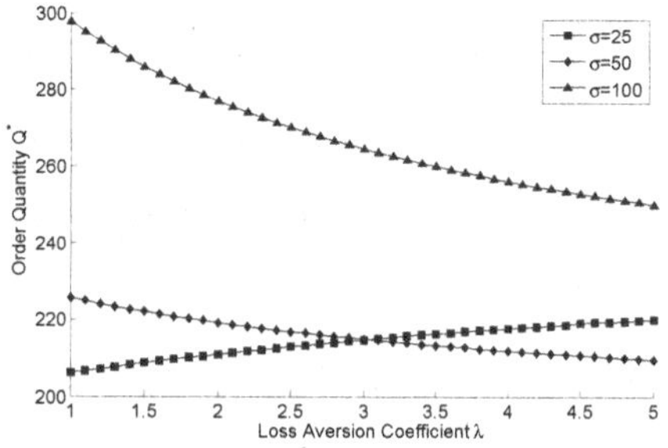

Fig. 3. Optimal policies with respect to different loss aversion coefficients and levels of demand variation. Demand follows a truncated normal distribution.

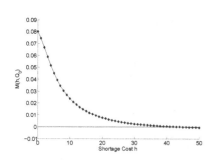
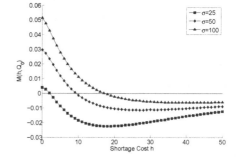

Fig. 4. The values of $M(h, Q_0)$ with respect to different shortage costs. Demand follows an exponential distribution.

Fig. 5. The values of $M(h, Q_0)$ with respect to different shortage costs and levels of demand variation. Demand follows a truncated normal distribution.

demand variation. Figures 4 and 5 show that $M(h, Q_0) > 0$ when the shortage cost h is less than a value, and thus the loss-averse retailer's order quantity is less than the risk-neutral retailer's. This result is consistent with Corollary 1. Furthermore, it follows from Fig. 5 that the higher the level of demand variation, the larger this value.

4 Conclusion

In this paper, we adopt loss-averse preferences to describe the retailer's decision-making behavior and study a single-period inventory problem with random yield and demand. We demonstrate that there exists a unique optimal order quantity and then compare it with the optimal order quantity of the risk-neutral retailer. If the shortage cost is small enough, especially if the unsatisfied demand is lost, the loss-averse retailer will always order less than the risk-neutral one. Moreover, we find that the loss-averse retailer's optimal order quantity may be decreasing in shortage cost and selling price, and increasing in purchasing cost. These will never occur in the risk-neutral case. A piecewise-linear loss aversion utility function is used in this paper. Future research may consider using a more general utility function to investigate the inventory problem with random yield and demand.

Acknowledgements. The paper is supported by Research Fund for the Doctoral Program of Higher Education (No. 20090002110035) and Natural Science Foundation of China (No. 61273233).

References

1. Grasman, S.E., Olsen, T.L., Birge, J.R.: Setting Basestock Levels in Multi-product Systems with Setups and Random Yield. IIE Trans. 40, 1158–1170 (2008)
2. Yano, C.A., Lee, H.L.: Lot Sizing with Random Yields: a Review. Oper. Res. 43, 311–334 (1995)
3. Wright, C.M., Mehrez, A.: An Overview of Representative Research of the Relationships between Quality and Inventory. Omega-Int. J. Manage. S. 26, 29–47 (1998)
4. Gerchak, Y., Vickson, R.G., Parlar, M.: Periodic Review Production Models with Variable Yield and Uncertain Demand. IIE Trans. 20, 144–150 (1988)
5. Wang, Y., Gerchak, Y.: Periodic Review Production Models with Variable Capacity, Random Yield, and Uncertain Demand. Manage. Sci. 42, 130–137 (1996)
6. Keren, B.: The Single-Period Inventory Problem: Extension to Random Yield from the Perspective of the Supply Chain. Omega-Int. J. Manage. S. 37, 801–810 (2009)
7. Inderfurth, K.: Analytical Solution for a Single-Period Production-Inventory Problem with Uniformly Distributed Yield and Demand. Cent. Eur. J. Oper. Res. 12, 117–127 (2004)
8. Fisher, M., Raman, A.: Reducing the Cost of Demand Uncertainty through Accurate Response to Early Sales. Oper. Res. 44, 87–99 (1996)

9. Kahn, J.A.: Why Is Production More Volatile than Sales? Theory and Evidence on the Stockout-Avoidance Motive for Inventory Holding. Q. J. Econ. 107, 481–510 (1992)
10. Kahneman, D., Tversky, A.: Prospect Theory: an Analysis of Decision under Risk. Econometrica 47, 263–291 (1979)
11. Schweitzer, M.E., Cachon, G.P.: Decision Bias in the Newsvendor Problem with a Known Demand Distribution: Experimental Evidence. Manage. Sci. 46, 404–420 (2000)
12. Wang, C.X., Webster, S.: The Loss-Averse Newsvendor Problem. Omega-Int. J. Manage. S. 37, 93–105 (2009)
13. Wang, C.X.: The Loss-Averse Newsvendor Game. Int. J. Prod. Econ. 124, 448–452 (2010)
14. Geng, W., Zhao, X., Gao, D.: A Single-Period Inventory System with a General S-shaped Utility and Exponential Demand. J. Syst. Sci. Syst. Eng. 19, 227–236 (2010)
15. Shen, H.C., Pang, Z., Cheng, T.C.E.: The Component Procurement Problem for the Loss-Averse Manufacturer with Spot Purchase. Int. J. Prod. Econ. 132, 146–153 (2011)
16. Wang, C.X., Webster, S.: Channel Coordination for a Supply Chain with a Risk-Neutral Manufacturer and a Loss-Averse Retailer. Decis. Sci. 38, 361–389 (2007)
17. Liu, W., Song, S., Wu, C.: Impact of Loss Aversion on the Newsvendor Game with Product Substitution. Int. J. Prod. Econ. (2012), doi:10.1016/j.ijpe.2012.08.017

A Novel Second Order Sliding Mode Control Algorithm for Velocity Control Permanent Magnet Synchronous Motor

Liang Qi and Hongbo Shi*

Key Laboratory of Advanced Control and Optimization for Chemical Processes
of Ministry of Education, East China University of Science and Technology,
200237 Shanghai, China
qi_liang999@yahoo.com.cn,
hbshi@ecust.edu.cn

Abstract. This paper focuses on performance improvements of the permanent magnet synchronous motor (PMSM) vector control. In this paper, a novel second order sliding mode control (SOSMC) algorithm is presented to accomplish velocity control of the PMSM. The integral manifold is utilized to avoid noise signals being amplified because of the acceleration information in control system, and the second order sliding mode control law is achieved by a Lyapunov function approach. The novel method can successfully eliminate the system chattering problem and improve the performance of the PMSM control system, such as fast response, high robustness and tracking speed performance. Meanwhile, an anti-windup control method is used to solve the problem of the windup phenomenon of the PMSM control system. The experimental results show that the proposed method is feasible and effective and is capable of controlling the permanent magnet synchronous motor.

Keywords: PMSM, SOSMC, Anti-Windup, Lyapunov function, Integral manifold.

1 Introduction

The permanent-magnet synchronous motors (PMSM) have many advantages of high power density, high efficiency, high reliability and fast dynamics, so they have been applied widely in factory automation, household appliances, computer, high-speed aerospace drives, and automobiles, etc. However, the control performance of PMSM is still influenced by the change of the mechanical parameter, the external load disturbances and perturbations in practical applications. It is difficult to attain the high-performance control of PMSM for the conventional PID-type control methods. Therefore, several modern and intelligent control techniques, such as sliding mode control(SMC)[1]-[7], adaptive control[[8], neural net control[9] and fuzzy control[10],[11], etc. have been studied to accomplish the control of PMSM. Sliding mode control is one of the effective control methods of PMSM since it owes several advantages, such as, insensitivity

---

* Corresponding author.

K. Li et al. (Eds.): ICSEE 2013, CCIS 355, pp. 213–220, 2013.
© Springer-Verlag Berlin Heidelberg 2013

to parameter variations, external disturbance rejection and fast dynamic response[1]-[4]. Consequently, SMC has been widely and successfully applied into the position and velocity control of PMSM. However, SMC has its own disadvantage, i.e., chattering phenomenon, which originated from the interaction between parasitic dynamics and high-frequency switching control. In order to avoid the phenomenon, several control methods were proposed to improve the sliding mode control, such as the saturation function or the sigmoid function instead of the Bang-Bang control[12], low-pass filter[13], hybrid SMC [14] and second order sliding mode control(SOSMC)[4]etc. But the methods bring some new problems. For example, the low-pass filter causes the phase lag, and hybrid SMC increases the computation load. Comparatively, SOSMC can not only eliminate the chattering but also preserve the main advantages of the SMC, i.e., robustness and precision of the SMC.

In this paper, a novel second order sliding mode control algorithm is proposed to accomplish velocity control of the PMSM owing to its system requirement of the fast response, robustness and good track performance and so on. The integral manifold is utilized to avoid the acceleration information required in control system. The control performance can be improved because of diminishing the differentiator to attain the acceleration information compared with others second order sliding mode control, which amplifies the noise signals. Meanwhile, the second order sliding mode control law is designed by the Lyapunov function approach and effectively eliminates the system chattering phenomenon. In addition, an anti-windup control method is used to address the windup phenomenon of the PMSM control system. The computer simulation results are presented to verify the feasibility of the method.

2   The Field-Oriented PMSM Control Scheme

In the stationary $(d - q)$ reference frame, the mathematics mode of permanent-magnet synchronous motor is shown as below:

$$\begin{cases} \dot{i}_d = -\frac{R_s}{L}i_d + p_n\omega i_q + \frac{u_d}{L} \\[2mm] \dot{i}_q = -p_n\omega i_d - \frac{R_s}{L}i_q - \frac{p_n\psi_f}{L}\omega + \frac{u_q}{L} \\[2mm] \dot{\omega} = \frac{p_n\psi_f}{J}i_q - \frac{B}{J}\omega - \frac{T_L}{J} \\[2mm] \dot{\theta} = \omega \end{cases} \tag{1}$$

where i_d, i_q and u_d, u_q are current and voltage (volt) of motor d axis and q axis respectively; R_s is stator resistance of motor (ohm); L is self inductance of motor stator (H); ψ_f is permanent magnet flux of motor (voltsec/rad); T_L is motor torque (Nm); p_n is pole-pairs of motor; J and B are the viscous friction coefficient and inertia constant of the motor; ω and θ are angular velocity (rad/sec) and rotor position of motor (rad).

The system is designed for the double close-loop control system to regulate the speed of motor by the field oriented control technology so that it obtains high performance.

Moreover, the method of the $i_d = 0$ vector control is utilized in order to simplify the control motor system.

The control system is regarded as linear system, and the overall configuration of the speed vector control of PMSM is shown in Fig.1, which consists of a PMSM, an SVPWM voltage source inverter, a power source rectifier, automatic current regulator(ACR) of the motor, an encoder used to detect speed and position , and a speed controller(SMC) based on the second order sliding mode control technology. The speed controller will be described in detail in the next section.

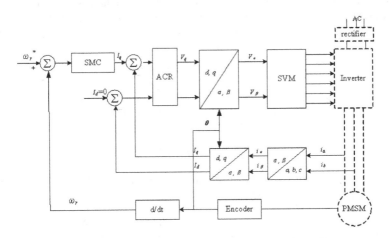

Fig. 1. The structure of a vector control system based on SMC

3   The SOSMC

Here, a novel second order sliding mode control algorithm is proposed to accomplish the speed control of PMSM. The chattering phenomenon can be avoided so as to improve the motor performance when compared to the conventional PID control.

The speed controller of the PMSM is designed to attain the highly precise speed track with reference speed and be robust to the disturbance of the load. Assume that the signal of the reference speed is ω^*, and it is sufficient smooth and has almost second order derivative everywhere. So define the error state equation as follows:

$$e_\omega = \omega^* - \omega \tag{2}$$

where ω^* is the reference speed of the motor, and ω is the actual speed of the motor. The motor speed error equation is obtained from (1) as follows:

$$\dot{e}_\omega = \dot{\omega}* - \dot{\omega} = \dot{\omega}* - \frac{p_n \Psi_f}{J} i_q + \frac{B}{J} \omega + \frac{T_L}{J} \tag{3}$$

where $\alpha = \frac{p_n \Psi_f}{J}$, $\beta = \frac{B}{J}$ and $\delta = \frac{T_L}{J}$, so the speed error equation(3) can be rewritten as:

$$\dot{e}_\omega = \dot{\omega}* - \dot{\omega} = \dot{\omega}* - \alpha i_q + \beta \omega + \delta \tag{4}$$

Because the error equation(4)is first order corresponding to the state equation(2), the chattering phenomenon is eliminated by the second order sliding mode control [15]-[16].

In order to achieve the good dynamic performance and improve the speed tracking precision, an integral manifold is designed as follows:

$$S = k_1 e_\omega + k_2 \int e_\omega dt \tag{5}$$

where k_1 and k_2 is the proportion gain and the integral gain, and $k_1 > 0$, $k_2 > 0$.

Theorem1: The speed error system 3 can converge to zero in finite time, while the integral manifold is chosen as 5, and the control law is designed as follows:

$$i_q = i_{eq} + i_{sw} \tag{6}$$

with

$$i_{eq} = \frac{1}{k_1 \alpha} (k_2 e_\omega + k_1 \beta \omega + k_1 \dot{\omega}*) \tag{7}$$

$$i_{eq} = \frac{1}{k_1 \alpha} (k_2 e_\omega + k_1 \beta \omega + k_1 \dot{\omega}*) \tag{8}$$

where λ_1, λ_2, γ_1 and γ_2 are designed parameters, and $\lambda_1 > 0, \lambda_2 = \lambda_{21} + \lambda_{22}$, $\lambda_{21} > k_1 \|\delta\|$ and $\lambda_{22} > 0$, $\gamma_1 > 0$, $\gamma_2 > 0$.

Proof: When the Lyapunov function $\dot{V} < 0$, the sliding mode control exists and the error system can converge. So the Lyapunov function can be defined as [17]:

$$V = \frac{1}{2} \gamma_1 S^T S + \frac{1}{2} \dot{S}^T \dot{S} + \lambda_1 \|S\| \tag{9}$$

Take the time derivative of (9), and obtain:

$$\dot{V} = \dot{S}^T (\gamma_1 S + \ddot{S} + \lambda_1 \text{sgn}(S)) \tag{10}$$

According to (5), the $\dot{S}$ and $\ddot{S}$ are rewritten as below:

$$\dot{S} = k_1 \dot{e}_\omega + k_2 e_\omega \tag{11}$$

$$\ddot{S} = k_1 (\dot{\omega}* - \alpha i_q + \beta \dot{\omega} + \dot{\delta}) + k_2 \dot{e}_\omega \tag{12}$$

The following equations is obtained from the (6), (7), (8) and (12):

$$\ddot{S} = k_1 (\dot{\omega}* - \alpha(i_{eq} + i_{sw}) + \beta \dot{\omega} + \dot{\delta}) + k_2 \dot{e}_\omega \tag{13}$$

$$i_{eq} = \frac{1}{k_1 \alpha} (k_2 \dot{e}_\omega + k_1 \beta \dot{\omega} + k_1 \ddot{\omega}*) \tag{14}$$

$$i_{sw} = \frac{1}{k_1 \alpha} (\lambda_1 sign(S) + \lambda_2 sign(\dot{S}) + (\gamma_1 S + \gamma_2 \dot{S})) \tag{15}$$

Hence:

$$\ddot{S} = -\lambda_1 \text{sgn}(S) - \lambda_2 \text{sgn}(\dot{S}) - \gamma_1 S - \gamma_2 \dot{S} + k_1 \dot{\delta} \tag{16}$$

As a result:

$$\dot{V} = \dot{S}^T(-\lambda_2 \text{sgn}(\dot{S}) - \gamma_2 \dot{S} + k_1 \dot{\delta}) \tag{17}$$

Owing to:$\lambda_{21} > k_1 \|\delta\|$ and $\lambda_{22} > 0$, $\gamma_2 > 0$ in the Theorem 1, so the inequation can be obtained as follows:

$$\dot{V} < \dot{S}^T(-\lambda_{22}\dot{S} - \gamma_2 \text{sgn}(\dot{S})) = -\lambda_{22}\|\dot{S}\|^2 - \gamma_2\|\dot{S}\| < 0$$

As the above formulas, the system states can reach the integral manifold $S = 0$ in finite time. Moreover, the system is moved around the sliding manifold, namely $S = \dot{S} = 0$, , the equation (5) can be obtained as:

$$k_1 e_\omega(t) + k_2 \int e_\omega dt = 0 \tag{18}$$

According to the principles of ordinary differential equations, the root of the equation (18) can be obtained as follows:

$$e_\omega = \exp(-k_i t) + \zeta \tag{19}$$

where $\zeta > 0$, $k_i = k_1/k_2$ is the positive constant.

From the equation (19), it is known that the tracking error e_ω can converge to zero exponentially if the constant coefficient k_1, k_2 is selected properly and equation (18) is strictly Hurwitz, which is a polynomial whose roots lie strictly in open left half of the complex plane, namely $\lim_{x \to \infty} e_\omega = 0$. Therefore, the tracking error system (4) can coverage to zero and is globally stable.

In order to avoid the windup phenomenon, the proposed algorithm is designed as follow (20) according to the anti-reset windup controller theory.

$$i_{sw} = \frac{1}{k_1 \alpha} \int (\lambda_1 sign(S) + \lambda_2 sign(\dot{S}) + (\gamma_1 S + \gamma_2 \dot{S}) - k_{c\omega}(i_q^* - i_q^r))dt \tag{20}$$

where $k_{c\omega}$ is compensate constant and i_q^r is the input of the limited amplitude.

4   Simulation and Experiment

In order to validate the feasibility and effectiveness of the proposed method, computer simulations are conducted. The computer simulation is mainly used to verify the performance of the novel second order sliding mode controller for PMSM. The simulation environment is the Matlab/Simulink. The Simulink model of the PMSM vector control system, which includes the proposed second order sliding mode controller, has been constructed. The parameters of PMSM are shown as the following table.

The results of the proposed sliding mode observer simulations are shown in Fig.2, Fig.3 and Fig.4. The parameters of the novel second order sliding mode controller are designed as follows: $\lambda_1 = 50$, $\lambda_2 = 1200$, $\gamma_1 = 15000$, $\gamma_2 = 30000$ and $k_1 = 1$, $k_2 = 14$.

Fig.2 shows the simulation experimental results of PMSM start response by two control algorithms. From the results, the system response is faster by the SOSMC than by PI control.

Table 1. Motor parameters in the simulation

Symbol	Quantity	Data
P	Power	1.0kW
T	Torque	1 Nm
V	Speed	1500 r/min
R_S	Stator resistor	2.875 ohm
Lq, Ld	Stator Inductor	0.0085 H
J	Rotor inertia	0.001 Kg· m^2
P	Poles	4
φ_f	Flux	0.175Wb

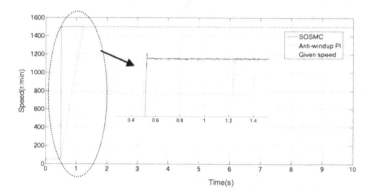

Fig. 2. The experimental results of PMSM start response

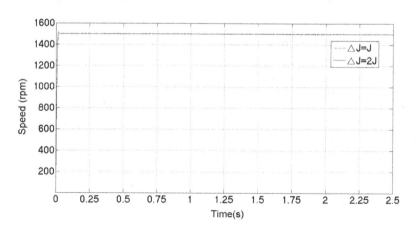

Fig. 3. Motor speed when J has a perturbation

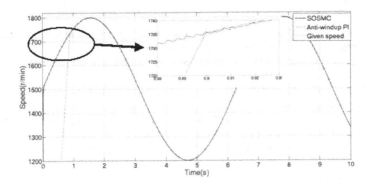

Fig. 4. Tracking curves of motor speed

Fig.3 shows the experimental results of PMSM speed control when J has a perturbation, i.e. $\triangle J=2J$. The experimental results in the Fig.3 testify that the SOSMC has good robustness when the system has the perturbation.

In order to evaluate the tracking performance of the second order sliding mode controller, the experiments are tested in the presence of the sinusoid external disturbances. Fig.3 shows the results of tracking speed in the condition of PI control and SOSMC when the given speed is $n^*=1500+300\sin(t)$ and the load torque is full load. As shown in Fig.4, the SOSMC provides better tracking performance than PI control.

5 Conclusion

This paper has proposed a novel second order sliding mode control algorithm to accomplish velocity control in the PMSM vector control system. The integral manifold is utilized to avoid the acceleration information amplifying the noise signals in control system. And the second order sliding mode control law is designed by the Lyapunov function approach. The algorithm can effectively eliminate the system chattering phenomenon compared with the conventional first order sliding mode control. In order to avoid the windup phenomenon of the PMSM control system, an anti-windup control method is used in the system. The feasibility and effectiveness of the novel algorithm has been testified through the computer simulation. The results indicate that the proposed second order sliding mode control algorithm can improve the performance of the PMSM control system such as fast response, high robustness and tracking speed performance.

Acknowledgments. We are very grateful to the editors and anonymous reviewers for their valuable comments and suggestions to help improve our paper. This work is supported by Research and Development Program of China (863 Program)(Grant no. 2011AA04A105), the Fundamental Research Funds for the Central Universities, and the Shanghai Commission of Science and Technology (Grant no. 11ZR1409800).

References

1. Uktin, V.I.: Sliding Mode Control Design Principles and Applications to Electric Drives. IEEE Trans. Industrial Electronics 40, 23–36 (1993)
2. Choi, H., Park, Y., Cho, Y.: Global Sliding-Mode Control Improved Design for a Brushless DC Motor. IEEE Control Systems Magazine 54, 27–35 (2001)
3. Shyu, K.K., Shieh, H.J.: A New Switching Surface Sliding-Mode Speed Control for Induction Motor Drive Systems. IEEE Trans. Power Electronics 11, 660–667 (1996)
4. Damiano, A., Gatto, G.L., Marongiu, I.: Second-Order Sliding-Mode Control of DC Drives. IEEE Trans. Industrial Electronics 51, 364–373 (2004)
5. Lin, F.J., Chiu, S.L., Shyu, K.K.: Novel Sliding Mode Controller for Synchronous Motor Drive. IEEE Trans. Aerospace and Electronic Systems 34, 532–542 (1998)
6. Baik, I.C., Kim, K.H., Youn, M.J.: Robust Nonlinear Speed Control of PM Synchronous Motor using Boundary Layer Integral Sliding Mode Control Technique. IEEE Trans. Control Systems Technology 8, 47–53 (2000)
7. Ghanes, M., Zheng, G.: On Sensorless Iinduction Motor Drives: Sliding-Mode Observer and Output Feedback Controller. IEEE Trans. Industrial Electronics 56, 3304–3413 (2009)
8. Rahman, M.A., Vilathgamuwa, M., Uddin, M.N.: Nonlinear Control of Interior Permanent-Magnet Synchronous Motor. IEEE Trans. Industry Applications 39, 408–416 (2003)
9. Wai, R.J., Chang, J.M.: Implementation of Robust Wavelet-Neural-Network Sliding-Mode Control for Induction Servo Motor Drive. IEEE Trans. Industrial Electronics 50, 1317–1333 (2003)
10. Cheng, K.Y., Tzou, Y.Y.: Fuzzy Optimization Techniques Applied to the Design of a Digital PMSM Servo Drive. IEEE Trans. Power Electronics 19, 1085–1089 (2004)
11. Lin, F.J., Su, K.H.: Self-Constructing Fuzzy Neural Network Speed Controller for Permanent-Magnet Synchronous Motor Drive. IEEE Trans. Fuzzy Systems 9, 751–759 (2001)
12. Chung, S.C.Y., Lin, C.L.: A Transformed Lure Problem for Sliding Mode Control and Chattering Reduction. IEEE Trans. Automatic Control 44, 563–568 (1999)
13. Paponpen, K., Konghirun, M.: An Improved Sliding Mode Observer for Speed Sensorless Vector Control Drive of PMSM. In: IEEE 5th International Conference on Power Electronics and Motion Control, pp. 1–5. IEEE Press, New York (2006)
14. Ho, H.F., Wong, Y.K., Rad, A.B.: Adaptive Fuzzy Sliding Mode Control with Chattering Elimination for Nonlinear SISO Systems. Simulation Modelling Practice and Theory 17, 1199–1210 (2009)
15. Bartolini, G., Ferrara, A., Usai, E.: Chattering Avoidance by Second-Order Sliding Mode Control. IEEE Trans. Automatic Control 43, 241–246 (1998)
16. Levant, A.: Principles of 2-sliding Mode Design. Automatica 43, 576–586 (2007)
17. Chiacchiarini, H.G., Desages, A.C.: Variable Structure Control with a Second-Order Sliding Mode Condition: Application to a Steam Generator. Automatica 31, 1157–1168 (1995)

Assessment of Texaco Syngas Components Using Extreme Learning Machine Based Quantum Neural Network

Wei Xu[1,*], Raofen Wang[2], Xingsheng Gu[3], and Youxian Sun[1]

[1] State Key Laboratory of Industrial Control Technology,
Zhejiang University, 310027 Hangzhou, China
xuwei0729@gmail.com
[2] Department of Automation, Shanghai University of Engineering Science,
201620 Shanghai, China
[3] Key Laboratory of Advanced Control and Optimization for Chemical Processes of
Ministry of Education, East China University of Science and Technology, 200237
Shanghai, China

Abstract. Quantum neural computing has nowadays attracted much attention, and tends to be a candidate to improve the computational efficiency of neural networks. In this paper, a new quantum neural network (QNN) is proposed based on quantum mechanics of superposition and collapse, etc. Instead of gradient descent methods and evolutionary algorithms, extreme learning machine (ELM) is introduced to analytically identify the parameters of the QNN. The ELM-QNN model is applied to the online and real-time assessment of the syngas components in a Texaco gasification process. The application would effectively avoid the problems of time delay and low accuracy which result from the manual analysis. In order to eliminate the redundant information stored in variables, principal component analysis (PCA) is adopted to reduce the number of input variables of ELM-QNN. The results indicate that ELM-QNN combined with PCA method has satisfied computational accuracy and efficiency. The PCA-ELM-QNN is very capable of being used for the real-time measurement of Texaco syngas components.

Keywords: Extreme learning machine, Quantum neural network, Texaco gasification, Syngas component, Principal component analysis.

1   Introduction

In recent years, ANN has been applied successfully to pattern recognition, automatic control, system modeling, signal processing, etc. However, it has the drawbacks of slow processing speed, limited memory storage and iterative learning. Owing to these, many researchers began to integrate other theories for improving the performance of ANN. The superposition of quantum mechanics provides

---

* Corresponding author.

K. Li et al. (Eds.): ICSEE 2013, CCIS 355, pp. 221–229, 2013.
© Springer-Verlag Berlin Heidelberg 2013

quantum computing with an advantage in processing huge data sets. Therefore, as an alternative, quantum computing has been introduced into the neural network. Professor Kak [1] first proposed the concept of "Quantum Neural Computation", and presented a new paradigm that combines neural computing and quantum computing. Since then, a variety of quantum neural models emerged. Perus discussed the mathematical analogies between the neural network theory and quantum mechanics [2]. Similar works have also been done in [3][4] to state new quantum neural networks. Learning is the fundamental feature of ANN. The adjustable parameters of the feedforward neural network are tuned by specific learning methods. In decades, gradient descent method is popular for learning the feedforward neural network. However, gradient descent method falls into local minima easily. Moreover, this method over-fits to the training data. In recent years, many researchers turn their attention to the evolutionary algorithms (EA). The global optimization capability enables them to be widely used for adjusting the parameters of the feedforword neural network. Many scholars [5][6] studied the applications of different EAs to optimizing the parameters. However, the slow learning speed weakens the optimization efficiency of EA. The iterative learning steps may take a lot of time to train neural networks in many applications, especially complex problems. Additionally, more time would be spent on choosing proper control parameters of EA. Different from the gradient descent method and EA, a new learning algorithm called extreme learning machine (ELM) [7][8] was proposed for single-hidden layer feedforward neural network (SLFN). ELM randomly chooses the input weights and the hidden layer thresholds, and then analytically identifies the output weights. It can be seen from many researches [9][10] that all the parameters of SLFN need not be tuned iteratively and would be obtained simply. Huang and Siew also extended ELM from SLFN to RBF network [11].

This paper focuses on the assessment of syngas components in the Texaco gasification process. A new quantum-inspired neural network is proposed, and the extreme learning machine is further applied to identifying all the adjustable parameters of the proposed QNN. The established ELM-QNN model is used for the real-time measurement of the Texaco syngas components under different operational situations.

2   Texaco Gasification Process

The Texaco gasification process [12] in a fertilizer plant of China is described as below. The coal water slurry is pumped into the Texaco gasifier. The gasifier is a two-compartment vessel, consisting of an upper refractorylined reaction chamber and a lower quench chamber. Oxygen and slurry flow through an injector nozzle into the reaction chamber. In the reaction chamber, they react to produce the raw syngas and the molten slag. Subsequently, the raw syngas and molten slag flow into the quench chamber where water cools and partially scrubs the raw syngas. The water quench also converts the molten ash into glass-like slag particles. The particles then pass down through the quench chamber/lockhopper system. The raw syngas leaving the quench chamber is then further scrubbed by using additional water in the Venturi scrubber and cooled to near ambient temperature. The gas flows to the bottom of raw gas scrubber. The solid particles in the gas are sunk into the water, and then the dust is removed from the gas at the center of the scrubber. Finally, the scrubbed raw syngas, i.e. Texaco syngas, including carbon monoxide (CO), hydrogen (H_2), carbon dioxide(CO_2), and a small amount of vapor (H_2O)), methane (CH_4), etc., is sent to the conversion procedure.

3 Extreme Learning Machine Based Quantum Neural Network

3.1 Quantum Neural Network

Quantum neural network (QNN) is a novel class of neural network models, which relies on the principles of quantum mechanics. In quantum computing, a set of qubits is used to process the information, corresponding to a sequence of bits in classical computers. For a qubit $|\varphi\rangle$, there are two ground states: $|0\rangle$ and $|1\rangle$. The qubit state $|\varphi\rangle$ can be represented as

$$|\varphi\rangle = \alpha|0\rangle + \beta|1\rangle \tag{1}$$

where α and β are the probability amplitudes of state $|0\rangle$ and $|1\rangle$ that satisfy

$$\alpha^2 + \beta^2 = 1 \tag{2}$$

When a qubit state is measured, only one value can be seen in the superposition. $|\varphi\rangle$ collapses into state $|0\rangle$ with probability α^2 or state $|1\rangle$ with probability β^2. The qubit state $|\varphi\rangle$ can be expressed in another way:

$$|\varphi\rangle = \cos\theta|0\rangle + \sin\theta|1\rangle \tag{3}$$

Based on the above quantum theory, we propose a novel quantum neural network. In our QNN, the firing and non-firing neuron states correspond to state $|1\rangle$ and $|0\rangle$ respectively. The inherent structure of qubit neuron is shown in Fig. 1.

In Fig. 1, the qubit neuron firstly collects the information from other neurons. As a control signal instead of the initial signal of a control qubit [13], the information is used to adjust the qubit state. For N arbitrary distinct samples $(\mathbf{x}_j, \mathbf{y}_j)$, where $\mathbf{x}_j=[x_{j1}, x_{j2}, ..., x_{jn}]^T \in \mathbf{R}^n$, and $\mathbf{y}_j=[y_{j1}, y_{j2}, ..., y_{jm}]^T \in \mathbf{R}^m$, the control signal u is given by

$$u = \sum_{k=1}^{n} w_{ik}x_{jk} + b_i = \mathbf{w}_i \cdot \mathbf{x}_j + b_i (j = 1, 2, ..., N) \tag{4}$$

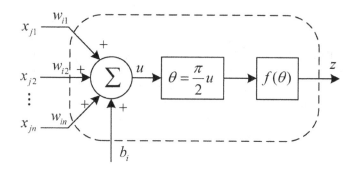

Fig. 1. The inherent structure of qubit neuron

where $\mathbf{w}_i=[w_{j1}, w_{j2}, ..., w_{jn}]^T$ is the input weight vector, and b_i is the threshold of the ith hidden node. $\mathbf{w}_i{\cdot}\mathbf{x}_j$ denotes the inner product of $\mathbf{w}_i$ and $\mathbf{x}_j$. The linear function is chosen as the activation function of the output nodes here.

And then, the state of the qubit is represented as

$$|\varphi\rangle = \cos(\theta)|0\rangle + \sin(\theta)|1\rangle = \cos(\frac{\pi}{2}u)|0\rangle + \sin(\frac{\pi}{2}u)|1\rangle \tag{5}$$

When the neuron is triggered, the qubit state collapses into state $|1\rangle$. The neuron state z is the probability with which the qubit will be found in the state $|1\rangle$.

$$z = f(\theta) = \sin^2(\theta) = \sin^2[\frac{\pi}{2}(\mathbf{w}_i \cdot \mathbf{x}_j + b_i)](j = 1, 2, ..., N) \tag{6}$$

According to Equations (4)-(6), the ith hidden neuron output is given by

$$HID_i = \sin^2[\frac{\pi}{2}(\mathbf{w}_i \cdot \mathbf{x}_j + b_i)](j = 1, 2, ..., N) \tag{7}$$

Finally, we obtain the network output for the jth sample:

$$o_j = \sum_{i=1}^{\tilde{N}} \beta_i HID_i = \sum_{i=1}^{\tilde{N}} \beta_i \sin^2[\frac{\pi}{2}(\mathbf{w}_i \cdot \mathbf{x}_j + b_i)](j = 1, 2, ..., N) \tag{8}$$

where $\beta_i = [\beta_{i1}, \beta_{i2}, ..., \beta_{im}]^T$ is the output weight vector.

3.2   ELM Algorithm for QNN

For the feedforward neural network, gradient descent-based methods like back-propagation (BP) algorithm [14] and evolutionary algorithms [15] are taken as the traditional learning rule. However, these learning methods are time-consuming. Comparatively, the ELM algorithm reaches the solutions straight-forwardly. And ELM is not to take much long time to train the feedforward network.

Here, extreme learning machine method is used to identify the parameters of the proposed QNN. Assume that the QNN can approximate the N samples with zero error which means that $\sum_{j=1}^{N} \|\mathbf{o}_j - \mathbf{y}_j\| = 0$. Thus, there exist β_i, $\mathbf{w}_i$ and b_i such that

$$\sum_{i=1}^{\tilde{N}} \beta_i \sin^2[\frac{\pi}{2}(\mathbf{w}_i \cdot \mathbf{x}_j + b_i)] = \mathbf{y}_j (j = 1, 2, ..., N) \tag{9}$$

The above N equations can be written compactly as $\mathbf{H}\beta = \mathbf{Y}$, where

$$\begin{aligned}
&\mathbf{H}(\mathbf{w}_1, ..., \mathbf{w}_{\tilde{N}}, b_1, ..., b_{\tilde{N}}, \mathbf{x}_1, ..., \mathbf{x}_N) \\
&= \begin{bmatrix} \sin^2[\frac{\pi}{2}(\mathbf{w}_1 \cdot \mathbf{x}_1 + b_1)] \cdots \sin^2[\frac{\pi}{2}(\mathbf{w}_{\tilde{N}} \cdot \mathbf{x}_1 + b_{\tilde{N}})] \\ \vdots \cdots \vdots \\ \sin^2[\frac{\pi}{2}(\mathbf{w}_1 \cdot \mathbf{x}_N + b_1)] \cdots \sin^2[\frac{\pi}{2}(\mathbf{w}_{\tilde{N}} \cdot \mathbf{x}_N + b_{\tilde{N}})] \end{bmatrix}_{N \times \tilde{N}}
\end{aligned} \tag{10}$$

After the input weights $\mathbf{w}_i$ and thresholds b_i of hidden layer nodes are chosen arbitrarily, the QNN can be simply considered as a linear system. The output weights β of QNN can be analytically determined according to Equation (11).

$$\beta = \begin{bmatrix} \sin^2[\frac{\pi}{2}(\mathbf{w}_1 \cdot \mathbf{x}_1 + b_1)] \cdots \sin^2[\frac{\pi}{2}(\mathbf{w}_{\tilde{N}} \cdot \mathbf{x}_1 + b_{\tilde{N}})] \\ \vdots \cdots \vdots \\ \sin^2[\frac{\pi}{2}(\mathbf{w}_1 \cdot \mathbf{x}_N + b_1)] \cdots \sin^2[\frac{\pi}{2}(\mathbf{w}_{\tilde{N}} \cdot \mathbf{x}_N + b_{\tilde{N}})] \end{bmatrix}_{N \times \tilde{N}}^{\dagger} \begin{bmatrix} \mathbf{y}_1^T \\ \vdots \\ \mathbf{y}_N^T \end{bmatrix}_{N \times m} \tag{11}$$

4   Results and Discussion

In the Texaco gasification process, the syngas components such as CO, H_2 and CO_2 are very critical for instructing the regular operations. However, in an application case of fertilizer plant, the syngas components are calculated through experiment analysis. Off-line manual computing is time-delay, and tends to greatly reduce the operational efficiency. Therefore, soft computing technique based on the proposed ELM-QNN method is adopted to execute the online assessment of the syngas components. Some measurable process variables are used to indirectly calculate the key unmeasurable variables, including CO, H_2 and CO_2 concentration in the Texaco syngas. The measurable variables in the gasification process include the characteristics of coal, slurry, oxygen, quenching water, etc. After smoothed and normalized, 253 groups of sample data are exposed to the ELM-QNN. Among the sample data, 200 groups are the training data for identifying the parameters of QNN, and the remaining are the testing data which are used to validate the generalization capability of ELM-QNN.

Though each of the measurable variables can reflect the process information partly, there exist correlativities between some of them. In order to eliminate the redundant information stored in variables, principal component analysis method is employed to reduce the number of input variables of ELM-QNN. Owing to the introduction of PCA, the number of input variables is reduced to 8.

We establish three ELM-QNNs with the same network structure to assess the CO concentration, H_2 concentration, and CO_2 concentration in the Texaco syngas, respectively. The number of the hidden nodes is set to 80. The training data are exposed to the PCA-ELM-QNNs. As shown in Fig. 2(a), Fig. 3(a) and Fig. 4(a), the training results of the three PCA-ELM-QNNs fit the analyzed values of CO, H_2, and CO_2 concentrations well. By using ELM method, the parameters of QNN are identified. Then, in Fig. 2(b), Fig. 3(b) and Fig. 4(b), the testing results indicate that PCA-ELM-QNNs provide the accurate measurements of the three components. The proposed models have good prediction capability.

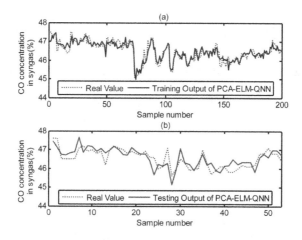

Fig. 2. Assessment of CO concentration in Texaco syngas (a) for training data and (b) for testing data

In a separate study, three other modeling methods, including SLFN (i.e. 3-layer BPNN), ELM-SLFN, and ELM-QNN, are adopted for comparison purpose. The network structures of the compared methods are the same as that of PCA-ELM-QNN. The comparison results of the four methods are shown in Table 1. For SLFN, the training results might over-fit to the training data. This would cause the bad generalization performance. For the PCA-ELM-QNN, the redundant information in the original sample data is eliminated by using PCA, and then ELM approach is used to identify the parameters of the proposed QNN. Generally, the generalization capability is the primary concern. In comparison with the three other methods on the testing sample, the mean square error (MSE) when using PCA-ELM-QNN is smaller for CO and H_2. Although the MSE for CO_2 is a little larger than that when using ELM-QNN, the concentration of CO and H_2 are the more significant components than CO_2 concentration. Thus, the performance degradation of the PCA-ELM-QNN for CO_2 is acceptable. The above results indicate that the proposed PCA-ELM-QNN performs better than

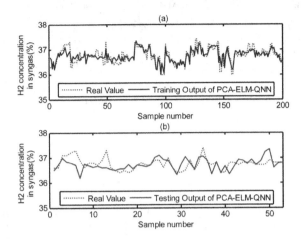

Fig. 3. Assessment of H_2 concentration in Texaco syngas (a) for training data and (b) for testing data

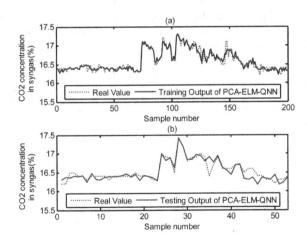

Fig. 4. Assessment of CO_2 concentration in Texaco syngas (a) for training data and (b) for testing data

Table 1. Comparisons of four methods

Methods	Train MSE			Test MSE		
	CO	H_2	CO_2	CO	H_2	CO_2
SLFN	0.038	0.036	0.027	0.505	0.352	0.205
ELM-SLFN	0.201	0.168	0.068	0.781	0.437	0.196
ELM-QNN	0.210	0.188	0.082	0.382	0.311	0.110
PCA-ELM-QNN	0.219	0.188	0.079	0.335	0.267	0.121

the other compared methods and can provide more reliable assessment of the Texaco syngas components.

5 Conclusions

A quantum neural network is proposed based on quantum mechanics. In the network, the states of quantum neurons and their interactions are investigated by using quantum theory of superposition and collapse. The extreme learning machine is employed as the learning algorithm of the proposed QNN. ELM randomly chooses the input weights, and then analytically identifies the output weights. The ELM-QNN model combined with principal component analysis is used to assess the syngas components in the Texaco gasification process. The results implied that PCA-ELM-QNN performs better than the other compared methods and is effective in the real-world application.

Acknowledgments. We are very grateful to the editors and anonymous reviewers for their valuable comments and suggestions to help improve our paper. This work is supported by the National Key Basic Research Program of China (No. 2012CB720500), the Key Program of National Natural Science Foundation of China (No. 60736021), the National Natural Science Foundation of China (No. 61273166), the Specialized Research Fund for Doctoral Program of Higher Education of China (No. 20100101110066), the Fundamental Research Funds for the Central Universities, and the Shanghai Commission of Science and Technology (Grant no. 11ZR1409800).

References

1. Kak, S.: On Quantum Neural Computing. Inform. Sci. 83, 143–160 (1995)
2. Perus, M.: Neuro-Quantum Parallelism in Brain-Mind and Computer. Informatica 20, 173–183 (1996)
3. Behrman, E.C., Niemel, J., Steck, J.E., Skinner, S.R.: A Quantum Dot Neural Network. In: Proceedings of the 4th Workshop on Physics of Computation, pp. 22–24 (1996)
4. Ventura, D., Martinez, T.: An Artificial Neuron with Quantum Mechanical Properties. In: Proceedings of the International Conference on Artificial Neural Networks and Genetic Algorithms, pp. 482–485 (1997)

5. Kim, B., Kim, S.: GA-optimized Backpropagation Neural Network with Multi-parameterized Gradients and Applications to Predicting Plasma Etch Data. Chemom. Intell. Lab. Syst. 79, 123–128 (2005)
6. Eberhart, R., Kennedy, J.: A New Optimizer using Particle Swarm Theory. In: Proceedings of the Sixth International Symposium on Micro Machine and Human Science, pp. 39–43. IEEE Press, New York (1995)
7. Huang, G.B., Zhu, Q.Y., Siew, C.K.: Extreme Learning Machine: A New Learning Scheme of Feedforward Neural Networks. In: 2004 Proceedings of IEEE International Joint Conference on Neural Networks, pp. 985–990. IEEE Press, New York (2004)
8. Huang, G.B., Zhu, Q.Y., Siew, C.K.: Extreme Learning Machine: Theory and Applications. Neurocomputing 70, 489–501 (2006)
9. Tamura, S., Tateishi, M.: Capabilities of a Four-layered Feedforward Neural Network: Four Layers versus Three. IEEE Trans. Neural Networks 8, 251–255 (1997)
10. Huang, G.B.: Learning Capability and Storage Capability of Two-hidden-layer Feedforward Networks. IEEE Trans. Neural Network 14, 274–281 (2001)
11. Huang, G.B., Siew, C.K.: Extreme Learning Machine: RBF Network Case. In: 8th International Conference on Control, Automation, Robotics and Vision, pp. 1029–1036. IEEE Press, New York (2004)
12. Rosenthal, S.: Site Technology Capsule: Texaco Gasification Process. EPA/540/R-94/514a, Environmental Protection Agency, United States (1995)
13. Kouda, N., Matsui, N., Nishimura, H., Peper, F.: Qubit Neural Network and its Learning Efficiency. Neural Comput. Appl. 14, 114–121 (2005)
14. Rumelhart, D.E., Hinton, G.E., Williams, R.J.: Learning Internal Representations by Error Propagation. In: Rumelhart, D.E., McClelland, J.L. (eds.) Parallel Distributed Processing: Explorations in the Microstructures of Cognition, pp. 318–362. MIT Press, Cambridge (1986)
15. Xu, W., Zhang, L., Gu, X.: Soft Sensor for Ammonia Concentration at the Ammonia Converter Outlet based on an Improved Particle Swarm Optimization and BP Neural Network. Chem. Eng. Res. Des. 89, 2102–2109 (2011)

2-Approximation and Hybrid Genetic Algorithm for Long Chain Design Problem in Process Flexibility

Yuli Zhang, Shiji Song, and Cheng Wu

Dept. of Automation, TNList, Tsinghua University, Beijing 100084, P.R. China
yl-zhang08@mails.tsinghua.edu.cn, {shijis,wuc}@mail.tsinghua.edu.cn

Abstract. Long chain flexibility strategy is an effective way to match the supply with the uncertain demand in manufacturing system. However there are few studies on the long chain design problem with nonhomogeneous link costs. This paper first presents a mixed 0-1 LP model and proves that it belongs to NP-complete. Then an approximation algorithm is proposed which includes three steps: 1) solve a relaxed LP; 2) generate a minimum spanning tree; 3) find the optimal local match. Under the quadrangle inequality assumption, we show that it is a 2-approximation algorithm. At last, based on another equivalent reformulation, we embed the 2-approximation algorithm and a 2-opt exchange local search into a hybrid genetic algorithm. By comparison with CPLEX solver, numerical experiments validate the effectiveness of the proposed algorithms.

Keywords: Long chain design, process flexibility, 2-approximation.

1 Introduction

In the increasingly competitive marketplace, there is an trend of shorter product upgrade cycle, more customized demand and higher demand variability. Besides the traditional operations strategies, such as inventory control and surplus capacity buffers, flexibility strategies have been proven as effective ways to decrease the mismatch between the supply and demand. It has been applied in many industry fields, such as automobile industry [1], textile industry [2] and semiconductor industry [3].

Tang and Tomlin [4] analyzed the potential supply chain risks and provided convincing arguments for deploying flexibility to mitigate supply chain risks. Process flexibility via flexible manufacturing process has been widely adopted to match the supply with the uncertain demand by shifting production quantities across internal plants or machines. Process flexibility is defined as the ability to build different types of products in the same plant or production facility at the same time [5]. Facing increasing deamnd uncertainty, process flexibility allows the company to reallocate its production capacity, workload and other flexible resources.

Although the total flexibility design [6], i.e., each plant build all products, provides the optimal performance of match, it suffers unacceptable expensive

K. Li et al. (Eds.): ICSEE 2013, CCIS 355, pp. 230–240, 2013.

setup cost. However, it has been shown that well designed limited flexibility, i.e., each plant builds only a few products, may yield most of the benefit of the total flexibility design. Furthermore, Jordan and Graves [5], through empirical analysis, have shown that the long chain configuration which chains products and plants together to the greatest extent possible, generates the greatest benefit.

In the case of symmetrical manufacturing system with n products and n plants, the long chain strategy requires that each plant produces exactly two products, demand for each product can be satisfied from exactly two plants and all plants and products are connected, directly or indirectly, by product assignment links. Motivated by the seminal work of [5], there are numerous empirical studies and analytical results on process flexibility. Graves and Tomlin [7] identified effective guidelines for process flexibility strategies in multi-stage supply chain, Gurumurthi and Benjaafar [8] extended this work to queuing systems, and Hopp et al. [9] to flexible workforce scheduling. Based on the set-theoretic methodology, Bassamboo et al. [10] analyzed newsvendor networks with multiple products and parallel resources, and characterized the optimal flexibility configuration for both symmetric and asymmetric systems. Akşin and Karaesmen [11] characterized the decreasing marginal value of flexibility and capacity in general process flexibility structures. Chou et al. [12] analyzed the worst-case performance of the flexible structure design problem using the graph expander structure, designed guidelines for general non-symmetrical systems and developed a simple and easy-to-implement heuristic to design flexible process structures.

To evaluate the efficiency of the long chain strategy in the symmetrical system, Chou et al [13] utilized the random walk and developed a system of equations to compute its performance. They showed that long chain structure performs well for a variety of realistic demand distributions, even when the system size is large. For manufacturing system with general demand and supply, they presented constraint sampling method to identify a sparse process flexibility structure within ϵ optimality of the total flexibility structure. David and Wei [14] provided the first non-asymptotic theory that explains the effectiveness of the long chain. Based on the supermodularity property of long chains, they showed that for any size system, not only large size, the long chain always maximizes expected sales among all 2-flexibility strategies.

Although long chain strategy has been proven as an effective guideline in process flexibility design, there are few studies on how to implement this strategy. In the literatures, especially under the symmetrical system assumption, each product is arbitrarily assigned to two different plants. However in reality the link costs between plants and products, which may contain the setup cost for one plant producing certain product and transportation cost to deliver the product to certain market, are usually quite different.

In this paper, we attempt to provide models and solutions on the implementation of the long chain strategy. First, under the bipartite graph representation of process flexibility, we present a mixed 0-1 linear programming model for the long chain design problem, and show that it belong to NP-complete. Second, although it can be transformed to a symmetrical Traveling Salesman Problem

(TSP), the triangle inequality, which is the key assumption for most of the effective TSP approximation algorithms, doesn't hold. We analyze the difficulties in implementing the well-known Christofides's algorithm [17],which is 3/2-approximation algorithm for TSP, in long chain design problem, and under the quadrangle inequality assumption we presents 2-approximation algorithm, i.e., the upper bound of its solution is no more than 2 time of the optimal solution. In the first step of the 2-approximation algorithm, a lower bound can be obtain by solving a linear programming. At last, based on the presented 2-approximation algorithm and 2-opt local search, a hybrid genetic algorithm (HGA) is presented to improved the solution quality. Numerical experiments and comparisons with the commercial solver CPLEX 11.1 validate the effectiveness of the presented 2-approximation algorithm and the HGA.

The paper proceeds as follows. Section 2 gives the mixed 0-1 linear programming model and analyzes the complexity of the long chain design problem. Section 3 analyzes the difficult in implementing the Christofides's algorithm and presents the 2-approximation algorithm. HGA is developed in section 4 and numerical experiments are carried out in section 5. Section 6 concludes this paper and give future research directions.

2   Long Chain Model

2.1   Formulation

A undirected bipartite graph $G = (I \cup J, E)$ represents the flexibility structures, where set I and J denote plant and product set, E denotes the link set. A link $(i, j) \in E$ means that plant i is assigned to produce j. We further define the link cost $d(i, j)$ for link (i, j) . The objective is determine the optimal long chain design with the minimal total link cost. See Fig. 1. The notation is summarized as follows:

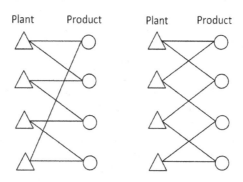

Fig. 1. Two example of long chains in bipartite graph with $n = 4$

The problem can be formulated as the following mixed 0-1 LP.

$$(P) \quad \min \quad f(x) = \sum_{i \in I} \sum_{j \in J} d_{i,j} x_{i,j}$$

$$\sum_{i \in I} x_{i,j} = 2 \quad j \in J \tag{1}$$

$$\sum_{j \in J} x_{i,j} = 2 \quad i \in I \tag{2}$$

$$\sum_{i \in U} \sum_{j \in \bar{V}} x_{i,j} \geq 1 \quad U \subset I, V \subset J, 1 < |U| = |V| < n \tag{3}$$

$$x_{i,j} \in \{0,1\} \quad i \in I, j \in J \tag{4}$$

Constraints (1) and (2) guarantee that each produce is satisfied exactly by two plants and each plant produce two products. Constraints (3) eliminate short chain solutions. This formulation contains $2n$ variables and $2n + \sum_{i=1}^{n-1}(C_n^i)^2$ constraints. Although, it can be transformed into a $2n$ nodes symmetrical TSP, the equivalent TSP formulation doesn't hold the triangle inequality property.

2.2   Problem Complexity

Proposition 1. *Long chain design problem is NP-complete.*

Proof. Let us first show that the long design problem is in NP, i.e., there is an certificate-checking algorithm such that for any yes instance of its recognition version problem, there exist a polynomial length certificate which can be checked by the certificate-checking algorithm in polynomial time. Since its recognition version problem is of the following form:

Given an instance with parameters $|I| = |J| = n$, *matrix* $[d_{i,j}]_{n \times n}$ *and an number* L, *is there a feasible solution* x *such that* $f(x) \leq L$?

Suppose we are given a yes instance $(n, [d_{i,j}]_{n \times n}, L)$ of the long chain problem, then we can construct the polynomial length certificate x as a list of the nodes which forms a long chain. This certificate can be checked efficiently for validity, because we only need to check whether n, $[d_{i,j}]_{n \times n}$ and L are appropriate, whether x forms a long chain, and whether its total length $f(x)$ is less than or equal to L. The certificate-checking algorithm will reach the answer yes at $O(n^2)$ steps. Hence, the long chain design problem is in NP.

Next we will show that any instance of Hamilton circuit problem can be polynomially transformed to a instance of long design problem. Given any instance of Hamilton circuit problem with graph $G = (V, E)$ and $|V| = n$, we construct an instance of long chain design problem by letting $|I| = |J| = n$, $E = (i, j), \forall i, j \in V$ and

$$d_{i,j} = \begin{cases} 0 & \text{if } i = j \ , \\ 1 & \text{if } (i, j) \in E \ , \\ M & \text{otherwise} \ . \end{cases}$$

where M is a big enough number. We shall argue that G has a Hamilton circuit if and only if the constructed long design problem has a long chain with length $L = n$. Suppose $v_1 v_2 \cdots v_n v_1, v \in V$ is a Hamilton circuit in G, then $i_1 j_1 i_2 j_2 \cdots i_n j_n i_1, i \in I, j \in J$ and $i_k = j_k = v_k, k = 1 \cdots n$ forms a long chain with $L = \sum_{k=1}^{n} d_{i_k j_k} + \sum_{k=1}^{n-1} d_{j_k j_{k+1}} + d_{j_n i_1} = 0 + (n-1) + 1 = n$.

For the *if* part, suppose that $i_1 j_1 i_2 j_2 \cdots i_n j_n i_1, i \in I, j \in J$ is a long chain with $2n$ links and $L = n$ in the constructed instance. By the definition of $d_{i,j}$, we must have $\{i_k = j_k, k = 1, \cdots n\}$ or $\{j_k = i_{k+1}, k = 1, \cdots n-1, j_n = i_1\}$. It is immediate that $i_1 i_2, \cdots i_n i_1$ is a Hamilton circuit in the original graph G. □

3   2-Approximation Algorithms

Although there are lots of good approximations for TSP, most of them are under the assumption of triangle inequality property. Furthermore, *unless P=NP, there is no (1 + ε)-approximation polynomail algorithm for the TSP for any $\epsilon > 0$*. As pointed in last section, the equivalent TSP for our problem doesn't hold the triangle inequality property, thus it seems harder.

First, we define the following quadrangle inequality (QI) property:

$$d_{i_1,j_1} + d_{i_2,j_1} + d_{i_2,j_2} \geq d_{i_1,j_2} \quad \forall i_1, i_2 \in I, j_1, j_2 \in J .$$

Note that if the matrix $\{d_{i,j}\}_{n \times n}$ satisfies the triangle inequality, the above inequality holds automatically.

The main idea of the presented algorithm can be summarized as:

1. Construct several short chains by solving a relaxed problem;
2. Find the "minimum spanning tree" for these short chains;
3. Find the bounded long chain by using the quadrangle inequality.

Specifically, in step 1, we relax constraints (3) in (P). Note that the following (RP) can be solved by LP because the constraint matrix is totally unimodular.

$$(RP) \quad \min \quad f(x) = \sum_{i \in I} \sum_{j \in J} d_{i,j} x_{i,j} \quad \text{s.t. (1), (2) and (4)}$$

Suppose the optimal solution of (RP) is $\{x_{i,j}^R : i \in I, j \in J\}$, the optimal value of (P) and (RP) are C^* and C_R respectively. The m short chains are $S_1, \cdots, S_m$, where $S_k = I_k \cup J_k$, $I_k \in I, J_k \in J, |I_k| = |J_k|$ for $1 \leq k \leq m$. Note that $I_k \cap I_l = J_k \cap J_l = \emptyset$ for $1 \leq k \neq l \leq m$, and $I_1 \cup I_2 \cup \cdots \cup I_m = I$, $J_1 \cup J_2 \cup \cdots \cup J_m = J$.

In step 2, by treating each short chain as a single "SC" node, define distance

$$D_{k,l} = \min \left\{ d_{i,j} : \{i \in I_k, j \in J_l\} \cup \{i \in I_l, j \in J_k\} \right\}$$

$$1 \leq k \neq l \leq m$$

For graph with m "SC" nodes and t, we can find "minimum spanning tree" in $O(m^2)$ time [16]. Suppose the link cost of the selected edges are $D^1, D^2, \cdots, D^{m-1}$.

Note that in the long chain each "SC" node must connect with at least two other "SC" nodes, we have $2 \sum_{k=1}^{m-1} D^k \leq C^*$.

In step 3, for given two short chains, if there are connected by link (i_1, j_2) in the minimum "spanning tree", we find another link (i_2, j_1) such that :

$$(i_2, j_1) = \arg_{i,j} \min \left\{ d_{i_1 j_2} + d_{ij} - d_{i_1 j} - d_{ij_2} : x^R_{i_1 j} = x^R_{ij_2} = 1 \right\} .$$

Then we merger these two short chains into one big chain by adding link (i_2, j_1) and deleting links (i_1, j_1), (i_2, j_2) as shown in Fig. 2.

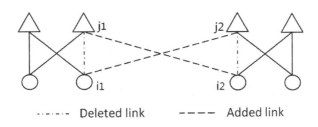

Deleted link   Added link

Fig. 2. Merger two short chains into one chain

Suppose that before merging these two short chains, the total cost is C, now the total cost is given as $C + d_{i_1 j_2} + d_{i_2 j_1} - d_{i_1 j_1} - d_{i_2 j_2} \leq C + 2 * d_{i_1 j_2}$, because of the quadrangle inequality assumption, i.e., $d_{i_2 j_1} \leq d_{i_1 j_2} + d_{i_1 j_1} + d_{i_2 j_2}$. The above analysis shows that the length of constructed long chain can be bounded by $C_R + 2 \sum_{k=1}^{m-1} D^k \leq C^* + C^* = 2C^*$. Thus, we give the following result.

Proposition 2. *The proposed approximation algorithm is 2-approximation algorithm for the long chain design problem under QI assumption.*

Proof. In the above analysis, we have shown that the proposed algorithm provides solution at worst 2 bounded by time optimal solution.

Next we only need to construct an worst-case instance. Consider the instance where all m "SC" nodes are evenly distributed on the circle with radius $R = m$. Suppose each "SC" node contains two plants and two products, which are close enough such that the total cost within each "SC" node is less than $\frac{1}{m^2}$. Figure 3. gives the "minimum spanning tree". For large enough m, the objective value obtained by the proposed algorithm can be estimated by

$$C(m) = 2(m - 1) \sin(\frac{\pi}{m}) + m \frac{1}{m^2} = 2(m - 1) \sin(\frac{\pi}{m}) + \frac{1}{m} .$$

On the other hand, for large enough m, the optimal long chain is obtained by connecting all adjacent "SC" node, thus its objective value can be estimated by

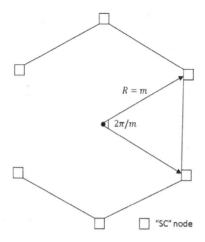

Fig. 3. "Minimum spanning tree" for the worst-case

$$C^*(m) = m\sin(\frac{\pi}{m}) + m\frac{1}{m^2} = m\sin(\frac{\pi}{m}) + \frac{1}{m} \ .$$

Let $m \to \infty$, we have

$$\lim_{m\to\infty}\frac{C(m)}{C^*(m)} = \frac{2(m-1)\sin(\frac{\pi}{m}) + \frac{1}{m}}{m\sin(\frac{\pi}{m}) + \frac{1}{m}} = 2 \ .$$

□

4 Hybrid Genetic Algorithm

In this section, we will improve the solution obtained by the proposed 2-approximation algorithm by hybrid genetic algorithm. Genetic algorithm (GA), presented by Holland [18], belongs to heuristic search methods, which are very useful when problems' search space is very large. It is not surprising that GA and its modification have been applied to the TSP problem, see [19], [20].

Here we attempt to design a hybrid genetic algorithm: 1) the proposed 2-approximation algorithm can be easily embedded into the HGA and the near optimal solution obtained can be used as the basis for further search. 2) Local search techniques in traditional TSP solution method is introduced to enhance the local search capacity of GA. In this section, we first present an equivalent formulation which is the basis for our HGA. Then we present the HGA based on the 2-approximation algorithm and 2-opt exchange local search.

4.1 Reformulation of the Long Chain Model

One drawback of the formulation in Section 2 is that there are exponential constrains in terms of n. Using the permutation matrix variable, we establish another mixed 0-1 formulation for the long chain design problem. This formulation contains the quadratic assignment problem as a special case. The equivalent reformulation is given as follows:

$$(EP) \quad minD \circ (XLY)$$

$$\sum_{i \in I} x_{i,j} = 1 \qquad j \in J \tag{5}$$

$$\sum_{j \in J} x_{i,j} = 1 \qquad i \in I \tag{6}$$

$$\sum_{i \in I} y_{i,j} = 1 \qquad j \in J \tag{7}$$

$$\sum_{j \in J} y_{i,j} = 1 \qquad i \in I \tag{8}$$

$$x_{i,j}, y_{i,j} \in \{0,1\} \qquad i \in I, j \in J \tag{9}$$

where matrix inner product $\circ$ is defined as $A \circ B = \sum_{i,j=1,\cdots,n} A_{i,j} B_{i,j} = Trace(A^T B)$, D is the constant distance matrix, i.e., $[D]_{i,j} = d_{i,j}$, and L can be chosen as any feasible solution in formulation (P).

Note that decision matrix variables $X = [x_{i,j}]_{n \times n}$ and $Y = [x_{i,j}]_{n \times n}$ are called permutation matrix variables because multiplying a matrix A from the left by a permutation matrix X results in a permutation of rows of A, and multiplying a matrix A from the right by X^T has the same effect on the columns of A. Thus for any solution of (P), i.e., any long chain, encoded by matrix L, we can still obtain a long chain by permutation operations XLY. Thus it is clear that (EP) is equivalent to (P).

4.2 Hybrid Genetic Algorithm

Note that for each permutation matrix $X \in \{0,1\}^{n \times n}$, we can encode it into a vector $x = (i_1 i_2 \cdots i_n)$, which is a permutation of the integers from 1 to n. Based on the reformulation (EP), we present a hybrid genetic algorithm which contains initialization, selection, crossover, mutation and 2-opt exchange local search.

The main body of HGA is given as follows:

1. Initialization. Run 2-approximation algorithm and let its solution be $L^* = [x_{i,j}^*]_{n \times n}$. Set $L = L^*$ and randomly generate x, y from the natural order.

2. Selection. Calculate objective value function for each individual, evaluate each individual by inverse of its objective value and select new generation by Roulette wheel selection method.

3. Crossover. According to the crossover probability, select certain individuals and exchange their permutation vector x.

4. Mutation. According to the mutation probability, select certain individuals and randomly change the permutation vectors x and y.

5. 2-opt Exchange. For permutation vectors x and y in each individual, exchange the order of any two elements until no improvement can be found. Note that there are many other available local search methods, see [21], [22].

5   Numerical Experiments

In this section, we test the effectiveness of the proposed 2-approximation algorithm (2-APP) and HGA on different test instances. Numerical experiments are implemented in Java and run on Intel (R) Xeon (R) CPU E5410.

The test instances are generated as follows: first, we randomly generate n points $(x_1, \cdots, x_n)$ for the plants and another n points $(y_1, \cdots, y_n)$ for products; then the link cost is defined by the p-norm, i.e., $d_{i,j} = \|x_i - y_j\|_p$, where $p = 1, 2, +\infty$. Other parameters are chosen as: population size $= 20$, mutation possibility $= 0.05$, crossover possibility $= 0.1$ and if no improvement is made in two consecutive iterations, then stop HGA. Value of RP is used as a lower bound for the original problem and we define the Relative Error (RE) of $(*)$ as

$$RE(\%) = \frac{\text{Value of } * - \text{Value of } RP}{\text{Value of } RP} \times 100\%.$$

Table 1 reports the relative error ($RE(\%)$) and CPU runtime ($T(s)$) of 2-APP, HGA and CPLEX solver for 10 randomly generated test instances. In this test, we set $n = 150$, $p = 2$. To compare the solution quality for given time limit, we set the time limit of CPLEX solver to 600 seconds. From Table 1, 2-App provides high quality solutions for the test instances and HGA further improves the solution quality by $1\% - 3\%$; however, even given 600 second, CPLEX solver can't find solution within relative error less than 40%.

Table 1. Performance of 2-APP, HGA and CPLEX when $n = 150, p = 2$

No.	2-APP RE(%)	T(s)	HGA RE(%)	T(s)	CPLEX RE(%)	T(s)
1	4.86	0.09	2.41	4.89	109	600
2	2.43	0.06	1.59	4.81	70.8	600
3	1.98	0.06	0.99	4.84	52.1	600
4	4.40	0.05	3.14	4.84	49.1	600
5	5.13	0.06	2.46	4.81	75.9	600
6	4.30	0.08	2.81	4.84	60.8	600
7	5.27	0.06	2.63	4.81	64.0	600
8	3.33	0.09	1.94	4.80	58.3	600
9	1.52	0.11	0.51	4.86	65.2	600
10	4.30	0.06	2.48	4.82	534	600

Next for problem parameter setting with problem size $n = 100, 150, 200$ and $p = 1, 2, +\infty$, we randomly generate 10 instances and report the average relative error and average CPU runtime in Table 2. Table 2 shows that for all the test instances, 2-APP can find solution with relative error less than 5% and HGA further decreases the relative error to less than 3% on average in reasonable time.

Table 2. Performance of 2-APP and HGA for different instances

| | 2-APP | | HGA | |
Parameters	$RE(\%)$	$T(s)$	$RE(\%)$	$T(s)$
$p = 1,\ n = 100$	5.41	0.03	2.95	2.19
$p = 1,\ n = 150$	2.43	0.06	1.59	4.81
$p = 1,\ n = 200$	4.51	0.14	2.72	8.69
$p = 2,\ n = 100$	4.54	0.02	2.55	2.19
$p = 2,\ n = 150$	4.14	0.05	2.44	4.87
$p = 2,\ n = 200$	4.09	0.17	2.40	8.68
$p = \infty,\ n = 100$	3.63	0.02	2.18	2.19
$p = \infty,\ n = 150$	3.84	0.08	2.51	4.82
$p = \infty,\ n = 200$	4.12	0.14	2.30	8.70

6   Conclusions and Future Works

This papers studies how to implement the long chain flexibility strategy when there are nonhomogeneous link costs between different plants and products. We first present a mixed 0-1 LP model and show that it belongs to NP-complete by transforming the NP-complete Hamilton circuit problem to the long chain design problem. However, the equivalent TSP doesn't holds the triangle equality property, thus we next present a new approximation algorithm which includes three steps: 1) solve a relaxed LP; 2) generate a minimum spanning tree; 3) find the optimal local match. Under the quadrangle inequality assumption, we show that it is a 2-approximation algorithm. Next a HGA is design to further improve the solution quality. At last, by numerical experiments we show that CPLEX solver fails to find good solution for given time while for all the test instances on average the 2-approximation algorithm and the HGA can obtain high quality solutions in very limited time.

For further research, other flexibility strategies, such as flexible supply contracts and postponement flexible pricing are interesting research topics on how to mitigate the supply chain risk by enhancing it flexibility.

Acknowledgment. This work was supported by Natural Science Foundation of China (No.61273233) and Research Fund for the Doctoral Program of Higher Education (No. 20090002 110035).

References

1. Wall, M.: Manufacturing flexibility. Automotive Industries 183(10), 44–45 (2003)
2. DesMarteau, K.: Leading the way in changing times. Bobbin 41(2), 48–54 (1999)
3. McCutcheon, D.: Flexible manufacturing: IBM's Bromont semiconductor packaging plant. Canadian Electronics 19(7), 26 (2004)
4. Tang, C., Tomlin, B.: The power of flexibility for mitigating supply chain risks. Int. J. Prod. Econ. 116(1), 12–27 (2008)
5. Jordan, W.C., Graves, S.C.: Principles on the benefits of manufacturing process flexibility. Manage. Sci. 41(4), 577–594 (1995)
6. Fine, C.H., Freund, R.M.: Optimal investment in product-flexible manufacturing capacity. Manage. Sci. 36(4), 449–466 (1990)
7. Graves, S.C., Tomlin, B.: Process flexibility in supply chains. Manage. Sci. 49(7), 907–919 (2003)
8. Gurumurthi, S., Benjaafar, S.: Modeling and analysis of flexible queueing systems. Nav. Res. Log. 51, 755–782 (2004)
9. Hopp, W., Tekin, E., Oyen, M.V.: Benefits of skill chaining in serial production lines with cross-trained workers. Manage. Sci. 50(1), 83–98 (2004)
10. Bassamboo, A., Randhawa, R.S., Mieghem, J.A.V.: Optimal flexibility configurations in newsvendor networks: going beyond chaining and pairing. Manage. Sci. 56(8), 1285–1303 (2010)
11. Akşin, O.Z., Karaesmen, F.: Characterizing the performance of process flexibility structures. Oper. Res. Lett. 35(4), 477–484 (2007)
12. Chou, M.C., Chua, G.A., Teo, C.P., Zheng, H.: Process flexibility revisited: The graph expander and its applications. Oper. Res. 59(5), 59–94 (2010)
13. Chou, M.C., Chua, G.A., Teo, C.P., Zheng, H.: Design for process flexibility: Efficiency of the long chain and sparse structure. Oper. Res. 58(1), 43–58 (2010)
14. David, S.L., Wei, Y.: Understanding the performance of the long chain and sparse designs in process flexibility. Oper. Res. (forthcoming)
15. Miller, C.E., Zemlin, R.A.: Integer programming formulation and traveling salesman problems. J. ACM 7, 326–329 (1960)
16. Christos, H.P., Kenneth, S.: Combinatorial optimization. Dover Publications Inc., Mineola (1998)
17. Christofides, N.: Worst-case analysis of a new heuristic for the traveling salesman problem. Report, Graduate School of Industrial Administration, Carnegie Mellon University, Pittsburgh, PA, USA (1976)
18. Holland, J.H.: Adaptation in natural and artificial systems. University of Michigan Press, Ann Arbor (1975)
19. Potvin, J.Y.: Genetic algorithms for the traveling salesman problem. Ann. Oper. Res. 63(3), 339–370 (1996)
20. Homaifar, A., Guan, S., Liepins, G.E.: A new approach on the traveling salesman problem by genetic algorithms. In: The Proceedings of the 5th International Conference on Genetic Algorithms, pp. 460–466. Morgan Kaufmann Publishers (1993)
21. Zhang, Y.L., Song, S.J., Zhang, H.M., Wu, C., Yin, W.J.: A hybrid genetic algorithm for two-stage multi-item inventory system with stochastic demand. Neural Comput. Applic. 21(6), 1087–1098 (2012)
22. Nilsson, C.: Heuristics for the traveling salesman problem, http://www.ida.liu.se/~TDDB19/reports_2003/htsp.pdf

Stochastic Gradient Algorithm for Hammerstein Systems with Piece-Wise Linearities*

Jing Chen[1,2] and Jia Chen[1]

[1] Wuxi Professional College of Science and Technology, Wuxi, P.R. China 214028
chenjing1981929@126.com
[2] Control Science and Engineering Research Center, Jiangnan University,
Wuxi, P.R. China 214122
302774418@qq.com

Abstract. This paper studies a stochastic algorithm for Hammerstein systems with piece-wise linearities. By using a switching function, the model of the nonlinear Hammerstein systems be changed to an identification model, then based on the derived model, a stochastic gradient identification algorithm is used to estimate all the unknown parameters of the systems. An example is provided to show the effectiveness of the proposed algorithm.

Keywords: Piece-wise linearity, Stochastic gradient, Parameter estimation, Hammerstein system.

1 Introduction

Hammerstein systems consist of a static nonlinear block followed by a linear dynamic block which are widely used in many areas, e.g., nonlinear filtering, actuator saturations, audio-visual processing, signal analysis. There exists a lot of work on identification of these nonlinear systems [1–6]. Some work assumed that the nonlinearity is the polynomial nonlinearity [6–8], others assumed that the nonlinearity is the hard nonlinearity [2,3,9–12,14]. Hard nonlinearity cannot be written as an analytic function of the input and is more common in engineering practice. Recently, identification of Hammerstein systems with hard nonlinearity has been received much attention [3,9,10,13–15]. For example, Bai used a deterministic approach and the correlation analysis method to estimate the parameters of systems with hard input nonlinearities [9]. Chen proposed a novel estimation algorithm for dual-rate Hammerstein systems with preload nonlinearity [13], and studied identification problems for Hammerstein systems with saturation and dead-zone nonlinearities [3].

This paper deals with the identification of Hammerstein systems with piece-wise linearities. By using the switching function, the model of the Hammerstein systems can be turned into an identification model, then based on the derived

---

* This work was supported by the National Natural Science Foundation of China.

K. Li et al. (Eds.): ICSEE 2013, CCIS 355, pp. 241–247, 2013.
© Springer-Verlag Berlin Heidelberg 2013

model, a stochastic gradient algorithm (SG) is proposed to estimate the unknown parameters of the systems.

Briefly, the paper is organized as follows. Section 2 describes the piece-wise linearities and derives an identification model. Section 3 studies estimation algorithms for the identification model. Section 4 provides an illustrative example. Finally, concluding remarks are given in Section 5.

2   The Piece-Wise Linearities

Consider a Hammerstein system

$$A(z)y(t) = B(z)f(u(t)) + v(t), \tag{1}$$

where $y(t)$ is the system output, $u(t)$ is the system input, and $v(t)$ is a stochastic white noise with zero mean, and $A(z)$ and $B(z)$ are polynomials in the unit backward shift operator $[z^{-1}y(t) = y(t-1)]$ and

$$A(z) := 1 + a_1 z^{-1} + a_2 z^{-2} + \cdots + a_n z^{-n},$$
$$B(z) := b_1 z^{-1} + b_2 z^{-2} + b_3 z^{-3} + \cdots + b_n z^{-n}.$$

The nonlinear input $f(u(t))$ is a piece-wise linearity which is shown in Figure 1 and can be expressed as

$$f(u(t)) = \begin{cases} m_1 u(t), & u(t) \geq 0, \\ m_2 u(t), & u(t) < 0, \end{cases}$$

where m_1 and m_2 are the corresponding segment slopes.

Define a switching function,

$$h(t) := h[u(t)] = \begin{cases} \frac{1}{2}, & u(t) \geq 0, \\ -\frac{1}{2}, & u(t) < 0. \end{cases}$$

Then the output $y(t)$ can be written as

$$f(u(t)) = (m_1 - m_2)u(t)h(u(t)) + \frac{1}{2}(m_1 + m_2)u(t), \tag{2}$$

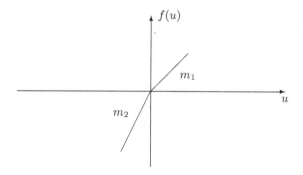

Fig. 1. The piece-wise linearity

and Equation (1) can be written as

$$A(z)y(t) = B(z)((m_1 - m_2)u(t)h(u(t))$$
$$+ \frac{1}{2}(m_1 + m_2)u(t)) + v(t). \tag{3}$$

From (3), we can see that the output $y(t)$ of the nonlinear block can be written as an analytic function of the input.

3   The Estimation Algorithms

Define the parameter vector $\boldsymbol{\theta}$ and the information vector $\boldsymbol{\varphi}(t)$ as

$$\boldsymbol{\theta} := [b_1(m_1 - m_2), b_2(m_1 - m_2), b_3(m_1 - m_2), \cdots ,$$
$$b_n(m_1 - m_2), \frac{1}{2}b_1(m_1 + m_2), \frac{1}{2}b_2(m_1 + m_2),$$
$$\frac{1}{2}b_3(m_1 + m_2), \cdots , \frac{1}{2}b_n(m_1 + m_2),$$
$$a_1, a_2, a_3, \cdots , a_n]^{\mathrm{T}} \in \mathbb{R}^{3n},$$
$$\boldsymbol{\varphi}(t) := [u(t - 1)h(t - 1), u(t - 2)h(t - 2),$$
$$u(t - 3)h(t - 3), \cdots , u(t - n)h(t - n),$$
$$u(t - 1), u(t - 2), u(t - 3), \cdots ,$$
$$u(t - n), -y(t - 1), -y(t - 2), \cdots ,$$
$$-y(t - n)]^{\mathrm{T}} \in \mathbb{R}^{3n},$$

gets

$$y(t) = \boldsymbol{\varphi}^{\mathrm{T}}(t)\boldsymbol{\theta} + v(t). \tag{4}$$

If $\boldsymbol{\theta}$ has been estimated, none of the identification schemes can distinguish $b_i, i = 1, 2, 3, \cdots , n$ and $m_i, i = 1, 2$ from the estimated $\boldsymbol{\theta}$. Therefore, to get a unique parameterization, in this paper, we adopt the assumption that the first coefficient b_1 equals 1, i.e., $b_1 = 1$.

The parameter vector $\boldsymbol{\theta}$ and the information vector $\boldsymbol{\varphi}(t)$ be defined as

$$\boldsymbol{\theta} := [(m_1 - m_2), b_2(m_1 - m_2), b_3(m_1 - m_2),$$
$$\cdots , b_n(m_1 - m_2), \frac{1}{2}(m_1 + m_2),$$
$$\frac{1}{2}b_2(m_1 + m_2), \frac{1}{2}b_3(m_1 + m_2),$$
$$\cdots , \frac{1}{2}b_n(m_1 + m_2), a_1,$$
$$a_2, a_3, \cdots , a_n]^{\mathrm{T}} \in \mathbb{R}^{3n}, \tag{5}$$
$$\boldsymbol{\varphi}(t) := [u(t - 1)h(t - 1), u(t - 2)h(t - 2),$$
$$u(t - 3)h(t - 3), \cdots , u(t - n)h(t - n),$$

$$u(t-1), u(t-2), u(t-3), \cdots, u(t-n),$$
$$-y(t-1), -y(t-2), \cdots,$$
$$-y(t-n)]^{\mathrm{T}} \in \mathbb{R}^{3n}, \tag{6}$$

Using the following SG algorithm to estimate the parameter vector $\boldsymbol{\theta}$ in (5):

$$\hat{\boldsymbol{\theta}}(t) = \hat{\boldsymbol{\theta}}(t-1) + \frac{\varphi(t)}{r(t)}(y(t) - \varphi^{\mathrm{T}}(t)\hat{\boldsymbol{\theta}}(t-1)), \tag{7}$$

$$\varphi(t) = [u(t-1)h(t-1), u(t-2)h(t-2),$$
$$u(t-3)h(t-3), \cdots, u(t-n)h(t-n)$$
$$, u(t-1), u(t-2), u(t-3), \cdots, u(t-n),$$
$$-y(t-1), -y(t-2), \cdots, -y(t-n)]^{\mathrm{T}}, \tag{8}$$

$$r(t) = r(t-1) + \|\varphi(t)\|^2, \quad r(0) = 1. \tag{9}$$

where $\frac{1}{r(t)}$ is the step-size and the norm of matrix X is defined by $\|X\|^2 :=$ tr$[XX^{\mathrm{T}}]$.

The convergence of the SG algorithm is relatively slower compared with the recursive least squares algorithm. In order to improve the tracking performance of the SG algorithm, we can introduce a λ in the SG algorithm to get the SG algorithm with a forgetting factor (the FF-SG algorithm for short) as follows:

$$\hat{\boldsymbol{\theta}}(t) = \hat{\boldsymbol{\theta}}(t-1) + \frac{\varphi(t)}{r(t)}(y(t) - \varphi^{\mathrm{T}}(t)\hat{\boldsymbol{\theta}}(t-1)), \tag{10}$$

$$\varphi(t) = [u(t-1)h(t-1), u(t-2)h(t-2),$$
$$u(t-3)h(t-3), \cdots, u(t-n)h(t-n),$$
$$u(t-1), u(t-2), u(t-3), \cdots, u(t-n),$$
$$-y(t-1), -y(t-2), \cdots, -y(t-n)]^{\mathrm{T}} \tag{11}$$

$$r(t) = \lambda r(t-1) + \|\varphi(t)\|^2,$$
$$0 < \lambda < 1, \quad r(0) = 1. \tag{12}$$

4   Example

Consider the following linear dynamic block,

$$[1 - 0.1q^{-1}]y(t) = [q^{-1} + 1.2q^{-2}]f(u(t)) + v(t),$$

the input $\{u(t)\}$ is taken as a persistent excitation signal sequence with zero mean and unit variance, and $\{v(t)\}$ is taken as a white noise sequence with zero mean and variance $\sigma^2 = 0.10^2$, the piece-wise linearity is shown in Figure 1 and with parameters: $m_1 = 1$, $m_2 = 0.8$. Then we have

$$\boldsymbol{\theta} = [m_1 - m_2, b_2(m_1 - m_2), 0.5(m_1 + m_2),$$
$$0.5b_2(m_1 + m_2), a_1]^{\mathrm{T}}$$

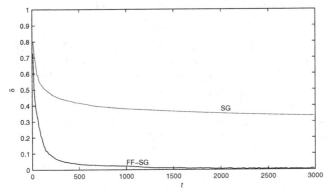

Fig. 2. The parameter estimation errors δ versus t

$$= [\alpha_1, \alpha_2, \alpha_3, \alpha_4, \alpha_5,]^{\mathrm{T}}$$
$$= [0.2, 0.24, 0.9, 1.08, -0.1]^{\mathrm{T}},$$
$$\varphi(t) = [h(u(t-1))u(t-1), h(u(t-2))u(t-2),$$
$$u(t-1), u(t-2), -y(t-1)]^{\mathrm{T}}.$$

Applying the proposed SG and FF-SG algorithms to estimate the parameters of this system, the parameter estimates and their errors are shown in Tables 1-2 and the parameter estimation errors $\delta := \|\hat{\theta} - \theta\| / \|\theta\|$ versus t are shown in Figure 2.

Table 1. The SG estimates and errors

t	α_1	α_2	α_3	α_4	α_5	δ (%)
100	-0.0422	0.0043	0.4938	0.5448	-0.2357	52.9384
200	-0.0291	0.0224	0.5536	0.6047	-0.2483	47.3742
300	-0.0180	0.0347	0.5844	0.6351	-0.2537	44.4015
500	-0.0108	0.0442	0.6168	0.6669	-0.2658	41.6292
1000	0.0009	0.0534	0.6589	0.7046	-0.2646	37.9839
1500	0.0055	0.0585	0.6771	0.7217	-0.2663	36.4216
2000	0.0102	0.0630	0.6906	0.7342	-0.2663	35.2145
2500	0.0135	0.0666	0.7024	0.7442	-0.2663	34.2374
3000	0.0160	0.0690	0.7093	0.7510	-0.2653	33.5802
True values	0.2000	0.2400	0.9000	1.0800	-0.1000	

Let $\hat{\alpha}_i$ be the ith element of the vector $\hat{\theta}$. From the definition of θ, we have: $\hat{a}_1 = \hat{\alpha}_5$, $\hat{b}_2 = \frac{\hat{\alpha}_2}{\hat{\alpha}_1}$. Furthermore, we can compute the estimates $\hat{m}_1 = \hat{\alpha}_3 + \frac{\hat{\alpha}_1}{2}$, $\hat{m}_2 = \hat{\alpha}_3 - \frac{\hat{\alpha}_1}{2}$.

Table 2. The FF-SG estimates and errors

t	α_1	α_2	α_3	α_4	α_5	$\delta\ (\%)$
100	0.1174	0.0619	0.8128	0.9035	-0.1812	20.0681
200	0.1927	0.1290	0.9055	1.0303	-0.1449	9.0010
300	0.2224	0.1689	0.8981	1.0440	-0.1097	5.7755
500	0.2285	0.1994	0.9009	1.0734	-0.1009	3.4647
1000	0.2206	0.2145	0.9003	1.0762	-0.1026	2.2923
1500	0.2105	0.2298	0.8962	1.0786	-0.1004	1.0502
2000	0.2033	0.2333	0.8983	1.0778	-0.1023	0.5736
2500	0.1952	0.2395	0.8952	1.0825	-0.0981	0.5179
3000	0.1980	0.2361	0.8958	1.0754	-0.1055	0.6487
True values	0.2000	0.2400	0.9000	1.0800	-0.1000	

5 Conclusions

An approach to identify Hammerstein systems with piece-wise linearity is presented in this paper. The model of the nonlinear system be turned into an identification model by using a switching function, then based on the identification model, we proposed an SG algorithm and an FF-SG algorithm to estimate all the parameters of the system. The simulation results verify the proposed algorithm.

References

1. Liu, Y., Bai, E.W.: Iterative identification of Hammerstein systems. Automatica 43(2), 346–354 (2007)
2. Ding, F., Liu, P.X., Liu, G.: Identification methods for Hammerstein nonlinear systems. Digital Signal Processing 21(2), 215–238 (2011)
3. Chen, J., Wang, X.P., Ding, R.F.: Gradient based estimation algorithm for Hammerstein systems with saturation and dead-zone nonlinearities. Applied Mathematical Modelling 36(1), 238–243 (2012)
4. Yu, L., Zhang, J.B., Liao, Y.W., Ding, J.: Parameter estimation error bounds for Hammerstein finite impulsive response models. Applied Mathematics and Computation 202(2), 472–480 (2008)
5. Wang, D.Q., Chu, Y.Y., Ding, F.: Auxiliary model-based RELS and MI-ELS algorithms for Hammerstein OEMA systems. Computers & Mathematics with Applications 59(9), 3092–3098 (2010)
6. Chen, J., Zhang, Y., Ding, R.F.: Auxiliary model based multi-innovation algorithms for multivariable nonlinear systems. Mathematical and Computer Modelling 52(9-10), 1428–1434 (2010)
7. Wang, D.Q., Chu, Y.Y., Ding, F.: Auxiliary model-based RELS and MI-ELS algorithms for Hammerstein OEMA systems. Computers & Mathematics with Applications 59(9), 3092–3098 (2010)
8. Ding, F., Shi, Y., Chen, T.: Auxiliary model-based least-squares identification methods for Hammerstein output-error systems. Systems & Control Letters 56(5), 373–380 (2007)

9. Bai, E.W.: Identification of linear systems with hard input nonlinearities of known structure. Automatica 38(5), 853–860 (2002)
10. Vörös, J.: Parameter identification of discontinuous Hammerstein systems. Automatica 33(6), 1141–1146 (1997)
11. Vörös, J.: Modeling and parameter identification of systems with multisegment piecewise-linear Characteristics. IEEE Transactions on Automatic Control 47(1), 184–188 (2002)
12. Vörös, J.: Parameter identification of Wiener systems with multisegment piecewise-linear nonlinearities. Systems & Control Letters 56(2), 99–105 (2007)
13. Jing, C., Lixing, L., Ruifeng, D.: Parameter Estimation for Dual-Rate Sampled Data Systems with Preload Nonlinearities. In: Zhang, T. (ed.) Mechanical Engineering and Technology. AISC, vol. 125, pp. 43–50. Springer, Heidelberg (2012)
14. Vörös, J.: Modeling and identification of systems with backlash. Automatica 46(2), 369–374 (2010)
15. Rochdi, Y., Giri, F., Gning, J.B., Chaoui, F.Z.: Identification of block-oriented systems in the presence of nonparametric input nonlinearities of switch and backlash types. Automatica 46(5), 864–877 (2010)

Robust Non-fragile Controllers Design for Uncertain Linear Switched Systems

Junyong Zhai*, Bin Wang, and Shumin Fei

Key Laboratory of Measurement and Control of CSE, Ministry of Education
Southeast University, Nanjing 210096, China
jyzhai@seu.edu.cn
http://www.springer.com/lncs

Abstract. This paper deals with the problem of how to design robust non-fragile controllers for linear switched systems with uncertainties which are included in both the system matrix and the input one. Combining the piecewise Lyapunov function with the average dwell-time method, we develop the sufficient conditions for the existence of non-fragile state feedback controllers, which guarantee that the switched system is exponentially stable and has a certain stability margin for all admissible uncertainties by means of switching among multiple models. The corresponding results are given in terms of linear matrix inequalities.

Keywords: Non-fragile, piecewise Lyapunov functions, average dwell-time, exponentially stable.

1 Introduction

A hybrid dynamical system consists of a family of continuous-time subsystems and a rule that orchestrates the switching between them. Due to their significance both in theory development and in practical applications, switched systems have been attracting considerable attention during the last decades [1–8]. Switched systems have numerous applications in control of mechanical systems, the automotive industry, aircraft and air traffic control, switching power converters, and many other fields. Hybrid control, which is based on switching between different models and controllers, has also received growing interest, due to its advantages, for instance, on achieving stability, improving transient response, and providing an effective mechanism to cope with highly complex systems and systems with large uncertainties. Two key problems in the study of switched systems are the stability analysis and control synthesis. It has been shown that average dwell- time approach is an effective tool for choosing certain switching laws, under which asymptotic and/or exponential stability can be obtained [9–11]. For the analysis of stability of the switched systems, there are mainly about common Lyapunov function, multiple Lyapunov functions and dwell-time method. In [12–15] have designed the controllers with appropriate switched rules using

---

* Corresponding author.

K. Li et al. (Eds.): ICSEE 2013, CCIS 355, pp. 248–256, 2013.

multiple Lyapunov functions. In [16–18], dwell- time have been used to study the stability of the switched systems. In [19–21] have analyzed the stability of the switched systems and designed the controllers by means of average dwell-time.

Robustness of control systems to uncertainties has always been the central issue in feedback control and therefore for uncertain dynamical systems, a large number of robust controller design methods have been presented [22] and [23]. Because controller implementation is subject to imprecision inherent in analog-digital and digital-analog conversion, finite word length, and finite resolution measuring instruments and roundoff errors in numerical computations and any useful design procedure should generate a controller which also has sufficient room for readjustment of its coefficients [24–26]. For linear continuous-time systems with structured uncertainties existing in the system matrix only, a design method of a robust non-fragile state feedback controller has been suggested [12]. Also, a design method of a H_1 controller for linear systems with additive controller gain variations has been derived [27]. Oya, Hagino and Mukaidani [25] considered the problem of robust non-fragile controllers for linear continuous-time systems. However, so far the design problem of robust non-fragile controllers for linear switched systems with uncertainties which are included in both the system matrix and the input one has not been discussed. From this viewpoint on the basis of the existing result for quadratic stabilization, we present a design method of a robust non-fragile controller for linear switched systems with structured uncertainties existing in both the system matrix and the input one. In this paper, we show that sufficient conditions for the existence of the robust non-fragile controller are given in terms of linear matrix inequalities (LMIs).

In this paper, we use $P > 0 (\geq, <, \leq 0)$ to denote a positive definite (positive semi-defined, negative definite, negative semi-definite) matrix. $\mathbf{R}^n$ is n-dimensional real space; $\mathbf{R}^{m \times n}$ is set of all the m by n matrices. For any vector or matrix A, A^T means the transpose of A; $He\{A\}$ donates $A + A^T$. In the symmetric matrix, $*$ means the symmetric part of the symmetric matrix. For the real symmetric matrices A and B, $A < B$ ($A \leq B$) means $A - B$ is negative definite (semi-definite) matrix. I represents the identity matrix. $\lambda_{min}(P), \lambda_{max}(P)$ denote minimal and maximal eigenvalues of matrix P. $\| \cdot \|$ denotes the usual 2-norm.

2   Preliminaries

Consider the following switched linear uncertain system

$$\dot{x}(t) = (A_\sigma + \Delta A_\sigma)x(t) + (B_\sigma + \Delta B_\sigma)u(t) \tag{1}$$

where $x(t) \in \mathbf{R}^n$ is the state, $u(t) \in \mathbf{R}^m$ is the control input, the right continuous function $\sigma(t) : [0, +\infty) \to M = \{1, 2, \cdots, m\}$ is the switching signal, corresponding to it, the switching sequence

$$\Sigma = \{x_0, (i_0, t_0), (i_1, t_1), \cdots, (i_j, t_j), \cdots | i_j \in M\}$$

means that the i_jth subsystem is active when $t \in [t_j, t_{j+1})$. A_σ, B_σ are constant matrices of appropriate dimensions. The uncertainties ΔA_σ and ΔB_σ are assumed to satisfy the following assumption.

Assumption 1. The uncertainties are in the form:

$$\Delta A_\sigma = D_\sigma \Delta_{A_\sigma}(t) L_\sigma, \quad \Delta B_\sigma = E_\sigma \Delta_{B_\sigma}(t) M_\sigma \tag{2}$$

where $D_\sigma, L_\sigma, E_\sigma$ and M_σ are the constant matrices with appropriate dimensions, and ΔA_σ and ΔB_σ are the unknown, real and possible time-varying matrices satisfying $\Delta_{A_\sigma}^T(t)\Delta_{A_\sigma}(t) \leq I, t \geq 0$ and $\Delta_{B_\sigma}^T(t)\Delta_{B_\sigma}(t) \leq I, t \geq 0$, respectively.

For the system (1), we design the state feedback controller as follows

$$u(t) = (K_\sigma + \Delta K_\sigma)x(t) \tag{3}$$

where $K_\sigma \in \mathbf{R}^{m \times n}$ denotes the state feedback gain matrix and ΔK_σ denotes the control gain perturbation matrix. In this paper, we consider the following two forms of the control gain perturbations.
(i) the additive form:

$$\Delta K_\sigma = F_\sigma \Delta_{K_\sigma}(t) N_\sigma \tag{4}$$

(ii) the multiplicative form:

$$\Delta K_\sigma = F_\sigma' \Delta_{K_\sigma}'(t) N_\sigma' K_\sigma \tag{5}$$

where $F_\sigma, N_\sigma, F_\sigma'$ and N_σ' are the known constant matrices, $\Delta_{K_\sigma}(t)$ and $\Delta_{K_\sigma}'(t)$ denote time-varying uncertainties satisfying

$$\Delta_{K_\sigma}^T(t)\Delta_{K_\sigma}(t) \leq I, \quad \Delta_{K_\sigma}'^T(t)\Delta_{K_\sigma}'(t) \leq I .$$

To conclude this section, we recall the following lemmas which will be used in the proof of our main results.

Lemma 1 (Schur Complement). *For any given constant real symmetric matrix* $P = \begin{bmatrix} P_{11} & P_{12} \\ P_{12}^T & P_{22} \end{bmatrix}$, *the following three arguments are equivalent*

$$(i) P < 0;$$
$$(ii) P_{22} < 0, P_{11} - P_{12}P_{22}^{-1}P_{12}^T < 0;$$
$$(iii) P_{11} < 0, P_{22} - P_{12}^T P_{11}^{-1} P_{12} < 0 . \tag{6}$$

Lemma 2. *Let* U, V *be real matrices of appropriate dimensions. Then, for any matrix* $Q > 0$ *of appropriate dimension and any scalar* $\varepsilon > 0$, *such that*

$$UV + V^T U^T \leq \varepsilon^{-1} U Q^{-1} U^T + \varepsilon V^T Q V . \tag{7}$$

3   Non-fragile Controller Design

In this section, we will show how to design state feedback gain K_i and switching law $\sigma(t)$ for switched linear uncertain system (1) to be exponentially stable.

Definition 1. *[19] For the switched signal σ and any $t \geq \tau \geq 0$, let $N_\sigma(t, \tau)$ denote the system switching times in the open interval (τ, t). If*

$$N_\sigma(t, \tau) \leq N_0 + \frac{t - \tau}{\tau_D} \tag{8}$$

holds for $\tau_D > 0$ and $N_0 \geq 0$, then τ_D is called average dwell-time and N_0 is said to be the chatter bound.

Definition 2. *[15] The switched system (1) is exponentially stable if all the state trajectories satisfy*

$$\|x(t)\| \leq k_1 e^{-k_2 t} \|x(0)\| \tag{9}$$

for some $k_1 > 0$ and $k_2 > 0$, k_2 is called stability margin.

We first consider the nominal system of the switched system (1). That is

$$\dot{x}(t) = A_\sigma x(t) + B_\sigma u(t) \tag{10}$$

and the state feedback controller

$$u(t) = K_\sigma x(t) . \tag{11}$$

Theorem 1. *Given a scalar $\lambda_0 > 0$, if there exist positive matrix P_i and matrix K_i, such that*

$$A_i^T P_i + P_i A_i + K_i^T B_i^T P_i + P_i B_i K_i + 2\lambda_0 P_i < 0, i \in M \tag{12}$$

holds, then the closed-loop system (10) and (11) is globally exponentially stable with stability margin $\lambda \in (0, \lambda_0)$ for any switching signal with average dwell-time satisfying

$$\tau_D \geq \tau_D^* = \frac{\ln \mu}{2(\lambda_0 - \lambda)}$$

where $a = \sup_{i \in M} \lambda_{max}(P_i), b = \inf_{i \in M} \lambda_{min}(P_i)$ and $\mu = \frac{a}{b}$.

Proof. Define the piecewise Lyapunov functional candidate $V(x(t)) = x^T P_\sigma x$, which is positive definite since $P_\sigma \in \mathbf{R}^{n \times n}$ is a positive definite matrix. When the i-th subsystem is activated, $V(x(t)) = V_i(x(t)) = x^T P_i x$. From the definition of a, b, μ, for $\forall i, j \in M$, we have the following two inequalities.

$$V_i(x(t)) = x^T P_i x \leq \mu x^T P_j x = \mu V_j(x(t)) \tag{13}$$

$$b\|x\|^2 \leq x^T P_i x \leq a\|x\|^2 \tag{14}$$

Assume that $[t_k, t_{k+1})$ is any switching interval, in which the i-th subsystem is activated, then the time derivative of the $V(x(t))$ along the trajectory of system (10) can be calculated as

$$\dot{V}(x(t)) = x^T (A_i^T P_i + P_i A_i + K_i^T B_i^T P_i + P_i B_i K_i)$$
$$\leq -2\lambda_0 x^T P_i x = -2\lambda_0 V(x(t)) \ . \tag{15}$$

From (15), it can be deduced that

$$V(x(t)) \leq e^{-2\lambda_0(t-t_k)} V(x(t_k)), \quad t \geq t_k \ . \tag{16}$$

Let $0 = t_0 < t_1 < \cdots < t_k = t_{N_\sigma(t,0)}$ denote the switching sequences in the interval $[0, t)$. Substituting (13) into (16) yields

$$V(x(t)) \leq e^{-2\lambda_0(t-t_k)} V(x(t_k)) \leq \mu e^{-2\lambda_0(t-t_{k-1})} V(x(t_{k-1})) \leq \cdots$$
$$\leq \mu^{N_\sigma(t,0)} e^{-2\lambda_0 t} V(x(0)) = e^{-2\lambda_0 t + N_\sigma(t,0) \ln \mu} V(x(0)) \ . \tag{17}$$

From (8) and definition of τ_D^*, it can be concluded that

$$N_\sigma(t,0) \ln \mu \leq (N_0 + \frac{t}{\tau_D^*}) \ln \mu \leq 2\alpha + 2(\lambda_0 - \lambda)t \tag{18}$$

where $\alpha = \frac{N_0 \ln \mu}{2}$. Substituting (18) into (17), it yields

$$V(x(t)) \leq V(x(0)) e^{-2\lambda t + 2\alpha} \tag{19}$$

From (14) and (19), we can obtain

$$\|x(t)\| \leq \sqrt{\mu} \|x(0)\| e^{\alpha - \lambda t} \ . \tag{20}$$

Therefore, the system (10) is exponentially stable with stability margin λ.

Next, we consider the system (1) with the uncertainties (2), and the actual controller (3) with perturbations (4)(or (5)). The closed-loop system can be described as

$$\dot{x}(t) = [A_i + \Delta A_i + (B_i + \Delta B_i)(K_i + \Delta K_i)]x(t), \quad \forall i \in M \ . \tag{21}$$

Theorem 2. *Given a scalar $\lambda_0 > 0$, if there exist positive matrix P_i and matrix $K_i, i \in M$, such that*

$$[A_i + \Delta A_i + (B_i + \Delta B_i)(K_i + \Delta K_i)]^T P_i + P_i[A_i + \Delta A_i +$$
$$(B_i + \Delta B_i)(K_i + \Delta K_i)] + 2\lambda_0 P_i < 0 \tag{22}$$

holds, then the closed-loop system (1) and (3) is globally exponentially stable with stability margin $\lambda \in (0, \lambda_0)$ for any switching signal with average dwell-time satisfying $\tau_D \geq \tau_D^*$.

Proof. The proof is very similar to the one for Theorem 1 with some modifications. For the sake of space, the detailed proof is omitted here.   □

In what follows, we will show that the design method of the robust non-fragile controller based on the LMI framework [28].

Firstly, we give the sufficient condition for the existence of non-fragile state feedback controller with additive control gain perturbations of form (4).

Theorem 3. *Consider system (1), for given scalar $\varepsilon_{1i} > 0, \varepsilon_{2i} > 0, \varepsilon_{3i} > 0, \varepsilon_{4i} > 0, \lambda_0 > 0, \forall i \in M$, if there exist $X_i > 0, W_i$, and the state feedback gain matrix $K_i = W_i X_i^{-1}$ (if it exists), such that the LMIs condition*

$$\begin{bmatrix} \Gamma_i & X_i L_i^T & X_i N_i^T & W_i^T M_i^T & 0 & E_i & X_i N_i^T \\ * & -\varepsilon_{1i} I & 0 & 0 & 0 & 0 & 0 \\ * & * & -\varepsilon_{2i} I & 0 & 0 & 0 & 0 \\ * & * & * & -\varepsilon_{3i} I & 0 & 0 & 0 \\ * & * & * & * & -\Xi(\varepsilon_{4i}) & 0 & 0 \\ * & * & * & * & * & -\varepsilon_{4i} I & 0 \\ * & * & * & * & * & * & -\varepsilon_{4i} I \end{bmatrix} < 0 \qquad (23)$$

holds, then the closed-loop system (21) with additive control gain perturbations (4) is exponentially stable with stability margin λ under arbitrary switching signal in terms of average dwell-time satisfying $\tau_D \geq \tau_D^*$. In (23), $\Gamma_i = He\{A_i X_i + B_i W_i\} + \varepsilon_{1i} D_i D_i^T + \varepsilon_{2i} B_i F_i F_i^T B_i^T + \varepsilon_{3i} E_i E_i^T + 2\lambda_0 X_i$, $\Xi(\varepsilon_{4i}) = I - \varepsilon_{4i}(I + M_i F_i F_i^T M_i^T)$.

Proof. Using Lemma 2, we can get

$$He\{P_i \Delta A_i\} \leq \varepsilon_{1i} P_i D_i D_i^T P_i + \varepsilon_{1i}^{-1} L_i^T L_i,$$

$$He\{P_i B_i \Delta K_i\} \leq \varepsilon_{2i} P_i B_i F_i F_i^T B_i^T P_i + \varepsilon_{2i}^{-1} N_i^T N_i,$$

$$He\{P_i \Delta B_i K_i\} \leq \varepsilon_{3i} P_i E_i E_i^T P_i + \varepsilon_{3i}^{-1} K_i^T M_i^T M_i K_i \ .$$

Due to $He\{P_i \Delta B_i \Delta K_i\} = He\{P_i E_i \Delta_{B_i}(t) M_i F_i \Delta_{K_i}(t) N_i\}$, one has

$$He\{P_i \Delta B_i \Delta K_i\} \leq (P_i E_i \Delta_{B_i}(t) + N_i^T \Delta_{K_i}^T(t) F_i^T M_i^T) \times$$
$$(P_i E_i \Delta_{B_i}(t) + N_i^T \Delta_{K_i}^T(t) F_i^T M_i^T)^T \ .$$

Substituting the above inequalities into (22), it yields

$$A_i^T P_i + P_i A_i + K_i^T B_i^T P_i + P_i B_i K_i + 2\lambda_0 P_i + \varepsilon_{1i} P_i D_i D_i^T P_i + \varepsilon_{1i}^{-1} L_i^T L_i +$$
$$\varepsilon_{2i} P_i B_i F_i F_i^T B_i^T P_i + \varepsilon_{2i}^{-1} N_i^T N_i + \varepsilon_{3i} P_i E_i E_i^T P_i + \varepsilon_{3i}^{-1} K_i^T M_i^T M_i K_i +$$
$$(P_i E_i \Delta_{B_i}(t) + N_i^T \Delta_{K_i}^T(t) F_i^T M_i^T)(P_i E_i \Delta_{B_i}(t) + N_i^T \Delta_{K_i}^T(t) F_i^T M_i^T)^T < 0 \ .$$
$$(24)$$

Let us introduce the matrix $X_i = P_i^{-1}$ and consider the change of variable $W_i = K_i X_i$. Then, pre- and post-multiplying (24) by P_i^{-1}, and using Lemma 1, it can be deduced that

$$
\begin{bmatrix}
\Gamma_i & X_i L_i^T & X_i N_i^T & W_i^T M_i^T & E_i \Delta_{B_i}(t) + X_i N_i^T \Delta_{K_i}^{\mathrm{T}}(t) F_i^T M_i^T \\
* & -\varepsilon_{1i} I & 0 & 0 & 0 \\
* & * & -\varepsilon_{2i} I & 0 & 0 \\
* & * & * & -\varepsilon_{3i} I & 0 \\
* & * & * & * & -I
\end{bmatrix} < 0 . \tag{25}
$$

Furthermore, by simply algebraic manipulation to (25) gives

$$
\begin{bmatrix}
\Gamma_i & X_i L_i^T & X_i N_i^T & W_i^T M_i^T & 0 \\
* & -\varepsilon_{1i} I & 0 & 0 & 0 \\
* & * & -\varepsilon_{2i} I & 0 & 0 \\
* & * & * & -\varepsilon_{3i} I & 0 \\
* & * & * & * & -I
\end{bmatrix} + He \left\{
\begin{bmatrix}
E_i & X_i N_i^T \Delta_{K_i}^T(t) \\
0 & 0 \\
0 & 0 \\
0 & 0 \\
0 & 0
\end{bmatrix}
\begin{bmatrix}
0 & 0 & 0 & 0 & \Delta_{B_i}(t) \\
0 & 0 & 0 & 0 & F_i^T M_i^T
\end{bmatrix}
\right\} < 0 . \tag{26}
$$

Using Lemma 2 to (26), we can get

$$
\begin{bmatrix}
\Gamma_i + \varepsilon_{4i}^{-1}(E_i E_i^{\mathrm{T}} + X_i N_i^T N_i X_i) & X_i L_i^T & X_i N_i^T & W_i^T M_i^T & 0 \\
* & -\varepsilon_{1i} I & 0 & 0 & 0 \\
* & * & -\varepsilon_{2i} I & 0 & 0 \\
* & * & * & -\varepsilon_{3i} I & 0 \\
* & * & * & * & -\Xi(\varepsilon_{4i})
\end{bmatrix} < 0 . \tag{27}
$$

By applying Lemma 1, it is easy to verify that (27) is equivalent to (23). □

Remark 1. In Theorem 3, there are many parameters to choose before solving the LMIs condition (23). As we known, λ_0 is the stability margin of single subsystem, which is determined prior. In (23), the inequality holds unless $\Xi(\varepsilon_{4i}) > 0$, then we will choose $0 < \varepsilon_{4i} < 1$. In general, the other parameters are always not too large, such as we can choose them as 1.

Next, the following theorem will give the sufficient condition for the existence of non-fragile state feedback controller with multiplicative control gain perturbations (5).

Theorem 4. *Consider system (1), for given scalar $\varepsilon_{1i} > 0, \varepsilon_{2i} > 0, \varepsilon_{3i} > 0, \varepsilon_{4i} > 0, \lambda_0 > 0, \forall i \in M$, if there exist $X_i > 0, W_i$, and state feedback gain matrix $K_i = W_i X_i^{-1}$ (if it exists), such that the LMIs condition*

$$
\begin{bmatrix}
\Gamma_i' & X_i L_i^T & W_i^T N_i'^T & W_i^T M_i^T & 0 & E_i & W_i^T N_i'^T \\
* & -\varepsilon_{1i} I & 0 & 0 & 0 & 0 & 0 \\
* & * & -\varepsilon_{2i} I & 0 & 0 & 0 & 0 \\
* & * & * & -\varepsilon_{3i} I & 0 & 0 & 0 \\
* & * & * & * & -\Xi'(\varepsilon_{4i}) & 0 & 0 \\
* & * & * & * & * & -\varepsilon_{4i} I & 0 \\
* & * & * & * & * & * & -\varepsilon_{4i} I
\end{bmatrix} < 0 \tag{28}
$$

holds, then the closed-loop system (21) with multiplicative control gain perturbations (5) is exponentially stable with stability margin λ under arbitrary

switching signal in terms of average dwell-time satisfying $\tau_D \geq \tau_D^*$. *In (28)*,
$\Gamma_i' = He\{A_iX_i+B_iW_i\}+\varepsilon_{1i}D_iD_i^{\mathrm{T}}+\varepsilon_{2i}B_iF_i'F_i'^{\mathrm{T}}B_i^{\mathrm{T}}+\varepsilon_{3i}E_iE_i^{\mathrm{T}}+2\lambda_0X_i$, $\Xi'(\varepsilon_{4i}) = I - \varepsilon_{4i}(I + M_iF_i'F_i'^{\mathrm{T}}M_i^{\mathrm{T}})$.

Proof. The proof is similar to that of Theorem 3. For the sake of space, the detailed proof is omitted here.   □

4 Conclusions

In this paper, based on average dwell-time method, the robust non-fragile state feedback controllers for a class of switched linear uncertain systems have been designed. The closed-loop systems can be exponentially stable via switching with the designed feedback controllers. The proposed method is feasible and convenient by just solving the LMIs condition. In addition, the proposed method can improve the controller performance. However, the method needs that every subsystem has some stability margin, which remains to be further investigated.

Acknowledgments. This work was supported in part by National Natural Science Foundation of China (61104068), Natural Science Foundation of Jiangsu Province (BK2010200), China Postdoctoral Science Foundation (2012M511176) and Research Fund for the Doctoral Program of Higher Education of China (20090092120027) and (20110092110021).

References

1. Hespanha, J.P.: Uniform stability of switched linear systems: extensions of LaSalles invariance principle. IEEE Trans. Autom. Control 49, 470–482 (2004)
2. Liberzon, D.: Switching in systems and control. Birkhauser, Basel (2003)
3. Liberzon, D., Morse, A.S.: Basic problem in stability and design of switched systems. IEEE Control Syst. Mag. 19, 59–70 (1999)
4. Morse, A.S.: Supervisory control of families of linear setpoint controllers–part I: exact matching. IEEE Trans. Autom. Control 41, 1413–1431 (1996)
5. Sun, Z.D., Ge, S.S.: Analysis and synthesis of switched linear control systems. Automatica 41, 181–195 (2005)
6. Xie, G.M., Wang, L.: Necessary and sufficient conditions for controllability and observability of switched impulsive control systems. IEEE Trans. Autom. Control 49, 960–966 (2004)
7. Zhang, L., Shi, P., Boukas, E.K., Wang, C.: Robust L_2-L_1 filtering for switched linear discrete time-delay systems with polytopic uncertainties. IET Control Theory Appl. 1, 722–730 (2007)
8. Zhao, J., Hill, D.J.: On stability, L_2 gain and H1 control for switched systems. Automatica 44, 1220–1232 (2008)
9. Hespanha, J.P., Morse, A.S.: Stability of switched systems with average dwell-time. In: 38th IEEE Conf. Decision and Control, pp. 2655–2660. IEEE Press, New York (1999)
10. Ishii, H., Francis, B.A.: Stabilizing a linear system by switching control with dwell-time. IEEE Trans. Autom. Control 47, 1962–1973 (2002)

11. Zhai, G.S., Hu, B., Yasuda, K., Michel, A.: Stability analysis of switched delayed systems with stable and unstable subsystems: an average dwell time approach. In: Proceedings of 2000 American Control Conference, pp. 200–204. IEEE Press, New York (2000)
12. Wang, R., Feng, J.X., Zhao, J.: Design methods of non-fragile controller for a class of uncertain switched systems. Control and Decision 21, 735–738 (2006) (in Chinese)
13. Branicky, M.S.: Stability of switched and hybrid systems. In: 33rd Conference on Decision and Control, pp. 3498–3503. IEEE Press, New York (1994)
14. Branicky, M.S.: Multiple Lyapunov functions and other analysis tools for switched and hybrid systems. IEEE Trans. Autom. Control 43, 475–482 (1998)
15. Ji, Z.J., Wang, L.: Disturbance attenuation of uncertain switched linear systems. In: 43rd IEEE Conference on Decision and Control, pp. 3708–3713. IEEE Press, New York (2004)
16. Ishii, H., Francis, B.A.: Stabilizing a linear system by switching control with dwell time. IEEE Trans. Autom. Control 47, 1962–1973 (2002)
17. Geromel, J.C., Colaneri, P.: H∞ and dwell time specifications of continuous-time switched linear systems. IEEE Trans. Autom. Control 55, 207–212 (2010)
18. Allerhand, L.I., Shaked, U.: Robust control of switched linear systems with dwell time. IEEE Trans. Autom. Control 56, 381–386 (2011)
19. Hespanha, J.P., Morse, A.S.: Stability of switched systems with average dwell-time. In: 38th Conference on Decision and Control, pp. 2655–2660. IEEE Press, New York (1999)
20. Zhai, G., Hu, B., Yasuda, K., Michel, A.N.: Stability analysis of switched systems with stable and unstable subsystems: an average dwell time approach. In: Proceedings of the 2000 American Control Conference, pp. 200–204. IEEE Press, New York (2000)
21. Yang, G.H., Ding, D.W.: H∞ static output feedback control for discrete-time switched linear systems with average dwell time. Control Theory and Applications 4, 381–390 (2010)
22. Dorato, J.: Non-fragile controller design: an overview. In: Proceedings of 1998 American Control Conference, pp. 2829–2831. IEEE Press, New York (1998)
23. Famularo, D., Dorato, P., Abdallah, C.T., et al.: Robust non-fragile LQ controllers: the static state feedback case. Int. J. Contr. 73, 159–165 (2000)
24. Jadbabaie, A., Abdallah, C.T., Famularo, D., Dorato, P.: Robust, non-fragile and optimal controller design via linear matrix inequalities. In: Proceedings of the 1998 American Control Conference, pp. 2842–2846. IEEE Press, New York (1998)
25. Oya, H., Hagino, K., Mukaidani, H.: Robust non-fragile controller for uncertain linear continuous-time systems. In: 31st Annual Conference of IEEE Iudustrial Electronics Society, pp. 1–6. IEEE Press, New York (2005)
26. Keel, L.H., Bhattacharyya, S.P.: Robust, fragile, or optimal. IEEE Trans. Autom. Control 42, 1098–1105 (1997)
27. Yang, G.H., Wang, J.L., Lin, C.: H_1 control for linear systems with additive controller gain variations. Int. J. Contr. 73, 1500–1506 (2000)
28. Yu, L.: Robust control–LMI, pp. 6–22. Tsinghua University Press, Beijing (2002)

An Intelligent Variable Spraying Decision-Making System Based on Fuzzy Neural Network for Greenhouse Mobile Robot

Guoqin Gao, Haiyan Zhou, and Xuemei Niu

School of Electrical & Information Engineering, Jiangsu University,
Zhen Jiang 212013, P.R. China
{gqgao,niuxm}@ujs.edu.cn, zhouhy_888@126.com

Abstract. To improve the effective utilization rate of pesticide and reduce the pesticide residues and chemical pollution during spraying process, an intelligent decision-making method for variable spraying based on fuzzy neural network is designed according to the feature of the mobile robot spraying in greenhouse, combined with the spraying principle of variable spraying system for row-walking mobile robot. The decision system of offline training fuzzy neural network is built by integrating the information of the level of plant diseases and insect pests, the distance and area of spraying target. The simulation results show that the fuzzy neural network intelligent decision-making method can realize real-time and quick decisions by off-line training. It has the greater decision accuracy than the fuzzy decision system on the samples not appearing in training because of its strong adaptability and generalization ability and has a good fit for the uncertain work environment in greenhouse.

Keywords: intelligence decision-making, fuzzy neural network, variable spraying, mobile robot.

1   Introduction

In the 21 century, great pressure in supplying agricultural and sideline products, low utilization of water and land resources and increasingly competition in international agriculture market require China to develop precise agriculture. Greenhouse is an important part of precise agriculture, so it is significant to study its effective variable spraying decision system.

Greenhouse environment is thick airtight, high temperature, high humidity, and easy to cause the spraying operator poisoning [1], thus it is important to study the variable spraying adapting to the greenhouse environment to improve the usage rate of spraying pesticide, and guarantee good health of operators. At present, greenhouse pesticides are sprayed in the extensive way. The spraying scheme is mostly decided by operator's experience, which is heavy workload, and influenced by subjective factors. For a long time, the technology of using pesticide still stays on the level of large capacity and large droplets, and the

K. Li et al. (Eds.): ICSEE 2013, CCIS 355, pp. 257–265, 2013.

pesticide utilization ratio is very low [2]. Therefore, there is an urgent need to research and apply the variable spraying technology to agricultural production.

The so-called variable spraying technology refers to the technology of adjusting the injection quantity with the change of the target information. Various variable spraying systems have been designed by scholars both at home and abroad for different demands. In [3], a multi-nozzle air blast spray was designed with three electromagnetic valves and three ultrasonic sensors, which implements the variable spraying according to the area change of grape leaves, but it lacks the ability of disease information acquisition. A fuzzy method was proposed in [4], for the lack of learning and self-adaption, the method wasnt applied to greenhouse with harsh environment. An adaptive neural decision method was put forward in [5], setting weed area and motion speed as inputs quantity and spray volume as output, and using the measured data to train the generated initial fuzzy system, however, the changes of the speed will lead to the changes of spraying system pressure, it is difficult to control the spraying performance. It is hard for the existing decision-making methods to get a desirable effect in the complicated uncertain greenhouse. Consequently, in this paper, an intelligent decision-making method combining artificial neural network with fuzzy control was put forward. An intelligent decision-making system of off-line training fuzzy neural network was constructed, and the spraying quantity was calculated according to the obtained information of target features. It offers a new decision-making method for the variable spraying researches of greenhouse mobile robots.

2   System Compositions

Greenhouse is a complex system. There are many factors influencing pesticide spraying, such as crop area, distance, the level of plant diseases and insect pests, temperature, humidity and so on. Theoretically, the more impact factors as inputs of the decision maker, the better effect of decision will be got. However, with the inputs increasing, the structure of decision will be more complex, and the fuzzy rule will be made more difficultly. Among the numerous factors, crop area, distance and the level of plant diseases and insect pests are the most influential factors. According to the cultivation method and growth features of band crop in modern greenhouse, combined with the spraying principle of mobile robots, the spraying quantity mainly decided by the crop area, distance and damage degree to implement the rapid and effective decision-making.

The variable spraying system of mobile robot is consisted of an image collector, an ultrasonic sensor, a decision maker, a controller, an electromagnetic valve, a flow sensor, a sprayer, etc. The system structure is shown in Fig. 1.

The spraying quantity is calculated by the decision maker according to the got target information, and is compared with the actual flow measured by the flow sensor. According to results of the comparison, the PWM (Pulse Width Modulation) [6] duty cycle is adjusted real-timely to drive the electromagnetic valve and realize the variable spraying.

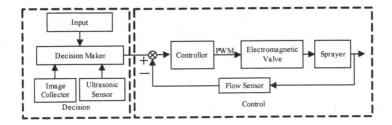

Fig. 1. Structure of the variable spraying system

3   Decision-Making System of Variable Spraying

3.1   Design of Decision-Making System

To realize the automatic variable spraying, the decision is the key link and the decision precision has a vital role in the implementation of variable spraying. In this paper, an intelligent decision-making system based on the offline training fuzzy neural network is innovative designed. According to the bond crop in greenhouse, it fuses not only the information of distance and area, but also the level of plant diseases and insect pests in order to make the uncertain greenhouse mobile robot spray efficiently and accurately, reduce the use of pesticides, improve the pesticides efficiency, and cuts down the pollution of environment. The decision-making system of variable spraying is shown in Fig. 2.

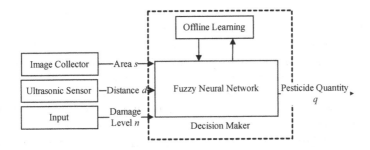

Fig. 2. Decision-making system of variable spraying

The camera sends the collected crop image to the computer through the USB interface, and computer extracts and calculates the crop row from the background to decide the spraying area. In addition, the ultrasonic sensor is used to measure the distance information of spraying target. Because of the complexity working environment, especially the existences of uncertainty illumination condition, and individual difference, it is difficult to distinguish and locate the damage position quickly by collecting and processing the images real-timely. In order to improve the efficiency and practicality of spraying robots, and considering the features of plant diseases and insect pests in common glasshouse, the level of plant diseases and insect pests are input through the computer user interface. On that basis, an offline training fuzzy neural network with self-learning and adaptive capacity is constructed, which fuses the information of area, distance and the level of plant diseases and insect pests, and implements quickly an intelligent decision for the spraying robot on pesticide quantity.

3.2   Design of Fuzzy Neural Network

The variable spraying process is complex non-linear, so it is hard to build the mathematical model. Multi-input fuzzy rules aimed at the distance, area, and damage level are not only difficult to establish, but also cant adaptively adjust with the change of greenhouse environment. Hence, the decision-making method based on fuzzy neural network is proposed to improve the ability of adapting to the change of environment and impact factors in the conventional fuzzy decision. The architecture of the fuzzy neural network is shown in Fig. 3.

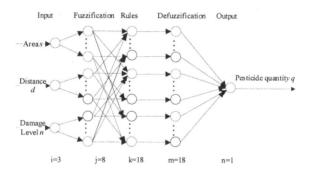

Fig. 3. Architecture of the fuzzy neural network

Use fuzzy neural network [7,8] structure to handle the input fuzzification, fuzzy inference, fuzzy rules, and defuzzification. There are five layers.

Layer 1 (input layer): there are three nodes representing the crops area 's', the spraying distance 'd', the damage level of disease and insect 'n'. In the layer, each neuron represents an input signal, the number of neurons is equal to the number of variables appeared in the premised fuzzy rules, it passes input vector directly to the next layer. The ith neuron connects with the ith unit of input variable X.

Layer 2 (fuzzification input layer): there are eight nodes, each node represents a language variable: area, the unit is m^2 , given three language variable values, namely, 'big', 'medium', 'small'. Distance, the unit is m, given three language variable values, namely, 'near', 'medium', 'far'. Damage level of plant disease and insect pest, given two language variable values, namely, 'serious', 'not serious'. In this layer, each neuron is used to simulate the membership function of input variables, and its role is to calculate the membership functions of input weight belonging to the language variable fuzzy set. Employ Gauss function as the membership function, which is expressed as:

$$y = e^{-\frac{(x-c)^2}{\sigma^2}} \tag{1}$$

$$I_{isi}^{(2)} = -\frac{(x_i^1 - c_{isi})^2}{\sigma_{isi}^2}, o_{isi}^{(2)} = \mu_i = e^{-\frac{(x_i^{(1)} - c_{isi})^2}{\sigma_{isi}^2}} \tag{2}$$

where, c_{isi} and o_{isi} represent the center and width of the membership function.

Layer 3 (fuzzification rules layer): There are 18 nodes representing 18 fuzzy rules used to complete the antecedent calculation of fuzzy rules. Each node represents one fuzzy rule, which is used to match the front layer of fuzzy rules and calculate the applicable degrees of each rule.

$$\alpha_j = \mu_1^{s_{1j}} \mu_2^{s_{2j}}, s_{1j} \in 1, 2, \ldots, m_1, s_{2j} \in 1, 2, \ldots, m_2 \tag{3}$$

$$I_k^3 = \mu_1^3 \mu_2^3 = \alpha_k, o_k^3 = I_k^3 \tag{4}$$

Layer 4 (fuzzification output layer): There are 18 nodes representing 18 applicable degrees of each rule. Each node is processed by normalization, and also can be used as the output layer weights of next network layer.

$$I_j^4 = \overline{\alpha_j} = \frac{\alpha_j}{\sum_{i=1}^{N_A} \alpha_i}, o_j^{(4)} = I_j^{(4)} = \alpha_j \tag{5}$$

Layer 5 (output layer): there is only one node representing the result of pesticide decision.

For the fuzzy segment value of input component has been determined, the parameters needed to learn are only the last layer of connection weights $\omega_{ij}(i = 1, 2, \ldots, r; j = 1, 2, \ldots, m)$, as well as the width $\sigma_{ij}(i = 1, 2, \ldots, r; j = 1, 2, \ldots, m)$ and center value c_{ij} of the second floor of the membership functions. The error cost function is defined as:

$$E = \frac{1}{2} \sum_{i=1}^{r} (y_{di} - y_i)^2 \tag{6}$$

where, y_{di} and y_i represent the output and desired output of network separately. BP algorithm is used to calculate $\frac{dE}{d\omega_{ij}}, \frac{dE}{dc_{ij}}, \frac{dE}{d\sigma_{ij}}$, and a gradient optimization algorithm is used to adjust $\omega_{ij}, c_{ij}, \sigma_{ij}$. Finally the learning algorithms of adjusting parameters are presented as:

$$\omega_{ij}(k+1) = \omega_{ij}(k) - \beta \frac{dE}{d\omega_{ij}}, j = 1, 2, \ldots, m_i \tag{7}$$

$$c_{ij}(k+1) = c_{ij}(k) - \beta \frac{dE}{dc_{ij}}, j = 1, 2, \ldots, m_i \tag{8}$$

$$\sigma_{ij}(k+1) = \sigma_{ij}(k) - \beta \frac{dE}{d\sigma_{ij}}, j = 1, 2, \ldots, m_i \qquad (9)$$

where, $i = 1, 2, \ldots, n, \beta > 0$ is the learning rate.

The training process of neural networks is that constantly adjust the weight coefficient between each layer until the anticipation error of the neural network output meets the application requirements and save the learning fuzzy neural network data. After learning, download the fuzzy neural network weights and thresholds to the decision maker.

4   Simulation Experiment and Analysis

4.1   Selection of Sample Data

The data are collected from real-time operating on greenhouse spraying site to constitute training samples and train the constructed fuzzy neural network.

Analyzing the real-time operating records, it is concluded that the actual quantity of spraying is determined by operators spraying experience combining with influence factors including the crop area, distance, damage level. The operation data, to some extent, reflects the operator's experience and strategy. Choose $\frac{2}{3}$ of the sample data to train the fuzzy neural network (the rest $\frac{1}{3}$ as the test samples), and make the trained network memory these experience and automatically generate a series of fuzzy rules, then the reasonable rate of spraying can be decided. Parts of sample dates are shown in Table 1.

4.2   Fuzzy Neural Network Training

In this paper, the model is built up by MATLAB neural network toolbox function. Select 7, 7, 7 as each input fuzzy division and Gauss function as membership function. In learning algorithm, the training goal error is defined as 0.001, and the training step is defined as 1000. The momentum and adaptive gradient descent training function Traindx is used as training function. Then the MATLAB model of fuzzy neural network is established.

Results of network training are shown in Fig. 4. It is indicated that the deviation between the actual training output and the expectancy output is very small, the fuzzy neural network reflects the decision rules well and the training result is more ideal. Target error of the network, after 191 steps training, meets the convergence requirement.

4.3   Experimental Verification

The spraying mobile robot system for greenhouse studied in this paper is shown in Fig. 5. The system is consisted of a mobile robot, a computer, a water tank for pesticide, an image collector, an ultrasonic sensor, a decision maker, a controller, an electromagnetic valve, a flow sensor, a sprayer, etc.

Table 1. Part of Sample Data

Sample Number	Area (m^2)	Distance (m)	Damage level	Pesticide Quantity $(L \cdot s^{-1})$
1	0.00656	0.2394	0	0.02125
2	0.00689	0.2315	0	0.02131
3	0.00736	0.2382	0	0.02370
4	0.02074	0.2458	0	0.02573
5	0.02579	0.3367	0	0.02948
6	0.02584	0.4285	0	0.0542
7	0.05891	0.2759	0	0.02917
8	0.08652	0.3756	0	0.05341
9	0.09010	0.4152	0	0.06324
10	0.01949	0.2521	1	0.02932
11	0.02198	0.3516	1	0.07301
12	0.01736	0.4156	1	0.05629
13	0.03403	0.2657	1	0.05438
14	0.01863	0.2981	1	0.04730
15	0.04517	0.4162	1	0.07307
16	0.06086	0.2429	1	0.04833
17	0.06103	0.3507	1	0.05924
18	0.09211	0.4128	1	0.06269

Validate the decision-making precision of the trained fuzzy neural network with the reserved test samples in the spraying mobile robot system. The error between the actual output and the sample output is shown in Fig. 6. In the same condition, the fuzzy decision system is simulated by using the same fuzzy division method and reserved test sample. The error is shown in Fig. 6.

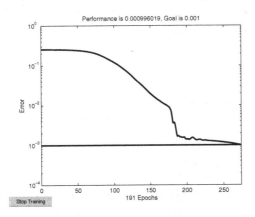

Fig. 4. Results of the network learning

Fig. 5. System of variable spraying mobile robot

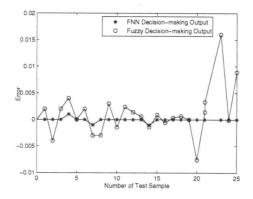

Fig. 6. Error output of test sample

The simulation experiment shows that the fuzzy neural network decision method proposed can implement high precision decision effectively and meet the requirements for mobile robot spraying pesticide in greenhouse.

5   Conclusions

In this paper, the model of variable spraying for a greenhouse mobile robot based on a fuzzy neural network is established, and an intelligent decision-making method of variable spraying based on the fuzzy neural network is proposed. The simulation results show that

1) Considering comprehensively cultivation method, growth features and the level of pests and diseases in modern greenhouse, the intelligent decision-making system proposed integrates not only the target information of distance and area, but also the damage level of plant diseases and insect pests.

2) The fuzzy neural network decision-making method proposed can realize the real-time and quick decision by offline training. It has the greater decision accuracy than the fuzzy decision system on the samples not appearing in training

because of its strong adaptability and generalization ability and has a good fit for the uncertain work environment in greenhouse.

3) Studies in this paper lay a foundation of theory and experiment for implementing the accurate spraying of greenhouse mobile robot with uncertainty so as to reduce the quantity of greenhouse pesticide, improve the effective utilization rate, and reduce the pollution of the environment and so on.

Acknowledgments. This work was financially supported by the Priority Academic Program Development of Jiangsu Higher Education Institutions (NO. 6, 2011) and Zhenjiang Municipal Agricultural Key Technology R&D Program (Grant No. NY2011013).

References

1. Nuyttens, D., Windey, S., Sonck, B.: Biosystems Engineering 89(4), 417–423 (2004)
2. Zheng, W., Ying, X.: Journal of Agricultural Mechanization Research (5), 219–221 (2008) (in Chinese)
3. Gil, E., Escola, A., Rosell, J.R.: Crop Protection 26(8), 1287–1297 (2007)
4. Shao, L., Dai, Z., Cui, H.: Transactions of The Chinese Society of Agricultural Machinery 36(11), 11011 (2005) (in Chinese)
5. Chen, S., Yin, D., Wei, X.: Journal of Drainage and Irrigation Machinery Engineering 29(3), 272–276 (2011) (in Chinese)
6. Deng, W., Ding, W.: Transactions of the Chinese Society for Agricultural Machinery 39(6), 77–80 (2008) (in Chinese)
7. Nauck, D.: International Journal of Approximate Reasoning 32(2/3), 103–130 (2003)
8. Li, M., Sethi, I.K.: Confidence-based active learning. IEEE Transaction on Pattern Analysis and Machine Intelligence 28(8), 1251–1261 (2005)

Research on Vector Control and PWM Technique of Five-Phase PMSM

Jianyong Su[1,2], Hongwei Gao[1], Pingzhi Zhao[3], and Guijie Yang[1]

[1] School of Electrical Engineering and Automation, Harbin Institute of Technology, Harbin, China
[2] Shenzhen Academy of Aerospace Technology, Shenzhen, Guangdong, China
[3] Beijing Leader Harvest Technologies Co, Ltd, Beijing, China

Abstract. The mathematic model that based on vector space for five-phase permanent magnet synchronous motor (PMSM) is established. The five-phase PMSM is designed with five sets of windings in Y-connected mode shifted by 72 electrical degrees. In the model, the variables of the motor are mapped to the $\alpha1$-$\beta1$ fundamental subspace associated with the electromechanical energy conversion and $\alpha3$-$\beta3$ third harmonic subspace. Different from six-phase PMSM, the five-phase model implies that the harmonic current in $\alpha3$-$\beta3$ subspace have positive effect on electromechanical energy conversion. The vector control scheme for five-phase PMSM is presented, in which the currents in the subspace $\alpha1$-$\beta1$ and $\alpha3$-$\beta3$ are controlled in closed loop. A space vector pulse width modulation (SVPWM) algorithm of five-phase voltage-source inverter based on near two vectors is discussed. The voltage in $\alpha1$-$\beta1$ and $\alpha3$-$\beta3$ can be modulated simultaneously. Through the simulation and experiment analysis, the vector control method and PWM technique for five-phase PMSM are proved to be feasible and effective.

Keywords: Five-phase PMSM, Modeling, Vector control, PWM.

1 Introduction

In order to achieve the demand of low-voltage high-power output, there are many advantages about multi-phase motor, such as small torque ripple and suitable for fault-tolerant operation, etc [1,2]. The current is sharing to more bridges among each so that the current could be reduced on the each power device. Therefore, multi-phase motor could be used with a wide range of applications in the field that limiting output voltage and requiring high reliability [3,4].

According to the traditional three-phase PMSM, the advantages of five -phase motor are: (1)increase the torque ripple frequency and reduce the torque ripple amplitude, (2) run in the case of losing a phase, (3) reduce the harmonic current in the DC bus [5-7]. The windings structure of five-phase PMSM is separated by 72 electrical angle, as shown in Fig.1. The five-phase PMSM can be supplied by the five-phase voltage source inverter (VSI), as shown in Fig.2. Compared with the three-phase PMSM, modeling and control of five-phase motor is more complex [8,9].

K. Li et al. (Eds.): ICSEE 2013, CCIS 355, pp. 266–274, 2013.
© Springer-Verlag Berlin Heidelberg 2013

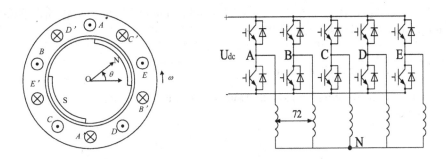

Fig. 1. Windings structure of five-phase **Fig. 2.** Five-phase voltage source inverter
PMSM

In this paper, the method of decoupling vector space is taken to establish
the mathematical model of five-phase PMSM, and the respective vector control
algorithm are discussed. Based on the structure of five-phase VSI, the SVPWM
modulation with near two vectors is applied. This method achieves that the
voltage of each subspace can be outputted at the same time, the low order
harmonics is eliminated, and the five-phase motor achieves the same control
features with the traditional three-phase PMSM.

2   Mathematic Model of Five-Phase PMSM

2.1   Model in Natural Coordinate

The voltage equation in natural coordinate of five-phase PMSM is

$$u_s = R_s i_s + \frac{d\Psi_s}{dt} \tag{1}$$

Where u_s is is the voltage matrix, $u_s = [u_A \ u_B \ u_C \ u_D \ u_E]^T$,
R_s is is the resistance matrix, $R_s = r_s \times I_s$,
i_s is is the current matrix, $i_s = [i_A \ i_B \ i_C \ i_D \ i_E]^T$,
Ψ_s is is the flux matrix, $\Psi_s = [\psi_A \ \psi_B \ \psi_C \ \psi_D \ \psi_E]^T$.
    The flux equation is

$$\Psi_s = L_s i_s + \Psi_m \tag{2}$$

Where L_s is is the inductance matrix.
    Ψ_m is the flux linkage of permanent magnet,

$$\Psi_m = [\Psi_{mA} \ \Psi_{mB} \ \Psi_{mC} \ \Psi_{mD} \ \Psi_{mE}]^T$$

The flux linkage of permanent magnet Ψ_m is

$$\Psi_m = \Psi_{m1} \begin{bmatrix} cos\theta_0 \\ cos\theta_1 \\ cos\theta_2 \\ cos\theta_3 \\ cos\theta_4 \end{bmatrix} + \Psi_{m3} \begin{bmatrix} cos3\theta_0 \\ cos3\theta_1 \\ cos3\theta_2 \\ cos3\theta_3 \\ cos3\theta_4 \end{bmatrix} \tag{3}$$

Where $\theta_i = \theta - i\alpha$, $\alpha = 2/5\pi$. Ψ_{m1} is the amplitude of fundamental flux, Ψ_{m3} the amplitude of third harmonic flux.

2.2   Transformation Matrix

The static decoupling transformation matrix on vector space of the five-phase motor is

$$
T(0) = \sqrt{\frac{2}{5}}
\begin{bmatrix}
1 & cos\alpha & cos2\alpha & cos3\alpha & cos4\alpha \\
0 & sin\alpha & sin2\alpha & sin3\alpha & sin4\alpha \\
1 & cos3\alpha & cos6\alpha & cos9\alpha & cos12\alpha \\
0 & sin3\alpha & sin6\alpha & sin9\alpha & sin12\alpha \\
\sqrt{1/2} & \sqrt{1/2} & \sqrt{1/2} & \sqrt{1/2} & \sqrt{1/2}
\end{bmatrix}
\tag{4}
$$

The variables are mapped to the static coordinate $\alpha1$-$\beta1$ subspace and $\alpha3$-$\beta3$ subspace. The amplitude doesn't change after the transformation. The fifth variable is zero because the neutral point is isolated.

The variables in the static coordinate are needed to be transformed into ones in rotating coordinate. The rotating transformation matrix is

$$
R(\theta) =
\begin{bmatrix}
cos\theta & sin\theta & 0 & 0 & 0 \\
-sin\theta & cos\theta & 0 & 0 & 0 \\
0 & 0 & cos3\theta & sin3\theta & 0 \\
0 & 0 & -sin3\theta & cos3\theta & 0 \\
0 & 0 & 0 & 0 & 1
\end{bmatrix}
\tag{5}
$$

From the $T(0)$ and $R(\theta)$, the rotating transformation matrix between natural coordinate and rotating coordinate is

$$
T(\theta) = R(\theta)T(0) = \frac{2}{5}
\begin{bmatrix}
cos\theta_0 & cos\theta_1 & cos\theta_2 & cos\theta_3 & cos\theta_4 \\
-sin\theta_0 & -sin\theta_1 & -sin\theta_2 & -sin\theta_3 & -sin\theta_4 \\
cos3\theta_0 & cos3\theta_1 & cos3\theta_2 & cos3\theta_3 & cos3\theta_4 \\
-sin3\theta_0 & -sin3\theta_1 & -sin3\theta_2 & -sin3\theta_3 & -sin3\theta_4 \\
1/2 & 1/2 & 1/2 & 1/2 & 1/2
\end{bmatrix}
\tag{6}
$$

2.3   Model in Rotating Coordinate

The voltage equation in rotating coordinate of five-phase PMSM is

$$
u_{dqs} = T(\theta)u_s = R_s i_{dqs} + \frac{d}{dt}\Psi_{dqs} - \Omega_{dqs}\Psi_{dqs}
\tag{7}
$$

where u_{dqs} is the rotating voltage matrix,

$$
u_{dqs} = T(\theta)u_s = [u_{d1}\ u_{q1}\ u_{d3}\ u_{q3}\ u_0]
$$

i_{dqs} is the rotating current matrix,

$$
i_{dqs} = T(\theta)i_s = [i_{d1}\ i_{q1}\ i_{d3}\ i_{q3}\ i_0]
$$

$\boldsymbol{\Psi}_{dqs}$ is the rotating flux matrix

$$\boldsymbol{\Psi}_{dqs} = \boldsymbol{T}(\theta)\boldsymbol{\Psi}_s = \boldsymbol{L}_{dqs}\boldsymbol{i}_{dqs} + \boldsymbol{\Psi}_{dqm} \tag{8}$$

$\boldsymbol{\Omega}_{dqs}$ is the speed matrix.

$$\boldsymbol{\Omega}_{dqs} = \frac{d\boldsymbol{T}(\theta)}{dt}\boldsymbol{T}^{-1}(\theta) \tag{9}$$

In (8), the inductance matrix is

$$\boldsymbol{L}_{dqs} = \boldsymbol{T}(\theta)\boldsymbol{L}_s\boldsymbol{T}^{-1}(\theta) = \begin{bmatrix} L_{d1} & 0 & L_{13} & 0 & 0 \\ 0 & L_{q1} & 0 & L_{13} & 0 \\ L_{13} & 0 & L_{d3} & 0 & 0 \\ 0 & L_{13} & 0 & L_{q3} & 0 \\ 0 & 0 & 0 & 0 & L_{ls} \end{bmatrix} \tag{10}$$

The flux matrix of magnet is

$$\boldsymbol{\Psi}_{dqm} = \boldsymbol{T}(\theta)\boldsymbol{\Psi}_m = \begin{bmatrix} 0 & \psi_{m1} & 0 & \psi_{m3} & 0 \end{bmatrix}^T \tag{11}$$

The speed matrix is

$$\boldsymbol{\Omega}_{dqs} = \frac{d\boldsymbol{T}(\theta)}{dt}\boldsymbol{T}^{-1}(\theta) = \omega \begin{bmatrix} 0 & 1 & 0 & 0 & 0 \\ -1 & 0 & 0 & 0 & 0 \\ 0 & 0 & 0 & 3 & 0 \\ 0 & 0 & -3 & 0 & 0 \\ 0 & 0 & 0 & 0 & 0 \end{bmatrix} \tag{12}$$

2.4   Electromagnetic Torque Equation

With the virtual displacement method in the electric motor theory, the electromagnetic torque is

$$T_e = P\left[\frac{1}{2}\boldsymbol{i}_s^T\frac{\partial \boldsymbol{L}_s}{\partial \theta}\boldsymbol{i}_s + \boldsymbol{i}_s^T\frac{\partial \boldsymbol{\Psi}_m}{\partial \theta}\right] \tag{13}$$

From (7)-(12), the electromagnetic torque T_e is calculated,

$$T_e = \frac{5P}{2}\left[\begin{matrix} \psi_{m1}i_{q1} + 3\psi_{m3}i_{q3} + (L_{d1} - L_{q1})i_{d1}i_{q1} \\ + 3(L_{d3} - L_{q3}) + 2L_{13}(i_{d1}i_{q3} - i_{q1}i_{d3}) \end{matrix}\right] \tag{14}$$

When the field oriented control, $i_d = 0$ is adopted, as

$$T_e = \frac{5P}{2}(\psi_{m1} + 3\psi_{m3}i_{q3}) \tag{15}$$

The torque is composed of traditional foundational torque $5P/2\psi_{m1}i_{q1}$ and three harmonic torque $15P/2\psi_{m3}i_{q3}$. So the three harmonic can be applied to increase the output torque.

3   Mathematic Model of Five-Phase PMSM

With the transformation matrix (6), the voltage space vector in the $\alpha1$-$\beta1$ sub-space and $\alpha3$-$\beta3$ subspace is defined.

$$V_{\alpha1\beta1} = v_{\alpha1} + jv_{\beta1} = \frac{2}{5}(V_{AN} + e^{j\alpha}V_{BN} + e^{j2\alpha}V_{CN} + e^{j3\alpha}V_{DN} + e^{j4\alpha}V_{EN}) \quad (16)$$

$$V_{\alpha3\beta3} = v_{\alpha3} + jv_{\beta3} = \frac{2}{5}(V_{AN} + e^{3j\alpha}V_{BN} + e^{j\alpha}V_{CN} + e^{j4\alpha}V_{DN} + e^{j2\alpha}V_{EN}) \quad (17)$$

There are total $32(2^5)$ voltage space vectors with different combination of five-phase VSI. These vectors are divided into four groups according to vector amplitude.

The vector mapping is different in $\alpha1$-$\beta1$ subspace and $\alpha3$-$\beta3$ subspace. The vector distribution in $\alpha1$-$\beta1$ subspace is shown in Fig.3. The vector distribution in $\alpha3$-$\beta3$ subspace is shown in Fig.4.

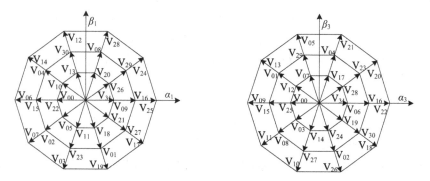

Fig. 3. Voltage vectors in $\alpha1$-$\beta1$ subspace   **Fig. 4.** Voltage vectors in $\alpha3$-$\beta3$ subspace

In the SVPWM algorithm of three-phase VSI, the voltage vector is implemented by the near two vectors, according to the volt-second balance principle. In the five-phase VSI, the voltage vector can be implemented by the near two largest vectors and two zero vectors.

In a modulation section, the period is defined as T_s. The two largest vectors are V_k and V_{k+1}. The equations are established.

$$T_s V_{ref} = T_k V_k + T_{k+1} V_{k+1}$$
$$T_s = T_k + T_{k+1} + T_0 \quad (18)$$

Where T_k is the action time of V_k in one modulation period. T_{k+1} is the action time of V_{k+1} in one modulation period. The T_k and T_{k+1} are calculated from

$$T_k = \frac{|V_{ref}|sin(k\pi/5 - \gamma)}{|V_l|sin(k\pi/5)}T_s$$

$$T_k = \frac{|V_{ref}|sin(\gamma - (k-1)\pi/5)}{|V_l|sin(k\pi/5)}T_s \qquad (19)$$

$$T_0 = T_s - T_k - T_{k+1}$$

where $|V_l|$ is the amplitude of the largest vector.

4   Vector Control of Five-Phase PMSM

Based on decoupled mathematic model, the scheme of rotor field oriented control for five-phase PMSM is presented, shown in Fig.5.

In the Fig.5 there is the current control in the two subspaces which are $\alpha 1$-$\beta 1$ and $\alpha 3$-$\beta 3$. The field oriented control $i_d = 0$ is adopted. The $i_{d1}^* = 0$ and $i_{d3}^* = 0$ are zero. The $i_{q1}^* = 0$ is the output of speed PI regular similar to the traditional vector control of three-phase PMSM.

The difference between the three-phase and five-phase is appliance of the three harmonic torque according to (5). In the Fig.5, the $i_{q3}^* = 0$ reference is calculated according to the $i_{q1}^* = 0$. The coefficient K_{q3} from $i_{q1}^* = 0$ to $i_{q3}^* = 0$ is dependent on the rotor design and the limitation of VSI current. In most field, can be set as zero. At this time K_{q3} is zero.

After the calculation of four PI regulators, the voltage reference $i_{q3}^* = 0$ is obtained in the corresponding subspace. Before the SVPWM calculation, these reference need to be converted in the stationary coordinate by the $T^{-1}(\theta)$ module.

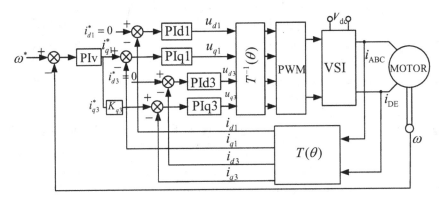

Fig. 5. Rotor field oriented control of five-phase PMSM

5   Simulation and Experiment Analysis

The SVPWM waveform in the No.1 section is shown in Fig.5. The action time T_0 and T_2 is calculated by (19). In one switching period, each device switches on and switches off only once.

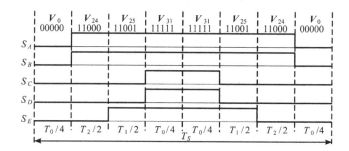

Fig. 6. Switching signals of VSI in No.1 section

The algorithm raised in the paper is experimentally verified. In the experiment, the PMSM parameters are: $Rs = 17\Omega$, $Ld1 = 55mH$, $Lq1 = 81mH$, $Ld3 = 45mH$, $Lq3 = 32mH$, $L13 = 11mH$, $J = 0.006Kgm^2$, $P = 4$. The PWM frequency is 10 kHz.

The current of phase A and B is shown in Fig.7 when K_{q3} is zero in steady state. The current of phase A and B when $K_{q3} = 0.15$ is shown in Fig.7 in steady state.

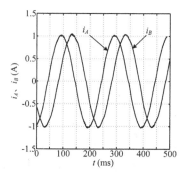
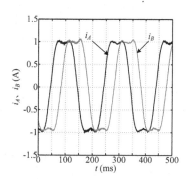

Fig. 7. Phase A and B when K_{q3} is zero

Fig. 8. Phase A and B when K_{q3} isn't zero

In Fig.7, the current only contains the fundamental component. In Fig.8, the current also contains the three harmonic component, which can increase the electromagnetic torque because of the three harmonic flux of the air gap. And the amplitude of phase current doesn't need to increase in this condition.

The dynamic responses of electromagnetic torque are shown in Fig.9 and Fig.10. The Fig.9 corresponds with $K_{q3} = 0$, the fig.10 corresponds with $K_{q3} = 0.15$. In these experiments, the current references i_{q1}^* and i_{q3}^* varies in step mode. And the amplitude of phase current is as same as Fig.7 and Fig.8.

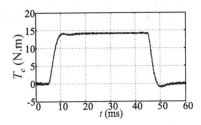

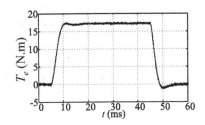

Fig. 9. Step response of Te when $K_{q3} = 0$

Fig. 10. Step response of Te when $K_{q3} \neq 0$

The electromagnetic torque when K_{q3} isnt zero is larger than the one when K_{q3} is zero. The comparison of Fig.9 and Fig.10 proves previous conclusion about the torque increase due to the three harmonic current injection.

6   Conclusion

The five-phase permanent magnet synchronous motors could be used with a wide range of applications in the low voltage and high power fields. In this paper,

(1) The vector space decoupling is applied respectively to establish mathematical model of the five-phase PMSM. Associated physical quantities are converted to the fundamental $\alpha 1$-$\beta 1$ subspace and three harmonic $\alpha 3$-$\beta 3$ subspace. And the three harmonic current injection can increase the electromagnetic torque.

(2) The control method of the near two largest vectors is used, then the voltage of the $\alpha 1$-$\beta 1$ subspace and the $\alpha 3$-$\beta 3$ subspace can also be converted at the same time.

(3) Experimental results are consistent with the theoretical analysis and verify the effectiveness of the algorithm.

References

1. Sign, G.K.: Multi-Phase Induction Machine Drive Research-A Survey. Electric Power Systems Research 147, 61–147 (2002)
2. Leila, P.: On Advantages of Multi-Phase Machines. In: 31st Annual Conference of IEEE Industrial Electronics Society, Raleigh, USA, pp. 1574–1579 (2005)
3. Leila, P., Hamid, A.: Five-phase permanent-magnet motor drives. IEEE Transactions on Industry Applications 41, 30–37 (2005)
4. Toliyat, H.A.: Analysis and simulation of five-phase variable-speed induction motor drives under asymmetrical connections. IEEE Transactions on Power Electronics 13, 748–756 (1998)
5. Levi, E.: Multiphase Electric Machines for Variable-Speed Applications. IEEE Transactions on Industrial Electronics 55, 1893–1909 (2008)

6. Yu, F., Zhang, X., Li, H., Xiang, D.: Discontinuous space vector PWM control of five-phase inverter. Transactions of China Electrotechnical Society 21, 26–30 (2006)
7. Yu, F., Zhang, X., Li, H., Song, Q.: Space vector PWM control of five-phase inverter. Proceedings of the CSEE, 40–46 (2005)
8. Zhao, P., Yang, G., Liu, C.: Optimal SVPWM algorithm for five-phase VSI. Electric Machines and Control 13, 516–522 (2009)
9. Xu, Q., Jia, Z., Xiong, Y.: A fast algorithm of space vector PWM. Power Electronics 3, 46–48 (2000)

A Heuristic Approach Based on Shape Similarity for 2D Irregular Cutting Stock Problem

Yanxin Xu[*], Genke Yang, and Changchun Pan

Department of Automation, Shanghai JiaoTong University, Dongchuan Road. 800,
200240 Shanghai, China
iamxuyanxin@sjtu.edu.cn

Abstract. Cutting stock problem is an important problem that arises in a variety of industrial applications. This research constructs an irregular-shaped nesting approach for two-dimensional cutting stock problem. The techniques of shape similarity are utilized, drawn from computer vision and artificial intelligence. These techniques enable the approach to find potential matches of the unplaced pieces within the void regions of the sheet, and thus the packing density and the performance of solutions are highly improved. The proposed approach is able to deal with complex shapes in industrial application and achieve high-quality solution with shorter computational time. We evaluate the proposed method using 15 established benchmark problems available from the EURO Special Interest Group on Cutting and Packing. The results demonstrate the effectiveness and high efficiency of the proposed approach.

Keywords: Cutting Stock Problem, Grid Approximation, Shape Similarity, Fourier Descriptor.

1 Introduction

Cutting and Packing Problem are a large family of problems arising in a wide variety of industrial applications, including the cutting of standardized stock units in the wood, steel and glass industries, packing on shelves or truck beds in transportation and warehousing, and the paging of articles in newspapers. There are many classic cutting and packing problems, including knapsack problem, bin packing and cutting stock problem etc. In this paper, we focus on the cutting stock problem (CSP). In CSP, a number of two-dimensional pieces must be cut from a couple of same stocks. The objective is to minimize the number of stocks. Using the typology of Wäscher, this is a Two-Dimensional Single Stock-Size Cutting Stock Problems (2DCSP) [1]. The 2DCSP has been proved to be NP-hard [2].

The 2DCSP can be defined as follows: Given a set L = (a1, a2, ..., an) of regular and irregular pieces to be cut (size of each piece $s(a_i) \in (0, A_o]$) from a set of rectangular stock-cutting sheets (objects) of size A_o (with fixed Length L and Width W), the CSP is to find cutting patterns to minimize the number of objects used. Typical assumptions are summarized as:

---

[*] Corresponding author.

K. Li et al. (Eds.): ICSEE 2013, CCIS 355, pp. 275–283, 2013.
© Springer-Verlag Berlin Heidelberg 2013

1. All pieces must be within the object (closeness);
2. Pieces must not overlap with each other (disjoint);
3. In this research, rotation of piece is allowed. And each piece has eight orientations: 0°, 45°, 90°, 135°, 180°, 225°, 270° and 315°.

This paper is organized as follows. A brief review of previous work in the field is presented in Section 2. The outline of our method is introduced and a novel heuristic placement strategy based on shape similarity is presents in detail in section 3. Section 4 gives experimental results on benchmark problems from the literatures that demonstrate the capabilities of the proposed approach. In Section 5 the research is concluded and possible issues for future work are suggested.

2   Literature Review

The Cutting and Packing problem involving irregular shape is also called nesting problem. A number of approaches have been proposed to tackle different nesting problems. We only give a brief discussion on some of the most interesting approaches previously presented in the literature.

The most visible attribute of nesting problems and the first obstacle researchers come up against is the geometry. As a result, developing a set of tools to assimilate the geometry is a non-trivial task and potentially a barrier that stifles academic research in this area. There exist a couple of solutions to this problem such as Rectangle Enclosure, Orbital Sliding of No Fit Polygon, Minkowski Sum of No Fit Polygon, and Phi Function [3-7].

In terms of solution methods a number of approaches are proposed depending on the type and the size of the problem. There have been many different strategies for producing solutions to the irregular cutting stock problem. These include exact algorithm (e.g. linear programming, dynamic programming, column generation etc.), heuristic placement strategy, meta-heuristic search techniques, and other novel approaches. For less constrained and simpler packing tasks, exact algorithms are developed along with problem-specific heuristic procedures [8]. For more complex packing tasks, heuristic search methods have been applied successfully for their solution. Heuristic placement strategy such as bottom-left (BL), bottom-left-fill is proposed to supply a rule for pieces to be placed on sheet [9-11]. Furthermore, this research utilizes techniques of shape similarity drawn from computer vision and artificial intelligence, and achieves high-quality solutions with shorter computational times [12].

The 2DCSP is NP hard due to the combinatorial explosion encountered as the size of problem increases. As a result, a number of published solution approaches focus on heuristic and meta-heuristics methodologies. Meta-heuristics are general frameworks for heuristics in solving combinatorial optimization problems. These meta-heuristic approaches include simulated annealing, tabu search, neural networks and genetic algorithms [13-16].

Although numerous approaches based on computational geometric description have obtained good performance, computational complexity for large and complex data sets is yet a huge difficulty. We adopt another method to represent shapes called grid approximation, in which pieces are represented by two-dimensional matrices. With use of grid approximation, it's not necessary to introduce additional routines to identify enclosed areas and geometric tool to detect overlap. The grid approximation is used in several literatures [17, 18].

3   Methodology

In 2DCSP, two kinds of heuristics are generally used, namely the selection heuristic for selecting pieces and objects, and the placement heuristic for placing pieces and objects. Many heuristics have been studied in the literatures and have their own superiority in some instances. In this research, Best Fit and Bottom Left irregular are respectively chosen to be the selection heuristic and placement heuristic. Firstly, it is decided by heuristic Best Fit that which stock sheet is chosen for the allocating piece. Then a hybrid approach combining heuristic Bottom Left irregular and other placement strategy is proposed to place the piece on this stock.

The proposed approach is composed of three steps. Firstly, the grid approximation is used to represent irregular-shaped pieces in two-dimensional matrices. The geometry of the stocks and pieces are converted into discrete form in order to make the nesting process faster and the actual geometry of the sheets and pieces independent. Secondly, a hybrid approach combining with heuristic Bottom Left irregular and a two-stage placement strategy is used to pack the ordered pieces. Finally, void regions are generated between the packed pieces and stock sheet after some pieces are placed. Then these void regions are matched with the unpacked pieces, according to their similarity of shape. Among the pieces that can be allocated in the void region, the most similar one will be allocated on the void region. A detailed description of every step is showed in following sections.

In order to evaluate the resulting solutions, we consider the ratio of the total area of all the allocated shapes to the area of the occupied stocks as a performance measure called packing density (PD). Its maximum value is 1, when there is no waste of resource material.

$$PD = \frac{\text{Area of Allocated Pieces}}{\text{Area of Occupied Stocks}}.$$

3.1   Pre-layout Phase

The operation of the method is divided into two phases: the pre-layout phase and the layout phase. In the pre-layout phase, the pieces are represented as a matrix by taking the grid approximation method and the initial sequence and orientation of the pieces based on their geometrical features are determines.

In this research, the grid approximation, a digitized representation technique, is used to represent multiple-shaped pieces including convex and concave. The matrix representation approach was proposed by Dagli 1990[19]. Each piece is represented by a matrix of size K by L which is the smallest rectangular enclosure of the irregular-shaped piece. The detailed technique applied in the paper is referred to W.K. Wong and Z.X. Guo 2009. By using this technique, each piece is enclosed by an imaginary rectangle for the sake of obtaining the reference points during the nesting process. Then, this particular rectangular area is divided into a uniform grid of 1mm×1mm size. In the case, the value of a pixel is '1' when the material of the sheet is occupied, otherwise the value of the pixel is '0'. $P_L^{(k)}$ and $P_W^{(k)}$ denote the length and the width of an enclosing rectangle corresponding to the piece p_k. R=1 mm, denotes the square side of a piece. Fig.1 gives an example of grid approximation.

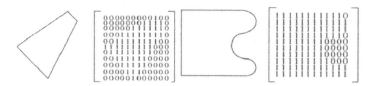

Fig. 1. Binary representations

Similar to the piece representation, the object is discretized as a finite number of equal-size pieces of size R^2. The object with a two-dimensional matrix of size $A_W^{(k)} \times A_L^{(k)}$ is represented as follows:

$$A^{(k)} = \begin{pmatrix} a_{11}^{(k)} & a_{12}^{(k)} & \cdots & a_{1A_L^{(k)}}^{(k)} \\ a_{21}^{(k)} & a_{22}^{(k)} & \cdots & a_{2A_L^{(k)}}^{(k)} \\ \vdots & \vdots & \ddots & \vdots \\ a_{A_W^{(k)}1}^{(k)} & a_{A_W^{(k)}1}^{(k)} & \cdots & a_{A_W^{(k)}A_L^{(k)}}^{(k)} \end{pmatrix}, \text{ where } A_W^{(k)} = \frac{P_W^{(k)}}{R} \text{ and } A_L^{(i)} = \frac{P_L^{(k)}}{R}$$

For each entry, $\quad a_{i,j}^{(k)} = \begin{cases} 1 \text{ , if pixel } (i,j) \text{ is occupied} \\ 0 \text{ , otherwise} \end{cases}$.

The initial sequence of the pieces is determined according to their area. The initial orientation for each packing piece is confirmed by the MRE (Minimum Rectangular Enclosure) of each piece. It is worth mentioning that the method does not replace the piece shapes by their MREs. Rather, it uses MREs as additional information for orienting the pieces.

3.2   Layout Phase

The placement of the pieces follows a single-pass placement strategy and takes place in a sequential manner. The manner entails that the method considers only one layout. In layout phase, a hybrid approach combining with heuristic Bottom Left irregular and two-stage placement strategy is proposed to construct a packing pattern according to the sequence and orientation of the packing pieces. Furthermore, a packing approach based on shape similarity draw from computer vision and artificial intelligence is adopted to increase the packing density of the nesting piece at a particular stage by reusing the complements of the allocated pieces on the sheet.

3.2.1 Hybrid Approach Combining with Heuristic Bottom Left Irregular and Two-Stage Placement Strategy

The heuristic Bottom Left irregular is used to allocate the current pieces on the stock sheet without affecting any other allocated pieces. The purpose is to minimize the total required area and the objective value can be stated as:

$$\text{objective value of sliding part} = x * W + y$$

Where x and y is respectively the coordinate of the piece along X axis and Y axis of the matrix representation of the stock sheet; W is the width of the stock sheet. By using this approach, the first packing piece is placed at the lower left-hand corner of the empty stock sheet. The following pieces are allocated along the Y axis until there is no enough space. Then a new row of pieces along X axis is formed. The two-stage placement strategy is referred to W.K. Wong and X.X. Wang 2009. In this case, the enclosing rectangles of the packing pieces are first examined, and then the packing pieces are compacted directly. The differences are that the compaction routine is done when each enclosing rectangle is placed rather than implement it in a single step after all the enclosing rectangles are allocated. This compaction routine is able to obtain a tighter packing pattern and provide more space for the unpacked pieces. The hybrid approach improves the quality of packing pattern without more computational effort.

3.2.2 Packing Approach Based on Shape Similarity

After pieces are placed, a number of void regions are generated by previous allocations, namely the complements of the allocated pieces. These void regions may be reused to increase the packing density at a particular stage. The packing approach is based on shape similarity draw from computer vision and artificial intelligence. The characteristic is the utilization of a shape similarity criterion for matching void regions of the layout and remaining pieces. As the grid of the allocated pieces in sheet partially or wholly covered is assigned by '1', then the void region between these pieces is the region assigned by '0' among '1's. The void regions are extracted from the stock sheet by scanning each row of the matrix representation of the stock plate. The result of scanning the sheet is a pool of line segments as sections of void regions. These line segments are later combined and form void regions.

Whenever a new row along X axis of the stock sheet appears as described in 3.2.1, the rectangular area with the size of the width of the stock sheet at end of the new row is considered as the active portion of the stock sheet at each allocation stage. Then the scanning operation as described above is done in the active portion.

Having defined what constitutes a void region in the layout, we can now describe how the approach attempts to find effective placements of the remaining pieces. This approach is referred to Alexandros and Murray 2007, but this research employs another method to solve it. In order to determine whether it is appropriate to place a piece to a void region, the approach considers the ratio of their areas. Firstly, the packing piece must be able to be placed into the void region. Then the similarity of shapes between the piece and void region is under considered. Thus it's effective to decrease computational time. The following equation is employed as an evaluation function (EF) for each possible placement

$$
EF = \begin{cases} M, & \text{if } 1 \leq \dfrac{\text{area of void region}}{\text{area of part}} \leq 2.5 \\ 1000, & \text{otherwise} \end{cases}.
$$

Where, M is a measure that expresses the similarity between the piece's shape and the region's shape. When the area ratio of the void region and the piece is not between the thresholds we have defined, the EF takes a great value in order to prefer other possible solutions. However, the evaluation function is just a quick and approximate measure and the actual quality of a potential placement still need to consider the performance measure to the layout by first placing the piece.

3.2.3   Fourier Descriptors Method for Shape Similarity

The problem of shape similarity has been well studied in the literature of computer vision, and a number of different methods have been proposed [20, 21]. In this paper, the Fourier descriptors method is adopted as it can achieve both good representation and easy normalization. The normalization can make the similarity measure of two shapes is invariant under translation, rotation, and change of scale.

For a given shape defined by a closed curve C. At every time t, there is a complex u(t), $0 \leq t < T$. Since u(t) is periodic, $u(t + nT) = u(t)$ can be implied. Therefore, it is possible to expand u(t) into Fourier series.

The discrete Fourier transform is given by

$$
a_n = \frac{1}{N} \sum_{t=0}^{N-1} u(t) \exp(-j2\pi nt / N) \quad n = 0,1,\dots N\text{-}1
$$

u(t) is often called a shape signature which is any one dimensional function representing shape boundary. It has been shown that FDs derived from centroid distance function outperforms FDs derived from other shape signatures. The centroid distance function r(t) is expressed by the distance of the boundary points from the centroid (x_c, y_c) of the shape and N is the number of boundary points.

$$
r(t) = ([x(t) - x_c]^2 + [y(t) - y_c]^2)^{1/2}, \text{ where } x_c = \frac{1}{N} \sum_{t=0}^{N-1} x(t),\ y_c = \frac{1}{N} \sum_{t=0}^{N-1} y(t)
$$

Since shapes generated through rotation, translation and scaling of a same shape are similar shapes, shape descriptors should be invariant to these operations. Based on the above analysis of FD properties, it is possible to normalize FDs into shape invariants. Now considering the following expression

$$
b_n = \frac{a_n}{a_0} = \frac{\exp(jn\tau) \cdot \exp(j\phi) \cdot s \cdot a_n^{(0)}}{\exp(j\tau) \cdot \exp(j\phi) \cdot s \cdot a_0^{(0)}} = \frac{a_n^{(0)}}{a_0^{(0)}} = b_n^{(0)} \exp[j(n-1)\tau)]
$$

Where b_n and $b_n^{(0)}$ are normalized Fourier coefficients of the derived shape and the original shape respectively. The set of magnitudes of the normalized Fourier coefficients of the shape $\{|b_n|, 0 < n < N\}$ can now be used as shape descriptors, denoted as $\{FD_n, 0<n<N\}$. The similarity between the query shape Q and the target shape T is given by the Euclidean distance d between their FDs.

$$d = (\sum_{i=1}^{N} \left| FD_i^Q - FD_i^T \right|^2)^{\frac{1}{2}}.$$

4   Performance Evaluation

All algorithms are implemented in Visual C++ and we also use the library CGAL to do some geometrical operations. The tests are performed on a computer with processor Intel Pentium Dual 1.8 GHz, 2 GB of RAM, and Windows XP operating system.

Few work for packing problems with pieces of irregular shape are done in the literature, and especially we can hardly find any related work for the 2DCSP and 2DBPP when pieces have irregular shape. As a result, we adapt some other known instances for packing problems with one open dimension to test our algorithms. The generated instances are adapted from the Two-Dimensional Irregular Strip Packing problem, and they can be found at the ES-ICUP website. In these instances, the following information is available: the quantity of pieces; the set of allowable rotations for these pieces; and, the size of the stock.

The instances from AM Del Valle .et .al 2012 are adapted and adopted in this research [22]. Table 1 presents solutions computed by our algorithm. We use the total area of the pieces divided by the area of one stock as a lower bound for the optimal solution. The rows in this table contain the following information: instance name (Name); solution value computed by our algorithm (Solution); the lower bound (LB); the difference (in percentage) on number of stocks computed by Solve 2DCSP and LB; the time spent in seconds (Time).

From Table 1 we can see that the value of the solutions returned by the algorithm is on average 38.487% larger than the lower bound (in the worst case), However, for instance DIGHE1, it is 84.375% larger. The results of the algorithm should be much closer to the optimal solutions and these differences are mainly due to the weakness of the lower bound. The instance SHIRTS took 1,232,468.32 s ($\approx$14 days) of CPU processing. The CPU time spent is on average 140896.16s. Since the complexity of the working with pieces of irregular shape is largely increased, the problem becomes harder than the working with rectangular pieces and some instances even take a long time to be solved.

It is worth to mention that all the instances are executed 30 times, except SHIRTS (2 time), SWIM (20 times) and TROUSERS (10 times) due to the high CPU time required by them. All the results presented on Table 1 are the average values of all executions. Such results show that the algorithm returns good solutions for the cutting stock problem with pieces of irregular shape. However, it requires high CPU time when solving instances with several pieces of completely irregular shape.

Table 1. Results obtained for the 2CS

Name	Solution	LB	Difference(%)	Time(s)
FU	72	56	22.222	208.82
JACKOBS1	48	38	26.315	6706.62
JACKOBS2	45	30	50.000	6877.99
SHAPES0	53	30	76.666	26681.03
SHAPES1	54	32	68.750	59581.59
SHAPES2	61	50	22.000	9393.39
DIGHE1	59	32	84.375	232.29
DIGHE2	44	29	51.724	25.07
ALBANO	82	65	26.153	6730.06
DAGLI	56	43	30.232	9284.62
MAO	47	33	42.424	7053.12
MARQUES	51	44	15.909	8278.96
SHIRTS	43	37	16.216	1232468.21
SWIM	58	34	70.588	466799.09
TROUSERS	51	42	21.428	273121.59

5   Conclusion

Cutting and Packing problems exist almost everywhere in real world situation. In this paper, a novel heuristic approach for two-dimensional irregular cutting stock problem is presented, based on grid approximation and shape similarity. The approach is mainly drawn on techniques from computer vision and artificial intelligence and has shown its capability of finding high-quality solutions. Specifically, the advantages include the following aspects: firstly, the placement approach based on grid approximation provides the system designers with an easier way to detect whether overlap occurs. Secondly, the two-stage placement strategy improves the quality of packing pattern without compromising the computational effort. Thirdly, the packing approach based on shape similarity gets higher occupancy rate and better performance compared with conventional methods. The proposed method is assessed on 15 established benchmark problems and performs very well.

Acknowledgments. The work of this paper was supported by National Natural Science Foundation of China, No.61074150 and No. 61203178.

References

1. Wäscher, G., Haußner, H., Schumann, H.: An improved typology of cutting and packing problems. Eur. J. Oper. Res. 183(3), 1109–1130 (2007)
2. Garey, M., Johnson, D.: Computers and Intractability: A Guide to the Theory of NP-Completeness. W. H. Freeman and Company, New York (1979)
3. Jakobs, S.: On genetic algorithms for the packing of polygons. European Journal of Operational Research 88(1), 165–181 (1996)

4. Gomes, A.M., Oliveira, J.F.: A 2-exchange heuristic for nesting problems. European Journal of Operational Research 141(2), 359–370 (2002)
5. Burke, E.K., Hellier, R.S.R., Kendall, G., Whitwell, G.: Irregular Packing Using the Line and Arc No-Fit Polygon. Oper. Res. 58(4), 1–23 (2010)
6. Bennell, J., Scheithauer, G., Stoyan, Y., Romanova, T.: Tools of mathematical modelling of arbitrary object packing problems. Annals of Operations Research 179, 343–368 (2010)
7. Chernov, N., Stoyan, Y.G., Romanova, T.: Mathematical model and efficient algorithms for object packing problem. Computational Geometry: Theory and Applications 43(5), 535–553 (2010)
8. Gomes, A.M., Oliveira, J.F.: Solving irregular strip packing problems by hybridizing simulated annealing and linear programming. European Journal of Operational Research 171(3), 811–829 (2006)
9. Oliveira, J.F., Gomes, A.M., Ferreira, J.S.: TOPOS – a new constructive algorithm for nesting problems. OR Spektrum 22(2), 263–284 (2000)
10. Burke, E.K., Hellier, R.S.R., Kendall, G., Whitwell, G.: A new bottom-left-fill heuristic algorithm for the two-dimensional irregular packing problem. Oper. Res. 54(3), 587–601 (2006)
11. Bouganis, A., Shanahan, M.: A vision-based intelligent system for packing 2-D irregular shapes. IEEE Transactions on Automation Science and Engineering 4(3), 382–394 (2007)
12. Bennell, J.A., Oliveira, J.F.: The geometry of nesting problems: A tutorial. Eur. J. Oper. Res. 184, 397–415 (2008)
13. Wong, W.K., Zeng, X.H., Au, W.M.R.: A decision support tool for apparel coordination through integrating the knowledge-based attribute evaluation expert system and the T-S fuzzy neural network. Expert Systems with Applications 36(2), 2377–2390 (2009)
14. Yuen, C.W.M., Wong, W.K., Qian, S.Q., Chan, L.K., Fung, E.H.K.: A hybrid model using genetic algorithm and neural network for classifying garment defects. Expert Systems with Applications 36(2), 2037–2047 (2009a)
15. Guo, Z.X., Wong, W.K., Leung, S.Y.S.: A genetic-algorithm-based optimization model for scheduling flexible assembly lines. International Journal of Advanced Manufacturing Technology 36(1-2), 156–168 (2008a)
16. Yuen, C.W.M., Wong, W.K., Qian, S.Q., Chan, L.K., Fung, E.H.K.: Fabric stitching inspection using segmented window technique and BP neural network. Textile Research Journal 79(1), 24–35 (2009b)
17. Poshyanonda, P., Dagli, C.H.: Genetic neuro-nester. Journal of Intelligent Manufacturing 15(2), 201–218 (2004)
18. Wong, W.K., Wang, X.X., Mok, P.Y., Leung, S.Y.S., Kwong, C.K.: Solving the two-dimensional irregular objects allocation problems by using a two-stage packing approach. Expert Systems with Applications 36, 3489–3496 (2009)
19. Dagli, C.H., Hajakbari, A.: Simulated annealing approach for solving stock cutting problem. In: Proceedings of IEEE International Conference on Systems, Man, and Cybernatics, pp. 221–223 (1990)
20. Sajjanhar, A., Lu, G.: A grid based shape indexing and retrieval method. Austral. Comput. J. 29, 131–140 (1997)
21. Zhang, D., Lu, G.: Content-based shape retrieval using different shape descriptors: a comparative study. In: Proc. IEEE Int. Conf. Multimedia and Expo, pp. 317–320 (2001)
22. Del Valle, A.M., De Queiroz, T.A., Miyazawa, F.K., Xavier, E.C.: Heuristics for two-dimensional knapsack and cutting stock problems with items of irregular shape. Expert Systems with Applications 39(16), 12589–12598 (2012)

Analysis and Modeling of GTAW Weld Pool Geometry

XueWu Wang

Key Laboratory of Advanced Control and Optimization for Chemical Processes,
Ministry of Education, East China University of Science and Technology,
200237 Shanghai, P.R. China
wangxuew@ecust.edu.cn

Abstract. A three dimensional vision sensing system was used to mimic the human vision system to observe the three-dimensional weld pool surface in pipe Gas Tungsten Arc Welding (GTAW) process. Novel characteristic parameters containing information about the penetration state specified by its back-side weld pool width and height were proposed based on the reconstructed three dimensional weld pool surfaces. In order to obtain the penetration status in real time conveniently, a neural network model was established to estimate the penetration based on the proposed characteristic parameters. It was found that the top-side characteristic parameters proposed can reflect the back-side weld pool parameters accurately and the neural network is capable of predicting the penetration status in real time by observing the three-dimensional weld pool surface which is beneficial for penetration control of GTAW process.

Keywords: GTAW, Three-dimensional, Characterization, Model, PCA-BP.

1   Introduction

In manual welding process, skilled welders can ensure the welding quality through compensation for deviation in the process. This can be achieved by observing the weld pool surface, which contains sufficient information about the weld quality. In contrast, more rapid and accurate welding can be achieved by automatic welding machines, which are more convenient to adjust several welding parameters simultaneously. Besides, welding manipulators and robots can also substitute for workers to finish the task in some severe environments. However, machines typically lack the visual feedback ability welders possess. Hence, a capable sensing system is the first step to realize intelligent penetration control. Some previous works were done to meet this increasing demand. These methods can be divided into indirect methods, two dimensional vision methods, and three dimensional vision methods.

Indirect methods include line scanner [1], ultrasonic [2], pool oscillation [3], infrared [4]-[5], X-ray [6], and non-transferred arc methods [7], etc. Information acquired by above indirect methods can only reflect one characteristic of weld pool, and it is often not sufficient. Besides, high temperature, contamination, extensive arc light, and noise in welding process will make most above sensing methods invalid in some conditions. Vision sensing method is more similar to the welder's visual system,

K. Li et al. (Eds.): ICSEE 2013, CCIS 355, pp. 284–293, 2013.
© Springer-Verlag Berlin Heidelberg 2013

and can reflect the status of the welding process from the weld pool. In addition, it includes sufficient information about weld pool which is useful for process modeling and control. Many researches were conducted to apply vision method in welding process monitoring.

In Ref. [8]-[16], two dimensional measurements of weld pool were obtained, and model establishments, penetration control were conducted. By sensing two dimensional weld pool images, certain characteristic parameters and penetration control can be achieved. Furthermore, three dimensional weld pool shapes presents more sufficient information for determining the state of weld pool, analyzing pool defect, conducting penetration control, and studying pool physical process.

The weld pools were monitored and reconstructed in three dimensional [17]-[23]. Besides, multi-variable process model were established, and advanced control were conducted to obtain desired penetration control. Recently, a novel three-dimensional monitoring system was developed and the observed three-dimensional weld pool was characterized by its width, length and convexity [24]. An optimal linear model was also obtained in steady state.

In this paper, an advanced three dimensional monitoring system developed at the University of Kentucky [24] was utilized to sense the three-dimensional weld pool surface in GTAW process. Then, novel characteristic parameters were defined and welding experiments were conducted to obtain the process data. At last, a neural network model was established to predict the back side parameters precisely. Based on the proposed intelligent sensing model, the 3D monitoring system can functionalize as skilled welders to observe the weld pool and predict the penetration status.

2   System Structure and Image Processing

System structure for GTAW control system is shown in Fig.1 [24]. Based on the three dimensional sensing system and weld pool reconstruction, the weld pool shape and relative characteristic parameters were obtained. Then, the process soft-sensing model was established after the process dynamic characteristic was observed. This model can be used to estimate the penetration status online and provide necessary feedback for the advanced controller.

The experiment system is shown in Fig.2 [24]. Gas tungsten arc welding without filler for pipe welding was exploited in the system. A 20mw illumination laser at a wavelength of 685nm with variable focus was used to project a 19-by-19 dot matrix structured light pattern (Model SNF-519X (0.77)-685-20) on the weld pool surface in the pipe. An imaging plane was fastened coaxially to the laser to intercept the reflected laser rays. The intensity of laser almost remained as the same when the rays intercepted by the imaging plane while the intensity of the arc light decays significantly by distance. Then, the image was captured by the camera in real time at certain frequency. Furthermore, the geometry of weld pool surface could be extracted from the reflected points in captured images after image processing and reconstruction algorithm was applied [23]. The welding torch can be controlled by the computer to move up and down for adjusting the arc length. The current and the welding speed can also be controlled conveniently. The material used in this paper was stainless steel 304, and the thickness of the pipe was 2.03 mm.

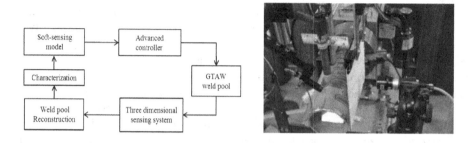

Fig. 1. GTAW control system structure   **Fig. 2.** Experiment system for pipe GTAW

After the reflected points were captured, the captured two dimensional images on the plane were processed to obtain three dimensional geometry of the weld pool surface. The process includes calibration, image processing, and reconstruction. The original reflected image, processed image, points after image processing, reconstructed points, points in horizontal plane, and three dimensional weld pool shape are given in the Fig. 3 where welding current was set at 60 A, welding speed was 1.5 mm/s and arc length was 4mm.

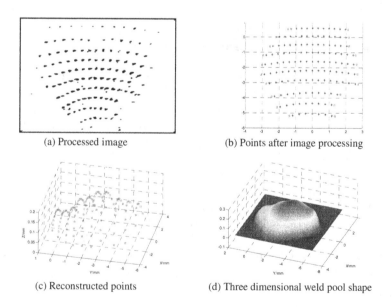

(a) Processed image   (b) Points after image processing

(c) Reconstructed points   (d) Three dimensional weld pool shape

Fig. 3. Weld pool reconstruction results

3   Characteristic Parameters Definition

In previous works, some characteristic parameters were given to describe the weld pool [12], [16], [17], [25]. All the parameters mentioned in above works are in two dimension plane. In [24] the weld pool width, length, and convexity were proposed as the characterization parameters of the weld pool surface. Some parameters reflecting more detailed three dimensional characteristics are presented in this paper.

In Fig.4, L and W give the basic description of weld pool in two dimensional, L_h and W_h present more information about the rear part of weld pool. L, W, L_h and W_h are the length, width, half length, half width of the weld pool respectively. H, S, and ΔS describe the characteristic in $Y - Z$ plane. H, S, and ΔS denote the height of the weld pool, the cross-section area, and the pool concavity area. Based on Fig.4 (a), the area of the weld pool in horizon can be calculated. Here the area is considered as circular, and radius of the circle can be obtained, which is denoted with R. Among these parameters, S, ΔS and R are introduced to characterize the weld pool for their correlation with the penetration at the first time.

Based on the above analysis, $L, W, L_h, W_h, H, S, \Delta S, R$ are selected as characteristic parameters to describe the weld pool. H_b, W_b are parameters describing the penetration status, where H_b is the backside height, and W_b is the backside width. Normally, a camera will be applied to obtain the value of W_b in real time, but the cost is high and it is difficult to place the vision system in some conditions (e.g., the accessibility is poor). H_b can be measured when the weld pool solidifies, but some delay will exist. In order to sense these two variables in real time, soft-sensing model thus is presented as an alternative to estimate them online. Based on the three dimensional weld pool sensing technology and weld pool reconstruction, L, W, L_h, W_h, H, S, ΔS, R of the weld pool shape will be calculated in real-time. L, W, L_h, W_h, and R are calculated based on Fig.4 (a); $H, S, \Delta S$ are obtained from Fig.4(b). H_b, W_b are measured after welding. Then the correlation between the characteristic parameters and H_b, W_b will be studied, and the soft sensing model will be established.

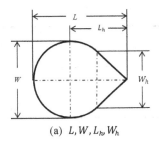

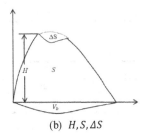

(a) L, W, L_h, W_h (b) $H, S, \Delta S$

Fig. 4. Weld pool characteristic parameters description

From the above analysis, characteristic parameters of the front weld pool were defined to correlate with the penetration status. Experiments were conducted to obtain relative data, and the relationships between the characteristic parameters and penetration status in different welding parameters were analyzed. Analysis results show that H_b and W_b correlate with the selected characteristic parameters. Hence, the relationships between the characteristic parameters and H_b, W_b are desired to be obtained. The characteristic parameters thus can be used to describe the penetration status, and this makes real time sensing of penetration status become possible. Models describing the relationships between the characteristic parameters and H_b, W_b will be discussed in next section.

4   Model Establishment

Section 3 illustrates that a specific characteristic parameter may vary with the penetration status in different ways depending on the welding conditions. This suggests that characteristic parameters are coupled in determining the weld penetration and the model that relates the weld penetration must be relatively complex. In order to obtain the model which can account for these variations, I, U, V are considered when establishing the model. At last, the inputs of model are determined as, $W, L_h, W_h, H, S, \Delta S, R, I, U, V$, the outputs are H_b, W_b . Table 1 and Table 2 show the auxiliary and estimation variables.

Table 1. Auxiliary variables

Variable	Variable name	Unit	Range
L	Length	mm	3~6
W	Width	mm	2~5
L_h	Half Length	mm	2~4
W_h	Half Width	mm	2~4.5
H	Height	mm	0~0.3
S	Section Area	mm^2	0~1
ΔS	Concavity	mm^2	0~0.03
R	Radius	mm^3	0~2
I	Current	A	50~72
L_a	Arc Length	mm	4~5.5
V	Speed	mm/S	1~2

Table 2. Estimation variables

Variable	Variable name	Unit	Range
H_b	Backside Height	mm	0~0.5
W_b	Backside Width	mm	0~6

After the characteristic parameters were defined, and enough experiments were conducted using various welding parameters, it was found that the top-side characteristic parameters correlate with the back-side parameters. Then the experiments data can be applied to establish the model which describes the relationships between these characteristic parameters, welding parameters and back-side parameters. The data used for model establishment were got from the 3D sensing system, the data from 1 to 38 are training data, and other 10 data are test data.

In the welding process, the auxiliary variables can be acquired through the monitoring system and image processing algorithm. Prior to establishing the model, the data have to be pretreated. Some apparently wrong data will be cancelled in this step to improve the accuracy of the acquired data. Because different variables have different units, a variable may differ from other variables in several orders. It is thus necessary to normalize all the data in order to map all the data into [0, 1] because the modeling results can be improved by using normalized data to fit a model with multiple inputs. Normalizing can be described as follows. If data sample space of a variable is $x = (x_1, x_2, \ldots x_n)$, then normalized data is:

$$x_i' = (x_i - min\,(x))/(max(x) - min\,(x)) \tag{1}$$

Back-propagation neural network (BPNN) model will be established to obtain the relationships between the welding parameters, characteristic parameters and estimation parameters in this section. After the model is established, it will be used to estimate the back-side parameters of the weld pool based on welding parameters and characteristic parameters in real time.

H gives the pool characteristic in Z axis, S shapes the pool characteristic in $Y - Z$ plane, and R holds pool information in horizontal plane. Hence, we select H, S and H, R as auxiliary variables respectively to establish the model. The structure of the BPNN is 2-5-1. The weights of the BPNN model are shown as follows, and the upper-right corner of the weights denotes its layer,

$$[w_{1,1}^1, w_{1,2}^1, \ldots, w_{1,n}^1, b_1^1, w_{2,1}^1, w_{2,2}^1, \ldots, w_{2,n}^1, b_2^1, \ldots,$$
$$w_{m,1}^1, w_{m,2}^1, \ldots, w_{m,n}^1, b_m^1, w_1^2, w_2^2, \ldots, w_m^2, b^2]$$

The amount of weights for the BPNN model is

$$M = m \times (n + 1) + m + 1 \tag{2}$$

where n is the number of input-layer neuron nodes, and m is the number of hidden-layer neuron nodes.

$$y = \sum_{m=1}^{5} w_m^2 \times f\left(\sum_{n=1}^{2} w_{mn}^1 x_n + b_m^1\right) + b^2 \tag{3}$$

$$f(x) = 1/(1 + e^{-x}) \tag{4}$$

After the structure of BPNN was determined, the data was used to train the BPNN, and the weights were obtained. Then Eq. (3) can be used to estimate the penetration in real-time. Here, x_1 and x_2 denote auxiliary variables, and y denotes dominant variables. Based on the real-time estimation of penetration, better analysis of GTAW process will be conducted which will in turn benefit the control system for it to achieve desired results.

Fig. 5 is the model simulation results based on the BPNN model where S and H are selected as auxiliary variables. As shown in Fig. 5 the data from 1 to 38 are the training data, and other 10 data are the test data. Table 3 shows the simulation errors, and it could be concluded that the estimation effect is accurate enough to realize the real-time estimation.

Fig. 6 is the model simulation results based on the BPNN model where R and H are selected as auxiliary variables. Table 4 shows the simulation errors, and it was concluded that the estimation result is acceptable to realize the real-time estimation. It can be found that the model based on S and H gives more precise estimation for H_b, but W_b can be estimated more accurately with model based on R and H. The reason is that the auxiliary variable S includes information of L and H, and auxiliary variable R contains information of L and W.

In above discussion, 11 auxiliary variables are presented and they are all related to the primary variables. Hence, all auxiliary variables will be applied to establish the model in this section. Every auxiliary variable reflects part information of the process in different degree, but variables are coupled and would provide redundant information. Principle component analysis (PCA) method can simplify the data matrix through dimension reduction. PCA make use of unrelated variables to reflect most information of original correlated auxiliary variables and delete redundant variables, which can improve calculation speed of model and simplify the data analysis. PCA is a well-developed statistic method, and has been widely applied in soft-sensing field.

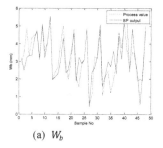 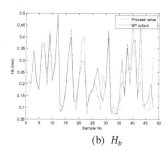

(a) W_b (b) H_b

Fig. 5. Simulation results of BPNN model based on S and H

Table 3. Simulation errors of BPNN model based on S and H

	Training MSE	Training ABE	Training RE	Testing MSE	Testing ABE	Testing RE
W_b	0.4761	0.4819	0.2203	0.7636	0.7121	0.2470
H_b	0.0018	0.0296	0.1743	0.0147	0.0925	0.4687

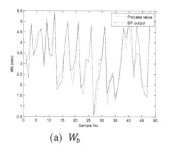 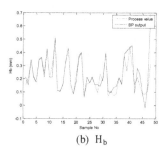

(a) W_b (b) H_b

Fig. 6. Simulation results of BPNN model based on R and H

Table 4. Simulation errors of BPNN model based on R and H

	Training MSE	Training ABE	Training RE	Testing MSE	Testing ABE	Testing RE
W_b	0.1766	0.2931	0.1272	0.4303	0.5531	0.1809
H_b	0.0011	0.0208	0.1433	0.0216	0.1075	0.5385

In this paper, PCA method was used to conduct dimension reduction from 11 to 4, and obtain the soft-sensing model. Fig. 7 is the model simulation results based on the PCA-BPNN model.

Table 5 shows the simulation errors. It was concluded that PCA-BPNN model is better in sensing W_b and H_b based on the top-side characteristic parameters. This is because all eleven auxiliary variables are used here, and more information about penetration status is included in the model. Hence, the PCA-BPNN model can be recommended as the preferred method in this paper to obtain real-time estimation for W_b and H_b in GTAW process.

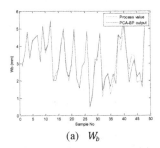
(a) W_b

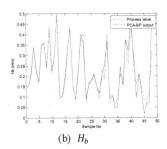
(b) H_b

Fig. 7. Simulation results of PCA-BPNN model based on all 11 inputs

Table 5. Simulation errors of BPNN model based on all 11 inputs

	Training MSE	Training ABE	Training RE	Testing MSE	Testing ABE	Testing RE
W_b	0.0613	0.1448	0.0688	0.2602	0.4679	0.1561
H_b	0.0005	0.0144	0.0929	0.0036	0.0470	0.2759

5   Conclusion

Based on the newly developed three dimensional vision monitoring system for the GTAW process, 304stainless steel pipe with 2.03 mm thickness was used to conduct experiments. The proposed novel top-side characteristic parameters are found to be closely related to the back-side weld pool parameters that directly quantify the weld penetration. Developed neural network models are effective in estimating the weld penetration specified by the back side bead width and height. With the proposed top-side characteristic parameters and neural network models, weld pool observation and penetration prediction can be realized in real time conveniently.

References

1. Vorman, A.R., Brandt, H.: Feedback control of GTA welding using puddle width measurement. Weld. J. 9, 742–749 (1976)
2. Graham, G.M., Ume, I.C.: Automated system for laser ultrasonic sensing of weld penetration. Mechatronics 8, 711–721 (1997)
3. Ju, J.B., Suga, Y., Ogawa, K.: Penetration Control by Monitoring Molten Pool Oscillation in TIG Arc Welding. In: 20th International Offshore and Polar Engineering Conference, Kitakyushu, Japan, pp. 241–246 (2002)
4. Hartman, D.A., Delapp, D.R., Cook, G.E., Barnett, R.J.: Intelligent fusion control throughout varying thermal regions. In: The IEEE Industry Applications Conference, AZ, Phoenix, pp. 635–644 (1999)
5. Beardsley, H.E., Zhang, Y.M., Kovacevic, R.: Infrared sensing of full penetration state in gas tungsten arc welding. Int. J. Mach. Tool. Manu. 8, 1079–1090 (1994)
6. Rokhlin, S.I., Guu, A.C.: A study of arc force, pool depression, and weld penetration during gas tungsten arc welding. Weld. J. 8, 381s–390s (1993)
7. Lu, W., Zhang, Y.M.: Robust sensing and control of the weld pool surface. Meas. Sci. Technol. 17, 2437–2446 (2006)
8. Luo, H., Devanathan, R., Wang, J., Chen, X., Sun, Z.: Vision based neurofuzzy logic control of weld pool geometry. Sci. Technol. Weld. Joi. 5, 321–325 (2002)
9. Wang, J.J., Lin, T., Chen, S.B.: Obtaining weld pool vision information during aluminum alloy TIG welding. Int. J. Adv. Manuf. Technol. 3, 219–227 (2005)
10. Fan, C.J., Lv, F.L., Chen, S.B.: Visual sensing and penetration control in aluminum alloy pulsed GTA welding. Int. J. Adv. Manuf. Technol. 1-2, 126–137 (2009)
11. Chen, B., Wang, J.F., Chen, S.B.: Prediction of pulsed GTAW penetration status based on BP neural network and D-S evidence theory information fusion. Int. J. Adv. Manuf. Technol. 1-4, 83–94 (2010)
12. Zhang, Y.M., Li, L., Kovacevic, R.: Dynamic estimation of full penetration using geometry of adjacent weld pools. J. Manuf. Sci. E-T. ASME 4, 631–643 (1997)
13. Kovacevic, R., Zhang, Y.M.: Neurofuzzy model-based weld fusion state estimation. IEEE Contr. Syst. Mag. 2, 30–42 (1997)
14. Zhang, Y.M., Kovacevic, R.: Neurofuzzy model based control of weld fusion zone geometry. IEEE T. Fuzzy Syst. 3, 389–401 (1998)
15. Kovacevic, R., Zhang, Y.M., Li, L.: Monitoring of weld penetration based on weld pool geometrical appearance. Weld. J. 10, 317s–329s (1996)
16. Wu, C.S., Gao, J.Q.: Vision-based neuro-fuzzy control of weld penetration in gas tungsten arc welding of thin sheets. Int. J. Modelling, Identification and Control 2, 126–132 (2006)
17. Zhao, D.B., Chen, S.B., Wu, L., Dai, M., Chen, Q.: Intelligent Control for the Shape of the Weld Pool in Pulsed GTAW with Filler Metal. W. R. C. 11, 253s–260s (2001)
18. Chen, S.B., Zhao, D.B., Lou, Y.J., Wu, L.: Computer Vision Sensing and Intelligent Control of Welding Pool Dynamics. In: Tarn, T.-J., Zhou, C., Chen, S.-B. (eds.) Robotic Welding. Intelligence and Automation. LNCIS, vol. 299, pp. 25–55. Springer, Heidelberg (2004)
19. Zhang, Y.M., Wu, L., Walcott, B.L., Chen, D.H.: Determining joint penetration in GTAW with vision sensing of weld-face geometry. Weld. J. 10, 463s–469s (1993)
20. Zhang, Y.M., Kovacevic, R., Wu, L.: Dynamic analysis and identification of gas tungsten arc welding process for full penetration control. J. Eng. I-T. ASME 1, 123–136 (1996)
21. Zhang, Y.M., Kovacevic, R., Li, L.: Adaptive control of full penetration GTA welding. IEEE T. Contr. Syst. T. 4, 394–403 (1996)

22. Saeed, G., Zhang, Y.M.: Weld pool surface depth measurement using calibrated camera and structured-light. Meas. Sci. Technol. 18, 2570–2578 (2007)
23. Song, H.S., Zhang, Y.M.: Measurement and analysis of three-dimensional specular gas tungsten arc weld pool surface. Weld. J. 4, 85s–95s (2008)
24. Zhang, W.J., Liu, Y.K., Wang, X.W., Zhang, Y.M.: Characterization of three dimensional weld pool surface in GTAW. Weld. J. 7, 195s–203s (2012)
25. Zhang, Y.M., Kovacevic, R., Li, L.: Characterization and real-time measurement of geometrical appearance of weld pool. Int. J. Mach. Tools Manufact. 7, 799–816 (1996)

An Improved Discrete Artificial Bee Colony Algorithm for Hybrid Flow Shop Problems

Zhe Cui and Xingsheng Gu*

Key Laboratory of Advanced Control and Optimization for Chemical Process of
Ministry of Education, East China University of Science and Technology,
Shanghai 200237, China
xsgu@ecust.edu.cn

Abstract. Being a typical NP-hard combinatorial optimization problem, the hybrid flow shop (HFS) problem widely exists in manufacturing systems. In this paper, we firstly establish the model of the HFS problem by employing the vector representation. Then an improved discrete artificial bee colony (IDABC) algorithm is proposed for this problem to minimize the makespan. In the IDABC algorithm, a novel differential evolution and a modified variable neighborhood search are studied to generating new solutions for the employed and onlooker bees. The destruction and construction procedures are utilized to obtain solutions for the scout bees. The simulation results clearly imply that the proposed IDABC algorithm is highly effective and efficient as compared to six state-of-the-art algorithms on the same benchmark instances.

Keywords: hybrid flow shop problem, mathematical model, artificial bee colony, differential evolution, scheduling.

1 Introduction

Production scheduling is a decision-making process that plays a crucial role in manufacturing and service industries [1]. As industries are facing increasingly competitive situations, the classical flow shop model is not applicable to some practical industry processes. As a result, the hybrid flow shop (HFS) problem in which a combination of flow shop and parallel machines operate together arises. Although the HFS problem, also called multi-processor or flexible flow shop, widely exists in real manufacturing environments, e.g., in chemical, oil, food, tobacco, textile, paper, and pharmaceutical industries, there is no effective method to solve this problem.

As to the computational complexity, since even the two-stage HFS problem is strongly NP-hard [2] by minimizing the maximum completion time (makespan), the multi-stage HFS problem is at least that difficult. Despite of the intractability, the HFS problem has great significance in both engineering and theoretical fields. Thus, it is meaningful to develop effective and efficient approaches for such the problem considered.

---

* Corresponding author.

K. Li et al. (Eds.): ICSEE 2013, CCIS 355, pp. 294–302, 2013.

Compared with a large number of literatures on the classic flow shop scheduling problem, the HFS problem has not been well studied. Santos et al. [3] presented a global lower bound for makespan minimization that has been used to analyze the performance of other algorithms. Neron et al. [4] used the satisfiability tests and time-bound adjustments based on the energetic reasoning and global operations to enhance the efficiency of another kind of B&B method proposed in Carlier and Neron [5]. With advanced statistical tools, Ruiz et al. [6] tested several heuristics in the realistic HFS problem and suggested that the modified NEH heuristic [7] outperformed the other dispatching rules. The genetic algorithm (GA) was applied by researchers to solve the HFS problem under the criterion of makespan minimization [8]. On the basis of vertebrate immune system, Engin and Doyen [9] proposed the artificial immune system (AIS) technique that incorporated the clonal selection principle and affinity maturation mechanism. Inspired by the natural mechanism of the ant colony, Alaykyran et al. [10] introduced an improved ant colony optimization (ACO) algorithm. Niu et al. [11] presented a quantum-inspired immune algorithm (QIA) for the HFS problem to minimize makespan. Liao et al. [12] developed a particle swarm optimization (PSO) algorithm. This algorithm hybridized the PSO and bottleneck heuristic to fully exploit the bottleneck stage, and further introduced simulated annealing to help escape from local optima. In addition, a local search is embedded to further improve its performance.

The artificial bee colony (ABC) algorithm, simulating the intelligent foraging behaviors of honey bee colonies, is one of the latest population-based evolutionary meta-heuristics [13]. Basturk and Karaboga suggested that the ABC algorithm has a better performance than the other population-based algorithms for solving continuous problems [14], [15]. Nevertheless, on account of its continuous nature, the studies on the ABC algorithm for combinatorial optimization problems is very limited.

As we know, there is no published work using the ABC-based algorithm for the HFS problem. In this paper, we firstly establish the model of the HFS problem by employing the vector representation, and then an improved discrete artificial bee colony (IDABC) algorithm is proposed for the problem to minimize the makespan. The rest of the paper is organized as follows. In Section 2, the model of the HFS problem is formulated. Section 3 presents the details of the proposed IDABC algorithm. The simulation results are provided in Section 4. Finally, conclusions are drawn in Section 5.

2 Problem Statement

2.1 Description of the Problem

The HFS problem can be described as follows. There are n jobs J=$\{1,2,...,i,...,n-1,n\}$ that have to be performed on s stages S=$\{1,2,...,j,...,s-1,s\}$, and each stage s has m_s identical machines. These identical machines are continuously available from time zero and have the same effect. At least one stage j must has more than one machine. Every job has to visit all of the stages in the same order

string from stage 1 through stage s and is processed by exactly one machine at every stage. A machine can process at most one job at a time and a job can be processed by at most one machine at a time. The processing time $p_{i,j}$ is given for each job at each stage. The scheduling problem is to choose a machine at each stage for each job and determine the sequence of jobs on each machine so as to minimize the makespan.

2.2 Mathematical Model

As we know, on the research of the HFS problem, there are two formats to represent a solution, namely the matrix representation and the vector representation. In this paper, we employ the vector representation [16], which considers the sequence of jobs only at the stage one. This subset sequence should contain the collection of all potentially good solutions for the problem and is a one-to-one correspondence. Most importantly, it is very convenient to design and operate by using this format. A subset sequence is decoded to a complete schedule by employing a generalization of the List Scheduling (LS) algorithm to incorporate the jobs at other stages [17], [18]. For scheduling jobs at each stage, the LS algorithm is based on the first-come-first-service rule, in which the jobs with the shortest completion time from the previous stage should be scheduled as early as possible. It could result in a non-permutation schedule, that is, the sequence of jobs at each stage may be different. The model of the HFS problem can be formulated as follows in terms of this representation:

Minimize:
$$C_{\max}(\pi_1) = \max_{i=1,2,\ldots n} \{C_{\pi_s(i),s}\}$$

Subject to:

$$\begin{cases} C_{\pi_1(i),1} = p_{\pi_1(i),1} \\ IM_{i,1} = C_{\pi_1(i),1} \end{cases} i = 1, 2, \ldots m_1$$

$$\begin{cases} C_{\pi_1(i),1} = \min_{k=1,2,\ldots m_1} \{IM_{k,1}\} + p_{\pi_1(i),1} \\ NM_1 = \arg\min_{k=1,2,\ldots m_1} \{IM_{k,1}\} \\ IM_{NM_1,1} = C_{\pi_1(i),1} \end{cases} i = m_1 + 1, m_1 + 2, \ldots n$$

$$\pi_j(i) = g(C_{\pi_{j-1}(i),j-1})\ i = 1, 2, \ldots n;\ j = 2, 3, \ldots s$$

$$\begin{cases} C_{\pi_j(i),j} = C_{\pi_j(i),j-1} + p_{\pi_j(i),j} \\ IM_{i,j} = C_{\pi_j(i),j} \end{cases} i = 1, 2, \ldots m_1;\ j = 2, 3, \ldots s$$

$$\begin{cases} C_{\pi_j(i),j} = \max\{C_{\pi_j(i),j-1}, \min_{k=1,2,\ldots m_j}\{IM_{k,j}\}\} + p_{\pi_j(i),j} \\ NM_j = \arg\min_{k=1,2,\ldots m_j}\{IM_{k,j}\} \\ IM_{NM_j,j} = C_{\pi_j(i),j} \end{cases}$$
$$i = m_j + 1, m_j + 2, \ldots n;\ j = 2, 3, \ldots s$$

where π_j is the job permutation at the stage j; $\pi_k(i)$ is the ith job in the π_k; $C_{\pi_k(i),j}$ is the completion time of job $\pi_k(i)$ at the stage j; $IM_{i,j}$ represents the idle moment of machine i at the stage j; NM_j denotes the serial number of earliest available machine at the moment at the stage j; the function

$S_1(i) = g(S_2(i))$ $(i=1,2,...n)$ means that $S_1(i)$ is the permutation of $(1,2,...n)$ based on the ascending order of $S_2(i)$ $(i=1,2,...n)$; and $\arg\min_k\{IM_k\}$ stands for the argument of the minimum, i.e. the set of points of the given argument for which the given function attains its minimum value.

In the above recursive equations, we firstly calculate the completion time of the jobs at the stage one, then that of the stage two, until the last stage.

3   The Improved Discrete Artificial Bee Colony (IDABC) Algorithm for the HFS Problem

3.1   Individual Representation and Initialization

Owing to the continuous nature of the ABC algorithm, it can not be directly used for the HFS problem. So it is important to find a suitable mapping which can conveniently convert a harmony to a solution. The model of the HFS problem is formulated by using the vector representation in last section. As a result, we adopt this representation in the proposed IDABC algorithm. The individual in the IDABC algorithm is represented by a permutation of jobs at the stage one $= \{\pi(1), \pi(2), ..., \pi(n)\}$.

To guarantee the initial population with a certain quality and diversity, it is constructed randomly except that one is established by the aforementioned NEH heuristic [7]. According to [6], the NEH heuristic, a typical constructive method for the permutation flow shop scheduling problem, is also very robust and well performing for the HFS problem.

3.2   Employed Bee Phase

In the original ABC algorithm, the employed bees exploit the given food sources in their neighborhood. Here we propose a novel differential evolution scheme for the employed bees to generate neighboring food sources. The differential evolution scheme consists of three steps: mutation, crossover, and selection.

In the mutation part, two parameters mutation rate (MR), insert times (IT) are introduced. For each incumbent individual, a uniformly random number is generated in the range of $[0,1]$. If it is less than MR, the mutant individual is obtained by operating the insert operation on the best individual π best in the population IT times; otherwise, the mutant individual is gained by operating the insert operation on a randomly selected individual IT times.

Next, partially mapped crossover (PMX) [19], a widely used crossover operator for permutation-based, is used in the crossover part. The incumbent individual and the mutated individual will undergo the PMX operation with a crossover rate (CR) to obtain the two crossed individuals. On the other hand, the two crossed individuals are the same as the mutated individual with about a probability of $(1-CR)$.

Following the crossover operation, the selection is conducted. The one with the lowest value of the objective function among the two crossed individuals and the incumbent individual will be accepted. In other words, if either of these two crossed individuals yields a better makespan than the incumbent individual, then the better individual will replace the incumbent one and become a new member in the population; otherwise, the old individual is retained.

3.3 Onlooker Bee Phase

There are NP onlooker bees and each communicates with its corresponding employed bee. After the probability selection, also called the wheel selection, a modified variable neighborhood search (VNS) [20] is incorporated into our algorithm as a hybrid strategy to further improve the performance. We use two structures of neighborhoods, which are referred to as the insert local search and the swap local search. The procedures of the insert local search and the swap local search are given in Fig. 1, where u and v are two positive integers chosen randomly in the range of $[1,n]$. The local search combining both the insert local search and the swap local search is illustrated as follows:

Step1. Perform the insert local search. If the individual is improved, go back to step 2; otherwise end the procedure.
Step2. Perform the swap local search. If the individual is improved, go back to step 1; otherwise end the procedure.

$loop=0$	$loop=0$
randomly generate u, v and $u{\neq}v$	randomly generate u, v and $u{\neq}v$
π'=insert(π, u, v)	π'=swap(π, u, v)
do{	do{
randomly generate u, v and $u{\neq}v$	randomly generate u, v and $u{\neq}v$
π''=insert(π', u, v)	π''=swap(π', u, v)
if($C_{max}(\pi'')<C_{max}(\pi')$)	if($C_{max}(\pi'')<C_{max}(\pi')$)
$\pi'=\pi''$	$\pi'=\pi''$
endif	endif
$loop{+}{+}$	$loop{+}{+}$
}while($loop<n{\times}(n{+}1)$)	}while($loop<n{\times}(n{+}1)$)
if($C_{max}(\pi){\leq}C_{max}(\pi)$)	if($C_{max}(\pi'){\leq}C_{max}(\pi)$)
$\pi=\pi'$	$\pi=\pi'$
endif	endif
(a)the insert local search	(b)the swap local search

Fig. 1. The procedure of modified variable neighborhood search

If the new food source obtained is better than or equal to the incumbent one, the new food source will be memorized in the population. The onlooker bee phase in the IDABC algorithm provides the intensification of the local search on the relatively promising solutions.

3.4 Scout Bee Phase

As it has been stated in the basic ABC algorithm, the scout bees search randomly in the predefined space. This procedure will increase the population diversity and avoid getting trapped in local optima, whereas this will also decrease the search efficacy. The new food source of a scout bee is produced as follows. Firstly, a tournament selection with the size of two is applied due to its simplicity and efficiency. That is, a scout bee selects two individuals π_a and π_b randomly from the population, and compares them with each other. If the makespan of π_a is smaller than that of π_b, π_a wins the tournament and π_b loses. Then, the scout bee generates a new solution new by employing the destruction and construction procedures of the iterated greedy (IG) algorithm [21]. The destruction and construction procedures are performed on the better individual π_a in the tournament selection and it has one parameter: destruction size (d). After that, the new solution π_{new} becomes a new member in the population and the worse one π_b is discarded. In this phase, the number of scout bees is ten percent of that of food sources.

4 Simulation Results and Comparisons

4.1 Experimental Setup

To fully examine the performance of the IDABC algorithm, an extensive experimental comparison with other powerful methods are provided. The IDABC algorithm was coded in Visual C++ and run on an Intel Pentium 3.06 GHz PC with 2 GB RAM under Windows 7 operating system.

The test problems used in experiments are the 98 different benchmark problems which are presented in [5]. The sizes of these problems vary from 10 jobs and 5 stages to 15 jobs and 10 stages. The processing times of the operations in these 98 instances are uniformly distributed between 3 and 20. Three characteristics that define a problem are the number of jobs, the number of stages and the number of identical machines at each stage. Therefore, we use the notation of j10c5b1 for instance, which means a 10-job, 5-stage problem. There are 55 easy problems and 43 hard problems. The problems with a and b machine layouts are easy problems. The problems with c, d, and e machine layouts are relatively harder to solve, so they are mostly grouped as hard problems.

The benchmark problems taken from Carlier and Neron [5] are relative simple. Thus, another 10 benchmark problems generated by Liao et al. [12] recently are also utilized in this section. In each problem, there are 30 jobs and 5 stages. At each stage, the machine number has a uniform distribution in the range of [3,5]. The processing times in these problems are within [1,100].

In the proposed IDABC algorithm, there are five main parameters: NP, MR, CR, IT, and d. We set the parameters $NP=n$ (the number of jobs), $MR=0.8$, $CR=0.8$, $IT=4$, and $d=4$ in the following experiments.

4.2 Computational Results

Comparison of Carlier and Neron's Benchmarks. Several meta-heuristics have been applied to Carlier and Neron's benchmark problems. To evaluate the performance of the proposed IDABC algorithm in solving the HFS problem under the criterion of makespan minimization, the IDABC algorithm was compared with a B&B method [4], an AIS [9], an ACO [10], a GA [8], a QIA [11], and a PSO algorithm [12]. The maximum run time of the algorithm was set at 1600s or until the lower bound (LB) [3], [4] was reached. If the LB was not found within this time limit, the search was stopped and the best solution was accepted as the final solution.

In order to establish more accurate and objective comparisons, the computational results of these compared algorithms are obtained from their original papers. For each test problem, the proposed IDABC algorithm was run independently twenty times and the performance of all the compared algorithms was summarized in Table 1. In Table 1, Solved means the number of problems which the algorithm can solve, and Deviation denotes the average relative percentage error to LB.

Table 1. Comparison results on Carlier and Neron's benchmark problems

Algorithm	Easy problems		Hard problems	
	Solved	Deviation	Solved	Deviation
B&B	53	2.17%	24	6.88%
AIS	53	0.99%	24	3.13%
ACO	45	0.92%	18	3.88%
GA	53	0.95%	24	3.05%
QIA	29	0	12	5.04%
PSO	53	0.95%	24	2.85%
IDABC	55	0.94%	43	2.82%

As it can be noticed from Table 1, the machine layouts have an important effect on the complexity of problems that affects solution quality. In the 55 easy problems and 43 hard problems, B&B, AIS, GA, and PSO can solve 53 easy problems and 24 hard problems, ACO can solve only 45 easy problems and 18 hard problems, QIA can solve only 29 easy problems and 12 hard problems, while the proposed IDABC algorithm can solve all the 98 problems. The average percentage deviation values of the easy and hard problems generated by IDABC are equal to 0.94% and 2.82%. For the 55 easy problems, QIA has a zero deviation value but it can solve only 29 of the 55 easy problems. The performance of PSO is comparable with the proposed IDABC, but it still cannot solve problems as many as the proposed IDABC. Thus, it is concluded that the IDABC algorithm is more effective and efficient in comparison with other algorithms for Carlier and Neron's benchmark problems.

Comparison of Liao's Benchmarks. For Liao's benchmark problems, two meta-heuristics: AIS and PSO, have been applied to the problems in [12]. The computational results are shown in Table 2, where the experimental data of AIS and PSO were obtained from their original papers, and the IDABC algorithm was run twenty independent replications for each problem. The execution time for each problem is limited to 200 seconds. In Table 2, AVE, MIN, and STD indicate the values of average, minimum, and standard deviation, respectively. T presents the average computation time (given in seconds) that the solution converges to the final solution.

Table 2. Comparison results on Liao's benchmark problems

Problem	AIS				PSO				IDABC			
	AVE	MIN	STD	T(s)	AVE	MIN	STD	T(s)	AVE	MIN	STD	T(s)
j30c5e1	485.35	479	2.58	99.44	474.70	471	**1.42**	96.16	**465.15**	463	1.50	**56.81**
j30c5e2	620.70	619	1.63	80.24	616.25	**616**	0.44	55.28	**616**	**616**	**0**	**1.51**
j30c5e3	625.70	614	4.81	116.70	610.25	602	4.70	64.56	**596.4**	593	**1.70**	49.14
j30c5e4	588.55	582	3.38	108.63	577.10	575	1.52	86.98	**566.2**	565	**1.20**	39.29
j30c5e5	618.75	610	3.42	101.19	606.80	605	**1.11**	79.84	**602**	600	1.56	57.67
j30c5e6	625.75	620	3.01	100.47	612.50	605	3.49	67.99	**603.05**	601	**1.47**	55.02
j30c5e7	641.30	635	4.67	93.56	630.60	629	0.75	87.18	**626**	626	**0**	18.68
j30c5e8	697.50	686	5.14	100.68	684.20	678	2.50	97.67	**674.65**	674	**0.88**	55.18
j30c5e9	670.20	662	3.85	100.75	654.65	651	1.87	83.80	**643.65**	642	**1.04**	67.49
j30c5e10	613.45	604	5.33	89.29	599.75	594	5.28	77.46	**576.25**	573	**1.52**	76.05
Average	618.73	611.10	3.78	99.09	606.68	602.60	2.31	79.69	**596.94**	**595.3**	**1.08**	**47.69**

In Table 2, the smallest values of AVE, MIN, STD, and T in the rows are shown in bold, respectively. It can be noted that the overall mean values of AVE, MIN, and STD yielded by the IDABC algorithm are equal to 596.94, 595.3, and 1.08, respectively, which are much better than those generated by AIS and PSO. Besides, the average computation time T of IDABC: 47.69 seconds is much shorter than that generated by AIS and PSO. From these observations, it is shown that the IDABC algorithm can obtain a better solution than AIS and PSO in an obviously shorter computational time. This means that the IDABC algorithm can converge to the good solutions faster than AIS and PSO. Also, it can be seen that the IDABC algorithm is more robust than both AIS and PSO for Liao's benchmark problems.

5   Conclusions

This paper establishes the model of the HFS problem by employing the vector representation and presents an improved discrete artificial bee colony (IDABC) algorithm for the HFS problem to minimize the makespan. Our future work is to extend the IDABC algorithm to other kinds of scheduling problems such as stochastic scheduling and multi-objective scheduling.

Acknowledgments. The authors are grateful to Carlier, Neron, and Liao for making the benchmark set available and to the anonymous reviewers for giving us helpful suggestions. This work was supported by National Natural Science Foundation of China (Grant no. 61174040, 61104178), the Fundamental Research Funds for the Central Universities.

References

1. Pinedo, M.: Scheduling: theory algorithms and systems. Prentice-Hall, Englewood Cliffs (2002)
2. Gupta, J.N.D.: Two-stage hybrid flowshop scheduling problem. J. Oper. Res. Soc. 39, 359–364 (1988)
3. Santos, D.L., Hunsucker, J.L., Deal, D.E.: Global lower bounds for flow shops with multiple processors. Eur. J. Oper. Res. 80(1), 112–120 (1995)
4. Neron, E., Baptiste, P., Gupta, J.N.D.: Solving hybrid flow shop problem using energetic reasoning and global operations. Omega 29(6), 501–511 (2001)
5. Carlier, J., Neron, E.: An exact method for solving the multi-processor flow-shop. RAIRO-Oper. Res. 34(1), 1–25 (2000)
6. Ruiz, R., Serifoglu, F.S., Urlings, T.: Modeling realistic hybrid flexible flowshop scheduling problems. Comput. Oper. Res. 35(4), 1151–1175 (2008)
7. Nawaz, M., Enscore, E., Ham, I.: A heuristic algorithm for the m-machine, n-job flow shop sequencing problem. Omega 11, 91–95 (1983)
8. Kahraman, C., Engin, O., Kaya, I., Yilmaz, M.K.: An application of effective genetic algorithms for solving hybrid flow shop scheduling problems. Int. J. Comput. Intell. Sys. 1(2), 134–147 (2008)
9. Engin, O., Doyen, A.: A new approach to solve hybrid flow shop scheduling problems by artificial immune system. Future Gener. Comp. Sys. 20, 1083–1095 (2004)
10. Alaykyran, K., Engin, O., Doyen, A.: Using ant colony optimization to solve hybrid flow shop scheduling problems. Int. J. Adv. Manuf. Tech. 35, 541–550 (2007)
11. Niu, Q., Zhou, T., Ma, S.: A quantum-inspired immune algorithm for hybrid flow shop with makespan criterion. J. Univers. Comput. Sci. 15, 765–785 (2009)
12. Liao, C.J., Tjandradjaja, E., Chung, T.P.: An approach using particle swarm optimization and bottleneck heuristic to solve hybrid flow shop scheduling problem. Appl. Soft. Comput. 12(6), 1755–1764 (2012)
13. Karaboga, D.: An idea based on honey bee swarm for numerical optimization. Technical report, Computer Engineering Department, Engineering Faculty, Erciyes University (2005)
14. Karaboga, D., Basturk, B.: A powerful and efficient algorithm for numerical function optimization: artificial bee colony algorithm. J. Global. Optim. 39(3), 459–471 (2007)
15. Karaboga, D., Basturk, B.: On the performance of artificial bee colony (ABC) algorithm. Appl. Soft. Comput. 8(1), 687–697 (2008)
16. Wardono, B., Fathi, Y.: A tabu search algorithm for the multi-stage parallel machine problem with limited buffer capacities. Eur. J. Oper. Res. 155, 380–401 (2004)
17. Oguz, C., Zinder, Y., Do, V.H., Janiak, A., Lichtenstein, M.: Hybrid flow-shop scheduling problems with multiprocessor task systems. Eur. J. Oper. Res. 152, 115–131 (2004)
18. Oguz, C., Ercan, M.F.: A genetic algorithm for hybrid flow shop scheduling with multiprocessor tasks. J. Scheduling 8, 323–351 (2005)
19. Goldberg, D.E., Lingle, R.J.: Alleles, loci and the traveling salesman problem. In: 1st International Conference on Genetic Algorithms and their Application, pp. 154–159. Lawrence Erlbaum (1985)
20. Mladenovic, N., Hansen, P.: Variable neighborhood search. Comput. Oper. Res. 24, 1097–1100 (1997)
21. Ruiz, R., Stutzle, T.: A simple and effective iterated greedy algorithm for the permutation flowshop scheduling problem. Eur. J. Oper. Res. 177(3), 2033–2049 (2007)

Circulating Fluidized Bed Coal-Saving Optimization Control Method

Tengfei Jiang[1], Debin Yin[2], Dewei Li[1,*], Yugeng Xi[1], and Wu Zhou[1]

[1] Department of automation, Shanghai Jiao Tong University,
Key Laboratory of System Control and Information Processing,
Ministry of Education, Shanghai, 200240
dwli@sjtu.edu.cn
[2] Shanghai Xinhua Control Technology (Group) CO., LTD, Shanghai 200241

Abstract. The circulating fluidized bed boiler is widely used in thermal power plants. With the proposal of energy-saving emission reduction, how to reduce coal consumption while ensure the output steam quality at the same time has become an important topic. This paper combines the technology of RTO (real-time optimization) and zone control in DMC (dynamic matrix control) to achieve this goal. The proposed method adds the coal consumption into the objective function of DMC controller and the operation point of the boiler is permitted to change within a zone which can be set according to the actual requirements of the circulating fluidized bed boiler. The zone control in DMC provides the freedom to reduce the coal consumption and achieves the economic optimal target. Compared to the simple use of constrained DMC control, the proposed method is verified to be remarkable coal-saving by the case study of a 150 t/h boiler of a power plant in Sichuan.

Keywords: Circulating fluidized bed boiler, Dynamic Matrix Control, Real Time Optimization, Zone Control, Coal-saving Control.

1 Introduction

Compared with the conventional boiler, the circulating fluidized bed (CFB) boiler can be universal to bituminous coal, lignite, anthracite, coal gangue and other fuels. Meanwhile, the unique hydrodynamic properties and structure of CFB boilers improve the combustion efficiency. These advantages make the CFB boiler widely[1].

The CFB boiler is a MIMO thermodynamic system with large inertia and large time delay, strong coupling and nonlinear characteristics. Hence, it is difficult to design the automatic control system for the CFB boiler. [2] studied the

---

* This work is supported in part by the National Science Foundation of China (Grant No. 60934007, 61074060), State Key Laboratory of Synthetical Automation for Process Industries, and Innovation Program of Shanghai Municipal Education Commission (Grant No. 11CXY08), Shanghai Jiao Tong University graduate innovative ability special fund(Grant No. Z-030-008).

circulating fluidized bed boiler mechanism and the linear model, and proposed multivariable decoupling PID control method. Since then, fuzzy-PID control (such as [3] and [4]), BP neural network control ([5]), and other control method have been proposed. In order to handle the physical constraints, the predictive control strategy in [6] has been applied in the control of CFB boilers.

In recent years, with the increasing energy demand, policies of energy saving and environmental protection are promulgated. How to reduce coal consumption, or improve the efficiency of coal has become the industrial focus of attention. By the experiments of raw materials, Yanxiao Zhong and his team studied influences of pulverized coal fineness on the combustion efficiency and then proposed to improve the economizer effect by pretreatment of coal in [7]. In terms of process, [8] suggested to use the cycle steam pipe to dryer pretreatment coal to improve coal-fired efficiency and then to save coal. On the other hand, Lv Wenxiang and his colleague proposed adaptive optimization method based on intelligent optimal control strategy to ensure that the boiler runs with a high combustion efficiency to achieve both normal operation and environmental protection[9].

Pretreatment of coal can contribute greatly to the combustion efficiency. But it also results in the cost for equipment investment to the end user. For the control of a CFB boiler, to achieve the dual purpose of circulating fluidized bed boiler coal-saving and the normal operation is a multi-objective optimization control problem. For the similar purpose, a real time optimization (RTO) layer can be designed for system operating point optimization. As mentioned in [10], the combination of the technique of RTO and DMC can be expected to reduce more than 30% of fuel in the industrial fuel gas system.

In order to reduce the coal consumption of a CFB boiler, this paper refers the idea of RTO to design the DMC controller where the coal consumption is added into the objective function of DMC. Meanwhile, in order to achieve the goal of saving coal further, the DMC adopts the zone control strategy. That means, the operation point of circulating fluidized bed boiler can be adaptively chosen near the ideal operating point. This provides the freedom to reduce the coal consumption. The permitted zone of operation point can be chosen according to the actual requirements of the circulating fluidized bed boiler. The final simulation verifies that the proposed method can achieve a remarkable reduction of coal consumption.

This paper is organized as follows. Section 2 introduces the related technology, including the CFB boiler, DMC, and RTO. Section 3 proposes the coal-saving control algorithm. And Section 4 gives a case study about a 150 t/h boiler of a power plant in Sichuan, which verifies the effectiveness of the proposed method.

2   The Problem Formulation and Background Technology

2.1   The CFB Boiler

The circulating fluidized bed boiler is a typical MIMO thermal system, whose main input and output relationships are shown in Fig. 1.

Fig. 1. The input-output relationships of a circulating fluidized bed boiler

Fig. 1 shows the bed temperature is influenced by the feeding coal amount, the primary air and the secondary air. On the other hand, the three inputs can also influence the pressure and temperature of steam, flue gas oxygen content. As pointed out by [11], the response time delay from fuel to the steam pressure is about 45s, and the rise time is about 350s. Thus, the circulating fluidized bed boiler is also a large time delay system subjected to physical constraints from feeding coal, primary air and secondary air.

From the process, the temperature control of a CFB boiler is a zone control. That is, the bed temperature is allowed to fluctuate within a threshold range of settings which can ensure the burn rate without coking risks. Similarly, the main steam pressure, temperature, and flue gas oxygen content are also allowed to fluctuate around the set values.

2.2 Dynamic Matrix Control[12]

Dynamic matrix control (DMC) is one of model predictive control (MPC) algorithms. Since the circulating fluidized bed boiler is a constrained multi-variable system with large time delay, DMC can be used as a better choice of control strategy.

Suppose there are m outputs, p inputs. $\mathbf{a_{ij}}$ is model vector from u_j to y_i. The control horizon and predictive horizon of DMC are denoted as M and P, respectively. Then, dynamic matrix is

$$\mathbf{A}=\begin{bmatrix}\mathbf{A}_{11} \cdots \mathbf{A}_{1m} \\ \vdots \qquad \vdots \\ \mathbf{A}_{p1} \cdots \mathbf{A}_{pm}\end{bmatrix}, \mathbf{A}_{ij}=\begin{bmatrix} a_{ij}(1) & & 0 \\ \vdots & \ddots & \\ a_{ij}(M) & \cdots & a_{ij}(1) \\ \vdots & & \vdots \\ a_{ij}(P) & \cdots & a_{ij}(P-M+1) \end{bmatrix}$$

$\tilde{\mathbf{y}}_M(k)$ is the predictive output at time k and $\tilde{\mathbf{y}}_{P0}(k)$ is the reference of system output when $\Delta u(k) = 0$. Thus, $\tilde{\mathbf{y}}_M(k) = \tilde{\mathbf{y}}_{P0}(k) + \mathbf{A}\Delta\mathbf{u}(k)$. The optimized performance of Dynamic Matrix Control is

$$\min_{\Delta\mathbf{u}_M} \|\omega(k) - \tilde{\mathbf{y}}_{P0}(k) - \mathbf{A}\Delta\mathbf{u}_M(k)\|_{\mathbf{Q}}^2 + \|\Delta\mathbf{u}_M(k)\|_{\mathbf{R}}^2 \qquad (1)$$
$$s.t. \ \ \Delta\mathbf{u}_{M,\min} \leq \Delta\mathbf{u}_M(k) \leq \Delta\mathbf{u}_{M,\max}$$

Where $\mathbf{\Delta u}_M(k) = \begin{bmatrix} \mathbf{\Delta B}_M(k) \\ \mathbf{\Delta Q}_{1M}(k) \\ \mathbf{\Delta Q}_{2M}(k) \end{bmatrix}$, $\mathbf{\Delta B}_M(k)$ is the change to the amount of coal(t/h), $\mathbf{\Delta Q}_{1M}(k)$ is the change to the primary air $(\mathrm{Nm}^3/\mathrm{h})\mathbf{\Delta Q}_{2M}(k)$ is the change of the secondary air $(\mathrm{Nm}^3/\mathrm{h})$. Denote $\tilde{\mathbf{y}}_{N1}(k)$ as the $k+1$ system output predicted at time k. $\mathbf{e}(k+1)$ is error between the actual output and the predicted output at time $k+1$. Then the modified vector of the predictive output is $\tilde{\mathbf{y}}_{cor}(k+1) = \tilde{\mathbf{y}}_{N1}(k) + \mathbf{H}\mathbf{e}(k+1)$.

2.3   Industrial RTO

In large-scale industrial process control, RTO is usually used together with predictive control. RTO optimizes operating points based on economic objective of the process, while MPC implements control based on the operating point optimized by RTO. According to [13], there are several kinds of expression of RTO, among which one widely used RTO used in practical applications. the RTO with linear objective function is expressed as the following optimization problem:

$$
\begin{aligned}
\min \ & \mathbf{c_y}\mathbf{y}^{TG} + \mathbf{c_u}\mathbf{u}^{TG} \\
s.t. \ & \mathbf{u}_{\min}^{TG} \leq \mathbf{u}^{TG} \leq \mathbf{u}_{\max}^{TG}, \\
& \mathbf{y}_{\min}^{TG} \leq \mathbf{y}^{TG} \leq \mathbf{y}_{\max}^{TG}, \\
& \mathbf{y^{TG}} = f(y, u)
\end{aligned}
\tag{2}
$$

where $\mathbf{y}^{TG}$ is real-time optimization relative vector from output, and $\mathbf{u}^{TG}$ is real-time optimization relative vector from input. The optimal operation point is achieved via linear optimization of RTO layer.

For the implementation of RTO, there are two general ways: one is constructing a dedicated RTO layer, optimizing the steady-state value for fast systems; another is combining the RTO layer and MPC to achieve the dual purposes of optimal control and operation point (see detail in [14]). As shown in [15], direct combination of RTO and MPC simplifies the control structure.

3   The Optimal Control of Coal-Saving of CFB Boiler (RTO/DMC)

For the optimal control of the CFB boiler, MPC control can achieve good control performance, but contribute very little on coal-saving. On the other hand, operation point optimization by RTO needs the additional devices and communication, which increase the cost of system. Hence, we proposed a control structure shown in Fig. 2, which embeds the RTO into the DMC controller as in [15]. The control structure integration coal-saving and dynamic optimization and make it possible to set the CFB boiler online according to the practical conditions. Based on the integration of coal and dynamic control optimization shown in Fig. 2, the predictive control optimization problem is formulated as

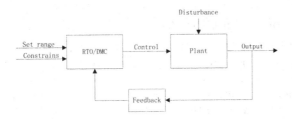

Fig. 2. RTO / DMC control structure

$$\min_{\Delta\mathbf{u}_M,\delta(k)} \quad \|\omega(k) - \tilde{\mathbf{y}}_{P0}(k) - \mathbf{A}_\delta\delta(k) - \mathbf{A}_{\Delta\mathbf{u}}\Delta\mathbf{u}_M(k)\|_{\mathbf{Q}}^2 + \|\Delta\mathbf{u}_M(k)\|_{\mathbf{R}}^2$$
$$+ \|\delta(k)\|_{\lambda_1}^2 + \lambda_2\mathbf{C}_c\mathbf{B}_c\Delta\mathbf{B}_M(k) \tag{3}$$
$$s.t. \ \underline{\delta} \le \delta(k) \le \bar{\delta},$$
$$\Delta\mathbf{u}_{M,\min} \le \Delta\mathbf{u}_M(k) \le \Delta\mathbf{u}_{M,\max}$$

where $\delta(k) = \begin{bmatrix} \delta_{T_b}(k) \\ \delta_{P_0}(k) \\ \delta_{T_0}(k) \\ \delta_{Y_{O_2}}(k) \end{bmatrix}$, $\mathbf{A}_\delta = blkdiag(\begin{bmatrix} 1 \\ \vdots \\ 1 \end{bmatrix}_{P\times 1}, \begin{bmatrix} 1 \\ \vdots \\ 1 \end{bmatrix}_{P\times 1}, \begin{bmatrix} 1 \\ \vdots \\ 1 \end{bmatrix}_{P\times 1}, \begin{bmatrix} 1 \\ \vdots \\ 1 \end{bmatrix}_{P\times 1})$,

$\Delta\mathbf{B}_M(k) = \begin{bmatrix} \mathbf{I}_{M\times M} & \mathbf{0}_{M\times M} & \mathbf{0}_{M\times M} \end{bmatrix} \Delta\mathbf{u}_M(k)$. Here, we adopt the zone control to the operation point, which can ensure the operation point within a permitted zone and provide the freedom to reduce the consumption of coal. In optimization problem (3), $\delta_{T_b}(k)$, $\delta_{P_0}(k)$, $\delta_{Y_{O_2}}(k)$, and $\delta_{T_0}(k)$ are zone variables about bed temperature set point, steam pressure set point, steam temperature set point, and flue gas oxygen content point, respectively. $\delta(k)$ is optimized under certain constraints which should be under allowed values of real process. Weighting matrix λ_1 is introduced as a regulated parameter to reduce the fluctuation of the operation point. In addition, output constrains can be introduced to ensure process requirement when the system is running near the operation point.

In order to achieve the purpose of saving coal, the penalty for the prediction of coal consumption is added to optimization problem (3), i.e. the integral of $\Delta\mathbf{B}_M(k)$ in the control domain $\sum_{k=1}^{M} \Delta\mathbf{B}_M(k) = \mathbf{C}_c\mathbf{B}_c\Delta\mathbf{B}_M(k)$. Here

$\mathbf{C}_c = [1 \cdots 1]_{1\times M}$, $\mathbf{B}_c = \begin{bmatrix} 1 & 0 & \cdots & 0 \\ 1 & 1 & \ddots & \vdots \\ \vdots & \vdots & \ddots & 0 \\ 1 & 1 & \cdots & 1 \end{bmatrix}_{M\times M}$, which is equivalent to the $\mathbf{u}^{TG}$ in

RTO layer and λ_2 is the equivalent weighting parameter as $\mathbf{c}_u$ in RTO layer. $\Delta\mathbf{B}_M(k)$ can be positive or negative. It is easy to find that the closer $\mathbf{B}_M(k)$ meets the lower bound, more coal consumption is reduced. Hence, the linear item of coal consumption is adopted rather than the quadratic item.

4   Case Study

4.1   System Model of a CFB Boiler

Consider a 150 t/h power plant circulating fluidized bed boiler system in Sichuan, China, whose structure is shown in Fig. 3. It can be seen from Fig. 3. that there are three entrances as coal feeding, two entrances at bottom of the CFB tank as primary air, and two entrances of secondary air from different height of bed tank. Manipulated variables of the CFB boiled are bed temperature (°C), main steam pressure (MPa), the main steam temperature (°C) and flue gas oxygen content (%).

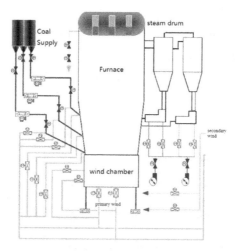

Fig. 3. The circulating fluidized bed boiler system of a power plant

The parameters of the considered 150 t/h circulating fluidized bed boiler are Bed temperature: 850 °C; Steam pressure: 9.8MPa; Steam temperature: 540 °C; Flue gas oxygen content: 3% to 5%. The allowed range of zone control of temperature is 800 to 900°C, the range of steam pressure is 0.01Mpa, the range of steam temperature is 5 °C, and the range of flue gas oxygen-containing is 1%. According to identification method given in [16], equation (4) is identified to shows the linear model of the CFB boiler. The corresponding step response is shown in Fig. 4. Choose sampling period $T = 10s$, the model length $N = 500$. Then, the dynamic matrix can be constructed.

$$
\begin{bmatrix} T_b \\ P_0 \\ T_0 \\ Y_{O_2} \end{bmatrix} =
\begin{bmatrix}
\frac{44.1(1-39.6s)}{(117.43s+1)(878.33s+1)}e^{-150s} & \frac{0.00041(623.26s-1)}{(625s+1)^2(13.3s+1)}e^{-30s} & \frac{0.0024(425.11s-1)}{(323.56s+1)^2}e^{-5.14s} \\
\frac{0.05}{(397.24s+1)^2}e^{-40s} & \frac{0.00015(20.19s+1)}{(173.67s+1)(154.27s+1)} & \frac{0.0011}{(93.774s+1)^2}e^{-80s} \\
\frac{20.158}{(1701.6s+1)(1.475s+1)}e^{-37s} & \frac{0.00051(864.8s+1)}{(296.77s+1)^2}e^{-100s} & \frac{0.00115}{(1270s+1)(250s+1)}e^{-39.43s} \\
\frac{-0.511(211.01s+1)}{(252.8s+1)^2(1000s+1)} & \frac{0.00008(138.6s+1)}{(673.14s+1)^2} & \frac{0.00045}{(74.32s+1)(9.93s+1)}
\end{bmatrix}
\begin{bmatrix} B \\ Q_1 \\ Q_2 \end{bmatrix} \tag{4}
$$

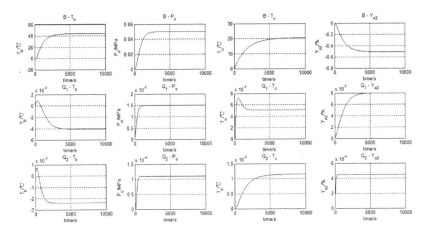

Fig. 4. The step responses of the CFB boiler

4.2 Constrained DMC

Consider online computation, we choose control horizon as $M = 1$ optimization horizon $P = 100$, and the weighting matrices $\mathbf{Q} = blkdiag(\mathbf{Q_c}, \mathbf{Q_c}, \mathbf{Q_c}, \mathbf{Q_c})$, $\mathbf{Q_c} = diag(1, ..., 1, 100) \in R^{100 \times 100}$. Here, we impose the terminal weighting parameter of the terminal output of the optimization horizon as 100 to improve the system stability. In addition, weighting matrix $\mathbf{R} = blkdiag(0.1\mathbf{R_c}, \mathbf{R_c}, 0.1\mathbf{R_c})$, $\mathbf{R_c} = [1]_{1 \times 1}$ The physical constraints of the system include: the amount of coal feeding should be within $[10, 30](t/h)$; the primary air should be within $[10000, 80000](Nm^3/s)$; the secondary air should be within $[10000, 40000](Nm^3/s)$. The control goal is to steer the boiler from the current operating point to a new one as $\omega_r = [800, 4, 470, 3.7]^T$, where the data represents the bed temperature, steam pressure, steam temperature, and flue gas oxygen content. The control results of constrained DMC control are shown in Fig. 5. It can be seen from Fig. 5 that from a given initial state to a new one, the amount of coal consumption, primary air and secondary air eventually converge to a steady vector. Bed temperature, steam temperature and steam pressure reached stable, and basically reached the preset reference value. There is gap between flue gas oxygen content and the preset reference value, which is within 3% to 5%. This gap is resulted by the character of the three-input and four-output system.

4.3 RTO/DMC Optimal Control

According to the settings and operation requirements, set $\underline{\delta} = [-20, -0.01, -5, -1.5]^T$, $\bar{\delta} = [20, 0.01, 5, 0.8]^T$, $\lambda_1 = diag(0.01, 0.01, 0.01, 0.01)$. For the same control goal as Section 4.2, we choose the same parameters as the constrained DMC to achieve a fair comparison and let $\lambda_2 = 100$ to design the RTO/DMC optimal controller. The control results are shown in Fig. 6. As shown in Fig. 6, the dynamic performance of the RTO/DMC control is

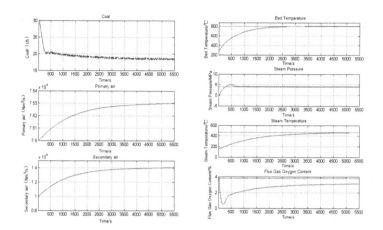

Fig. 5. Control inputs and system output of constrained DMC

similar to that of the constrained DMC. However, as the RTO/DMC controller online optimizes the operation point within an allowed zone according to the coal-saving goal, the controlled variable eventually stabilize within the allowable range of a given operating point of a constant value and the amount of coal feeding is less than that of constrained DMC.

In order to investigate the RTO/DMC coal-saving result further, random disturbance is added to the output of the system. Repeat the test for five times, the statistics amount of coal is given in Table 1. Table 1 tells us that compared to the constrained DMC controller, RTO/DMC controller can save more than 7% of coal consumption, which is significant coal-saving.

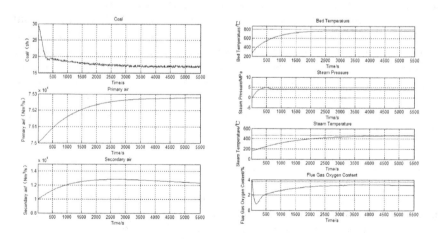

Fig. 6. Control inputs and system output of RTO/DMC

Table 1. Coal Consumption Comparison between Constrained DMC and RTO/DMC

Number	Constrained DMC $B_{CDMC}(t)$	RTO/DMC $B_{RTO/DMC}(t)$	Coal-saving Percentage $\frac{B_{CDMC}-B_{RTO/DMC}}{B_{RTO/DMC}}$ (%)
1	31.1590	28.6834	8.630776
2	31.1551	28.7779	8.260505
3	30.9670	28.8656	7.279946
4	30.9674	28.7987	7.530548
5	31.1106	28.9495	7.465068
Average	31.07182	28.81502	7.833369

5   Conclusion

For the CFB boiler, this paper proposed the optimal control strategy to reduce the coal consumption. By combining the RTO technique and DMC with zone control, the proposed method not only ensures the normal process conditions but also achieve the coal-saving. This proposed method is verified by a case study of a 150 t/h circulating fluidized bed boiler of the power plant in Sichuan.

References

1. Jiang, Y., et al.: Application of Advanced Process Control in Circulating Fluidized Bed Boiler. Applied Mechanics and Materials. Advances in Science and Engineering II, 305–308 (2011)
2. Ma, S.-X., Yang, X.-Y.: Study on Dynamic Behavior of the Combustion System of a Circulating Fluidized Bed Boiler. Proceedings of the Chinese Society for Electrical Engineering 26(9), 1–6 (2006)
3. Yu, X.: Application of fuzzy-PID control algorithm in bed temperature control of Circulation fluidized bed boiler. Journal of North China Electric Power University 32(3), 43–47 (2005)
4. Liu, C.Y., et al.: The Study of the Control of the Bed Temperature in the Circulating Fluidized Bed Boiler Based on the Fuzzy Control System. In: International Conference on Computer and Communication Technologies in Agriculture Engineering, pp. 285–288 (2010)
5. Wu, J., et al.: The Application of BP Neural Network to Bed Temperature Control System of CFB Boiler. Intelligent Systems and Applications, 1–4 (2009)
6. Wang, X.: Dynamic Modeling and Model Predictive Control of the Combustion System of a Circulating Fluidized Bed Boiler. In: 2008 China Doctor Academic Forum (2008)
7. Yan, X.-Z.: Experimental Investigation of the Influence of Coal Powder Fineness on Combustion Characteristic. Journal of Power Engineering 27(5), 684–686 (2007)
8. Xu, Z.: Process for reducing coal consumption in coal fired power plant with steam piping drying. United States Patent Application Publication. US2011/0214427 A1 (2011)
9. Lv, W.: Study on intelligent optimal control of Furnace thermal. Chinese Journal of Scientific Instrument 30(6), 108–113 (2009)

10. Muller, C.J., et al.: Modeling, validation, and control of an industrial fuel gas blending system. Journal of Process Control 21(6), 852–860 (2011)
11. Yang, J.-Q.: Analysis and Design of Control Systems of Circulating Fluidized Bed Boilers. Chinese Journal of Power Engineering 25(4), 517–522 (2005)
12. Xi, Y.: Model Predictive Control. National Defense Industry Press, Beijing (1993)
13. Darby, M.L., et al.: RTO: An overview and assessment of current practice. Journal of Process Control 21(6), 874–884 (2011)
14. Adetola, V., Guay, M.: Integration of real-time optimization and model predictive control. Journal of Process Control 20(2), 125–133 (2010)
15. De Souza, et al.: Real time optimization (RTO) with model predictive control (MPC). Computers and Chemical Engineering 34(12), 1999–2006 (2010)
16. Wu, Z.: Modeling study of 150 t/h circulating fluidized bed boiler combustion system. Control Engineering (to be published)

Optimal Reactive Power Planning
Based on Quadratic Programming Function
of MATLAB Optimization Toolbox

Yang Wu[1], Jinlu Li[2], and Jian Zhang[1]

[1] School of Electrical Engineering of Zhengzhou University, Zhengzhou, China, 450001
[2] North China University of Water Resources and Electric Power, Zhengzhou, China
wuyang3434@qq.com

Abstract. This paper puts forward a method of using the quadratic programming function based on the MATLAB Optimization Toolbox to solve the optimal reactive power planning of the distribution network. The article uses the annual economic benefit maximization for the objective function, establishes the mathematical model of optimal reactive power planning of distribution network and considers the power factor constraints and bus voltage amplitude constraints. Through the practical example, the optimization results prove that this mathematical model and algorithm are effective and efficient.

Keywords: Optimal reactive power planning, quadratic programming, MATLAB Optimization Toolbox.

1   Introduction

With the increase of power and distribution network load, the economic operation of power grid will be the important decision factor in the economic development. The operation and management of the power system should not only pay attention to the safety and reliability, but also consider the economy [1]. The structure of distribution network and power distribution of power system also have changed with the increase of power and load, cause that the reactive power distribution of power system is not reasonable, there are plenty of reactive power current flow in the low voltage circuit, lead to the power supply capability of distribution transformer and distribution line are reduced, so, decrease the network loss, improve the transmission efficiency and the economy of power system are the practical problems of power system operation department must face, these problems also are the main research directions in the power system now.

Although the reactive power does not directly constitute available energy, in the power system it plays an important role [2]. If the reactive power is excess, will make the voltage increase, exceed the voltage upper limits and harm the operation of the power grid and the insulation of equipments, if the reactive power is shortage, can make the voltage is too low, can not meet the load demand, the serious situation can cause voltage collapse.

K. Li et al. (Eds.): ICSEE 2013, CCIS 355, pp. 313–321, 2013.

Optimal reactive power planning is an important way to solve the above problems, it researches that optimal compensation position, capacity and the input time of the newly increased reactive power compensation equipment during a period(generally in the future for 1-3 years). Optimal reactive power planning contains two aspects: from the point of view of operation, should minimize the active power loss and optimize the system operation economy; from the point of view of investment, should minimize the investment of the newly increased reactive power compensation equipment, can described by a objective function and a set of constraints in mathematics. In the area of reactive power optimization, the common optimization methods include: nonlinear programming, linear programming and mixed integer programming. Because of reactive power optimization is a nonlinear problem, so nonlinear programming is used to reactive power optimization firstly, the representative methods include simplified gradient method, Newton method optimal flow algorithm and quadratic programming, and quadratic programming is a mature branch in the field of mathematical programming.

For the reactive optimization problem, the objective function and constraints often have a form of quadratic function, so quadratic programming can be used to solve the problem of reactive power optimization [3]. The advantage of quadratic programming is the better accuracy and reliability, but the computing time increases sharply with the increased of problem scale and can appear the no convergence phenomenon when solve the critical feasibility problems. In fact, for the optimal reactive power planning of the electric power system, the request of computing time is not strict, so quadric programming completely suits for optimal reactive power planning.

This study discusses optimal reactive power planning based on quadratic programming function in MATLAB Optimization Toolbox, the work procedures are shown as following:

1. To establish the mathematical model of the optimal reactive power planning: determine the decision variable, construct the objective function and select the constraint conditions.
2. To program in MATLAB and use quadratic programming function to find the optimal decision variable and the optimal solution of objective function.
3. To analysis and verify the rationality and effectiveness of the mathematical model and optimization method through the actual examples.

2   Mathematical Model of Optimal Reactive Power Planning

Reactive power planning mainly considers the following three aspects: the active power loss of transmission lines, the installation and maintenance costs of reactive compensation devices and the deviation degree of voltage. Through the optimization planning calculation, determine the optimal compensation position and capacity.

2.1   Objective Function

Different distribution network, according to the requirements of the planning and improvement, the objective function can have a variety of forms.

The article uses the annual economic benefit maximization for the objective function which contains two parts: the annual saving power costs after compensating the reactive power Z_e and the annual comprehensive costs of the reactive compensation equipments Z_c.

$$\max \quad F = Z_e - Z_c = (\Delta P_0 - \Delta P_\Sigma) \tau_{max} \beta - K_a K_c \left(\sum_{i=1}^{NC} Q_{ci} \right) \tag{1}$$

where, Q_{c1}, Q_{c2}, ... Q_{cNC} —the compensation capacities of NC compensation points(decision variable).

ΔP_0—the active power loss in the situation of peak load before compensating the reactive power.

ΔP_Σ—the active power loss in the situation of peak load after compensating the reactive power.

τ_{max}—the hours of the annual peak load loss.

β—the active power price.

K_a—the maintenance and depreciation rate of compensation device.

K_c—the comprehensive cost of unit compensation capacity.

If $Z_e > Z_c$, it means that the saving power cost is more than the comprehensive investment of compensation device, so the investment can be considered cost-effective, the difference between the saving cost and the investment is the annual economic benefit, this is the optimization objective of the paper and the optimization purpose is maximum annual economic benefit.

2.2   Constraints

In order to guarantee the quality of electric energy, need to prescribe the upper and lower limits of node voltage, the voltage amplitude constraints are shown as following:

$$V_{k\,min} \leq V_k \leq V_{k\,max} \tag{2}$$

where, V_k —voltage of the node.

$V_{k\,min}$, $V_{k\,max}$ — upper and lower limits of node voltage.

$k = 1, 2, ... KV$, $KV \leq N$, N is the node number, KV is the voltage constraint node number.

The power factor reflects the efficiency of power transmission, constraints are shown as following:

$$\cos \phi_{l\,min} \leq \cos \phi_l \leq \cos \phi_{l\,max} \tag{3}$$

where, $l = 1, 2, ... NF$, $NF \leq NL$, NL is the system branch number, NF is the power factor constraints branch number.

3   Quadratic Programming

Because of the paper's optimization process is achieved with the help of quadratic programming function in MATLAB Optimization Toolbox, so should convert the objective function and constraints of the above section into the standard form of quadratic programming.

The standard form of quadratic programming:

$$
\begin{cases}
\min f(x) = \dfrac{1}{2} x^T H x + f^T x \\
A \cdot x \leq b \\
A_{eq} \cdot x = b_{eq} \\
lb \leq x \leq ub
\end{cases}
\tag{4}
$$

where, H is a $m \times m$ matrix, A is a $n \times m$ matrix, A_{eq} is a $n \times m$ matrix, $x, f, lb, ub \in R^m$, b, $b_{eq} \in R^n$.

3.1   A Method of Solving ΔP_0

For the optimal reactive power planning, this paper uses a simplified method to obtain the branch power according to the load forecast data and the relationship of node-branch. Now, through a simple example to illustrate this method:

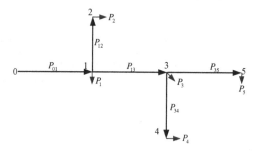

Fig. 1. Simplified illustration of a power distribution network

In the Fig.1, 0-5 are the node numbers (0 is the power supply point), P_{01} - P_{35} is the branch power of each branch which are unknown parameters(the subscript numerical sequence means the direction of power flow), P_1 - P_5 is the load power of each node (if the node is not load node, then the corresponding power is zero), the node-branch association matrix A of Fig.1 is shown as following:

$$A = \begin{bmatrix} 1 & -1 & -1 & 0 & 0 \\ 0 & 1 & 0 & 0 & 0 \\ 0 & 0 & 1 & -1 & -1 \\ 0 & 0 & 0 & 1 & 0 \\ 0 & 0 & 0 & 0 & 1 \end{bmatrix} \tag{5}$$

where, for example, $A_{n,l} = 1$, it means that the direction of branch power l is a input power for the node n, $A_{n,l} = -1$, it means that the branch power l is a output power for the node n, the number 0 means the node and branch don't have the direct relationship, the meanings of other elements are same as above.

When the matrix A is obtained, then, can get the power of each branch according the following equation:

$$A \cdot \begin{bmatrix} P_{01} \\ P_{12} \\ P_{13} \\ P_{34} \\ P_{35} \end{bmatrix} = \begin{bmatrix} P_1 \\ P_2 \\ P_3 \\ P_4 \\ P_5 \end{bmatrix} \tag{6}$$

The active power loss in the situation of peak load before compensating the reactive power also can calculated by the branch power, is shown as following[5]:

$$\Delta P_0 = \sum_{l=1}^{NL} \frac{P_l^2 + Q_l^2}{V_N^2} R_l \tag{7}$$

where, V_N is the rated voltage, R_l is the resistance of branch.

3.2   A Method of Solving ΔP_Σ

The specific expression of ΔP_Σ is shown as following:

$$\Delta P_\Sigma = \sum_{l=1}^{NL} \frac{P_l^2 + \left(Q_l - Q_{l,c}\right)^2}{V_N^2} R_l \tag{8}$$

$$Q_{l,c} = \sum_{i=1}^{NC} \alpha_{l,ci} Q_{ci} \tag{9}$$

where, $Q_{l,c}$ is the sum of the branch l downstream nodes' compensation capacities, α is an $NL \times NC$ adjacency matrix, its meaning can be explained by the Fig.1, if the nodes 2, 4 and 5 are the planned reactive compensation nodes, then, the adjacency matrix α of the Fig.1 is shown as following:

$$\alpha = \begin{bmatrix} 1 & 1 & 1 \\ 1 & 0 & 0 \\ 0 & 1 & 1 \\ 0 & 1 & 0 \\ 0 & 0 & 1 \end{bmatrix} \tag{10}$$

where, the value of the variable $\alpha_{l,ci}$ is only 0 or 1, $\alpha_{l,ci} = 0$, it means that the set of downstream nodes of the branch l don't contain the capacitor node ci ; $\alpha_{l,ci} = 1$, it means that the set of downstream nodes of the branch l contain the capacitor node ci .

3.3   A Method of Solving V_k

The specific expression of V_k is shown as following:

$$V_k = V_s - \sum_{l=1}^{NL} \beta_{k,l} \Delta V_l = V_s - \sum_{l=1}^{NL} \beta_{k,l} \frac{P_l R_l + (Q_l - Q_{l,c}) X_l}{V_N} \tag{11}$$

where, V_s is the power supply point voltage, ΔV_l is the voltage drop of the branch after the reactive power compensation, β is a $KV \times NL$ adjacency matrix [4], it means that the channel from the power supply point s to the node k go past which branches.

In the Fig.1, if the nodes 2, 4 and 5 are the nodes of voltage constraints, then, the adjacency matrix β of the Fig.1 is shown as following:

$$\beta = \begin{bmatrix} 1 & 1 & 0 & 0 & 0 \\ 1 & 0 & 1 & 1 & 0 \\ 1 & 0 & 1 & 0 & 1 \end{bmatrix} \tag{12}$$

The value of the variable $\beta_{k,l}$ is 0 or 1, $\beta_{k,l} = 1$, it means that the channel from the power supply point s to the node k go past the branch l , $\beta_{k,l} = 0$, it means that the channel from the power supply point s to the node k don't go past the branch l .

3.4   Quadratic Expressions of Mathematical Model

Because of ΔP_0 don't contain the decision variable, so don't consider it in the optimization calculation, and quadratic programming function returns the minimum value of the objective function, this paper uses the annual economic benefits as the objective function, the optimization purpose is to get the maximum annual economic benefits, so the objective function should convert into the form of equation (13).

$$\min \quad F = \Delta P_\Sigma \tau_{max} \beta + K_a K_c (\sum_{i=1}^{NC} Q_{ci}) \tag{13}$$

Based on the analysis of the section 3.2 and 3.3, then, the adjusting variables of the formula (13) only are the compensation capacities of the reactive compensation nodes (decision variable), after the formula derivation, the standard quadratic programming form of objective function and constraints are shown as following:

$$obj. \quad \min F = \frac{1}{2} Q_{ci}^T H Q_{ci} + f^T Q_{ci}$$

$$s.t. \quad \begin{cases} -P_l \cdot tg\phi_{l\,min} + Q_l \le \sum_{i=1}^{NC} \alpha_{l,ci} Q_{ci} \le -P_l \cdot tg\phi_{l\,max} + Q_l \\ V_{k\,min} \le V_s - \sum_{l=1}^{NL} \beta_{k,l} \Delta V \le V_{k\,max} \end{cases} \tag{14}$$

where, Q_{ci} is a NC dimensions column vector which contains the reactive compensation capacity of each reactive compensation node, Q_{ci}^T is the transposition of Q_{ci}; H is an $NC \times NC$ matrix, the specific equation is shown as following:

$$H = \frac{2}{V_N^2} \tau_{max} \beta \alpha^T diag(R_l) \alpha \tag{15}$$

where, $diag(R_l)$ is a $NL \times NL$ diagonal matrix, the diagonal elements are the resistances of each branch.

f is an NC dimensions column vector, f^T is the transposition of the vector f, the specific equation is shown as following:

$$f = K_a K_c - \frac{2}{V_N^2} \alpha^T diag(R_l) Q_l \tag{16}$$

Through the above analysis, then it's ready for the next optimization calculation, the specific calculation process is achieved by programming in MATLAB, next, can verify the reasonable and the effective of the mathematical model and algorithm which the paper proposed through a practical example.

4 Actual System Analysis

In a actual distribution network, the parameters are shown as following: $N = 24$, $NL = 23$, $KV = 23$, $NF = 15$, $NC = 15$ (low-voltage buses of 15 distribution transformers as the reactive compensation nodes), $\beta = 0.5$ RMB/kwh ,15 load nodes, the investment cost of the unit capacity of the compensation device is 7000 RMB/Mvar, the fixed cost is 10000 RMB, both are included in the comprehensive cost of unit compensation capacity K_c, $K_a = 11\%$, $V_N = 110KV$, $\tau_{max} = 3000h$.

The results of the compensation capacities of the reactive power compensation points are shown as following Table 1.

Table 1. Reactive Power Compensation Capacities

Node Number	Compensation Capacities of Reactive Device （Mvar）
3	10.3776
5	14.4522
7	6.5662
9	5.3522
10	7.6382
11	7.4143
13	8.8902
15	9.3802
16	4.5822
17	6.1843
19	15.8703
20	15.5853
22	4.1690
23	4.9553
24	3.7553

The results of comparison between the voltages of nodes before the reactive compensation and the voltage of nodes after the reactive compensation are shown as following in Table 2.

Table 2. Results of Voltage Optimization

Node Number	Node Voltage before Reactive Power Compensation	Node Voltage after Reactive Power Compensation
3	0.9445	0.9876
5	0.9245	0.9795
7	0.9598	0.9884
9	0.9626	0.9828
10	0.9542	0.9824
12	0.9467	0.9783
14	0.9551	0.9864
15	0.9671	0.9829
16	0.9602	0.9806
18	0.9096	0.9741
19	0.9094	0.9728
21	0.9604	0.9770
22	0.9564	0.9761
23	0.9609	0.9760
24	0.9480	0.9754

The compensation effect of distribution power network is shown in Table 3.

Table 3. Compensating Effect of Distribution Power Network

Compensating Effect	Ten Thousand RMB
Active Power Loss before Compensating	308.4
Active Power Loss after Compensating	265.2
Z_e	43.2
Z_c	11.3
Annual Economic Benefits	31.9

The optimization results show that, through the reactive power compensation, the voltage quality of nodes and the power factor of branches all have been improved, the reduced operation cost and the annual economic benefits are obvious.

5   Conclusion

Through the practical example, the results prove that the new increased reactive compensation devices can decrease the network loss and improve the quality of voltage, the economic benefits are very obvious, and the mathematical optimization model and quadratic programming algorithm is reasonable and effective.

References

1. Liu, H., Sheng, Z.-J.: Research on reactive power optimization in the power market. Electric Power Automation Equipment 24(12), 33–35 (2004)
2. Mamandur, K.R.C., Chenweth, R.D.: Optimal control of reactive power flow for improvement in voltage profiles and for real power loss minimization. IEEE Trans. PAS-100(7), 3185–3194 (1981)
3. Shen, R.-G.: Power system reactive power integrated optimization-quadratic programming algorithm. Proceedings of the CSEE 6(5), 40–48 (1986)
4. Wang, X.-Z., Li, X.-L.: Topology identification of power network based on incidence matrix. Power System Technology 25(2), 10–13 (2001)
5. Chen, H.: Power system steady analysis. China Electric Power Press (1995)

Research on Zero-Wait Scheduling Problems in Multiproduct Processes with Due Dates

Zhenhao Xu and Xingsheng Gu

Key Laboratory of Advanced Control and Optimization for Chemical Processes, Ministry of Education, East China University of Science and Technology, Shanghai 200237, China
{xuzhenhao,xsgu}@ecust.edu.cn

Abstract. Production scheduling is an important aspect of batch process operations to achieve high productivity and operability. In this paper, we considered a zero-wait multiproduct scheduling with due dates under uncertainty, where the total weighted earliness/tardiness penalty is to be minimized. The imprecise processing time is expressed with the triangle fuzzy variable and the model has been established based on fuzzy cut-set theory. A new improved shuffled frog-leaping algorithm (ISFLA) is introduced to search optimal objective for the given problem, which has a new updating rule in memeplexes. In order to enhance the ability to search the global optimum, the strategy of Forced Moving of the worst frog in each sub-memeplex is proposed to increase the diversity. Simulated results demonstrate that ISFLA has outperformed the conventional SFLA, which is effective and robust in solving the zero-wait scheduling with due-date window.

Keywords: zero-wait, uncertainty, SFLA, forced moving.

1 Introduction

Products manufactured by batch processes are commonly of high-value and low-volume, such as specialty chemicals, pharmaceuticals, agricultural, food and consumer products, and most recently the constantly growing spectrum of biotechnologyenabled products [1]. Batch plants are frequently required to handle the manufacturing of more than one product and are flexibly designed to meet varying product requirements due to market fluctuations. So, scheduling in batch processes with constrained resources is a very complex but important problem.

In a batch plant, intermediate storage between processing units is also important to maintain smooth flow of materials and to satisfy the requirements of processing recipes. The different types of intermediate storage policies which have been frequently studied [2]. A ZW scheduling problem occurs in a production environment in which a product must be processed until completion without any interruption either on or between processing units. This situation is encountered in a variety of industrial processes, for example, in chemical and pharmaceutical industries where a series of processes must follow one another immediately owing to the chemical instability of intermediate products [3]. The ZW scheduling belongs to the NP-complete class of

K. Li et al. (Eds.): ICSEE 2013, CCIS 355, pp. 322–332, 2013.
© Springer-Verlag Berlin Heidelberg 2013

problems [4]. Ryu et al. developed a new multiproduct batch scheduling model with penalty functions for earliness and tardiness [5]. Cafaro and Cerda developed a new MILP continuous-time framework for periodically updating the work schedule of a single unidirectional multiproduct pipeline over a multiperiod rolling horizon [6]. Subbiaha et al. discussed multi-product batch scheduling problems with intermediate due dates [7]. The model formulation is extended to include release dates of the raw materials and due dates of the production orders. And the meeting of due dates is modeled by causing additional costs.

In the paper, the zero-wait scheduling problem in multiproduct processes with the due-date window under uncertainty is discussed, and a newly evolutionary algorithm named Improved Shuffled Fog-Leaping Algorithm (ISFLA) is proposed to solve it. The remaining sections of this paper are organized as follows. In section 2, an *NP-hard* problem, the earliness and tardiness multiproduct scheduling under zero-wait strategy is presented. Section 3 proposes an improved shuffled frog-leaping algorithm and section 4 summarizes experiments and results. The remaining section presents the conclusion.

2   Problem Definition

In a multiproduct process, all products to be produced have the same production route. Considering some assumptions regarding the multiproduct process, multiproduct batch processes scheduling problems can be described as the following.

2.1   Model Formulation

Parameters

N number of products, $(i = 1,2,\cdots,N)$

M number of processing units, $(j = 1,2,\cdots,M)$

$\tilde{O}_{ij}$ fuzzy processing time of product i on processing unit j, which includes the transfer time, the set-up time and the clean-up time, etc.

$[e_i\ t_i]$ due-date window of product i, the earliness due-date and the tardiness due-date are denoted by e_i, t_i respectively.

h_i the weight of earliness,

w_i the weight of tardiness, normally $h_i < w_i$.

Decision Variables

$\tilde{S}_{i_k j}(\tilde{C}_{i_k j})$ starting (completion) time of the product i in the k th position of processing sequence on unit j.

$\tilde{S}_{iM}\left(\tilde{C}_{iM}\right)$ starting (completion) time of the last operation of product i.

$\sigma\left(\sigma_1,\sigma_2,\cdots\sigma_N\right)$ processing sequence under the minimum of *makespan*.

$\tilde{P}$ penalty of products between the due-data window.

$\tilde{d}_{ij}$ the delay between products i and j on the first unit.

A product batch must be scheduled without any wait as soon as it finishes processing on one unit under ZW policy. We used the procedure of calculating delays presented by Reddi and Ramamurthy [8]. Let $\tilde{d}_{ij}$ define the delay between products i and j, when product j follows product i in the sequence. It is given by

$$\tilde{d}_{ij} = \max_{m=2,M}\left\{0, \sum_{k=2}^{m}\tilde{O}_{ij} - \sum_{k=1}^{m-1}\tilde{O}_{ij}\right\} \tag{1}$$

Objective Function

Using these donations, the fuzzy ZW scheduling model is formulated as following,

Minimize:

$$\tilde{P} = \sum_{i=1}^{N}\left[h_i \max\left(0, e_i - \tilde{C}_{iM}\right) + w_i \max\left(0, \tilde{C}_{iM} - t_i\right)\right] \tag{2}$$

Subject to:

$$\tilde{C}_{i_k j} = \tilde{C}_{i_k(j-1)} + \tilde{O}_{i_k j}, \quad \forall j > 1$$

$$\tilde{C}_{i_k j} \geq \tilde{C}_{(i_k-1)j} + \tilde{d}_{(i_k-1)i_k} + \tilde{O}_{i_k j}, \quad i_k \in N, j \in M$$

$$\tilde{C}_{i_k j} = \tilde{S}_{i_k j} + \tilde{O}_{i_k j}$$

$$\tilde{S}_{ij} \geq 0$$

2.2   Solution Description

We used the Triangular Fuzzy Numbers (TFN) to represent the fuzzy processing time of operations, which is determined by a triplet $\tilde{O} = \left(O^L, O^M, O^U\right)$.

A fuzzy set $\tilde{O}$ of the universe X can be specified by a membership function $\mu_{\tilde{O}}(x)$, which takes its value in the interval [0, 1]. It is completely characterized by the set of ordered pairs:

$$\tilde{O} = \left\{\left(x, \mu_{\tilde{O}}(x)\right) \mid x \in X\right\} \tag{3}$$

The concept of fuzzy α-cuts is extended in this paper. It is also called the level of probability. The fuzzy scheduling objective can be transformed into the two different programming function based on the concept of the fuzzy cut-set.
The α-level optimal programming model,

$$min\left\{P_\alpha^L = \sum_{i=1}^{N}\left[h_i\,max\left(0, e_i - C_{iM}{}^L_\alpha\right) + w_i\,max\left(0, C_{iM}{}^L_\alpha - t_i\right)\right]\right\} \qquad (4)$$

$$s.t. \quad C_{i_k j}{}^L_\alpha = C_{i_k(j-1)}{}^L_\alpha + O_{i_k j}{}^L_\alpha, \quad \forall j > 1$$

$$C_{i_k j}{}^L_\alpha \geq C_{(i_k-1)j}{}^L_\alpha + d_{(i_k-1)i_k}{}^L_\alpha + O_{i_k j}{}^L_\alpha, i_k \in N, j \in M$$

$$C_{i_k j}{}^L_\alpha = S_{i_k j}{}^L_\alpha + O_{i_k j}{}^L_\alpha$$

$$S_{ij}{}^L_\alpha \geq 0$$

The α-level worst programming model,

$$min\left\{P_\alpha^R = \sum_{i=1}^{N}\left[h_i\,max\left(0, e_i - C_{iM}{}^R_\alpha\right) + w_i\,max\left(0, C_{iM}{}^R_\alpha - t_i\right)\right]\right\} \qquad (5)$$

$$s.t. \quad C_{i_k j}{}^R_\alpha = C_{i_k(j-1)}{}^R_\alpha + O_{i_k j}{}^R_\alpha, \quad \forall j > 1$$

$$C_{i_k j}{}^R_\alpha \geq C_{(i_k-1)j}{}^R_\alpha + d_{(i_k-1)i_k}{}^R_\alpha + O_{i_k j}{}^R_\alpha, i_k \in N, j \in M$$

$$C_{i_k j}{}^R_\alpha = S_{i_k j}{}^R_\alpha + O_{i_k j}{}^R_\alpha$$

$$S_{ij}{}^R_\alpha \geq 0$$

It can be got that an interval $\left[P_\alpha{}^L, P_\alpha{}^R\right]$ of the penalty objective $\tilde{P}$ at the α-level set, which may help decision-makers gain the variable range of the optimal objective under a certain possible extent.

3   The Improved Shuffled Frog-Leaping Algorithm (ISFLA)

3.1   Shuffled Frog-Leaping Algorithm in General

The shuffled frog-leaping algorithm (SFLA) is a meta-heuristic optimization method that mimics the memetic evolution of a group of frogs when seeking for the location that has the maximum amount of available food. It is based on evolution of memes carried by the interactive individuals, and a global exchange of information among themselves [9]. A number of researchers have studied on the SFLA and its application.

3.2   The Improved Shuffled Frog-Leaping Algorithm (ISFLA)

Based on the general SFLA, the improved Shuffled Frog-Leaping Algorithm (ISFLA) for zero-wait multiproduct scheduling under uncertainty is proposed. There are some key notes in the evolution mechanism of the proposed Shuffled Frog-Leaping Algorithm.

Encoding Scheme. Taking into account the production characteristics of the zero-wait multiproduct batch processes, every product is represented by a character, and it can appear only once in the coding. Then, the sequence of the characters in the coding is the job sequence.

New Updating Rule in Memeplexes. By using the frog-leaping rule of SFLA, it is often produced illegal solution for multi-product problems. And if the model has higher dimension, a feasible solution is impossible by applying the updating rule of SFLA. The original SFLA will always generate a random solution, which makes the algorithm more random search.

Therefore, a modified updating rule is devised to ensure that can both produce the feasible solution and maintain the structure characteristic of the SFLA. This will be clearly illustrated in Fig. 1.

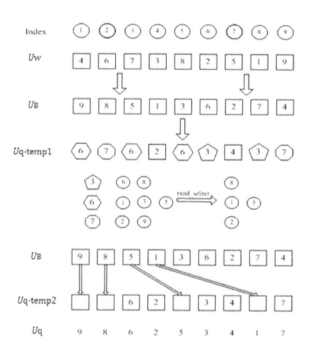

Fig. 1. The illustration of new frog-leaping rule

Forced Moving of the Worst Frog in Each Memeplex. In order to improve the performance of the SFLA and prevent premature convergence, the following strategy of Forced Moving is developed, which is operated by forcing the worst frog to be mutated in sub-memeplex. In the local exploration, when the fitness value of new frog generated randomly is not better than that of the worst frog, the worst frog will be forced to jump by applying the mutation operation. Mutation is just used to produce perturbations on frogs in order to maintain the diversity of memeplexes. In this paper, a mutation operation related with the maximum step size Smax is designed.

4   Experimental Analysis

The first experiment is the problem considered is with five processing units and ten products under zero-wait strategy. Table 1 lists the uncertain processing time of various products expressed by the Triangular Fuzzy Numbers. And the due-date windows, penalty weights of products are also in it. Fig. 2 and 3 are respectively the evolution curves of the models based on the fuzzy-set theory when α is 0.7.

Table 1. The fuzzy processing times (unit), due-date (unit) and penalty weights of products

Products	UNIT 1	UNIT 2	UNIT 3	UNIT 4	UNIT 5	Due-date Windows $[E, T]$	Weights of E/T (h, w)
1	(37 40 43)	(13 15 18)	(11 12 13)	(22 25 28)	(8 10 11)	[350,450]	(2, 3)
2	(6 7 9)	(39 41 44)	(21 22 23)	(32 36 38)	(7 8 9)	[300,400]	(2, 3)
3	(37 41 43)	(136 155 166)	(30 33 35)	(106 121 131)	(145 160 174)	[850,1000]	(3, 4)
4	(10 12 14)	(69 74 82)	(20 24 27)	(44 48 53)	(72 78 82)	[550,650]	(5, 5)
5	(5 7 8)	(82 95 101)	(67 72 78)	(48 52 54)	(140 153 166)	[700 ,800]	(4, 4)
6	(11 12 14)	(11 14 15)	(60 62 68)	(29 32 35)	(131 162 176)	[250,300]	(2, 2)
7	(9 11 14)	(6 7 8)	(27 31 33)	(23 26 28)	(29 32 35)	[300,350]	(4, 4)
8	(28 31 35)	(37 39 41)	(123 141 158)	(5 6 8)	(17 19 21)	[400,500]	(3, 4)
9	(28 32 33)	(88 92 95)	(11 12 13)	(11 14 16)	(93 102 112)	[350,450]	(3, 5)
10	(23 27 29)	(105 114 121)	(19 21 22)	(84 90 96)	(48 52 59)	[500,600]	(2, 3)

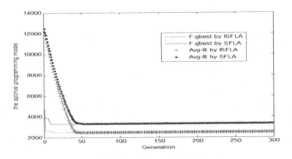

Fig. 2. Convergence characteristic of the optimal programming model $(\alpha = 0.7)$

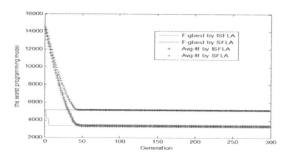

Fig. 3. Convergence characteristic of the worst programming model ($\alpha = 0.7$)

Fig. 2 and 3 illustrate the evolution curve of the two models. In each figure, curve *F_gbest* is the optimal value of objective, which shows the best frog in each iteration; curve *Avg_fit* is the mean value of the objective of all frogs in memeplexes. Along with the increases of evolutionary generation, the curve gradually tends to a stable value, which indicating that the convergence of the method. When α is 0.7, the optimal objective of the optimal programming is 2280.5, and the processing sequence is [6 8 9 2 1 7 4 10 5 3]; the optimal objective of the worst programming is 3569.5, and the processing sequence is [7 2 1 6 8 9 4 10 5 3].

The parameter selection is critical to ISFLA performance. ISFLA has four parameters: the number M of memeplexes, the number N of frogs in a sub-memeplex, the number Nmax of local evolution in a sub-memeplex and the maximum step size Smax. In the test, the maximum step size Smax is devised to be the same with the number of products, so the other three parameters will be determined through extensive experiments. Some benchmark instances with different sizes have been selected to test, which have the objective of makespan. For each test problem, a total of 10 run with different combinations of the parameters. Firstly, the frog memeplex is fixed to 10, and each sub-memeplex has 20 frogs. The local exploration for each sub-memeplex is set to 15, 30, 50, 75, and 100. And we defined the maximal deviation MDP in this paper as follows,

$$MDP = \frac{opt\_max - opt\_min}{opt\_min} \times 100\% \tag{6}$$

Supposed *opt_max* be the maximum objective value found in the optimization computation, *opt_min* be the minimum objective. MDP can indicate the extent of scatter of the objective set. Table 2 illustrates the comparison of different local exploration iteration for every sub-memeplex. From the table, when Nmax is set to 15 or 75,100, the MDP value is not better than that of 30.

Table 2. Comparimion of *N*max Value of Different Size

	Nmax				
	15	30	50	75	100
Ta001(20x5)	0	0	0	2.32	1.41
Car1(11x5)	0	0	0	0	0
Car3(12x5)	1.79	1.5	1.79	3.0	1.79
Car7(7x7)	2.87	0.8	0.8	0.8	0.8
reC03(20x5)	2.06	1.97	2.51	2.43	1.98
Ta001(20x5)	0	0	0	2.32	1.41

Then, the frog memeplex is fixed to 10, and the local exploration is set to 20. The number of frogs in each sub-memeplex is set to 12, 25, 33, 50, and 66. Fig. 4 shows the variance of MDP with frogs' amount in sub-memeplexes. At last, the number of frogs in sub-memeplexes is fixed to 30, the number of local iteration for each sub-memeplex is set to 20, and the number of sub-memeplex is set to 5, 10, 15, 20, and 30. We can get the comparison of different size problems in the test.

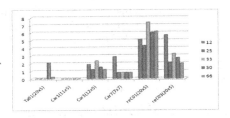

Fig. 4. Comparison of the different number of frogs in sub-memeplexes

Based on those experimental results, the suitable values are found as follows: the number of memeplexes and the number of frogs in each sub-memeplex are 15 and 25, respectively, number of processing cycles on each memeplex before shuffling is 30.

To further analyze the performance of the proposed method ISFLA, we compared our proposed algorithm on the same type of scheduling problems with the general SFLA. In the simulation, we used some benchmarks which would be fuzzed as the test examples. The triangular number $\tilde{O}(O^L, O^M, O^U)$ is applied to denote the processing time of experimental job, where O^M is data of some typical hard benchmarks, O^L and O^U can be got by $O^M - rand(0, O^M * 0.2)$, $O^M + rand(0, O^M * 0.2)$. The due-date window is assumed by $[ms - rand(0, ms * 0.2), ms + rand(0, ms * 0.2)]$. The earliness weights h_i and tardiness weights w_i are uniformly generated in the interval [1, 6], and each instance was randomly performed 10 times. Statistic results are shown in Table 3 and Table 4.

Table 3. Comparison of the Two Algorithms for Some Fuzzed Benchmarks (SFLA)

Fuzzed Benchmarks	α	The optimal model	The worst model
		Best/Avg/Worst	*Best/Avg/Worst*
reC03(20x5)	0.1	10426/**11749**/13173	14304/**16404**/18477
	0.3	10664/**12319**/13964	13693/**15281**/17692
	0.5	10719/**12102**/13717	12537/**14462**/15900
	0.7	10726/**12259**/13744	12817/**14184**/15268
	0.9	11646/**13400**/ 15161	11663/**13497**/15795
Hel2(20x10)	0.1	1379.7/**1536.0**/1698.2	2043.2/**2383.4**/ 2779.5
	0.3	1331.3/**1558.2**/1787.9	1817.5/**2225.1**/2567.8
	0.5	1339.5/**1604.5**/1730.5	1621.5/**2033.2**/2354.5
	0.7	1378.7/**1657.1**/1834.4	1708.9/**2024.9**/2411.8
	0.9	1551.4/**1738.0**/2234.8	1506.1/**1931.6**/2156.2
reC21(30x10)	0.1	39558/**44539**/47597	61499/**66039**/ 72406
	0.3	42984/**46845**/52274	58897/**64931**/72591
	0.5	43151/**48965**/51901	58787/**62089**/65979
	0.7	46060/**50726**/53355	52888/**57325**/61852
	0.9	48685/**53119**/55610	52006/**56309**/62575
reC31(50x10)	0.1	165230/ **169730**/175730	236840/**245410**/255170
	0.3	155370/**174740**/182410	217930/**231780**/240690
	0.5	167140/**184650**/195610	216790/**222720**/228130
	0.7	178310/**191910**/200740	194600/**212680**/220960
	0.9	186430/**197210**/207700	191480/**204290**/215380
hel1(100x10)	0.1	62233/**64763**/67278	88020/**91471**/93879
	0.3	63604/**67107**/69009	86517/**88259**/90413
	0.5	66723/**69200**/71155	81993/**84475**/87200
	0.7	68372/**72274**/74323	78347/**81186**/82925
	0.9	71579/**74719**/78392	73668/**77380**/80358

Table 4. Comparison of the Two Algorithms for Some Fuzzed Benchmarks (ISFLA)

Fuzzed Benchmarks	α	The optimal model	The worst model
		Best/Avg/Worst	*Best/Avg/Worst*
reC03(20x5)	0.1	8174.4/**8782**/9223	11169/**12235**/13608
	0.3	8823.5/**9471**/10439	10590/**11415**/11919
	0.5	8527.0/**9327**/10329	10819/**11309**/12019
	0.7	8699.2/**9643**/10207	10097/**11073**/12723
	0.9	9233.2/**10187**/11263	9534/ **10474**/11257
Hel2(20x10)	0.1	993.9/**1064.3**/1179.9	1342.9/**1576.3**/1737.0
	0.3	1044.0/**1158.3**/1266.3	1313.4/**1526.2**/1797.3
	0.5	1072.0/**1137.5**/1220.0	1296.0/**1446.3**/1674.5
	0.7	1103.3/**1210.3**/1316.9	1217.9/**1348.5**/1514.0
	0.9	1139.9/**1283.3**/1418.6	1211.5/**1337.5**/1490.0
reC21(30x10)	0.1	28703/**31182**/35876	41404/**45752**/50503
	0.3	29115/**31966**/34746	37934/**43180**/47480
	0.5	31954/**33857**/36886	39115/**41652**/49139
	0.7	34014/**35780**/38743	35857/**40090**/44067
	0.9	32916/**37150**/40407	34987/**38091**/40371

Table 4. (*Continued*)

reC31(50x10)	0.1	118850/**127140**/137360	174640/ **197070**/212550
	0.3	127040/**135530**/154350	163770/**180080**/191750
	0.5	129470/**141750**/155910	148820/**172950**/183590
	0.7	132280/**145110**/154160	135600/**159930**/178970
	0.9	141350/**153370**/166680	147570/**160000**/178390
hel1(100x10)	0.1	54972/**59251**/62630	81188/**83072**/84658
	0.3	60079/**61488**/63151	78055/**80143**/83015
	0.5	61398/**62902**/65205	73912/**76444**/78183
	0.7	63148/**65391**/66380	70617/**73361**/77638
	0.9	63853/**68679**/70562	66468/**70733**/73478

From the tables, we can see that the best solution found by ISFLA is better than that of SFLA. Even the worst value is better than that of SFLA. For different the fuzzy α-cuts, the average value of the objective obtained by the proposed method ISFLA in the optimal programming model is smaller than that of the SFLA method, which is also appeared in the worst programming model. Even for the large-scale scheduling problems, it also performed well than SFLA. From those above comparison results, the proposed ISFLA had superior performance than SFLA with respect to convergence and stability.

5   Conclusion

This paper proposes an improved evolutionary algorithm based on the general SFLA method to solve the fuzzy zero-wait scheduling problem with due dates in multi-product batch processes. In the ISFLA, a new frog leaping rule is considered for modification of the general SFLA. And the strategy of Forced Moving of the worst frog in each sub-memeplex is proposed to increase the diversity of memeplexes. The fuzzy mathematical model is proposed to denote the uncertain zero-wait scheduling problems based on the fuzzy cut-set theory. According to simulation results, the proposed algorithm reaches a much better optimal solution in comparison with the SFLA, and proved the feasibility. Furthermore, the proposed method can also be used to solve other complex problems.

Acknowledgments. This work was supported   by National Natural Science Foundation of China (Grant No. 61174040, 61104178), Fundamental Research Funds for the Central Universities.

References

1. Mokeddem, D., Khellaf, A.: Multicriteria Optimization of Multiproduct Batch Chemical Process Using Genetic Algorithm. J. Food Process Eng. 33, 979–991 (2010)
2. Kim, M., Jung, J.H., Lee, I.B.: Optimal Scheduling of Multiproduct Batch Processes for Various Intermediate Storage Policies. Ind. Eng. Chem. Res. 35, 4058–4066 (1996)

3. Lee, D.S., Vassiliadis, V.S., Park, J.M.: List-based Threshold-accepting Algorithm for Zero-wait Scheduling of Multiproduct Batch Plants. Ind. Eng. Chem. Res. 41, 6579–6588 (2002)
4. Papadimitriou, C.H., Kanellakis, P.C.: Flowshop Scheduling with Limited Temporary Storage. In: Proc. 16th Annu. Allerton Conf. Communication, Control, and Computing, Monticello, pp. 214–223 (1978)
5. Ryu, J.H., Lee, H.K., Lee, I.B.: Optimal Scheduling for a Multiproduct Batch Process with Minimization of Penalty on Due Date Period. Ind. Eng. Chem. Res. 40, 228–233 (2001)
6. Cafaro, D.C., Cerda, J.: Dynamic Scheduling of Multiproduct Pipelines with Multiple Delivery Due Dates. Comput. Chem. Eng. 32, 728–753 (2008)
7. Subbiaha, S., Tometzkia, T., Panekb, S., Engell, S.: Multi-product Batch Scheduling with Intermediate Due Dates Using Priced Timed Automata Models. Comput. Chem. Eng. 33, 1661–1676 (2009)
8. Reddi, S.S., Ramamoorthy, C.V.: On the Flow-shop Sequencing Problem with No Wait In the Process. Opl. Res. Q. 23, 323–331 (1972)
9. Eusuff, M., Lansey, K., Pasha, F.: Shuffled Frog-leaping Algorithm: A Memetic Meta-heuristic for Discrete Optimization. Eng. Optim. 38, 129–154 (2006)

The Optimization Algorithm of Torque Compensation for PMSM Systems with Periodic-Nonlinear Load

Ning Chen, Yong Fan, Weihua Gui, Hao Zhang, and Shouyi Yu

School of Information Science and Engineering,
Central South University, Changsha, 410083, China
ningchen@csu.edu.cn

Abstract. Non-linear variations of load torque cycles in the low frequency operation of interior permanent magnet synchronous motor lead to rotor imbalance. To deal with this issue, we apply a low-frequency torque compensation algorithm based on control parameterization method. Given the load torque, the control variable is approximated by a piecewise constant function whose magnitudes are taken as decision vectors. The control problem is thus transferred into a mathematical programming problem , which can be solved by the Sequential Quadratic Programming (SQP) algorithm. The simulation results show that, the state variables are close to their target values. Thus, the method avoids chattering and ensures system stability.

Keywords: IPMSM, control parameterization, sequential quadratic programming algorithm.

1 Introduction

Permanent magnet synchronous motor (PMSM) is becoming more and more popular in the field of Robot, NC Machine Tool, Electric Vehicle, Compressor, because of its high power density, high efficiency, low loss and easy to control. Due to system nonlinearity and instability of control parameters in practice especially in low frequency operations, the reciprocating compressor performs three procedures including suction, compression and exhaust. The speed of rotor is affected by load torque and electromagnetic torque, i.e., when load torque is large but electromagnetic torque is small, the rotor speed is low. Variations in load torque and electromagnetic torque lead to fluctuations of the rotor speed and cause chattering. The cyclical fluctuations of rotor speed within a cycle will lead to system speed imbalances and significantly reduce system performance. It may also cause pipeline rupture when Inverter Air-Conditioner operates in low frequency state for a long time. During the low frequency process, periodic-nonlinear variations of the compressor load torque have fatal consequences to the stable operation of the system. The traditional FOC control strategy is effective for real-time speed control, but the speed loop is slower and the reference torque

K. Li et al. (Eds.): ICSEE 2013, CCIS 355, pp. 333–341, 2013.

current regulated by speed PI is hard to follow the load fluctuation on time(see, for example, [1]). A method of low frequency torque compensation is proposed in[2], but it is a forward-feed compensation, which makes it difficult to accurately locate the compensation point. In[3], an automatic torque estimating schemes is proposed, which is based on current feed forward compensation and load torque estimation. However, it is difficult to accurately estimate the load in actual system . As far as we know, there is no existing work on applying control parameter optimization techniques to attack the torque compensation problem for PMSM systems with periodic-nonlinear load. Therefore, aimed at solving periodic-nonlinear variations of the load torque under low frequency operations of the compressor, we present a method in this paper based on control parameter optimization techniques.

With the increasing development of science and technology, industrial production on the energy saving requirements are extremely urgent, the optimization method has become indispensable to modern theoretical research and industrial production of energy saving methods(see, for example, [4]). Analytical methods and numerical methods are used to solve these optimal control problems. For the analytical method, it includes the variation formulation (see, for example, [5]) and dynamic programming technique (see, for example, [6]). The algorithms developed in this paper belongs to the numerical method, including control parameterization (see, for example, [7]), orthogonal collocation (see, for example, [8]), iterative dynamic programming (see, for example, [9]) and intelligent optimization (see, for example, [10]). The control parameterization has low computationally complexity and does not have instability problems as in intelligent optimization methods. The control parameterization technique divides the time interval [0, T] into several subintervals. Then, the control is approximated by piecewise constant function with possible discontinuities at the partition points, which are called switching times. So the infinite dimensional complex control problem can be transformed into finite dimensional optimization problems with constant control. Based on IPMSM motor mathematical models, we use the sequential quadratic programming algorithm to find the optimal numerical solution of the actual control of the system state equations, thus obtaining the optimal control parameters. The rest of the paper is organized as follows: Sec. II formulates the problem and Sec. III presents the algorithm. Simulations are given in Sec. IV. Finally, a conclusion is give in Sec .V.

2   Problem Description

Unlike surface mount permanent magnet synchronous motor, the inductance of d-coordinate (L_d) of the interior permanent magnet synchronous motor (IPMSM) is not equal to the q-coordinate (L_q). By analyzing the mathematical model of the IPMSM (see, for example, [13]), we obtain the model used in the optimal control algorithm. The motor stator voltage equations in the d-q rotating coordinate system are as below:

$$\begin{cases} L_d \frac{di_d}{dt} = -R_s i_d + n_p \omega L_q i_q + u_d \\[2mm] L_q \frac{di_q}{dt} = -R_s i_q - n_p \omega L_d i_d - n_p \omega \phi + u_q \\[2mm] J \frac{d\omega}{dt} = \frac{3}{2} n_p [(L_d - L_q) i_d i_q + \phi i_q] - \tau_L \end{cases} \tag{1}$$

where i_d and i_q are direct axis and quadrature axis current, respectively. J is the Moment of Inertia, τ_L is the load moment, ϕ is the permanent magnetic flux, R_s is the Stator Resistance, n_p is the role pair number of motor, u_d and u_q are direct axis and quadrature axis voltage, L_d and L_q, respectively, d-q axis stator inductance.

x and u are defined as follows.

$$x = \begin{pmatrix} x_1 \\ x_2 \\ x_3 \end{pmatrix} = \begin{pmatrix} i_d \\ i_q \\ \omega \end{pmatrix}, u = \begin{pmatrix} u_1 \\ u_2 \end{pmatrix} = \begin{pmatrix} u_d \\ u_q \end{pmatrix}. \tag{2}$$

(1) can be rewritten as:

$$\frac{d\mathbf{x}(t)}{dt} = f(t, x(t), u(t)) \tag{3}$$

where $\mathbf{x} = [x_1, \cdots, x_n]^T \in \mathbb{R}^n$, and $\mathbf{u} = [u_1, \cdots, u_r]^T \in \mathbb{R}^r$ are the state and control vectors, respectively and $\mathbf{f} = [f_1, \cdots, f_n]^T \in \mathbb{R}^n$ is as follows.

$$f(t, x(t), u(t)) = \begin{bmatrix} -\frac{R_s}{L_d} x_1 + \frac{n_p L_q}{L_d} x_2 x_3 + \frac{1}{L_d} u_1 \\[2mm] -\frac{R_s}{L_q} x_2 - \frac{n_p L_d}{L_q} x_1 x_3 - \frac{n_p \phi}{L_q} x_3 + \frac{1}{L_q} u_2 \\[2mm] -\frac{3n_p(L_d - L_q)}{2J} x_1 x_2 + \frac{3n_p \phi}{2J} x_2 - \frac{1}{J} \tau_L \end{bmatrix}. \tag{4}$$

here τ_L is a function with a constant period, the fitting curve in a cycle shown in Fig. 1.

In one period, τ_L can be expressed by:

$$\begin{aligned} \tau_L = 2.29 &- 1.75\cos(131.26t) - 3.38\sin(131.26t) - 0.53\cos(262.51t) \\ &+ 0.58\sin(262.51t) \end{aligned} \tag{5}$$

The initial value of the state variables x for differential equations (3) is denoted by:

$$x(0) = x^0 \tag{6}$$

We also define U_1 and U_2 as follows

$$U_1 = \left\{ v = [v_1 \cdots v_\gamma]^T \in \mathbb{R}^n : (E^i)^T v \leq b_i, i = 1, \cdots, q \right\} \tag{7a}$$

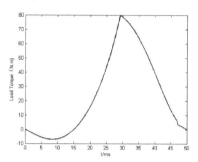

Fig. 1. Load torque curve fitting by a period

$$U_2 = \left\{ v = [v_1 \cdots v_\gamma]^{\mathrm{T}} \in \mathbb{R}^n : c_i \leq v_i \leq d_i, i = 1, \cdots, r \right\} \tag{7b}$$

where E^i is the vector with *gamma* dimenions, b_i, c_i and d_i are real numbers.

Assume that $U = U_1 \cap U_2$, obviously, U is a compact convex subset satisfying $U \in \mathbb{R}^n$. Given interval $t \in [0, T]$, if $u(t) \in U$, then the boundary measurable function u is called admissible control within the interval $[0, T]$. Let $\mathcal{U}$ is a collection of admissible control. For each control variable $\mathbf{u} \in \mathcal{U}$, let $\mathbf{x}(\cdot|\mathbf{u})$ be the corresponding vector-valued functions, which is absolutely continuous and satisfies the differential equation (3) in $[0, T]$. Obviously, this function is also a dynamic system (3) satisfying the initial conditions (6) corresponding to the $\mathbf{u} \in \mathcal{U}$ state solution, so the optimal control problem can be expressed as follows:

Problem 1. Given the system (1), find a control $\mathbf{u} \in \mathcal{U}$ such that the cost function

$$g_0(u) = \int_0^T [(x_1(t) - \bar{x}_1(t))^2 + (x_2(t) - \bar{x}_2(t))^2] dt \tag{8}$$

is minimized subject to the following inequality constraints:

$$\begin{aligned} a \leq x_1 \leq b, \\ c \leq x_2 \leq d \end{aligned} \tag{9}$$

where $\bar{x}_1(t)$ and $\bar{x}_2(t)$ are the targets of the state variable x_1, x_2.

3   Solution Algorithm

This paper will use parameterization to solve optimal control problem with inequality constraints. The method is to set up state variable of system dynamic model as piecewise constant functions with respect to time. The time interval $[0, T]$ is divided into some sub-interval, and each constant control to instantly switch at split point. In order to achieve optimal control of dynamic systems, many numerical optimization methods are used to solve these control parameters. In [7] and [14], the convergence of parametic approach is proved. We use mathematical programming techniques to solve problem 1. We solve the gradient formula of the objective function $g_0(u)$ on $\mathbf{u}$ before solving the original dynamic system optimal control problem. When the state values, the co-state values, the objective function values and the corresponding gradient values of dynamic system are known, the sequential quadratic programming (see, for example, [15-21]) is applied to obtain the optimal solution.

3.1 Gradient Formulae

Consider the following system, which is known as the corresponding co-state system:

$$\frac{d\lambda^0(t)}{dt} = -\left[\frac{\partial H_0(t, [\mathbf{x}, \mathbf{u}, \lambda^0](t))}{\partial \mathbf{x}}\right]^{\mathrm{T}} \tag{10a}$$

Boundary conditions are given as:

$$\lambda^0(t) = 0, t > T \tag{10b}$$

H_0 is the corresponding Hamiltonian function defined by:

$$H_0(t, x, u) = (x_1(t) - \bar{x}_1(t))^2 + (x_2(t) - \bar{x}_2(t))^2 + (\lambda^0)^{\mathrm{T}} f(t, x, u) \tag{11}$$

To solve the dynamical system (3), we divide the interval $[0, T]$ into $[0, h_l]$, $[kh_l, (k+1)h_l], k = 1, \ldots, \eta - 1, [\eta h_l, T]$, where $\eta = \mathrm{INT}(T/h_l)$. The state differential equation in each subinterval is solved individually.

To solve the co-state system (10), we subdivide the interval $[0, T]$ into $[T - h_l, T], [T-(k+1)h_l, T-kh_l], k = 1, \ldots, \eta-1$ and $[0, T-\eta h_l]$, where $\eta = Int\,(T/h_l)$. Then, the co-state system (10) can be solved on each shorter interval.

Theorem 1. Let $\mathbf{u}$ be any control in $\mathcal{U}$ and let $\Delta\mathbf{u}$ be any bounded measurable function defined in $[-h_l, T]$ such that $\Delta\mathbf{u}(t) \in \mathbb{R}^r$ for each $t \in [0, T]$ and $\Delta\mathbf{u}(t) = 0$ for all $t \in [-h_l, 0)$, the directional derivative for the function g_0 is:

$$\Delta g_0(\mathbf{u}) = \lim_{\varepsilon \to 0}\left\{\frac{g_0(\mathbf{u}+\varepsilon\Delta\mathbf{u}) - g_0(\mathbf{u})}{\varepsilon}\right\} = \frac{dg_0(\mathbf{u}+\varepsilon\Delta\mathbf{u})}{d\varepsilon}\bigg|_{\varepsilon=0}$$

$$= \frac{\partial g_0(\mathbf{u})}{\partial \mathbf{u}}\Delta\mathbf{u} \tag{12}$$

$$= \int_0^T \frac{\partial H_0(t)}{\partial \mathbf{u}}\Delta\mathbf{u}(t)dt$$

where $H_0(t) = H_0(t, \mathbf{x}, \mathbf{u})$. In [7] we can get the proof of this theorem.

3.2 Control Parameterization

Using the control parameterization method (see, [7] and [12]), we can partition the time interval $[0, T]$ into q subintervals $[\tau_{j-1}, \tau_j)$, where $j = 1, \ldots, q$. The control variable can be approximated as:

$$\mathbf{u}^q(t) = \sum_{j=1}^{q} \sigma^{q,j} \chi_{[\tau_{j-1}, \tau_j)}(t) \tag{13}$$

where τ_j $(j = 0, 1, \ldots, q-1)$ is interval endpoints. $\sigma^{q,j} \in U$, while $I = \chi_{[\tau_{j-1}, \tau_j)}$ denotes the indicator function of the interval $[\tau_{j-1}, \tau_j)$ which is defined as:

$$\chi_I(t) = \begin{cases} 1, & t \in I \\ 0, & \text{elsewhere} \end{cases} \tag{14}$$

Define $\sigma^q = [(\sigma^{q,1})^\mathrm{T}, \ldots, (\sigma^{q,q})^\mathrm{T}]^\mathrm{T}$ and $\sigma^{q,j} = [\sigma_1^{q,j}, \ldots, \sigma_r^{q,j}]^\mathrm{T}$, $j = 1, \ldots, q$. Furthermore, $\sigma^{q,j} \in U, j = 1, \ldots, q$. Let Ω be the set in $\mathbb{R}^{rq}$ such that $\sigma^q \in \Omega$. Apparently, $\mathbf{u}^q$ depends on the control parameter σ^q. For each control variable $\sigma^q \in \mathcal{U}^q$, the control parameter vector $\sigma^q \in \Omega$ has only a single satisfies (14). On the contrary, the control parameter vector $\sigma^q \in \mathcal{U}^q$ has only a unique $\sigma^q \in \Omega$.

So the **Problem1** can be approximated by the following: Minimizing

$$g_0(\sigma^q) = g_0(u^q(t)) \tag{15}$$

For $\sigma^q \in \Omega$.

To solve this problem, the following conclusion can be obtained:

For each $i = 0, 1, \ldots, N$ and each $j = 1, \ldots, q$, the gradient of the function $g_0(\sigma^q)$ with respect to $\sigma^{q,j}$ is:

$$\frac{\partial g_0(\sigma^q)}{\partial \sigma^{q,j}} = \int_I \frac{\partial H_0(t, x, u)}{\partial \mathbf{u}} dt \tag{16}$$

Thus, the above approximation problem can be transformed into the finite dimension optimal control problem.

4   Simulation Analysis and Verification

In order to verify the feasibility of optimal control parameters approach, we use a Panasonic 5RS102ZAA01 compressor motor for simulations, which is run on a 1.5 air-conditioning compressor-driven experimental platform. The compressor motor parameters are shown in Table 1.

Table 1. Panasonic 5RS102ZAA01compressor motor parameters

MOTOR PARAMTER	Parameter values
Permanent Magnetic Flux	0.105Wb
Number of Pole Pairs	3
Stator Winding	0.49Ω
Direct-axis Inductance	6.3mH
Quadrature axis Inductance	11.8mH
Moment of inertia	0.00063kg.m^2

Differential equations (4) can be expressed as:

$$f(t, x, u) = \begin{cases} -77.78x_1 + 5.62x_2x_3 + 158.73u_1 \\ -41.53x_2 - 1.6x_1x_3 - 26.69x_3 + 84.75u_2 \\ -39.29x_1x_2 + 750x_2 - 1587.3\tau_L \end{cases} \tag{17}$$

subject to the following inequality constraints:

$$-1 \leq x_1 \leq 1,$$
$$0 \leq x_2 \leq 3 \tag{18}$$

Set the time interval $[0, 2]$ (unit: s), the direct axis current i_d target equal to 0A, the i_q target value of 2A. Therefore, the objective function (8) can be expressed as:

$$g_0(u) = \int_0^2 [(x_1(t))^2 + (x_2(t) - 2)^2]dt \tag{19}$$

According to Theorem 1 and (16) and the actual motor parameters given by Tab.1, we can deduce gradient σ_1^j of the objective function $g_0(u)$ on the control parameters σ_1^j and σ_2^j $j = 1, 2, \cdots, 20$:

$$\begin{cases} \frac{\partial g_0(\sigma^q)}{\partial \sigma_1^j} = \int_I 158.73\lambda_1 dt \\ \\ \frac{\partial g_0(\sigma^q)}{\partial \sigma_2^j} = \int_I 84.75\lambda_2 dt \end{cases} \tag{20}$$

Using the obtained model parameters and selected simulation parameters, we use sequential quadratic programming algorithm to calculate the optimal control parameters corresponding to U_1 and U_2. Control curves are shown as Fig.2 and Fig. 3.

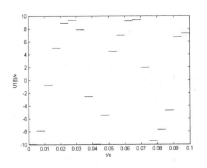

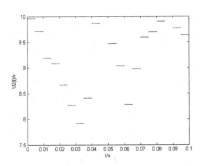

Fig. 2. U_1 **Fig. 3.** U_2

From Figure 2 and Figure 3, we can observe that U_1 and U_2 are much smaller than the actual values after optimization Furthermore, during the process of calculating the optimal control parameters, a new set of state variables are calculated that are shown in Figure 4, Figure 5 and Figure 6. They are closer to the actual state values, which also verifies the effectiveness of the parametric approach.

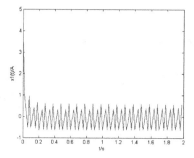

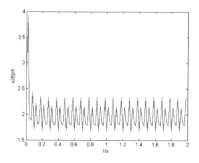

Fig. 4. D axis current state of the curve   **Fig. 5.** Q axis current state of the curve

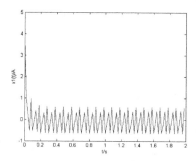

Fig. 6. Angular velocity state curve

5   Conclusion

This paper presents a torque compensation optimization algorithm based on control parameterization to interior permanent magnet synchronous motor control system. Firstly, by solving the control parameters of the gradient formula, the control problem is transformed into a mathematical programming problem, and the optimal solution is obtained. Simulation results show that the control parameters are much smaller than the actual value. Thus, the problem of rotor speed fluctuations in low frequency torque control is successfully solved. It can also be seen from Figure (5) that, the rotor speed closely follows the target speed after the control parameter. Hence system jitter is reduced and the accumulation of errors in the actual process is avoid. Therefore, our results demonstrate that the optimal control method can be a very promising tool for torque compensation in IPMSM system with periodic-nonlinear load.

References

1. Yuan, D.K., Tao, S.G.: AC permanent magnet motor speed regulation system. China Machine Press, Beijing (2011)
2. Chen, W.X.: Research and implementation of inverter air conditioner compressor drive control system. Central South University, Changsha (2011)

3. Huang, H., Ma, Y.J., et al.: Method to reduce the speed ripples of single rotator compressor of air conditioner in low frequency. Electric Machines and Control 15(3), 98–102 (2011)
4. Biegler, L.T., Grössmann, I.E.: Retrospective on optimization. Computers and Chemical Engineering 28(8), 1169–1192 (2004)
5. Chachuat, B., Mitsos, A., Barton, P.I.: Optimal design and steady-state operation of micro power generation employing fuel cells. Chemical Engineering Science 60(16), 4535–4556 (2005)
6. Michaei, B., Jesus, R.G.: Optimal control for linear systems with multiple time delays in control input. IEEE Transactions on Automatic Control 51(1), 91–96 (2006)
7. Teo, K.L., Goh, C.J., Wong, K.H.: A Unified Computational Approach to Optimal Control Problems. Longman Scientific and Technical, New York (1991)
8. Biegler, L.T.: Solution of dynamic optimization problems by successive quadratic programming and orthogonal collocation. Computers and Chemical Engineering 8(3), 243–248 (1984)
9. Luus, R.: Optimal control by dynamic programming using accessible grid points and region contraction. Hungarian Journal of Industrial Chemistry 17(4), 523–543 (1989)
10. Rajesh, J., Gupta, K., Kusumakar, H.K., et al.: Dynamic optimization of chemical processes using ant colony framework. Computers and Chemical Engineering 25(6), 583–595 (2001)
11. Wang, L.Y.: Modelling and optimization method based on control parameterization in purification process of zinc hydrometallurgy. Central South University, Changsha (2009)
12. Chen, N., Shen, X.Y., Gui, W.H., et al.: Control parameterization method of linear systems with multiple time-delays. Control Theory and Applications 28(8), 1099–1104 (2011)
13. Chen, B.S.: Electric drive automatic control system (The third version). China Machine Press, Beijing (2003)
14. Goh, C.J., Teo, K.L.: Control parametrization: A unified approach to optimal control problem with general constraints. Automatica 24(1), 3–18 (1988)
15. Schittkowski, K.: NLPQLP: A fortran Implementation of a Sequential Quadratic Programming Algorithm with Distributed and Non-monotone Line Search. University of Bayreuth, Bayreuth (2004)
16. Schittkowski, K.: NLPQLP: A fortran subroutine solving constrained nonlinear programming problems. Annals of Operations Research 5(4), 485–500 (1986)
17. Schittkowski, K., Zillober, C., Zotemantel, R.: Numerical comparison of nonlinear programming algorithm for structural optimization. Structural Optimization 7(1), 1–19 (1994)
18. Schittkowski, K.: Solving nonlinear programming problems with very many constraints. Optimization 25(2), 179–196 (1992)
19. Schittkowski, K.: Parameter estimation in systems of nonlinear equations. Numerische Mathematik 68(1), 129–142 (1994)
20. Sirisena, H.R., Chou, F.S.: An efficient algorithm for solving optimal control problems with linear terminal constraints. IEEE Transactions on Automatic Control 21(2), 275–277 (1976)
21. Sirisena, H.R.: Computation of optimal controls using a piecewise polynomial parameterization. IEEE Transaction on Automatic Control 18(4), 409–411 (1973)

Boundary Feedback Control Design
and Stability of Open-Channel Networks

Lihui Cen[1,2], Yugeng Xi[2], Dewei Li[2], and Yigang Cen[3]

[1] Department of Control Engineering, Central South University, Changsha, China
[2] Key Laboratory of System Control and Information Processing,
Ministry of Education, and Department of Automation,
Shanghai Jiaotong University, Shanghai, China
[3] School of Computer and Information Technology, Beijing Jiaotong University,
Beijing, China
lhcen@csu.edu.cn, {ygxi,dwli}@sjtu.edu.cn, ygcen@bjtu.edu.cn

Abstract. This paper proposed a boundary feedback control design for open-channel networks with trapezoidal cross sections by using a Riemann invariants approach. The open-channel network is well modeled by the nonlinear Saint-Venant equations. Based on the characteristic form in terms of Riemann invariants, the stabilizing boundary control is developed for a single canal. The stability condition and the boundary control design are subsequently generalized to open-channel networks composed by multireaches in cascade. The design of the boundary feedback control laws either for a single canal or for the cascaded networks is illustrated in a unified design framework.

Keywords: Open canals, Saint-Venant equations, Riemann invariants, boundary feedback control, stability analysis.

1 Introduction

The well-known Saint-Venant equations are commonly used in hydraulics to describe the water flow in open channels [1]. As a standard tool for solving engineering problems regarding the dynamics of open channels, these coupled quasi-linear hyperbolic partial differential equations play a key role in modeling and control of irrigation canals and rivers.

The control of open channels has attracting more increasing attention in the last decade. Many regulation methods about boundary control have been proposed ranging from the simplest to the most sophisticated ones in the last decades. In [2], a simplified irrigation model was proposed for controller design. Decentralized predictive control was applied to the linear discrete time model of irrigation canals in [3] and [4]. Litrico linearized the Saint-Venant equations and designed the controllers using control policies [5].

On the other hand, the boundary feedback control of open canals attracts increasing attention of scientists and engineers. The common method uses a Riemann invariant or Lyapunov approaches to investigate the existence of the

K. Li et al. (Eds.): ICSEE 2013, CCIS 355, pp. 342–350, 2013.

continuously differential solutions and the closed-loop stability based on the Saint-Venant equations. Some results also arise in recent years. [6] discussed the boundary feedback control of a single canal and a multi-reach canal in cascade with rectangular cross section by means of Riemann invariants. Coron [7], Li [8] and Cen and Xi [9] proposed a Lyapunov approach to study the boundary feedback control of a single canal and its stability. Litrico proposed a frequency domain approach to investigate boundary control of the linear Saint-Venant equations [10].

Boundary feedback control indicates that the control laws are elementary functions of the boundary water levels as the feedback. However, all the results mentioned above are limited to canals with rectangular cross sections or linearized Saint-Venant equations. For open canals with some particular cross sections such as the trapezoidal one that is the mostly common case, the above results fails to give the boundary feedback controller by only taking the boundary water levels as the feedback. This paper solves this problem. We start from the nonlinear Saint-Vennant equations without any approximation, linearization and discretisation. The aim is to propose a unified framework for the boundary feedback control design not only for a single canal but also for canal networks with multireaches in cascade.

2   Modelling in Open Channels

2.1   Saint-Venant Equations

Let us consider one horizontal canal with trapezoidal cross section delimited by two gates without friction, as shown in Fig. 1.

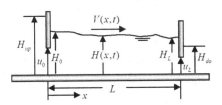

Fig. 1. A canal delimited by two underflow gates

In the following, regardless of a single canal or multi-reach canal we always assume that [11]: (A) The water flow satisfies the sub-critical condition. (B) The water levels at the gate boundaries can be measured online. (C) The gate openings, as the physical control actions, are elementary functions of the water heights $(H(0,t), H(L,t))$ at the gate boundaries.

From [11], the dynamics of such a canal described by Saint-Venant equations with their perturbation forms, the initial conditions and boundary conditions have the following forms respectively

$$\frac{\partial}{\partial t}\begin{pmatrix} a \\ v \end{pmatrix} + \begin{pmatrix} \bar{V}+v & \bar{A}+a \\ g/b(\bar{A}+a) & \bar{V}+v \end{pmatrix}\frac{\partial}{\partial x}\begin{pmatrix} a \\ v \end{pmatrix} = \begin{pmatrix} 0 \\ 0 \end{pmatrix}, (x,t)\in[0,L]\times[0,+\infty) \quad (1)$$

$$\begin{cases} a(x,t)|_{t=0} = a^0(x) \\ v(x,t)|_{t=0} = v^0(x) \end{cases} \quad (2)$$

$$\begin{cases} v(x,t)|_{x=0} = f^0(a(0,t)) \\ v(x,t)|_{x=L} = f^L(a(L,t)) \end{cases} \quad (3)$$

where L is the length of the canal, $A = A(x,t) = \bar{A}+a(x,t)$ denotes the wet cross section at x occupied by water at time t, $a = a(x,t)$ is the perturbation around the equilibrium cross section $\bar{A}$, $V = V(x,t) = \bar{V}+v(x,t)$ is the water velocity at point x at time t, $v = v(x,t)$ is the perturbation around the equilibrium water velocity $\bar{V}$, $b(\bar{A}+a)$ is the width of the water surface at section A, and g is the gravity acceleration constant. f^0 and f^L are continuously differentiable functions of a in a neighborhood of $0 \in R$ and satisfy $f^0(0) = 0$ and $f^L(0) = 0$.

By differentiating boundary conditions (3) with respect to t and using (1), the initial functions $a^0(x)$ and $v^0(x)$ should satisfy the boundary compatibility conditions [12].

In order to find the boundary feedback control laws, it is only necessary to determine a set of boundary conditions (3) to guarantee the initial-boundary value problem described by (1), (2) and (3) has continuously differentiable solutions.

2.2   Models in Terms of Riemann Invariants

The eigenvalues of the matrix $\begin{pmatrix} \bar{V}+v & \bar{A}+a \\ g/b(\bar{A}+a) & \bar{V}+v \end{pmatrix}$ are computed as $\lambda_\pm(a,v) = \bar{V}+v \pm \beta(a)$, where $\beta(a) = \sqrt{g\frac{\bar{A}+a}{b(\bar{A}+a)}}$. From the sub-critical condition, we know $\lambda_-(a,v) < 0 < \lambda_+(a,v)$. Equation (1) is a strictly quasi-linear hyperbolic equation. Thus the Riemann invariants of (1) are computed by

$$\xi_\pm(a,v) = \frac{1}{2}\left(v \pm \int_0^a \sqrt{\frac{g}{(\bar{A}+\alpha)b(\bar{A}+\alpha)}}d\alpha\right) \triangleq \frac{1}{2}\left(v \pm \int_0^a p(\alpha)d\alpha\right) \quad (4)$$

With the new coordinates ξ_+ and ξ_- in (4), the Saint-Venant equations (1), initial conditions (2) and boundary conditions (3) are respectively equivalent to

$$\frac{\partial}{\partial t}\begin{pmatrix} \xi_+ \\ \xi_- \end{pmatrix} + \begin{pmatrix} \lambda_+(\xi_+,\xi_-) & 0 \\ 0 & \lambda_-(\xi_+,\xi_-) \end{pmatrix}\frac{\partial}{\partial x}\begin{pmatrix} \xi_+ \\ \xi_- \end{pmatrix} = \begin{pmatrix} 0 \\ 0 \end{pmatrix}, (x,t)\in[0,L]\times[0,+\infty)$$
$$(5)$$

$$\begin{cases} \xi_+(x,t)|_{t=0} = \xi_+^0(x) \\ \xi_-(x,t)|_{t=0} = \xi_-^0(x) \end{cases} \quad (6)$$

$$\begin{cases} \xi_+(x,t)|_{x=0} = g^0(\xi_-(0,t)) \\ \xi_-(x,t)|_{x=L} = g^L(\xi_+(L,t)) \end{cases} \tag{7}$$

where $\xi_+^0(x) = \frac{1}{2}(v^0(x) + \int_0^{a^0(x)} p(\alpha)d\alpha)$, $\xi_-^0(x) = \frac{1}{2}(v^0(x) - \int_0^{a^0(x)} p(\alpha)d\alpha)$, $g^0(u)$ and $g^L(u)$ are continuously differentiable functions of u in a neighborhood of $0 \in R$, and the initial functions $\xi_+^0(x)$ and $\xi_-^0(x)$ also satisfy the corresponding boundary compatibility conditions.

2.3 Stability Analysis

Theorem 1. Consider the Saint-Venant equations (5) with initial conditions (6) satisfying the boundary compatibility conditions and the boundary conditions selected as

$$\xi_+(x,t)|_{x=0} = m_0\xi_-(0,t),$$
$$\xi_-(x,t)|_{x=L} = m_1\xi_+(L,t). \tag{8}$$

Suppose that
$$|m_0 m_1| < 1 \tag{9}$$

Then if $\|(a^0(x), v^0(x))\|_1$ is sufficiently small, there exists a unique continuously differentiable solution $(a(x,t), v(x,t))^T$ on $(x,t) \in [0,L] \times [0,+\infty)$ to the problem that is defined for all positive t and satisfies the estimate

$$\|(a(\cdot,t), v(\cdot,t))\|_1 < Ce^{-\alpha t} \|(a^0(x), v^0(x))\|_1$$

where $C > 0$ and $\alpha > 0$.

Proof: This Theorem is a generalized result of Theorem 2 in [12].
According to Theorem 1, we can develop the boundary control laws for the trapezoidal canal as follows.

Conclusion 1. If the boundary conditions are designed by the following boundary conditions

$$V(0,t) = \bar{V} - \frac{1+m_0}{1-m_0}\sqrt{\frac{g}{\bar{A}b(\bar{A})}} a(0,t),$$

$$V(L,t) = \bar{V} + \frac{1+m_1}{1-m_1}\sqrt{\frac{g}{\bar{A}b(\bar{A})}} a(L,t), \tag{10}$$

$$|m_0 m_1| < 1,$$

the result of Theorem 1 still holds.

3 Modelling and Control of Multi-reach Open Channels

3.1 Modelling

The aim of this section is to generalize the idea of the boundary control of a single canal to the open-channel networks made up of multireaches in cascade.

Without loss of generality, we take the simple case: two reaches in cascade. These reaches are assumed to have the same length L. As depicted in Fig. 2, $x = 0$ and $x = L$ corresponds to the left and right side of each reach respectively.

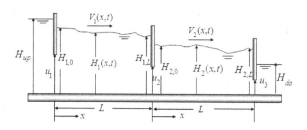

Fig. 2. A canal system consisting of 2 reaches

The Saint-Venant equations are written for each reach i, $(i = 1, 2)$

$$\frac{\partial}{\partial t}\begin{pmatrix} a_i \\ v_i \end{pmatrix} + \begin{pmatrix} \bar{V}_i + v_i & \bar{A}_i + a_i \\ g/b_i(\bar{A}_i + a_i) & \bar{V}_i + v_i \end{pmatrix} \frac{\partial}{\partial x}\begin{pmatrix} a_i \\ v_i \end{pmatrix} = \begin{pmatrix} 0 \\ 0 \end{pmatrix}, \tag{11}$$

$$(i = 1, 2), (x, t) \in [0, L] \times [0, +\infty).$$

Equation (11) has to be complemented by the initial conditions

$$\mathbf{a}(x, 0) = \mathbf{a}^0(x), \mathbf{v}(x, 0) = \mathbf{v}^0(x) \tag{12}$$

where $\mathbf{a}(x, t) = (a_1(x, t), a_2(x, t))^T$, $\mathbf{v}(x, t) = (v_1(x, t), v_2(x, t))^T$.
The boundary conditions are as follows.

$$\begin{aligned}
F_1(v_1(0, t), a_1(0, t)) &= v_1(0, t) - f_1^0(a_1(0, t)) = 0, \\
F_2(v_1(L, t), a_1(L, t)) &= v_1(L, t) - f_1^L(a_1(L, t)) = 0, \\
F_4(v_2(L, t), a_2(L, t)) &= v_2(L, t) - f_2^L(a_2(L, t)) = 0, \\
F_3(v_1(L, t), a_1(L, t), v_2(0, t), a_2(0, t)) &= (\bar{A}_1 + a_1(L, t)) \\
(\bar{V}_1 + v_1(L, t)) - (\bar{A}_2 + a_2(0, t))(\bar{V}_2 + v_2(0, t)) &= 0,
\end{aligned} \tag{13}$$

where $f_i^0(u)$ and $f_i^L(u)$, $(i = 1, 2)$ are continuously differentiable about u in a neighborhood of $0 \in R$. They satisfy $f_i^0(0) = 0$ and $f_i^L(0) = 0$.

The Riemann invariants are computed as $\xi_\pm^i(a_i, v_i) = \frac{1}{2}(v_i + \int_0^{a_i} p_i(\alpha)d\alpha)$, where $p_i(\alpha) = \sqrt{\frac{g}{(\bar{A}_i + \alpha)b_i(\bar{A}_i + \alpha)}}$. For the sake of simplicity, we denote the following vectors $\xi_-(x, t)$, $\xi_+(x, t)$ and $\xi(x, t)$ by

$$\xi_-(x, t) \stackrel{\Delta}{=} (\xi_-^1(x, t), \xi_-^2(x, t))^T,$$

$$\xi_+(x, t) \stackrel{\Delta}{=} (\xi_+^1(x, t), \xi_+^2(x, t))^T,$$

$$\xi(x,t) \triangleq \begin{pmatrix} \xi_-(x,t) \\ \xi_+(x,t) \end{pmatrix}.$$

With the new coordinates in terms of Riemann invariants, (11) is rewritten as

$$\frac{\partial \xi}{\partial t} + \Lambda(\xi) \frac{\partial \xi}{\partial x} = 0 \tag{14}$$

where $\Lambda(\xi) = diag(\lambda_-^1(\xi_+^1, \xi_-^1), \lambda_-^2(\xi_+^2, \xi_-^2), \lambda_+^1(\xi_+^1, \xi_-^1), \lambda_+^2(\xi_+^2, \xi_-^2))$.
The initial conditions (12) are equivalent to

$$\xi(x,0) = \xi^0(x) \tag{15}$$

The boundary conditions (13) are expressed in terms of Riemann invariants by

$$\begin{pmatrix} \xi_-(L,t) \\ \xi_+(0,t) \end{pmatrix} = \mathbf{g} \begin{pmatrix} \xi_-(0,t) \\ \xi_+(L,t) \end{pmatrix} \tag{16}$$

where $\mathbf{g} = (g_2, g_4, g_1, g_3)^T$ is a suitable function. Therefore (14), (15) and (16) constitute the characteristic form of (11), (12) and (13) for cascaded open channels.

3.2 Stability Analysis

Theorem 2 [6]. Consider the Saint-Venant equations (14) with initial conditions (15) satisfying the boundary compatibility conditions and the boundary conditions (16). If

$$\rho(abs(\nabla \mathbf{g}(\mathbf{0}))) < 1 \tag{17}$$

and $\left\| \xi^0 \right\|_1$ is sufficient small, then there exists a unique continuously differentiable solution on $(x,t) \in [0,L] \times [0,+\infty)$ which satisfies the estimate

$$\left\| \xi(\cdot,t) \right\|_{C^1[0,L]} < Ce^{-\alpha t} \left\| \xi^0 \right\|_{C^1[0,L]}, \forall t \geq 0$$

where $C > 0$ and $\alpha > 0$.

In Theorem 2, we observe that the condition (17) is only related to the Jacobian of $\mathbf{g}$ at the origin. Therefore we develop one specific form of the boundary conditions with stability guarantee as the following conclusion.

Conclusion 2. If the parameters m_i, $(i = 0, 1, 2)$ are chosen properly to satisfy the stability condition (17) and the boundary conditions (16) are replaced with the following explicit expressions

$$\begin{cases} V_1(0,t) = \bar{V}_1 - \frac{1+m_0}{1-m_0} \sqrt{\frac{g}{A_1 b_1(A_1)}}(A_1(0,t) - \bar{A}_1) \\ V_i(L,t) = \bar{V}_i + \frac{1+m_i}{1-m_i} \sqrt{\frac{g}{A_i b_i(A_i)}}(A_i(L,t) - \bar{A}_i), \quad (i = 1,2) \end{cases}, \tag{18}$$

the result of Theorem 2 still holds.

4 Design of Boundary Feedback Control

The boundary feedback control for a single canal and cascaded canal networks can be designed according to the following unified procedure

Step 1: Determine the function relationship between $A_1(0,t)$ and $H_1(0,t)$, and also the relationship between $A_i(L,t)$ and $H_i(L,t)$, $(i = 1, 2)$ according to the trapezoidal cross section, and substituting these relationships into (18). Particularly when $n = 1$, it is the case of a single canal.

Step 2: Select the parameters m_i, $(i = 0, 1, 2)$ such that $\rho(\text{abs}(\nabla \mathbf{g}(\mathbf{0}))) < 1$. Particularly when $n = 1$, select m_0 and m_1 such that $|m_0 m_1| < 1$, which is just the case of a single canal.

Step 3: Substitute the expressions achieved by Step 1 and Step 2 into the gate discharge relationships to obtain the boundary feedback control laws.

5 Application Examples

This section will show how to derive the boundary conditions according to the procedure. A canal with a trapezoidal cross section, as depicted in Fig. 3, is selected as an example to demonstrate the advantages of the proposed boundary feedback control.

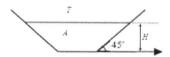

Fig. 3. Schematic of a canal with a constant trapezoidal cross section

Under this type of cross section, we use the boundary conditions of (10) to compute the boundary conditions as follows

$$\begin{cases} V(0,t) = \bar{V} - \frac{1+m_0}{1-m_0} \sqrt{\frac{g}{2\bar{A}\sqrt{\bar{A}+1}}} \left[H(0,t)(H(0,t)+2) - \bar{A} \right] \\ V(L,t) = \bar{V} + \frac{1+m_1}{1-m_1} \sqrt{\frac{g}{2\bar{A}\sqrt{\bar{A}+1}}} \left[H(L,t)(H(L,t)+2) - \bar{A} \right] \end{cases}, |m_0 m_1| < 1.$$

(19)

For the canal delimited by two underflow gates shown in Fig. 1, the gate discharge relationships are expressed by

$$\begin{aligned} x = 0 : A^2(0,t)V^2(0,t) = 2gu_0^2(H_{up} - H(A(0,t))), \\ x = L : A^2(L,t)V^2(L,t) = 2gu_L^2(H(A(L,t)) - H_{do}), \end{aligned}$$

(20)

where u_0 and u_L are respectively the gate opening heights. $H(A(0,t))$ and $H(A(L,t))$ are the upstream and downstream water heights inside the canal. H_{up} and H_{ab} are the water heights outside the canal.

According to Step 3, by substituting the expressions of (19) into (20), the boundary feedback control laws are obtained by

$$
\begin{cases}
u_0(t) = \dfrac{H(0,t)}{\sqrt{2g(H_{up}-H(0,t))}} \left| [\bar{V} - \frac{1+m_0}{1-m_0}\sqrt{\frac{g}{H}}(H(0,t)-\bar{H})] \right| \\
u_L(t) = \dfrac{H(L,t)}{\sqrt{2g(H(L,t)-H_{do})}} \left| [\bar{V} + \frac{1+m_1}{1-m_1}\sqrt{\frac{g}{H}}(H(L,t)-\bar{H})] \right|
\end{cases}, \ |m_0 m_1| < 1. \quad (21)
$$

6 Conclusion

This paper considered the boundary feedback control of canals with trapezoidal cross section, which is the common case in open canals. In such a case, the results achieved so far in the literature don't work well. Starting from the nonlinear Saint-Venant equations, this work investigates the boundary feedback control for a single canal as well as canal networks via a Riemann invariants approach. The main contribution is to propose a unified boundary feedback control design, which extends the results of the literature [6][7][8]. For a single canal, a set of boundary conditions in terms of Riemann invariants to guarantee the closed-loop system exponentially stable is developed. Based on these conditions, the stabilizing boundary control laws are derived. Then the stability condition is generalized to the cascaded case. The boundary control laws thus derived are elementary functions of the water levels at the gate boundaries. The advantage of our proposed approach lies in that the boundary feedback control laws are always elementary functions of the boundary water levels. Thus the boundary feedback control laws can be implemented only by taking the boundary water levels as the feedback.

Acknowledgments. This work was supported in part by the National Natural Science Foundation of China under Grant 61104078, 61074060, 61272028 and 61073079, the Foundation of Key Laboratory of System Control and Information Processing under Grant SCIP2011009, Ministry of Education, the Specialized Research Fund for the Doctoral Program of Higher Education under Grant 20110162120045 and the Fundamental Research Fund of Central South University under Grant 201012200159.

References

1. Chow, V.T.: Open Channel Hydraulic. Mac-Graw Hill Book Company, New York (1985)
2. Schuurmans, J., Clemmens, A.J., Dijkstra, S., Hof, A., Brouwer, R.: Modelling of Irrigation and Drainage Canals for Controller Design. J. Irrig. Drain. Eng. 125(6), 338–344 (1999)
3. Sawadogo, S., Fayer, R.M., Mora-Camino, F.: Decentralized Adaptive Predictive Control of Multireach Irrigation Canal. Int. J. Syst. Sci. 32(10), 1287–1296 (2001)
4. Gomez, M., Rodellar, J., Mantecon, J.A.: Predictive Control Method for Decentralized Operation of Irrigation Canals. Appl. Math. Model. 26(1), 1039–1056 (2002)

5. Litrico, X., Fromion, V.: Control of an Irrigation Canal Pool with a Mixed Control Politics. IEEE Trans. Control Syst. Tech. 14(1), 99–111 (2006)

6. de Halleux, J., Prieur, C., Coron, J.M., d'Andrea-Novel, B., Bastin, G.: Boundary Feedback Control in Networks of Open Channels. Automatica 39(8), 1365–1376 (2003)

7. Coron, J.M., d'Andrea-Novel, B., Bastin, G.: A Strict Lyapunov Function for Boundary Control of Hyperbolic Systems of Conservation Laws. IEEE Trans. Automatic Contr. 52(1), 2–11 (2007)

8. Li, L.: Decentralized Boundary Control of Irrigation Canal Networks via a Strict Lyapunov Method. In: 17th World Congress the International Federation of Automatic Control, Seoul, Korea, pp. 9970–9975 (2008)

9. Cen, L.H., Xi, Y.G.: Lyapunov-based Boundary Feedback Control in Multi-reach Canals. Science in China Series F: Information Sciences 52(7), 1157–1164 (2009)

10. Litrico, X., Fromion, V.: Boundary Control of Hyperbolic Conservation Laws Using a Frequency Domain Approach. Automatica 45(3), 647–656 (2009)

11. Cen, L., Xi, Y., Li, D.: A General Boundary Feedback Control Design for an Open Canal. In: 2011 International Conference on Modelling, Identification and Control, pp. 422–427. IEEE Press, New York (2011)

12. Leugering, G., Georg Schmidt, E.J.P.: On the modeling and stabilization of flows in networks of open canals. SIAM J. Control Optim. 41(1), 164–180 (2002)

Optimization of Train Energy-Efficient Operation Using Simulated Annealing Algorithm

Ting Xie[1], Shuyi Wang[2], Xia Zhao[1], and Qiongyan Zhang[3]

[1] School of Electronic Information and Electrical Engineering, Automation,
Shanghai JiaoTong University, Shanghai, 200240, China
hustxt@126.com, xiazhao@sjtu.edu.cn
[2] Department of Electrical and Computer Engineering, University of British
Columbia, Vancouver, V6T 1Z4, Canada
sdjksdafji@gmail.com
[3] Technical Center of Shanghai Shentong Metro Group, Shanghai, 201103, China
zhangqiongyan77@yahoo.com.cn

Abstract. Rail transit plays an increasingly important role in the public transportation system, and effectively reducing its huge energy consumption is of great practical significance. An optimization method is proposed to minimizes energy consumption by comprehensively considering speed limit, track alignment and running time. The objective function is total energy consumption. The decision variables are the location where train enters the state of coasting. A simulated annealing algorithm(SA) is developed to search for optimized coasting point. The developed model is applied to a particular segment of route in Shanghai. Experiment results demonstrate that, although there was a mite increase of running time, the method could effectively reduce energy consumption.

Keywords: rail, transitenergy, consumptionsimulated, annealingsimulation.

1 Introduction

Rail transit plays an increasingly important role in the urban public transportation system, but it is also energy-intensive. Developing an optimal train speed profile for energy-efficient train operation shows both theoretical and practical significance. Coasting, which is a train running state of removal of traction and brake, is considered to be one of the most efficient energy saving measures for train movements in an inter station run when the running time is guaranteed [1]. Under the constraints of train condition, speed limit, track alignment, and running time, an optimization method, which minimizes energy consumption by selecting appropriate coasting point, could increase efficient energy use and make rail transit system more sustainable.

In the field of energy-efficient control of rail transit, a lot of research has been developed by both domestic and international scholars [2-4]. However, previous studies mainly focused on a single coasting run(such as: traction-coasting-braking). The complicated track alignment and speed limits were not the main

K. Li et al. (Eds.): ICSEE 2013, CCIS 355, pp. 351–359, 2013.

focus. Time-efficient traction strategy is aimed at running at the whole section in the shortest possible time with the most traction and braking capacity [5,6]. An optimization method that the factors, such as speed limit, track alignment, and running time, are taken into account. A simulated annealing algorithm(SA) is developed to search for optimal coasting point. The developed model is applied to a particular segment of route in Shanghai. Experiment results demonstrate that, although there was a mite increase of running time, the method could effectively reduce energy consumption.

2   Energy Consumption

The total consumed energy for a train movement consists of traction energy and brake energy. In the Communication Based Train Control(CBTC) System, traction and brake systems receive command and amount of traction and brake from the on-board Automatic Train Controller (ATC) to produce the energy consumption. Fig. 1 describes the train traction characteristic in the overload situation.

The motion of a train is determined by the difference between tractive effort $F_{traction}$ and running resistance $F_{resistance}$. $F_{resistance}$ is a comprehensive resistance of a variety of resistance, including basic and additional resistance by its formation.

Basic resistance always exists in train movement. The acceleration of basic resistance can be determined by (1)

$$a_{\omega_0} = a + bv + cv^2.$$
(1)

where v is train speed; and a, b, c are basic resistance parameters, determined by operation conditions and train composition.

Additional resistance consists of ramp resistance, curve resistance and tunnel resistance, acceleration respectively:

$$a_{\omega_i} = \frac{[i_1 \times l + i_2 \times (L - l)]g}{10^3 L}.$$
(2)

where i_1, i_2 is track gradient; L is train length; l is the segment length of train on the i_1 ramp.

$$a_{\omega_r} = [\frac{600}{R_1} \times l + \frac{600}{R_2} \times (L - l)]\frac{g}{10^3 L}.$$
(3)

where R_1, R_2 is curve radius; L is train length; l is the segment length of train on the R_1 curve.

$$a_{\omega_s} = \frac{L_s v^2 g}{10^{10} \times 3.6^2}.$$
(4)

where L_s is the tunnel length; v is train speed in the tunnel.

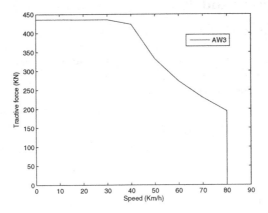

Fig. 1. Train traction curve

Accordingly, the total resistance acceleration(a_ω) is determined by

$$a_\omega = a_{\omega_0} + a_{\omega_i} + a_{\omega_r} + a_{\omega_s}. \tag{5}$$

At time t, the motion of a train can be described by

$$M_{train}(1 + \gamma)a^t = F_{traction}^t - F_{resistance}^t. \tag{6}$$

where M_{train} is train mass; γ is coefficient of rotating mass; $F_{traction}^t$, $F_{resistance}^t$ = tractive effort and resistance at time t respectively.

The consumed engine power P^t for for speed V^t at time t can be formulated as

$$P^t = \frac{F_{traction}^t V^t}{\eta}. \tag{7}$$

where η = power efficiency. e^t is the power consumed per interval Δt, which can be derived as

$$e^t = P^t \Delta t. \tag{8}$$

Accordingly, the total consumed energy(E) for a train movement between two stations is the sum of energy used in all time steps:

$$E = \sum_{t=1}^{N} e^t. \tag{9}$$

where N = total number of times steps needed to travel from one station to another.

3   Optimization Model

3.1   Objective Function

The optimization objective of tractive curve is to minimize energy consumption while keeping the running time under a given maximum value. Therefore, the objective function is formulated as follow:

$$minE = \sum_{t=1}^{N} e^t. \tag{10}$$

3.2   Assumption

To develop the model,some following assumptions are made:

1. The maximum tractive effort will be applied to move a train from standstill until it reaches the speed limit,and the maximum deceleration rate will be applied for braking to reach the next station.

2. In the time-efficient operation mode, the actual running curve of a train is tended to be as close as possible to the speed limit curve. If the speed limit of next section decreases than that of current section, train will have to brake to guarantee safety. In order to reduce braking energy as much as possible, here assume that train is only allowed to brake under specific conditions. These specific conditions are described in section C.

3.3   Decision Variables

The selection of decision variables mainly depends on the distribution of speed limit sections. Firstly, the speed limit sections between two stations are preprocessed to satisfy following rules [8,9]:

As shown in Fig. 2, for any three adjacent speed limit sections XZ1, XZ2 and XZ3, a train in the entrance of XZ2 executing emergency brake with min[V1,V2] should slow down to V3 before the train arrive at XZ3. Therefore, for any two adjacent speed limit sections, if the speed limit of second one is higher than the first one like XZ1 and XZ2, a train can move forward into the second section with the maximum allowed speed of the first section. Then there is no need to set a decision variable in the first section. But if the second one is lower like XZ2 and XZ3, there must exist a operating mode change point x1 in XZ2 to reduce speed to guarantee safely moving into XZ3. Then x1 is the decision variable in this section. When a train starts coasting from x1 to the entrance of next section, if the speed still exceeds the maximum allowed speed of next section, x1 will move back 3 steps automatically and repeat calculating again until the speed limit requirements are met; else if x1 move back to the entrance of the prior section(XZ1) and speed limit requirements are still not met, changes of speed limit are considered too frequently and brake is allowed in the inter-station. Accordingly range of x1 is changed from [S0,S3] to [S0,S2], where s2 is position of start braking. Then repeat optimization of x1.

3.4 Constraints

Constraints should include speed limits and maximum running time. The former has discussed in section C, the following is the constraint of maximum running time:

$$T \leq T_M. \tag{11}$$

In addition, the boundary conditions of variable should be set:

$$S0 < x1 < S3. \tag{12}$$

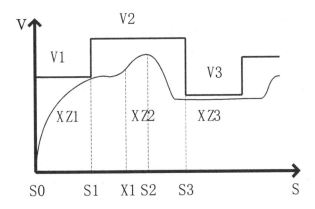

Fig. 2. Preprocessing of speed limit ranges

4   Solution Algorithm

The study's coasting point optimization problem is a large combinatorial optimization one. The classical mathematical method has great limitations for solving this problem.

Therefore, a metaheuristic method is desirable. Comparing to other heuristic search algorithms, Simulated Annealing(SA) algorithm has advantages of descriptive simplicity and efficiency, which is also less affected by initial conditions. Due to inequality constraints, the terminal solution in this study implies an acceptable and close to optimal solution within a reasonable time frame rather than the ideally optimal solution. The steps to execute the proposed SA are as follows:

Step1: Set the initial parameters, initial temperature $(T_{n=0}^{temp})$, the number of trials per temperature(L), temperature change function and minimum temperature;

Step2: Generate the initial solution, and set m=0, repeat step3 and step4 for m=1,2L;

Step3: reproduce a new solution, and set m=m+1, calculate ΔE;

Step4: If $\Delta E \leq 0$, accept the new solution; if $e^{-\frac{\Delta E}{k_B T^{temp}}} > random[0,1]$(where k_B = Boltzmann's constant), accept the new solution; otherwise, go to step3;

Step5: Reduce the system temperature according to the cooling schedule, $T_{n=n+1}^{temp} = T_n^{temp}\alpha^n$ (where n=0,1,2,...;α^n is cooling factor at iteration n; α is ranging between 0.8 and 0.99, here set 0.85);

Step6: If $T_n^{temp} < T_{min}^{temp}$, the algorithm terminates and outputs the optimized solution. Otherwise repeat Step 3 to 5 until the stop criterion is met.

5   Numerical Example

The developed model is applied to a particular segment of Shanghai Metro Line. The experiment segment is a 3200-m-long track route without tunnel, of which the maximum running time is 232s. The train characteristics and options of SA algorithm are listed in Table 1.The Track profile and the associated maximum operating speed are rendered in Table 2.

Table 1. Parameter Table

	Parameters	Values
Train characteristics(AW3)	Train mass	370000 kg
	Train rotating mass	2200 kg
	Train length	140.44 m
	Energy efficiency	0.85
Basic resistance parameter(AW3)	a	1.0079×10^{-2}
	b	0
	c	1.0293×10^{-4}
Simulated annealing options	Initial temperature	100
	Annealing function	Boltzman
	Temperature update	Metropolis rule
	Cooling schedule	Exponential
	Trials per temperature	100
	Minimum temperature	1×10^{-3}

6   Results Analysis

Optimized result is shown in Fig. 3. The solid blue line is track speed limit; the dashed green line is traction operation curve in time-efficient mode; the dotted red line is optimized energy-efficient operation curve. Obviously, in the time-efficient operation mode, the actual running curve of a train is very close to the speed limit curve, but there exists three coasting sectors in the optimized tractive curve, starting with X1, X2, X3 respectively. When determining the location of X2, even if the starting position is set to be the fourth speed limit section, the results still contradict with the requirements, therefore braking mode is started. X3 is in the last section and a train starts coasting for a while from X3 and changes to brake to stop in the station accurately.

Table 2. Track Alignmemt and Speed Limit

Distance(m)	Grade(%)	Radius(m)	V_{max}(Km/h)
0 -160	0	3000	50
160-205	0	3000	65
205-663	-2.74	400	65
663-818	-2	0	70
818-985	-2.7	1000	70
985-1266	-0.91	0	75
1266-1738	-1.01	600	65
1738-2032	-0.33	0	30
2032-3034	0.13	1500	70
3034-3060	0.32	0	70
3060-3200	0.2	0	55

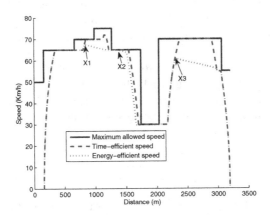

Fig. 3. Train traction operation curve under different modes

According to the simulation results in Table. 3, after optimization, though there is a slightly increase of running time by 4.56%, the energy consumption between two stations decreased significantly by 27.58%. In the previous study, Bocharnikov et al developed a model to optimize the timing of coasting associated with varying acceleration and deceleration rates with GA(2007). The method was applied to a 8.53-km track segment, accordingly 31.27% of energy consumption was saved albeit the travel time increased by 12.5%. It was found that here proposed optimization is very effective.

Table 3. Train operation indicators under time-efficient and optimized modes

Operation mode	Running time(s)	Energy consumption(KWh)	Coasting position(m)		
			X1	X2	X3
Time-efficient	2216	95.93			
Optimized	231.7	69.47	825	1363	2297

7 Conclusion

This paper proposed an optimization for train operation by selecting reasonable coasting point to reduce energy consumption, in which the combinatorial optimization problem used the simulated annealing algorithm. The developed model was applied for a particular track segment in Shanghai, in the given specific train characteristics and constraints(e.g., track alignment, speed limit and running time), There was a enormous decrease in energy consumption by 27.58%, At the same time, the total running time increased by 4.56%; however, this amount of increase is negligible because total running time of the train is allowed to increase by 5% –10% in practice. The evidence implies that the proposed method shows efficiency in consumption reducing and has pragmatic value in sustainability.

An immediate extension of this study is to enhance the developed model by investigating the impact of various track alignments and train characteristics on energy consumption. Additionally, the effect of regenerative braking system to the energy consumption and running time should also be taken into consideration, which will make the model more practical.

References

1. Ma, C., Mao, B., Liang, X., et al.: Coast control of urban train movement for energy efficiency. Journal of Transport Information and Safety 28(2), 37–42 (2010)
2. Kim, K., Chien, S.I.-J.: Optimal train operation for minimum energy consumption considering track alignment, speed limit and schedule adherence. Journal of Transportation Engineering 137(9), 665–674 (2011)
3. Ke, B.-R., Lin, C.-L., Yang, C.-C.: Optimisation of train energy-efficient operation for mass repid transit systems. IET Intelligent Transport Systems 6(1), 58–66 (2012)
4. Ding, Y., Liu, H., Bai, Y., et al.: A two-level optimization model and algorithm for energy-efficient urban train operation. Journal of Transportation Systems Engineering and Information Technology 11(1), 96–101 (2011)
5. Mao, B., He, T., Yuan, Z., et al.: A general-purposed simulation system on train movement. Journal of the China Railway Society 2(1), 1–5 (2000)

6. Wen, X., Du, Z., Yang, L.: Traction curve optimization of straddle-type monorail train based on genetic algorithm. Railway Locomotive & CAR 31(4), 83–87 (2011)
7. Chen, R., Zhu, C., Liu, L.: CBTC based urban rail transit train energy consumption algorithm and simulation. Application Research of Computers 28(6), 2126–2129 (2011)
8. Ye, H.: Introduction to UMT. China Railway Press, Beijing (2009)
9. Zeng, Y., Yu, W., Hu, H., et al.: Hierarchical restriction Method for traction calculation of high-speed railway. China Railway Science 30(6), 97 (2009)
10. Bocharnikov, Y.V., Tobias, A.M., Roberts, C., et al.: Optimal driving strategy for traction energy saving on DC suburban railways. IET Electic Power Applications 1(5), 675–682 (2007)

Artificial Cobweb Intelligent Routing Algorithm for Low-Voltage Power Line Communication

Liang Zhang, Xiaosheng Liu, Yan Zhou, and Dianguo Xu

Harbin Institute of Technology,
Harbin 150001, Heilongjiang Province, China
{LiangZhang,XiaoshengLiu,YanZhou,DianguoXu}xiaozhanghit@163.com

Abstract. This paper presents a novel artificial cobweb routing algorithm (ACRA) of tree-type physical topology for the low-voltage power line communication (LVPLC). The establishment, maintenance and reconstruction of routing are completed. The artificial cobweb routing algorithm is further improved and it has a broader general applicability in LVPLC of smart grid. Based on the cobweb structure, the proposed algorithm formulates communication timing which reduces the channel conflict sharply, and improves the channel utilization. The simulation experiment was based on MATLAB and Opnet14.5, the simulation results show that the ACRA has a greater advantage and practicality in guaranteeing the Quality of Service (QoS) and reliability of LVPLC.

Keywords: Smart Grid, Low-Voltage Power Line Communication, Artificial Cobweb, Routing Algorithm, Reliability.

1 Introduction

With the development of Smart Grid in China, the low-voltage power-line communication (LVPLC) has become one of the potential technologies to commute the information between the end users and the power provider, which has been attracting new attention and becoming one of the research hot topics [1]. However, the LV distribution networks are not designed for communications; they are therefore not favorable for data transmission. Furthermore, power line communication (PLC) systems have to operate with limited signal power in order to ensure low electromagnetic interference (EMI). In the case, the channel noise and signal attenuation are very significant and serious, which can affect the efficiency and reliability of PLC services[2].

To improve the reliability of PLC system, researchers in [3] introduced array codes into the PLC environment and studied the channel's performance, applied Middleton's model for the channel's background and impulsive noise, introduced a novel way of estimating the system's impulsive noise. Authors in [4–10] mainly focus on attenuation characteristics of the channel, channel noise characteristics,

---

[1] This work is supported by National Natural Science Foundation of China (Nos. 60972065, 51277042).

K. Li et al. (Eds.): ICSEE 2013, CCIS 355, pp. 360–372, 2013.
© Springer-Verlag Berlin Heidelberg 2013

impedance matching and coupling circuit design, the application of Orthogonal Frequency Division Multiplexing (OFDM) in PLC and frequency hopping modulation and demodulation techniques etc. These technologies improve the point to point communications of PLC essentially. The reliability of point to point communication does not mean the high reliability of PLC networks, so it is necessary to establish routing for the PLC. Q. Gao, J. Y. Yu etc [11] proposed clustered simple polling (CSP) and neighbor relay polling (NRP) algorithm, these schemes solve the "silent nodes" problem in an Automatic Meter Reading System Using Power-Line Communications and can be evaluated in terms of reliability and efficiency via selecting suitable relay node. But these two schemes did not consider the situation that some nodes in the network send information at the same time which can cause serious data conflict in the channel, the channel utilization will be very low and may not meet the QoS (quality of service) of communication system.

This paper presents an artificial cobweb routing algorithm (ACRA) for treetype physical topology of low voltage distribution based on artificial cobweb structure proposed in [12]. The artificial cobweb routing algorithm is further completed. This paper detailed describes the process of artificial cobweb algorithm, presents corresponding routing maintenance and restructuring strategy. Experimental platform is based on Opnet 14.5. The simulation results show that the artificial cobweb routing algorithm has a greater advantage and practicality in guaranteeing the QoS and reliability of power line communication.

2   Artificial Cobweb Routing Algorithm

2.1   Topology of LVPLC

Signal attenuation between phases is very large for the three phase power distribution grid of the secondary side in the transformer. Without phase coupling, each phase can be regarded as parallel and relatively independent. So that topology of any one phase can be the focus of research. According to the physical topology of low-voltage distribution proposed in reference [13], this paper establishes a single-phase power distribution network shown in Fig.1. BS (Base Station) is responsible for collection and coordination of the data of each terminal node in the network, and connected with wide area network (WAN) for information exchange with external. In practice, signal attenuates with the transmission distance [14], therefore only the terminal nodes, the physical distance of which are close to the BS, are able to ensure the reliable communication. In this case, traditional polling methods cannot guarantee the direct communication between BS and the nodes far away from BS, the success rate of communication may be very low [11]. In order to solve this problem, a novel method called artificial cobweb routing algorithm is proposed. In this method, firstly a communication route is established between BS and part of the nodes. Then the nodes are regarded as relays to expand the communication distance, and gradually all nodes are connected to the communication system. Finally the route between BS and all nodes is completed. Consequently it is necessary to solve the selection of relay nodes, the establishment, maintenance and reconstruction of route, etc.

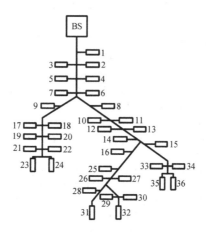

Fig. 1. Tree-type network

2.2   Artificial Cobweb Routing Initialization Algorithm

Reference [12] gives a detailed description of artificial cobweb structure so that this paper will not repeat them. In single-layer artificial cobweb structure, the central node is responsible to coordinate and deal with the information of its peripheral nodes within the subnet. In addition it is the relay for the communication between the central node and BS of other associated subnets. The communication route of BS to any one terminal node is exclusive in the network based on artificial cobweb network.

We assume that the physical link is connected shown in Fig.1, stipulate that:

1) All nodes in the networking process are responsible to record the physical signal strength of the received information.

2) Any node within the network can at least communicate with one other nod.

3) In maximum physical range the signal up to, the node is to find the possible number of nodes.

4) The logical ID of BS is 0, the logical ID acquired node is no longer involved in the new logical ID allocation process.

5) Upon completion of the networking, the central node will not become a bad node (Nodes can't communicate with other nodes caused by the changing of channel condition).

We illustrate the artificial cobweb routing initialization algorithm based on Fig.1:

1) When a networking broadcast is sent by the BS, assuming that it can only be intercepted by nodes 1 to 9, which have a close physical distance to the BS. These nodes send response to BS in turns and are assigned logical ID (1 to 9) by BS. Broadcast from BS is received by nodes 1 to 9, which indicates that the link in the physical range of these nodes is in good condition, and indicates that the 9 nodes can direct or indirect communicate with each other. BS selects a node h_1 from these 9 nodes as the central node.

2) When node h_1 sends a broadcast, the rest 8 nodes receive the broadcast and record the logical ID of h_1 (nodes without logical ID do not respond to the broadcast), and send response to h_1 . Central node h_1 record the logical ID of peripheral nodes to generate a routing table. The routing table will be sent to BS. So that the artificial cobweb topology network of subnet 1 is completed and records that the logical layer, where the 9 nodes locate, is layer 1.

3) Node h_1 sends a networking broadcast, assuming that the broadcast is received by node 10-24, setting that the logical layer, where these nodes locate, is layer 2. Repeat step 1), node h_1 assigns the logical ID and selects central node h_2 . Assuming that h_2 is any one from node 10-16. h_2 sends a broadcast, repeat step 2). Node 17-24, that locate in different branches, do not receive the broadcast from h_2 or receive a broadcast in weak intensity due to the signal attenuation. In this case, the composite index of signal strength β is introduced to make further judgment.

If $\beta_{h1} < \beta_{h2}$, the composite index of signal strength from h_1 recorded by node 10-16 is less than that from h_2 , repeat step 2), make h_2 as central node to form a new network. If $\beta_{h1} > \beta_{h2}$ or broadcast from h_2 is not received, node 17-24 send message to h_1 , repeat step 1), h_1 select a new central node h_3 from these nodes to form a new cobweb network. We assume broadcast from h_2 is not received by node 17-24 which locate in another branch, select h_3 as central node to form a new cobweb network. Stipulate that the logical layer of new cobweb network is layer 2. h_2 and h_3 send the routing table to BS by selecting h_1 as relay. $\beta = s \times \eta^l$, s is the characterization value of physical signal strength, l is the layer number of subnet to which information sources belong, η is the successful probability of communication between different nodes. η can be got by practical statistical values. The higher of η , the higher of reliability of communication between layers; contrary, the lower of the reliability.

4)h_2 and h_3 send a broadcast again, assuming that node 25-36 receive the broadcast from h_2 , repeat step 3), select central node h_4 and h_5 to form new cobweb network respectively in different branches. There are no nodes without logical ID response to broadcast from h_3 and get an empty response, then stop networking.

5)h_4 and h_5 send a broadcast, repeat step 3), there are not nodes without logical ID response to broadcast from h_4 and h_5, get empty response, stop networking. We stipulate that the cobweb logical layer which h_4 and h_5 locate is layer 3. h_4 and h_5 send the route table to BS by selecting h_2 , h_1 as relays. So far the initialization is completed.

If the central node h_4 and h_5 of the third layer still receive nodes without logical ID, then repeat the networking process, until all nodes get the logical ID and are connected into the network. At last, the communication route of all nodes within the network is established. Fig.2 shows the data frame format. Fra_H, Fra_E are frame head and frame end. Src is data source address. Dest is destination address. Con is control indicator, 0, 1 mean query, control respectively. Sta is status indicator, 0, 1 mean request, response respectively. Re1, Re2, Ren are relay nodes. Data is communication data.

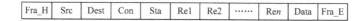

Fig. 2. Data frame format

2.3   An Example of Routing Algorithm

Single-phase in Fig.1 is consist of one BS and 36 terminal nodes. After artificial cobweb initialization algorithm, the MAC layer of the network constitutes the network structure in Fig.3. One broadcast of the BS is intercepted by 9 nodes, and these 9 nodes will then respond to the BS one by one to get the logical ID. The BS selects the node, of which the logical ID is 5, to be the central node of the layer 1, at last these nodes are composed of a cobweb in the MAC layer. The node, of which the logical ID is 5, continues to broadcast, and intercepted by 15 nodes of different physical branches, which are distributed logical ID 10-24. Node 5 selects the node of which the logical ID is 13 as the central node for networking. In case the broadcast of this central node 13 is not intercepted by nodes logical ID 17-24 in another branch, these nodes will send messages to node 5. Node 5 will then reselect node 19 as the central node to complete the networking of layer 2. Central node 13 and 19 search with the same algorithm until the rest of the nodes get logical ID to complete the networking of layer 3. After completing the cobweb route initialization, each central node generates a routing table to record the logical ID and layer of nodes in jurisdiction. The peripheral nodes record the logical ID and layer of its own and the central node. The complexity and length of the routing table is decided by the number of the nodes and the layers in the network. Fig.4 shows the example of data frame format. BS (address is 0) communicate with node 16 via the relay node 5 and relay node 13, BS communicate with node 31 via the relay node 5, relay node 13 and relay node 28.

In order to evaluate the effectiveness of ACRA, a simulation model has been carried out by MATLAB. In the range of 100m×100m, 40 meter units randomly distributed in this area constitute the tree topology of LVPLC, and there are no isolated nodes in it, as shown in Fig.5. Assume that BS located in the center of this area and was set as node1. The rest 39 nodes are the meters units distributed in LV power line network.

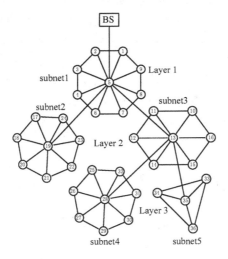

Fig. 3. Result of networking

Although the LVPLC environment is unpredictable, according to the actual point-to-point tests, the overall trend of power line signal transmission quality has a certain relationship with the communication distance due to the signal attenuation. It is assumed that effective communication is generated at random in a certain range, so that a dynamic communication distance could be set to imitate the changing signal transmission environment. In the simulation, the effective communication distance is generated randomly at 25-30m. The proposed ACRA of LVPLC is shown in Figure 6.

Fra_H	0	8	0	0	5	Data	Fra_E

Fra_H	0	16	0	1	5	13	Data	Fra_E

Fra_H	0	31	1	0	5	13	28	Data	Fra_E

Fig. 4. Example of data frame format

According to Figure 5 and Figure 6, each node in tree topology only has single point-to-point link. Whenever one communication link broke down, unless the whole network re-constructed again, the interrupted communication link won't be restored. In the ACRA network, the nodes in the same layer can communicate with other nodes. A meter unit may re-communicate with others due to the additional tangential connections when one communication link interrupted. After networking, the BS communicates with any node in the networking via the center node which it belongs to as relay node, example is shown in Table.1. It is obvious that communication quality of ACRA is more effective than that in

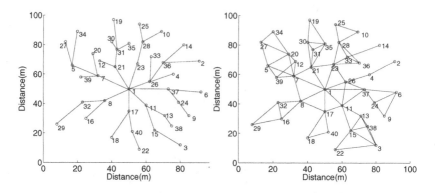

Fig. 5. Tree topology model   **Fig. 6.** Results of ACRA

Table 1. Example of Routing

Source node	Destination node	Relay/Center nodes
1	5	7-20
1	9	37-24
1	22	11-15

tree topology. The ACRA is reliable and strong enough to reduce the network reconfiguration frequency significantly.

3   Routing Maintenance and Reconstruction Algorithm

3.1   Routing Maintenance Principle

By the reasons of openness and time variability of the low-voltage power line network, bad nodes will appear in the operation of the network. In order to ensure the success rate of the data collection for BS, two principles to maintain the route are proposed:

1) After bad nodes caused by routing interrupt in the completed networking, the reconstruction process of local routing for the bad node is to start.

2) During the idle periods of the network, BS initiates the routing detection instructions. If any bad node is found, the principle 1) is in implementation to establish new communication routes.

3.2   Routing Reconstruction Algorithm

As shown in Fig.2, we assume that central node 13 in logical layer 2 has not received any information sent by node 11, so that it is confirmed to be a bad node, node 13 initiates the reconstruction process of local routing:

1) Central node 13 sends reconstruction broadcast to the nodes in its same subnet.

2) Nodes, which intercept the instructions of routing reconstruction, transmit the instructions in turns and add their own logical ID in the data packet.

3) Node 11 records the physical strength of reconstruction broadcast signals, calculate and select the node of max to be a relay node (node 12 in this paper), send response to node 12. Node 11 establishes communication route to central node 13 with the node 12 as relay.

4) Node 12 sends information to node 13, node 13 updates the routing table.

5) Node 13 informs the corresponding node to update routing table up layer by layer until the BS. With this the routing reconstruction process of bad node after networking is completed. When bad node appeared in the network, it is not necessary for all nodes to reconstruct route, only the nodes in the same subnet with bad nodes are involved, so that it highly shorten the route reconstruction and maintenance time, and improve the efficiency of the system.

3.3   Algorithm Timing

The PLC only allows one node to send data at any time, so it is necessary to present a reasonable timing strategy. Otherwise the channel conflict will seriously affect the performance of system. According to artificial cobweb routing algorithm (ACRA), a data transmission timing of nodes in subnet is established to make a comparative simulation with the clustered simple polling (CSP) algorithm mentioned in reference [11]. The interval time of communication between two meter nodes T_{WT} in ACRA is defined as (1). T_P represents the manage processing delay of one meter unit; T_T is the signal transmission time among nodes; T_R indicates the communication redundant time. Generally, the signal transmission time T_T in power line is fast enough to be ignored.

$$T_{WT} > T_P + T_T + T_R \tag{1}$$

According to the protocol and the experimental results, T_{WT} is taken as 150 ms in this paper. Finally the communication effectiveness of cobweb structure is confirmed from the aspect of data collision and channel utilization percent. Figure 7(a) is timing of ACRA, horizontal axis is time, single-layer cobweb is consist of n nodes, node n is the central node, the rest are peripheral nodes, every node sends data message according to the logical ID from small to large in turns to central node of its subnet with a interval T_{WT} . After receiving messages from all nodes, the central node then sends them to the central node or BS. So that it avoids the channel data collision caused by occupation of channel by several nodes at the same time. Figure 7(b) is the timing of CSP, it takes 5 nodes for instance. The node of medium distance will communicate with the BS through the node of close distance as a relay.

Without constant relay nodes and strict timing, the 5 nodes will find the close node by sending random messages, so as to result a large amount of collision of data in the channel. The corresponding simulation result is described in details in the fourth part of this paper.

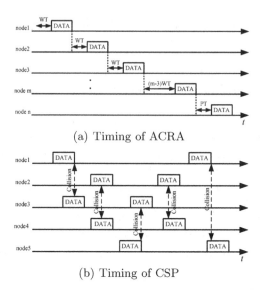

(a) Timing of ACRA

(b) Timing of CSP

Fig. 7. Timing of algorithm

4   Simulation Results and Analysis

4.1   Simulation Environment and Parameter

According to the distribution environment of actual low voltage distribution network, we set 21 terminal nodes and 1 BS node in the range of 50 meters in radius and use PC as simulation platform and Opnet14.5 to compilation and simulation environment. We assume that all nodes form three artificial cobwebs and they stay the same in simulation time. Fig.8 shows the topology structure after networking, subnet_0 represents BS node, subnet_1_0, subnet_2_0 and subnet_3_0 are their central node of each subnet respectively, the other nodes are terminal nodes. Based on the Konnex standard [15], we set the channel transmission rate 2.4Kbps and packet size 24bits. The transmission time of the packet t_x is calculated in formula (2)

$$t_{tx} = p/Baud \tag{2}$$

p is the packet size, is the communication rate.

4.2   Simulation Results and Analysis

According to the generating rate of each node is 24bits/s, the calculated throughput of link subnet_3_0-subnet_2_0 is 168 bits/s. The throughput of link subnet_2_0-subnet_1_0 includes the data of subnet3 and subnet 2, it is 336 bits/s. The throughput of link subnet_1_0-subnet_0 includes the data of all the 21 nodes, it is 504bits/s. The throughput simulation results of link subnet_3_0-subnet_2_0,

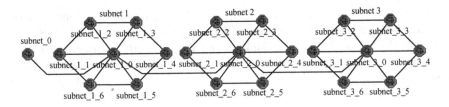

Fig. 8. Simulation model

subnet_2_0-subnet_1_0 and subnet_1_0-subnet_0 are shown in Fig.9(a). The figure shows that the throughput simulation results of every link are identical to theoretical calculation. According to the formula (2), the calculated transmission delay of each link is 0.07s, 0.14s, 0.21s. The simulation result of data transmission delay between each link is shown in Fig.9(b). The delay simulation results are also identical to theoretical calculation.

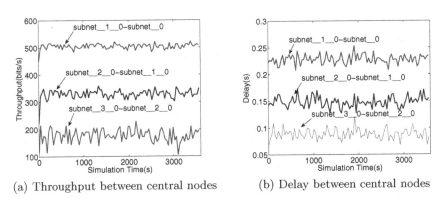

(a) Throughput between central nodes   (b) Delay between central nodes

Fig. 9. Simulation of links between center nodes

The simulation results indicate that in the aspects of throughput and transmission delay, the networking structure can guarantee the QoS of power line communication.

Figure 10(a) shows the simulation result of the throughput of link subnet_2_2-subnet_2_1 before and after routing reconstruction. While node subnet_2_2 operates normally, this link has no data throughput, after routing reconstruction at the time of 1200s, throughput of the link increases sharply and eventually stabilizes at 24 bits/s. And it is agreed with the throughput generated by node subnet_2_2, so that it reveals that this link undertakes the throughput of subnet_2_2 after routing reconstruction. Similarly, Figure 10(b) shows that, link subnet_2_1-subnet_2_0 undertakes the throughput of subnet_2_2 so that its throughput increases sharply after 1200s and eventually stabilizes at about 50 bits/s, the sum of throughput of node subnet_2_2 and subnet_2_1. Figure 10(c) shows the simulation result of throughput of link subnet_3_0 - subnet_2_0 before and after routing reconstruction when subnet_3_2 become blind node. It is clear from the figure that the throughput of this link is approximately constant after routing reconstruction. And it demonstrates that this kind of routing reconstruction method can well solve the problem of blind node, it improves the survivability of the system and it is somehow practically operational.

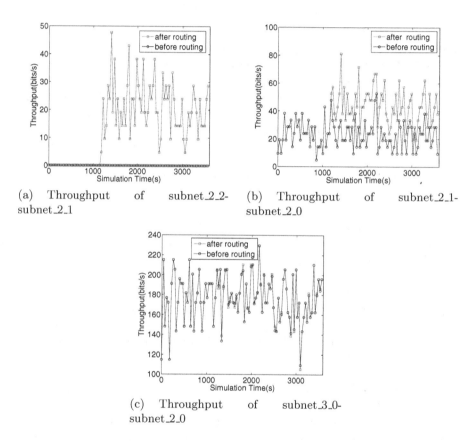

(a) Throughput of subnet_2_2-subnet_2_1

(b) Throughput of subnet_2_1-subnet_2_0

(c) Throughput of subnet_3_0-subnet_2_0

Fig. 10. Simulation results of routing reconstruction

Figure 11(a) is the simulation comparative results of channel conflict between ACRA and CSP. Vertical lines represent the occurrence of channel conflict at this moment. It can be seen from the figure that within the simulation time, conflict caused by ACRA (blue lines) is far less than that caused by CSP (green lines). ACRA divides the whole large network into several small networks, nodes of each one send messages to central node at some certain timing. It reduces the nodes in the channel at one moment to some extent, so that it reduces the conflict sharply, and improves the channel utilization shown in Figure 11(b).

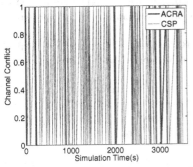

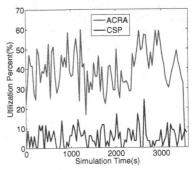

(a) Simulation results of channel conflict   (b) Simulation of channel utilization

Fig. 11. Simulation results of channel status

5 Conclusion

The artificial cobweb routing of tree-type physical topology of LVPLC is analyzed in this paper. The artificial cobweb routing algorithm is further improved. The conclusions are as follows:

1) The novel routing algorithm solves the problem of part of nodes can't access the PLC system caused by the limited data connectivity and improve the reliability of communication. Although the establishing of routing require some time, the principle of route maintenance keeps the process of route maintenance after networking is simple, efficient and easy to operate. But the process of algorithm initialization need some time, which should be improved in later work.

2) Routing reconstruction algorithm can solve the problems of bad nodes in local network, avoid all the nodes involved in the process of routing reconstruction, improve the time efficiency of communication system and enhance the anti-destroying ability of PLC system effectively.

3) Each node sends data according to strict time series, hence it reduces the channel collision rate sharply and improves the channel utilization. Therefore, it ensures the communication reliability of cobweb from the aspect of effectiveness. All these results prove that the cobweb structure is feasibility in routing method to improve the reliability of LVPLC and can guarantee the QoS of PLC. This routing algorithm may also provide a novel routing method for other network such as wireless sensor network, ad hoc network etc.

References

1. Mingyue, Z.: Transmission Characteristics of Low-Voltage Distribution Networks in China under the Smart Grids Environment. IEEE Transactions on Power Delivery 26, 173–180 (2011)
2. Sivaneasan, B., So, P.L., Gunawan, E.: A New Routing Protocol for PLC-Based AMR Systems. IEEE Transactions on Power Delivery 26, 2613–2620 (2011)

3. Nikoleta, A., Fotini-Niovi, P.: PLC Channel: Impulsive Noise Modeling and Its Performance Evaluation Under Different Array Coding Schemes. IEEE Transactions on Power Delivery 24, 585–595 (2009)
4. Zdravko, M., Zeljko, I., Alen, B.: Fixed-Data-Rate Power Minimization Algorithm for OFDM-Based Power-Line Communication Networks. IEEE Transactions on Power Delivery 25, 141–149 (2010)
5. Julian, M., Andrew, E.M.: Effective Communication Strategies for Noise-Limited Power-Line Channels. IEEE Transactions on Power Delivery 20, 887–892 (2007)
6. Van Rensburg, P.A.J., Ferreira, H.C.: Design Evaluation of a Dual Impedance-Adapting Power-Line Communications Coupler. IEEE Transactions on Power Delivery 25, 667–673 (2010)
7. Athanasios, G.L., Panayiotis, G.C.: Broadband Transmission via Underground Medium-Voltage Power Lines-Part I: Transmission Characteristics. IEEE Transactions on Power Delivery 25, 2414–2424 (2010)
8. Okazima, N., Baba, Y., Nagaoka, N., Ametani, A., Temma, K., Shimomura, T.: Propagation Characteristics of Power Line Communication Signals Along a Power Cable Having Semiconducting Layers. IEEE Transactions on Electromagnetic Compatibility 52, 756–759 (2010)
9. Tayyar, G., Eser, Ü., Hasan Basri, Ç., Hakan, D., Kivanç, M.: Noise Modeling and OFDM Receiver Design in Power-Line Communication. IEEE Transactions on Power Delivery 26, 2735–2742 (2011)
10. Jun, Z., Julian, M.: Robust Narrowband Interference Rejection for Power-Line Communication Systems Using IS-OFDM. IEEE Transactions on Power Delivery 25, 680–692 (2010)
11. Gao, Q., Yu, J.Y., Chong, P.H.J., So, P.L., Gunawan, E.: Solutions for the "Silent Node" Problem in an Automatic Meter Reading System Using Power-Line Communications. IEEE Transactions on Power Delivery 23, 150–156 (2008)
12. Xiao-sheng, L., Liang, Z., Yan, Z., Dian-guo, X.: Performance Analysis of Power Line Communication Network Model Based on Spider Web. In: 2011 IEEE 8th International Conference on Power Electronics-ECCE Asia, Jeju, pp. 953–959 (2011)
13. Sivaneasan, B., Gunawan, E., So, P.L.: Modeling and Performance Analysis of Automatic Meter-Reading Systems Using PLC Under Impulsive Noise Interference. IEEE Transactions on Power Delivery 25, 1465–1475 (2010)
14. Charles, J.K., Mohamed, F.C.: Attenuation Characteristics of High Rate Home-Networking PLC Signals. IEEE Transactions on Power Delivery 17, 945–950 (2002)
15. Konnex (KNX) standard: Konnex Assoc. (2002)

Trajectory Matching Algorithm
Based on Clustering and GPR
in Video Retrieval

Tianlu Wang and Xianmin Zhang

Department of Automation, Shanghai Jiao Tong University,
Image Processing and Pattern Recognition Institution of Shanghai Jiao Tong
University, Shanghai 200240
wtltina@yahoo.com.cn, zhangxm@sjtu.edu.cn

Abstract. This paper proposes an approach for trajectory matching in video retrieval. Algorithm is consist of three parts. First, a terse trajectory contains most important temporal and spatial characters are abstracted. Then, abstracted trajectories are classified into several classes by using reformed k-means cluster method according to their position and acceleration features. At the end, Gaussian Process regression model is built and trained using clustered trajectories and trajectories classes which are similar with the given retrieval targets are found out. Advantages of this algorithm include the possibility of generalized clustering for similar trajectories in different scales and partial trajectories matching.

1 Introduction

Current methods used in pictures and videos retrieval are mostly based on text tags, which are labeled manually. These manual operations waste too much time. We need more automatic video processing methods and constructing a new video retrieval system.

The application of moving object trajectory matching is adopted by many literatures. X. Li et al introduce a good two-step approach which makes clustering time effective and calculation efficient, but there is no specific method to solve the problem of limits on clustering trajectories with different scales and lengths [1]. O. Ossama et al adopt core-set which indicates the most essential partitions of a motion trajectory as cluster centroid [2]. K. Kim et al introduce an important and useful matching method which makes trajectory integrity unnecessary, but leave a puzzle on the collection of training samples [3].

There are also many related works about content based retrieval system. Specific methods and future research direction for video retrieve system are analyzed by W. Hu et al [4]. M. Broilo et al present the most recent motion trajectory representation and matching methods [5]. Z. Zhang et al and R. Hu et al show us the common useful distance measures, cluster algorithms and video data-sets [6] [7].

K. Li et al. (Eds.): ICSEE 2013, CCIS 355, pp. 373–381, 2013.
© Springer-Verlag Berlin Heidelberg 2013

R. Hu et al and L. Le Thi introduce retrieval methods using free-hand sketched queries as target ones [8] [9]. C. Williams et al adopt relevance feedback strategies to build a more user-centered retrieval system [10].

In the following parts, these methods are adopted: main segments abstraction, reformed k-means cluster and Gauss Process Regression model. Advantages of this processing framework are: (a) By extracting main features of trajectories, nonsignificant features like scale or length take no opposite effect on similar trajectories clustering and have less requirement on time synchronization. (b) Gaussian Process Regression model has no request on the integrity of an motion trajectory. In other words, even the similar trajectories of a partial one can be found out.

The rest of this paper is organized as follows: In section 2, algorithm framework is described in details and each part of this method is introduced successively. Simulation results are presented and explained in Section 3. Then the conclusion is shown in Section 4.

2   Framework for Trajectory Matching

The whole algorithm can be divided into three parts: trajectory disposing, trajectories clustering and particular trajectory matching.

2.1   Trajectory Disposing

Sampling. Moving objects are essential temporal-spatial features of videos. This paper starts from sampling points from available trajectories and these points contain features like position, velocity and acceleration. Sample points as below

$$Point(x, y, t, v_x, v_y, a_x, a_y) \tag{1}$$

(x, y) is the 2D position of the point. t is the instantaneous time when moving object goes through (x, y). v_x and v_y, a_x and a_y are corresponding instantaneous velocities and accelerations in each direction respectively.

Interval distance between $Point_{i+1}$ and $Point_i$ is constant denoted by l. And then, trajectories are divided into segments as below

$$Segment(PointS, PointE, \theta, a) \tag{2}$$

$PointS$ and $PointE$ are the starting point and ending point. θ is the angle that expresses the segment's motion direction. a expresses the average velocity change of moving object from $PointS$ to $PointE$.

θ and a are measured as

$$\theta = \begin{cases} \left(\frac{180}{\pi}\right)\left(arctan\left(\frac{S_{E_y}-S_{S_y}}{S_{E_x}-S_{S_x}}\right)\right), \\ \quad \text{if } S_{E_y} - S_{S_y} \geq 0 \text{ and } S_{E_x} - S_{S_x} > 0; \\ \\ 180 - \left(\frac{180}{\pi}\right)\left(arctan\left(\frac{S_{E_y}-S_{S_y}}{S_{E_x}-S_{S_x}}\right)\right), \\ \quad \text{if } S_{E_y} - S_{S_y} \geq 0 \text{ and } S_{E_x} - S_{S_x} < 0; \\ \\ 180 + \left(\frac{180}{\pi}\right)\left(arctan\left(\frac{S_{E_y}-S_{S_y}}{S_{E_x}-S_{S_x}}\right)\right), \\ \quad \text{if } S_{E_y} - S_{S_y} < 0 \text{ and } S_{E_x} - S_{S_x} \geq 0; \\ \\ 360 - \left(\frac{180}{\pi}\right)\left(arctan\left(\frac{S_{E_y}-S_{S_y}}{S_{E_x}-S_{S_x}}\right)\right), \\ \quad \text{if } S_{E_y} - S_{S_y} < 0 \text{ and } S_{E_x} - S_{S_x} < 0. \end{cases} \tag{3}$$

$$v = \sqrt{v_x^2 + v_y^2} \tag{4}$$

$$a = \frac{S_{E_v} - S_{S_v}}{S_{E_t} - S_{S_t}} \tag{5}$$

A moving object trajectory is a finite sequence of segments: $S_m = \{S_1, S_2, ..., S_m\}$ and points $Pt = \{Pt_1, Pt_2, ..., Pt_{m+1}\}$.

Extracting. In order to extract the important representative points and segments, we use angulation θ and acceleration a as constrains to detach those unimportant ones.

Assume that S_a is a main segment and S_b is a subsequent segment of S_a. If $|(S_a)_\theta - (S_b)_\theta| > \varepsilon$ or $|(S_a)_a - (S_b)_a| > \Delta$, S_b can be another important segment. The more specific algorithm is shown in Algorithm 1.

Then we can get main segments sequence $S_n = \{S_1^{'}, S_2^{'}, ..., S_n^{'}\}$.

2.2 Trajectory Clustering

Raw Classification. Trajectory direction histogram (TDH) expresses an approximative segments' direction distribution. It describes directional characters of trajectories. It's an effective method of pre-process classification for large amount of data.

First, in order to figure out trajectory direction histogram, let $(S_i^{'})_\theta \in [0, 360)$, $i = 1, 2, 3, ..., n$. And we divide $[0, 360)$ into N subintervals, each length of which is $aver = \frac{360}{N}$. If $javer \leq (S_i^{'})_\theta < (j+1)aver$, $1 \leq j \leq N$, $(S_i^{'})_\theta$ belongs to the jth subinterval. N_j is the number of segments whose angles belong to the jth subinterval. So, TDH of the trajectory can be calculated as

$$H = (h_1, h_2, ..., h_N), h_j = \frac{N_j}{N} \tag{6}$$

Algorithm 1. Extraction of main segments

Input:
S_m: the set of all segments in the given trajectory
ε: the maximum acceptable differences between adjacent segments' angular values
Δ: the maximum acceptable differences between adjacent segments' acceleration values
Output:
S_n: the set of extracted main segments in the given trajectory
Processing:
$S_a = S_1$;
for $S_b \in S_m$ **do**
   if $|(S_a)_\theta - (S_b)_\theta| > \varepsilon$ or $|(S_a)_a - (S_b)_a| > \Delta$ **then**
     S_b is another main segment;
     $S_a = S_b$; $S_b = S_{b+1}$;
   else
     S_b is not one of main segments;
     $S_b = S_{b+1}$;
   end if
end for

Algorithm 2. Raw classification

Input:
H: the TDH set of all trajectories
D: the TDH distance set between all trajectories
d_{min}: The min d value between TDH for trajectories be allowed in one class
Output:
C_L: the List of classes
Processing:
Sort set D and choose the trajectories whose d (distance between TDH) are the most small ones to be K centroid.
for $H_a \in H$ **do**
   $TempDis = 0$;
   for $i = 1; i \leq K; i++$ **do**
     if $TempDis < d(H_a, Cid_i)$ **then**
       $TempDis = d(H_a, Cid_i)$;
       $C_a = i$;
     end if
   end for
   if $TempDis < d_{min}$ **then**
     $K = K + 1$;
     H_a belongs to the new Kth class;
   else
     Then H_a belongs to the C_ath class, m_{C_a} is the number of trajectories belongs to $Class$;
     $Cid_{C_a} = \frac{m_{C_a}}{m_{C_a}+1}Cid_{C_a} + \frac{1}{m_{C_a}+1}H_a$
     $m_{C_a} = m_{C_a} + 1$;
   end if
end for

Trajectory direction histogram is a constrain for raw classification. Correlation distance between TDH is calculated as below

$$d(H_a, H_b) = \frac{\sum_I (H_a(I) - \bar{H}_a)(H_b(I) - \bar{H}_b)}{\sqrt{\sum_I (H_a(I) - \bar{H}_a)^2 \sum_I (H_b(I) - \bar{H}_b)^2}} \tag{7}$$

$$\bar{H}_k = \frac{1}{N} \sum_J H_k(J) \tag{8}$$

The smaller the value of d is the lower similarity between TDH is.

The number of trajectories is n, and the set of Trajectory Direction Histogram for n trajectories and Trajectory Direction Histogram distance between every two trajectories are $H = \{H_1, H_2, ..., H_n\}$, $D = \{d(H_i, H_j) | i \neq j; i, j = 1, 2, ...n\}$.

For raw classification using TDH similarity, a modified k-means method is adopted. Generally, the first step of k-means cluster is to confirm the k class centroid stochastically, but those centroid may be close to each other, which is not favorable for the further classification. So we'd better choose the ones between which the distance value d are the smallest in set D to be the k centroid. And a threshold d_{min} is settled. If the maximum distance value d between a trajectory and each class centroid is bigger than d_{min}, the trajectory can be the member of the class with the highest similarity. Otherwise the trajectory can be the centroid of a new class.

Class centroid is defined as

$$Cid = \frac{1}{m} \sum_{i=1}^m H_i; \tag{9}$$

where $H_i \in Class, i = 1, 2, ..., m$. And m is the number of trajectories belong to $Class$.

The more specific algorithm is shown in Algorithm 2.

Fine Classification. Segments in main sequence $S_n = \{S_1', S_2', ..., S_n'\}$ also have position and acceleration features. Through the new main sequence, we can pick out main points set $P_L = \{P_0, P_1, P_2, ..., P_n\}$. Main points are calculated as below

$$P_i = P_{i-1} + (S_i')_{E_{(x,y)}} - (S_i')_{S_{(x,y)}} \tag{10}$$

where, $P_0 = (0, 0)$.

The raw classification above have divided trajectories into k classes. We need to subdivide them by using more specific position and acceleration features to ensure an effective classification result. This time, position distance d_p and acceleration distance d_a between trajectories are defined as below

$$d_p = \frac{1}{n} \sum_{k=0}^n \sqrt{((P_a)_k^x - (P_b)_k^x)^2 + ((P_a)_k^y - (P_b)_k^y)^2} \tag{11}$$

$$d_a = \frac{1}{n} \sum_{k=1}^{n} |(S_a)_k^a - (S_b)_k^a| \tag{12}$$

where P_a and P_b are main points sets of two trajectories a and b, S_a and S_b are main segments sequences of them.

Fine classification also uses modified k-means method to subdivide classes. Position distance d_p and acceleration distance d_a are constrains of this fine classification. The more specific algorithm is shown in Algorithm 3.

Algorithm 3. Fine classification $(d_p + d_a)$

Input:
Class: Trajectories belong to this class
Output:
List of subclasses
Processing:
Find out trajectories whose $d_p + d_a$ are the most large ones to be K centroid.
for $T_a \in Class$ **do**
   $TempDis = d_p(T_a, Cid_1) + d_a(T_a, Cid_1);$
   for $i = 2; i \leq K; i++$ **do**
     if $TempDis > d_p(T_a, Cid_i) + d_a(T_a, Cid_i)$ **then**
       $TempDis = d_p(H_a, Cid_i) + d_a(T_a, Cid_i);$
       $C_a = i;$
     end if
   end for
   if $TempDis > d_{pamin}$ **then**
     $K = K + 1;$
     H_a belongs to the new Kth subclass;
   else
     Then T_a belongs to the C_ath subclass, m_{C_a} is the number of trajectories belongs to the C_ath subclass;
     position centroid:
     $CidP_{C_a} = \frac{m_{C_a}}{m_{C_a}+1} CidP_{C_a} + \frac{1}{m_{C_a}+1} P_{La}$
     acceleration centroid:
     $CidA_{C_a} = \frac{m_{C_a}}{m_{C_a}+1} CidA_{C_a} + \frac{1}{m_{C_a}+1} S_{Aa}$
     $m_{C_a} = m_{C_a} + 1;$
   end if
end for

2.3   Gaussian Process Regression for Matching

In order to find out similar ones for all kinds of trajectories, we adopt Gaussian Process Regression method [3] to build the model $a = f(q) + \tau$, where $\tau \sim \aleph(\mu, \sigma_2)$. q denotes the position of point (q_x, q_y). The position sequence is $q = \{q_1, q_2, ..., q_n\}$. a is the acceleration of the point, that includes a_x and a_y in different direction. The acceleration sequence is $a = \{a_1, a_2, ..., a_n\}$. In this part, we train Gaussian Progress Regression model for both acceleration a_x and

a_y. Given a test point q_*, we can calculate the posterior density $p(a_*|q_*, q, a)$ to determine whether the point fits Gaussian univariate normal distribution with mean $\bar{a}_*(q_*)$ and variance $\sigma^2_{\bar{a}_*}(q_*)$ [11] which are defined as

$$\bar{a}_*(q_*) = k^T(q_*)(K + \sigma^2 I)^{-1} a \tag{13}$$

$$\sigma^2_{\bar{a}_*}(q_*) = C(q_*, q_*) - k^T(q_*)(K + \sigma^2 I)^{-1} k(q_*) \tag{14}$$

where, $K_{ij} = C(q_i, q_j)$, $k(q_*) = (C(q_*, q_1), C(q_*, q_2), ..., C(q_*, q_n))^T$.

Key points for Gaussian Process Regression are the common covariance function and the calculation of their parameters. Main points of trajectories in classes are used for training.

The following is covariance function

$$C(q, q') = v_0 \exp\{-\frac{1}{2} \sum_{I=1}^{d} w_I(q_I - q'_I)^2\} + v_1 \tag{15}$$

where q_I is the Ith component of q. But the parameters $v_0, v_1, w_1, ..., w_d$ in above function are unknown.

The log likelihood $l = \log P(q, a|\phi)$

$$l = -\frac{1}{2} \log|K'| - \frac{1}{2} a^T K'^{-1} a - \frac{n}{2} \log 2\pi \tag{16}$$

where $K' = K + \sigma^2 I$, $\phi = (\log v_0, \log v_1, \log w_1, ..., \log w_d)$.

In order to obtain a local maximum likelihood, we can figure out ϕ by partial derivatives of l with respect to parameters.

$$\frac{\partial l}{\partial \phi_i} = -\frac{1}{2} tr\left(K'^{-1} \frac{\partial K'}{\partial \phi_i}\right) + \frac{1}{2} a^T K'^{-1} \frac{\partial K'}{\partial \phi_i} K'^{-1} a \tag{17}$$

In the end, the Gaussian mean and variance of each point in target trajectory are calculated. And the possibility for target trajectory belonging to the trained class are based on the number of points whose practical acceleration are in confidence interval.

3   Numerical Experiments

3.1   Trajectories Disposing and Clustering

Fig.1 is showing 9 raw trajectories which seemly have totally different position and acceleration characters. After disposing, main features are shown clearly in Fig.2. The final classification results are {Tra1,2,3,9}, {Tra4,5}, {Tra7,8}, {Tra6}.

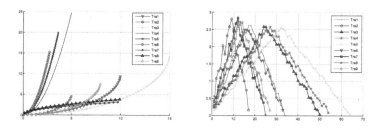

Fig. 1. position and acceleration feature of 9 raw trajectory

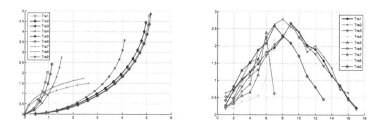

Fig. 2. position and acceleration feature of main segments

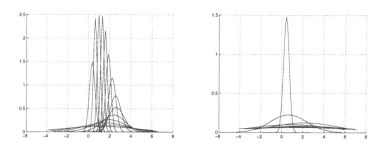

Fig. 3. probability density distribution of points in Tra10 (left) and Tra7 (right)

3.2   Trajectory Matching

Tra10 is the same as latter part of Tra9. Gaussian Process Regression model is built for Tra10 using main points extracted from Tra1,2,3,9 as training samples.

The left picture for Tra10 in Fig.3 shows that the most probability density distribution curve has a small variance which means a thick distribution around evaluated mean value. The corresponding actual acceleration value are in 90% confidence interval. The right one in Fig.3 shows the curve of the probability density distribution of points in Tra7. Those points variances are too big to have a dense distribution for their acceleration value. This is not an ideal Gaussian distribution though, the experiment result shows that their actual acceleration values are in 90% confidence interval. From this point of view, Tra7 is regarded as an unmatched trajectory with class {Tra1,2,3,9}.

4   Conclusion

In this paper, we introduce a new algorithm framework for motion trajectories cluster and match. Simulation results demonstrate that extracting main features and modified k-means cluster method approach good results on normal and abnormal trajectories classification. Gaussian Process Regression model makes partial trajectory match for the right class, which neglects trajectory integrity . Its training samples got from previous classification also get ride of surrounding constrains. But there are still some problems unthoughtful. The criterion setting of whether target trajectory match for one class in Gaussian Process Regression model needs more sample experiments.

References

1. Li, X., Hu, W., Hu, W.: A coarse-to-fine strategy for vehicle motion trajectory clustering. In: 18th International Conference on Pattern Recognition, ICPR 2006, vol. 1, pp. 591–594. IEEE (2006)
2. Ossama, O., Mokhtar, H.M.O., El-Sharkawi, M.E.: Clustering Moving Object Trajectories Using Coresets. In: Al-Majeed, S.S., Hu, C.-L., Nagamalai, D. (eds.) WiMoA 2011 and ICCSEA 2011. CCIS, vol. 154, pp. 221–233. Springer, Heidelberg (2011)
3. Kim, K., Lee, D., Essa, I.: Gaussian process regression flow for analysis of motion trajectories. In: 2011 IEEE International Conference on Computer Vision, ICCV, pp. 1164–1171. IEEE (2011)
4. Hu, W., Xie, N., Li, L., Zeng, X., Maybank, S.: A survey on visual content-based video indexing and retrieval. IEEE Transactions on Systems, Man, and Cybernetics, Part C: Applications and Reviews (99), 1–23 (2011)
5. Broilo, M., Piotto, N., Boato, G., Conci, N., De Natale, F.G.B.: Object Trajectory Analysis in Video Indexing and Retrieval Applications. In: Schonfeld, D., Shan, C., Tao, D., Wang, L. (eds.) Video Search and Mining. SCI, vol. 287, pp. 3–32. Springer, Heidelberg (2010)
6. Morris, B., Trivedi, M.: Learning trajectory patterns by clustering: Experimental studies and comparative evaluation. In: IEEE Conference on Computer Vision and Pattern Recognition, CVPR 2009, pp. 312–319. IEEE (2009)
7. Zhang, Z., Huang, K., Tan, T.: Comparison of similarity measures for trajectory clustering in outdoor surveillance scenes. In: 18th International Conference on Pattern Recognition, ICPR 2006, pp. 1135–1138. IEEE (2006)
8. Hu, R., Barnard, M., Collomosse, J.: Motion-sketch based video retrieval using a trellis levenshtein distance (2010)
9. Hu, R., James, S., Collomosse, J.: Annotated Free-Hand Sketches for Video Retrieval Using Object Semantics and Motion. In: Schoeffmann, K., Merialdo, B., Hauptmann, A.G., Ngo, C.-W., Andreopoulos, Y., Breiteneder, C. (eds.) MMM 2012. LNCS, vol. 7131, pp. 473–484. Springer, Heidelberg (2012)
10. Le Thi, L., Boucher, A., Thonnat, M.: Trajectory-based video indexing and retrieval enabling relevance feedback (2010)
11. Williams, C., Barber, D.: Bayesian classification with gaussian processes. IEEE Transactions on Pattern Analysis and Machine Intelligence 20(12), 1342–1351 (1998)

Optimal Operation for 3 Control Parameters of Texaco Coal-Water Slurry Gasifier with MO-3LM-CDE Algorithms

Cuiwen Cao, Yakun Zhang, and Xingsheng Gu

Key Laboratory of Advanced Control and Optimization for Chemical Processes,
Ministry of Education, East China University of Science and Technology,
200237 Shanghai, P.R. China
{caocuiwen,xsgu}@ecust.edu.cn

Abstract. Optimizing operation parameters for Texaco coal-water slurry gasifier with the consideration of multiple objectives is a complicated nonlinear constrained problem concerning 3 BP neural networks. In this paper, multi-objective 3-layer mixed cultural differential evolution (MO-3LM-CDE) algorithms which comprise of 4 multi-objective strategies and a 3LM-CDE algorithm are firstly presented. Then they are tested in 6 benchmark functions. Finally, the MO-3LM-CDE algorithms are applied to optimize 3 control parameters of the Texaco coal-water slurry gasifier in methanol production of a real-world chemical plant. The simulation results show that multi-objective optimal results are better than the respective single-objective optimal operations.

Keywords: Operation parameters, gasifier; multi-objective, MO-3LM-CDE.

1 Introduction

Gasification is a vital component of "clean coal" technology. [1] Texaco coal-water slurry gasifier is one of the most important upstream units in gasification process, and it generates synthesis gas which can be used to produce a wide range of chemical products. Optimal operation with the consideration of multiple objectives for the control parameters of gasifier can greatly increase the production efficiency and economic benefits.

In recent years, evolutionary algorithms (EAs) have been introduced to solve multi-objective optimization problem. Among them, differential evolution (DE) algorithm is one of the most efficient algorithms for problems over continuous space. It has been widely used to solve many optimization problems [2]-[4]. Enlightened by the far more rapid evolving velocity of human society, Reynolds [5] developed cultural algorithm (CA). CA is a dual inheritance system that models evolution at both the population space and the upper belief space. Quite a few EAs, such as DE [6]-[8], have been adopted in population space instead of evolutionary programming in standard CA, and multi-objective differential evolution algorithms (MODEs)

K. Li et al. (Eds.): ICSEE 2013, CCIS 355, pp. 382–391, 2012.

[9]-[15] show their outperform abilities. The paper is organized as follows. Section 1 is the introduction. In section 2, 4 multi-objective strategies are presented. The 3LM-CDE algorithm is in section 3. MO-3LM-CDE algorithms are tested in 6 benchmark functions in Section 4. The application of MO-3LM-CDE algorithms used in optimizing 3 control parameters of Texaco coal-water slurry gasifier is in section 5. The last section is the conclusion.

2 Multi-objective Optimization Strategies

The multi-objective problem considered in this paper has the following form:

$$\underset{X \in \mathbf{R}^n}{\text{Minimize}}: F(\mathbf{X}) = [f_1(\mathbf{X}), f_2(\mathbf{X}), ..., f_r(\mathbf{X})], \quad (r \geq 2)$$

$$\text{Subject to}: g_i(\mathbf{X}) \geq 0 \ (i = 1, 2, ..., k) \tag{1}$$

$$h_i(\mathbf{X}) = 0 \ (i = 1, 2, ..., l)$$

where the decision vector $\mathbf{X} = (x_1, x_2, ..., x_n)$ belong to the feasible n-dimensional space $\mathbf{R}_n$. The purpose is to search $\mathbf{X}^* = (x_1^*, x_2^*, ..., x_n^*)$, which can satisfy all constraints and at the same time make the objectives achieve optimum.

Some basic definitions [15] are given as follows.

Definition 1 (Pareto dominance): A vector $\mathbf{u} = (u_1, u_2, ..., u_r)$ is called to dominate another vector $\mathbf{v} = (v_1, v_2, ..., v_r)$ iff $\mathbf{u}$ is partially less than $\mathbf{v}$, i.e. $\forall i \in \{1, 2, ..., r\}, (u_i \leq v_i) \wedge (\exists j \in \{1, 2, ..., r\}: u_i < v_i)$.

Definition 2 (Pareto optimal solution): A solution $\mathbf{X_1}$ is said to be Pareto optimal solution with respect to feasible regions iff there is no $\mathbf{X_2}$ for $f(\mathbf{X_2})$ dominating $f(\mathbf{X_1})$.

Definition 3 (Pareto optimal set): The Pareto optimal set is defined as the set of all Pareto optimal solution.

In this paper, 4 strategies of transform multi-objective into single objective are used. They are listed below.

1. Main objective method: According to the general technical requirements of optimization design, the most important objective function is chosen from each branch objective function as the main objective function. At the same time, other branch objective functions are transformed into appropriate constraints.

2. Linear weighted method: According to the importance in the entire optimization design of each branch objective function, we can get a set of corresponding weighting factors $\omega_1, \omega_2, ..., \omega_r$. The new unified objective function is made up of the linear combinations of $f_j(\mathbf{X})$ and ω_j as equation (2).

$$f(\mathbf{X}) = \min[\sum_{j=1}^{r} \omega_j f_j(\mathbf{X})] \tag{2}$$

3. Max-min method: The value of evaluation function is the maximum value of each branch objective function shown in formula (3).

$$f(\mathbf{X}) = \min\{\max_{1 \leq j \leq r} [f_j(\mathbf{X})]\} \tag{3}$$

4. Ideal point method: $f_j^*(j = 1, 2,..., r)$ is the optimal value of each branch objective function. If the value of each branch objective function is as close as possible to the corresponding ideal value, we can get better non-inferior solutions. The evaluation function is as follows.

$$f(\mathbf{X}) = \min[\sum_{j=1}^{r}(f_j(\mathbf{X}) - f_j^*)^2] \qquad (4)$$

3   3LM-CDE Algorithm

3LM-CDE algorithm which has a population space, a medium space and a belief space is selected from our former work [16].The Pareto optimal set in section 2 is search by this algorithm and basic differential evolution (DE) algorithm.

4   Numerical Experiments

To validate and compare the performance of the 3LM-CDE and the 4 multi-objective optimization strategies, 6 test functions [15] (Appendix) are used and compared with standard DE algorithm which has the same value of each related parameter.

Fig.1-Fig.4 shows the Pareto solution vectors of Main Objective, Linear Weighted, Max-min, and Ideal Point methods of function g01, Fig.5-Fig.8 of g02, Fig.9-Fig.12 of g03, Fig.13-Fig.16 of g04, Fig.17-Fig.20 of g05, and Fig.21-Fig.24 of g06.

5   Optimal Operation for Texaco Coal-Water Slurry Gasfier

The input and output of Texaco water-coal slurry gasifier is shown in Fig.25. Under the stable production situation, there are 3 control parameters can be operated, which are the Total Oxygen feed flowrate (Fo) from the three-stream burner port on top of the gasfier (the oxygen stream in the outer annular tube), the Central Oxygen feed flowrate (Foc) from the tube of the three-stream burner port on the top of the gasfier (the oxygen flow in the central tube), and the Quench Water feed flowrate (Fw). Many types of factors influence the yield of the syngas and its effective components $(CO+H_2)$, and the relationships among them are very complicated which cannot be formulated.

Based on the work of reference [17], 18 variables of 3 BP neural network soft sensor models describing the relationship between the 15 inlet variables and 3 outlet variables were selected. Via data preprocessing, 310 groups' historical data of 18 variables during one month were firstly collected and normalized. Then, 15 inlet variables were reduced into 8 principal variables using principal component analysis. Finally, three 3-layer BP neural networks (Fig.26) were established, each of which had the same structure of 8 input nodes, 8 hidden nodes, and 1 output node. The outlet flowrate of synthesis gas (Fg), the CO volume% in the outlet syngas, and the H_2 volume% in the outlet syngas are the three output node variables. The actual values are gained via the denormalized process via equations (5).

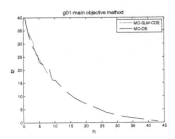

Fig. 1. Main objective method of g01

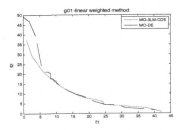

Fig. 2. Linear weighted method of g01

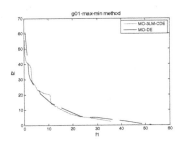

Fig. 3. Max-min method of g01

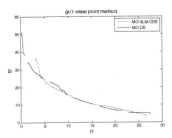

Fig. 4. Ideal point method of g01

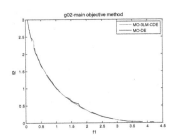

Fig. 5. Main objective method of g02

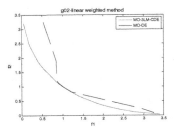

Fig. 6. Linear weighted method of g02'

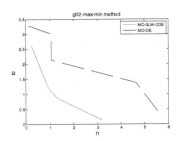

Fig. 7. Max-min method of g02

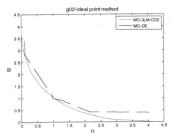

Fig. 8. Ideal point method of g02

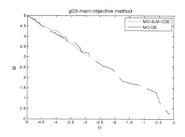

Fig. 9. Main objective method of g03

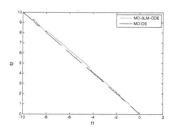

Fig. 10. Linear weighted method of g03

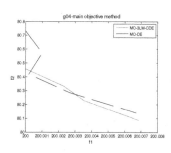

Fig. 11. Max-min method of g03

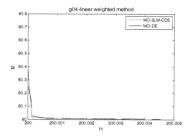

Fig. 12. Ideal point method of g03

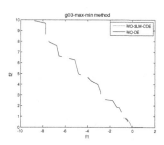

Fig. 13. Main objective method of g04

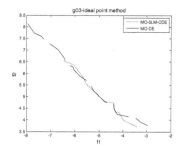

Fig. 14. Linear weighted method of g04

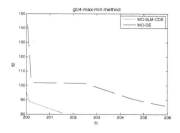

Fig. 15. Max-min method of g04

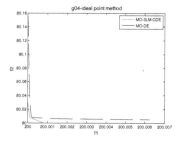

Fig. 16. Ideal point method of g04

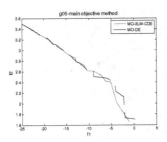

Fig. 17. Main objective method of g05

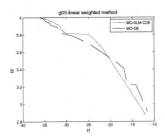

Fig. 18. Linear weighted method of g05

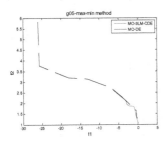

Fig. 19. Max-min method of g05

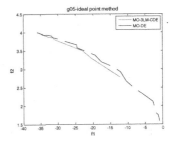

Fig. 20. Ideal point method of g05

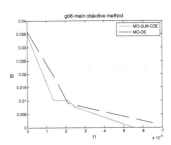

Fig. 21. Main objective method of g06

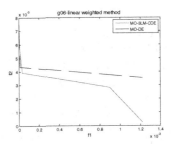

Fig. 22. Linear weighted method of g06

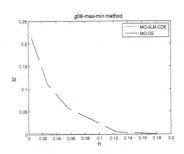

Fig. 23. Max-min method of g06

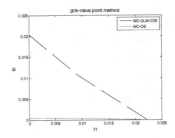

Fig. 24. Ideal point method of g06

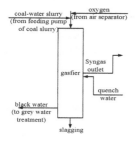

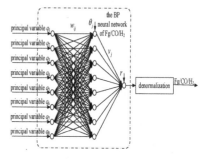

Fig. 25. The input and output of Texaco gasifier

Fig. 26. Structure of the 3 BP neural networks

$$f_{Fg} = Fg \times (250000 - 120000) + 120000$$
$$f_{CO} = Co \times (45 - 35) + 3 \qquad (5)$$
$$f_{H_2} = H_2 \times (45 - 35) + 35$$

The values of weights and thresholds of each of the 3 BP neural networks computing results are listed in Table 1. After finishing the former preparing work, we can begin the optimization task for the real-time operational parameters. The optimal model is presented as follows.

Objective function:
Single objective function
The single objective function is effective gas yield defined as equation (6). [17] The optimal algorithm is 3LM-CDE.

$$\max\{p = [f_{Fg} \times (f_{CO} + f_{H_2})]/(Fc^* \times Cc^* \times \rho_c)\} \qquad (6)$$

in which p, Fc^*, Cc^*, and ρ_c are the effective gas yield, current inlet flowrate of the coal slurry, current inlet concentration of the coal slurry, and the density of the coal slurry ($=1.086t/m^3$), respectively.

Multi-objective function
The multi-objective function is defined as equation (7). The 4 MO-3LM-CDE algorithms are used to optimize the results.

$$\textbf{maximize}\,[f_{Fg}, f_{CO}, f_{H_2}]$$
$$\textbf{Subject to}\, 15000 \le x_{Fo} \le 35000, 3000 \le x_{Foc} \le 6000, 250 \le x_{Fw} \le 400. \qquad (7)$$

in which x_{Fo}, x_{Foc}, and x_{Fw} are the 3 optimized control parameters of inlet Total Oxygen flowrate, inlet Central Oxygen flowrate and inlet flowrate of quench water.

Table 1. Weights and thresholds for each of the 3 BP neural networks

category	Fg	CO	H₂	category	Fg	CO	H₂	category	Fg	CO	H₂
w11	1.8635	4.9084	5.9429	w44	4.8276	1.7488	4.5775	w77	0.3240	0.4327	23.1049
w12	-0.1369	-9.6307	-13.5372	w45	-0.0668	9.9990	0.3142	w78	0.7294	-25.6689	20.2206
w13	-2.4829	-4.2817	-17.2574	w46	0.5664	-4.7937	-5.4420	w81	2.0962	24.6860	-10.0065
w14	-0.8504	-3.4585	2.0645	w47	0.0793	-3.0714	-26.4697	w82	-0.0461	-26.6032	37.3559
w15	1.6463	-7.3019	-1.5329	w48	0.7136	-0.7556	4.3870	w83	6.3219	17.1456	-8.0812
w16	1.5550	7.9417	1.4441	w51	-3.836	-25.859	22.3753	w84	4.6829	-14.7542	42.7188
w17	0.1801	4.3339	15.3921	w52	-7.4787	22.9181	8.9603	w85	-1.1910	2.6048	-18.3320
w18	2.6966	12.3964	-8.2909	w53	0.4707	10.7267	35.6822	w86	-1.8889	-17.7350	9.7700
w21	5.6230	-4.8583	23.8403	w54	0.3503	-0.9656	22.1330	w87	-1.2744	-4.3293	-0.5461
w22	-4.5553	-22.6512	-55.3841	w55	0.3606	-10.759	31.0260	w88	0.1974	11.3994	19.5003
w23	-1.7120	-1.1308	34.5353	w56	0.4855	22.9518	-17.9031	v1	-1.7597	-1.2467	9.2067
w24	-2.8146	-10.6592	-0.7447	w57	-3.8633	-12.909	0.7290	v2	2.6515	-1.2787	-1.2212
w25	-1.2761	42.0020	-22.4337	w58	1.1296	-8.6603	-19.1587	v3	-6.3735	12.3021	-15.3544
w26	-1.9847	-3.4065	-0.5483	w61	0.0474	0.3232	21.0601	v4	-4.4436	-2.8620	-1.2450
w27	-4.7681	1.4985	-18.7561	w62	1.7063	8.3583	-14.0994	v5	1.6059	-0.9401	-0.9907
w28	-0.2870	-26.2014	-14.2768	w63	0.9223	-3.8937	20.2876	v6	1.6943	1.0200	-1.1413
w31	-10.145	-17.5820	0.0019	w64	-1.0234	-7.8132	12.0555	v7	-3.0079	2.8675	1.0988
w32	1.7225	-22.9627	36.8186	w65	-0.7430	-10.303	23.5590	v8	1.5435	1.8162	1.3969
w33	4.1829	1.9811	9.9481	w66	-0.6885	-7.1607	9.1791	θ1	-1.8709	11.6277	4.1325
w34	-3.3798	0.8228	28.6763	w67	-0.2844	-3.9438	7.9657	θ2	1.1129	-17.6772	-21.5644
w35	1.4201	-11.6556	-25.4485	w68	-0.0763	19.6221	0.0733	θ3	-2.2206	-3.8848	-45.9344
w36	1.2433	-21.7318	6.4054	w71	-4.4433	-2.9771	13.8629	θ4	1.1723	-2.5505	9.0688
w37	4.5140	-15.2801	32.9958	w72	0.3643	16.5940	49.8957	θ5	1.8759	-19.0972	5.7394
w38	0.6301	28.0138	-3.5233	w73	7.0671	27.9733	29.1182	θ6	0.7607	14.6930	2.4959
w41	-3.1632	-19.4046	23.8918	w74	-1.3045	34.7179	49.9074	θ7	1.0704	10.2298	28.9257
w42	6.0327	-8.0793	-1.1947	w75	-0.7360	-13.746	31.4887	θ8	0.4504	17.4301	-15.3954
w43	2.9478	4.6934	33.4183	w76	-1.0396	26.2516	11.1552	r	1.9946	1.6019	-6.4239

Table 2. Optimal results between 3LM-CDE algorithm and MO-3LM-CDE algorithm

Algorithm	Optimize value of the 3 control parameters			Optimize results of the three outputs			Effective gas yield (m³/t)
	x_{Fo} (m³/h)	x_{Foc} (m³/h)	x_{Fw} (m³/h)	Fg(m³/h)	CO(V%)	H₃(V%)	
3LM-CDE	30297.852	4447.998	392.323	222782.338	40.555	39.673	3683.318
MO-3LM-CDE	34860.535	3622.009	374.148	212362.374	43.654	41.031	3706.047

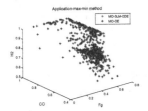

Fig. 27. Main objective method of application

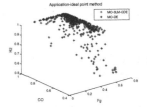

Fig. 28. Linear weighted method of application

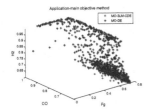

Fig. 29. Max-min method of application

Fig. 30. Ideal point method of application

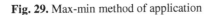

A real-world application of the Texaco gasfier is in its normal and stable working condition, and its 18 variables are M*=1.16%, A* = 7.68%, Va* = 41.39%, C* = 49.77%, ST* = 1210°C, Cc* = 65.05%, Fc* = 68.69 m^3/h, Pc* = 5.74MPa, Tc* = 42.91°C, Fo* = 28550.82 m^3/h, Foc* = 4340.00m^3/h, Po* = 7.40 MPa, To* = 33.55°C, Fw* = 337.55 m^3/h, Tw* = 229.45°C, Fg* = 210528.09 m^3/h, CO* = 39.96%, H$_2$* = 40.28%.[17]

MO-3LM-CDE and MO-DE algorithms with the 4 strategies in section 2 are used to optimize the 3 control parameters (x_{Fo}, x_{Foc}, and x_{Fw}) of the gasfier. The simulation results are shown in Fig.27-Fig.30. The Main Objective strategy achieved the best results. They are shown in Table 2. From Table 2 we can know that the optimal results of MO-3LM-CDE are more balanced and at the same time the effective gas yield is better than the single objective calculations.

6   Conclusion

Optimizing operation parameters for Texaco coal-water slurry gasifier with the consideration of multiple objectives was studied in this paper. 4 multi-objective strategies were developed. From the simulation results we can suggest that the strategies of finding the Pareto optimal sets have found better performances not only in the 6 test functions, but also in engineering practice problems.

Acknowledgments. Financial support from the National Natural Science Foundation of China (No.61174040), Shanghai commission of Nature Science (No. 12ZR1408100) and the Fundamental Research Funds for the Central Universities was grateful appreciated.

Appendix

Six test problems in reference [15] are listed below.

g01:

Minimize $[f_1(x) = x_1^2 + x_2^2, \ f_2(x) = (x_1 - 5)^2 + (x_2 - 5)^2]$

Subject to $-5 \le x_1, \ x_2 \le 10$

g02:

Minimize $[f_1(x) = x_1^2 + x_2^2, \ f_2(x) = (x_1 + 2)^2 + x_2^2]$

Subject to $-50 \le x_1 \le 50$

g03:

Minimize $[f_1(x) = -2x_1 + x_2, \ f_2(x) = 2x_1 + x_2]$

Subject to $0 \le x_1 \le 5, \ 0 \le x_2 \le 3$

$-x_1 + x_2 - 1 \le 0, \ x_1 + x_2 - 7 \le 0$

g04:

Minimize $[f_1(x) = 5x_1 + 3x_2; \ f_2(x) = 2x_1 + 8x_2]$

Subject to $0 \le x_1, \ x_2 \le 100$

$x_1 + 4x_2 - 100 \le 0, \ 3x_1 + 2x_2 - 150 \le 0$

$200 - 5x_1 - 3x_2 \le 0, 75 - 2x_1 - 8x_2 \le 0$

g05:

Minimize $[f_1(x) = -x_1^2 + x_2, \ f_2(x) = 0.5x_1 + x_2 + 1]$

Subject to $0 \le x_1, \ x_2 \le 7, (1/6)x_1 + x_2 - 6.5 \le 0$

$0.5x_1 + x_2 - 7.5 \le 0, 0.5x_1 + x_2 - 30 \le 0$

g06:

Minimize $[f_1(x) = x_1, \ f_2(x) = x_2]$

Subject to $0 \le x_1, x_2 \le 1, x_1^2 + x_2^2 \le 1$

References

1. Robinson, P.J., Luyben, W.L.: Simple dynamic gasifier model that runs in Aspen Dynamics. Industrial Engineering and Chemistry Research 47, 7784–7792 (2008)
2. Engelbrecht, A.P.: Computational Intelligence: an introduction. John Wiley & Sons, Ltd. (2007)
3. Storn, R., Price, K.: Differential Evolution-A simple and Efficient Heuristic for Global Optimization over Continuous Spaces. Global Optimization 11, 341–359 (1997)
4. Yan, J.F., Zhang, B.P., et al.: Application study on a novel differential evaluation algorithm. Computer Applications 3, 719–722, 725 (2008)
5. Reynolds, R.G.: An introduction to cultural algorithms. In: Proc. 3rd Annu. Conf. Evol. Program, pp. 131–139 (1994)
6. Ricardo, L.B., Coello Coello, C.A.: A cultural algorithm with differential evolution to solve constrained optimization problems. Computer Science 3315, 881–889 (2004)
7. Pu, Y.X., Liu, M.D.: Early Warning Technology for Cracking Severity Based on Improved Cultural. In: 8th World Congress on Intelligent Control and Automation, Jinan, pp. 5866–5871 (2010)
8. dos Santos, C.L., Clemente, T.S.R., Cocco, M.V.: Improved differential evolution approach based on cultural algorithm and diversity measure applied to solve economic load dispatch problems. Mathematics and Computers in Simulation 79, 3136–3147 (2009)
9. Coello Coello, C.A., Lamont, G.B., Van Veldhuizen, D.A.: Evolutionary algorithms for solving multi-objective problems, ch. 2. Springer (2007)
10. Deb, K., Pratap, A., Agarwal, S., Meyarivan, T.: A fast and elitist multi-objective genetic algorithm: NSGA-II. IEEE Trans. on Evolutionary Computation, 182–197 (2001)
11. Babu, B.V., Mathew Leenus Jehan, M.: Differential Evolution for Multi-objective Optimization. Evolutionary Computation 4, 8–12 (2003)
12. Knowles, J.D., Corne, D.W.: Approximating the Non-dominated Front Using the Pareto Archived Evolution Strategy. Evolutionary Computation 8, 149–172 (2000)
13. Kukkonen, Lampinen, J.: An extension of generalized differential evolution for multi-objective optimization with constraints. Computer Applications 3432, 752–761, 725 (2004)
14. Qin, H., Zhou, J.Z., Lu, Y.L., Li, Y.H., Zhang, Y.C.: Multi-objective Cultured Differential Evolution for Generating Optimal Trade-offs in Reservoir Flood Control Operation. Water Resour. Manage. 24, 2611–2632, 725 (2010)
15. Zheng, J.H.: Multi-objective evolutionary algorithms and its application, 1st edn., pp. 1–27. Science Press, Beijing (2007)
16. Cao, C., Zhang, Y., Gu, X., Zhong, X.: 3-layer mixed cultural evolutionary algorithms and its application in operation optimization of Texaco gasifier. Industrial & Engineering Chemistry Research (2nd review)
17. Sun, Y., Zhang, L.B., Gu, X.S.: Membrane computing based particle swarm optimization algorithm and its application. In: 2010 IEEE 5th International Conference on Bio-Inspired Computing, pp. 631–636 (2010)

Fuzzy Double Model Control for Air Supply on a PEM Fuel Cell System

Xiaohong Hao, Haochen Zhang, Aimin An, Xin Liu, and Liwen Chen

College of Electric and Information Engineering, Lanzhou University
of Technology, Lanzhou, Gansu, China, 730050
haoxh@lut.cn, {zhanghaochen1987,anaiminll,meteor7136}@163.com,
825781363@qq.com

Abstract. Oxygen excess ratio control is closely related to the performance and safety of a Proton Exchange Membrane fuel cell system. Some control strategies should be used to regulate the oxygen excess ratio at the suitable value in order to avoid stack starvation and damage. And in this paper, a simple fuzzy double model control has been proposed to adjust the oxygen excess ratio under variation load currents. The double model controller combines a PID controller and a fuzzy logic controller which can be switched based on the fuzzy inference rules during the regulation process. The simulation results demonstrate that the fuzzy double model control can adjust the oxygen excess ratio at the setting point when the current is changed, and improve the dynamic performance of oxygen excess ratio than fuzzy PID control.

Keywords: PEM fuel cell, oxygen excess ratio, fuzzy control, PID control, soft switching.

1 Introduction

With the serious environmental pollution and energy crisis around the world, it is very important for people to explore more efficient and clean energy. Proton Exchange Membrane (PEM) fuel cell is a kind of clean energy. PEM fuel cell has many advantages, such as high power density, clean, low operation temperature, and which is especially used in portable power supply [1], vehicles and distributed power plants [2]. The fuel and oxidant, usually hydrogen and oxygen or air, are supplied continuously to the anode and cathode by the gas supplied system.

On the anode side, hydrogen is fed to track a desired ratio of the air flow by a fast operation valve. And on the cathode side, the sufficient oxygen is needed to be supplied to the cathode channel by compressor or blower for fulfilling stoichiometric requirements to generate the demanded current. The insufficient oxygen supply results in the oxygen starvation phenomenon which implies a fast stack degradation and low power generation [3]. The oxygen starvation phenomenon occurs when the oxygen partial pressure falls below a critical level at the cathode. And the factor oxygen excess ratio has been introduced to describe the oxygen mass between the inlet and consumption. When the oxygen excess ration falls

K. Li et al. (Eds.): ICSEE 2013, CCIS 355, pp. 392–400, 2013.

below a suitable value, the oxygen starvation would occur. In several studies, controlling the oxygen excess ratio is proposed to prevent the oxygen starvation phenomenon [4]. The proposed control strategies include feedforward control [5, 6], Linear Quadratic Regulator (LQR) [7], dynamic matrix predictive control [8], model predictive control [9] and adaptive control [10]. These control methods have achieved well control effect. However, some of them are not suitable for the requirements of model parameters with time varying in the real system, and some strategies are more complex that it is difficult to apply. A simple control strategy has been appropriated in this work to regulate the oxygen excess ratio based on the fuzzy control and fuzzy inference.

The proposed control strategy in this paper is fuzzy double model control which is used to regulate the oxygen excess ratio at a desired level when the load current changes. And this control strategy contains the fuzzy control, PID control and the switching strategies between these two control methods within the adjusting process. The fuzzy logic control plays a major role at the beginning of regulation so as to increase the system response speed and enhance the system robustness, and then the PID control will play the leading role to achieve the purpose of eliminating the system steady state error when the system is above to reaching the steady state.

This paper is organized as follows: Section 2 presents a description of the PEM fuel cell and the mathematical definition of the oxygen excess ratio. Section 3 gives a detailed description of the fuzzy control, PID control and switching method based on the fuzzy inference of the fuzzy double model control. In section 4, the simulation results obtained with the proposed fuzzy double model control are shown and compared to the fuzzy PID control. In section 5 the major conclusions are drawn.

2   System Description

This section gives a brief introduction of the PEM fuel cell we studied in this work and exact mathematical definition of oxygen excess ratio process, and then an optimal control object is specified.

2.1   PEM Fuel Cell System

The PEM fuel cell can convert the chemical energy of hydrogen and oxygen into electricity through electrochemical reaction. A PEM fuel cell power system requires the integration of many components beyond the fuel cell stack itself, to allow efficient performance under different operating conditions. The PEM fuel cell system should include the fuel supply and air supply subsystems, the temperature regulation subsystem, the power conversion subsystem which is used to convert the generated DC power to regulated DC or grid quality AC power. And it is also necessary to require a controller to maintain efficiency and avoid degradation of available fuel cell voltage.

The PEM fuel cell model in this work is based on a 1.2 KW PEM fuel cell power system, The structure of the PEM fuel cell system studied in this work is shown in Figure 1.The supply of air/oxygen to the cathode is one of the key factors in operation of the PEM fuel cell stack and it is also the subject in this work.

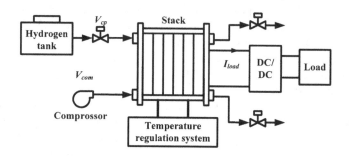

Fig. 1. The structures of a PEM fuel cell system

2.2   Oxygen Excess Ratio

The general proposed of this work is to design a control strategy to regulate the oxygen excess ratio λ_{O2} in the PEM fuel cell system. The oxygen excess ratio has been defined in work [11] as

$$\lambda_{O2} = w_{O2,cain}/w_{O2,react}. \tag{1}$$

where $w_{O2,react}$ is oxygen consumption rate. It is proportional to the current I

$$w_{O2,react} = nIM_{O2}/(4F). \tag{2}$$

with n is the number of cells in the fuel cell stack, and $n{=}46$, M_{O2} is the molar mass of oxygen and F is the Faraday constant.

The oxygen flow rate $w_{O2,cain}$ depends on the dry air flow rate $W_{a,cain}$:

$$w_{O2,cain} = \mu_{O2,cain}W_{a,cain}. \tag{3}$$

$\mu_{O2,cain}$is the oxygen mass fraction, and it is equal to 0.233. And the mass flow rate of dry air at the cathode inlet is described as

$$W_{a,cain} = W_{cain}/(1 + \omega_{ca}). \tag{4}$$

$$\omega_{ca} = M_v P_{v,cain}/(M_{a,ca}P_{a,cain}). \tag{5}$$

ω_{ca} is the humidity ratio, $M_v$and $M_{a,ca}$ are the molar mass of vapor and air at the cathode inlet. The vapor pressure $P_{v,cain}$ and the dry air pressure $P_{a,cain}$ are defined as

$$P_{v,cain} = \phi_{ca,in}p_{sat}|_T. \tag{6}$$

$$P_{a,cain} = P_{cain} - P_{v,cain}. \tag{7}$$

With $\phi_{ca,in}$ describes the relative humidity of air at the cathode inlet, and $p_{sat}|_T$ represent the vapor saturation pressure for temperature T[12].

In a real-time system, the size of the compressor outlet/cathode inlet complicates the installation of a pressure sensor and converts the cathode inlet pressure measuring into a difficult task. Therefore, a mathematical model which is used to express the pressure $P_{a,cain}$ as a function of air flow rate W_{cain} and the current I has been used[13]

$$P_{cain} = 1.0033 + 2.1 \times 10^{-4}W_{cain} - 4.757 \times 10^{-4}I. \tag{8}$$

The air flow rate W_{cain} depends on the voltage V_{com} applied to the compressor motor and the model can be described as the following transfer function:

$$P_{com} = 0.1/(0.01s^2 + 0.14s + 1). \tag{9}$$

It is clearly that the oxygen excess ratio depends on the oxygen consumption $w_{O2,react}$ and the oxygen mass flow rate $w_{O2,cain}$. And the oxygen consumption cannot be influenced as it depends only on the measurable disturbance current I, as the M_{O2}, n and F are constants. So the regulation of the oxygen excess ratio λ_{O2} has to be carried out by manipulating the oxygen mass flow rate which is regulated to the air flow rate. Hence, the compressor motor voltage V_{com} will be used as input signal in order to regulate the oxygen excess ratio λ_{O2}.

2.3   Control Object

As already mentioned above, the regulation of the oxygen excess ratio λ_{O2} is important in the fuel cell control, both for performance and safety. Tests showed that the oxygen starvation can reduce the performance of fuel cell, even causing the damage of the electro-catalyst [14]. To avoid this negative factor, the oxygen excess ratio has to satisfy $\lambda_{O2} > 1$, it means the mass of oxygen supply should greater than the consumption.

Several experiments have been carried out in order to choose a suitable oxygen excess ratio value to prevent the stack oxygen starvation and keep the fuel cell working at the high efficiency level. The experiment results show that the output power for a certain current will increase along with the oxygen excess ratio until reaching a maximum point. And in our work, a control objective of oxygen excess ratio is chosen to $\lambda_{O2ref} = 4$ through consulting the work [15]. This value can be considered as a tradeoff between efficiency and safety.

3   Fuzzy Double Model Control Design for Oxygen Excess Ratio

This section presents the design of a double model controller based on the fuzzy rules switching to regulate the oxygen excess ratio. First, we design a fuzzy controller and a PID controller, and then a switching method between two control strategies based on fuzzy rules in the double controller is proposed.

3.1   Fuzzy Control and PID Control Design for Oxygen Excess Ratio

The gas supply in the PEM fuel cell is the nonlinear and time-varying parameters system. And the system parameters and control effect would be impacted by the slight perturbation of sensors which is used to measure the air flow rate and load current. Hence, a controller with strong robust performance is needed, and the fuzzy control strategy is suitable for controlling this process. However, the process with fuzzy control has the steady state error. So it is necessary to join the integral elements to eliminate the steady state error, and the PID control strategy is considered naturally.

Fuzzy Control Design. In this work, the deviation e and deviation ration ec between the actual output and setting value are used to be the inputs of the fuzzy controller, and the controller output u_{fuzzy} is the voltage of compressor.

$$e(k) = \lambda_{O2,ref}(k) - \lambda_{O2}(k). \tag{10}$$

$$ec(k) = [e(k) - e(k-1)]/T_s. \tag{11}$$

where T_s is the sample time. The actual range of e and ec is [-4, 4] and [-40, 40], which can be converted to [-1, 1] by multiplying the factors. The output range is [-80, 80]. The fuzzy subsets of input is NB, NM, NS, ZO, PS, PM, PB, and the output fuzzy subsets is NB, N, ZO, P, PB. The basic form of fuzzy control rules is: "If the deviation e and the deviation ration ec is A and B, Then the fuzzy control output is u_{fuzzy}." The fuzzy control rules are obtained by Figure 2.

u_{Fuzzy}		e					
		NB	NM	NS	ZO	PS	PM	PB
	NB	NB	N	N	P	ZO	ZO	ZO
	NM	N	N	N	ZO	ZO	ZO	P
	NS	N	N	ZO	ZO	ZO	ZO	P
ec	ZO	N	ZO	ZO	ZO	ZO	P	P
	PS	N	ZO	ZO	ZO	ZO	P	P
	PM	ZO	ZO	ZO	ZO	P	P	P
	PB	ZO	ZO	ZO	N	P	P	PB

Fig. 2. Fuzzy rules table of the fuzzy controller. Each entry is in the format of fuzzy control output u_{fuzzy} with respect to the devision e and devision ration ec.

PID Control Design. The key of the PID control design is to select the appropriate PID controller parameters. Firstly, the mathematical model of the oxygen supply process can be identified by the least squares method, and then we could get the suitable parameters of PID controller by suboptimal reduction algorithm [16] and PID parameters setting algorithm described in work [17].

3.2 Double Model Controller Based on Fuzzy Rules Switching

As already mentioned above, both the fuzzy and PID control is required to improve the dynamic and steady state performance of the oxygen excess ratio regulation. The fuzzy controller should play a major role of regulation when the steady state error exists and error rate is big. Similarly, when the error rate is small and the steady state error exists, the PID control should play a major role instead of fuzzy control. Hence, a strategy based on the fuzzy rules is needed to determine when and how to switch the control model . The structure of the system is shown in Figure 3. In this paper, the switching strategy which we call

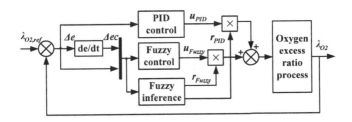

Fig. 3. Structures of the control system

as soft switching is proposed. r_{PID} and r_{fuzzy} are the adjustment coefficient of PID and fuzzy control, and the sum of r_{PID} and r_{fuzzy} is 1.

The range of r_{PID} and r_{fuzzy} are [0, 1]. And the output of the double model controller with fuzzy and PID control can be expressed as

$$u = r_{PID}u_{PID} + r_{fuzzy}u_{fuzzy}. \tag{12}$$

The fuzzy and PID controller can generate the appropriate amount of control through the error and error rate, and the control coefficients r_{PID} and r_{fuzzy} which can be obtained by fuzzy rules is used to determine which control strategy play a major role. And the control amount contains the corresponding proportion of PID and fuzzy control effect, which realize the soft switch between fuzzy control and PID control. In this work, a fuzzy inferior is proposed to choose the suitable control coefficient. The inputs of fuzzy inferior are e and ec, and the fuzzy subsets of input are N, Z, P. The output of fuzzy inferior is PID control coefficient r_{PID} , of which the range is [0, 1], and the output fuzzy subsets is PS, PB. And the fuzzy inference rules are shown in Figure 4. The membership functions of inputs and output are respectively showed in Figure 5.

4   Simulation and Analysis

The proposed control strategy was implemented as a Simulink model with the structure given in Figure 3. The sampling time for the simulation was set to be

r_{PID}		e		
		N	Z	P
	N	PS	PS	PS
ec	Z	PB	PS	PB
	P	PS	PS	PS

Fig. 4. Fuzzy rules table of the fuzzy inference

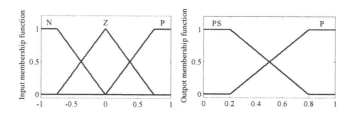

Fig. 5. The membership functions of inputs and output in the fuzzy inference

$T_s=10$ ms. The load current I_{load} varies in a wide range from 20 A to 35 A and load steps of different magnitudes were applied to the system. The compressor voltage amplitude was restricted at the range of [0, 80].

From the Figure 6, it is clearly that the load current changes impacts on λ_{O2}. When the load current increases, the consumption of oxygen is corresponding increase, and if the inlet mass of oxygen is constant, the oxygen excess ratio λ_{O2} will decrease; whereas the λ_{O2} will correspondingly increase in equal increments as the consumption of oxygen if the load current decrease with the condition of inlet oxygen is constant. From the work [15], we know that the output power of fuel cell is the maximum power output when the λ_{O2} is near 4. And if the oxygen excess ratio λ_{O2} is greater or less than 4, the output power of PEM fuel cell is not the maximum. Even the starvation phenomenon would occur when the oxygen excess ratio $\lambda_{O2} < 1$. As can be seen from the simulation, it is necessary to regulate the oxygen excess ratio at a suitable value.

It can be seen from the Figure 6 and 7 that the oxygen excess ratio is steadily regulated at the setting value with the load current changed under the action

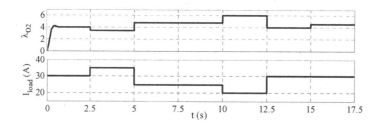

Fig. 6. Oxygen excess ratio curves with load current changed

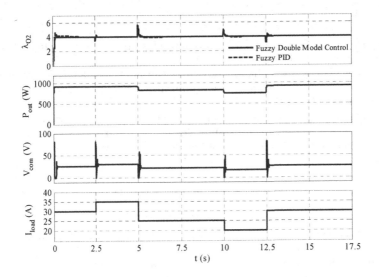

Fig. 7. Oxygen excess ratio, output power of PEM fuel cell, and compressor voltage curves under control with load current changed

of fuzzy model control. This is helpful for safety and efficient operation of the PEM fuel cell under the load current changes.

It is also shown form Figure 7 that the fuzzy double model control has the better control effect than traditional fuzzy PID control in regulation time, and which could make the fuel cell system into the optimal operational state within the less time.

5 Conclusions

In this paper an innovative fuzzy double model control for controlling the oxygen excess ratio has been proposed. The strategy contains the fuzzy control, PID control and the method used to switch the control model between PID and fuzzy control based on the fuzzy inference rules. The proposed fuzzy model control is relatively simple ,and the simulation results show that this control strategy has the better control effect than traditional fuzzy PID control in regulation time. However, the algorithm also needs to be evaluated and complemented in the real PEM fuel cell system, which is our future work to be done.

Acknowledgments. This research has been supported by the Nation Natural Science Foundation of China under Grant No. 61064003 and the Project of Commission of Finance of Gansu Province, China under Grant No. 1103ZTC143.

References

1. Stefanopoulou, A.G., Suh, K.W.: Mechatronics in fuel cell system. Control Enigneering Practice 15, 277–289 (2007)
2. Li, Q., Chen, W., Dai, C., Jia, J., Han, M.: Proton exchange membrane fuel cell model optimization based on seeker optimization algorithm. Proceeding of the CSEE 28, 119–124 (2008)
3. Yousfi-Steiner, N., Moçotéguy, P., Candusso, D., Hissel, D.: A review on polymer electrolyte membrane fuel cell catalyst degradation and starvation issue: cause, consequence and diagnostic for mitigation. Journal of Power Sources 194, 130–145 (2009)
4. Ramos-Paja, C.A., Bordons, C., Romero, A., Giral, R., Martines-Salamero, L.: Minimum fuel consumption strategy for PEM fuel cells. IEEE Transactions on Industrial Electronics 156, 685–696 (2009)
5. Pukrushpan, J.T., Stefanopoulou, A., Varigonda, S., Eborn, J., Haugstetter, C.: Control oriented model of fuel processor for hydrogen generation in fuel cell applications. Control Engineering Practice 14, 277–293 (2006)
6. Prayiton, A., Kubumwe, O., Bosma, H., Tazelaar, E.: Efficiency of polymer electrolyte membrane fuel cell stack. Telkomnika 9, 303–310 (2011)
7. Jong-Woo, A., Song-yul, C.: Coolant controls of a PEM fuel cell system. Journal of Power Sources 179, 252–264 (2008)
8. Feroldi, D., Serra, M., Riera, J.: Performance improvement of a PEM fuel cell system controlling the cathode outlet air flow. Journal of Power Sources 169, 205–212 (2007)
9. Gruber, J.K., Doll, M., Bordons, C.: Design and experimental validation of a constrained MPC for the air feed of a fuel cell. Control Engineering Practice 17, 874–885 (2009)
10. Zhang, J., Liu, G., Yu, W., Ouyang, M.: Adaptive control of the airflow of a PEM fuel cell system. Journal of Power Sources 179, 649–659 (2009)
11. Pukrushpan, J.T., Stefanopoulou, A.G., Peng, H.: Control of Fuel Cell Power Systems: Principles, Modeling, Analysis and Feedback Design. Springer, London (2004)
12. Nguyen, T.V., White, R.E.: A water and heat management model for proton exchange membrane fuel cells. Journal of Electrochemical Society 140, 2178–2186 (1993)
13. Peng, H., Stefanopoulou, A.G.: Control oriented modeling and analysis for automotive fuel cell systems. Journal of Dynamic Systems, Measurement, and Control 126, 14–25 (2004)
14. Thounthong, P., Sethakul, P.: Analysis of a fuel starvation phenomenon of a PEM fuel cell. In: 2007 Power Conversion Conference, pp. 731–738. IEEE Press, Nagoya (2007)
15. Gruber, J.K., Bordons, C., Oliva, A.: Nonlinear MPC for the airflow in a PEM fuel cell using a volterra series model. Control Engineering Practice 20, 205–217 (2012)
16. Xue, D., Atherton, D.P.: A suboptimal reduction algorithm for linear systems with a time delay. International Journal of Control 60, 181–196 (1994)
17. O'Dwyer, A.: Handbook of PI and PID controller tuning rules. Imperial College Press, London (2003)

Non-negative Mutative-Sparseness Coding towards Hierarchical Representation

Jiayun Cao and Lanjuan Zhu

Automation Department, Shanghai Jiaotong University, China

Abstract. In this work, the problem of data representation is studied. Many existing works focus on how to learn a set of bases, sparsely and effectively representing the data. However, as reaching a consensus, people always arrange the data in the hierarchical structure so that a clearer data framework and interaction can be got, just like patent documents listed by different layers of classes. Thus, different from existing works, we target at discovering the hierarchical representation of data. Non-negative mutative-sparseness coding (NMSC) is a method for analyzing the non-negative sparse components of multivariate data and representing the data as hierarchical structure. Specifically in a subsequent layer, the sparseness of each data is adjusted according to the corresponding hidden components in the upper layer. Our experimental evaluations show that the NMSC possesses great efficiency in clustering and sufficient merit in hierarchical organizing the observed document data.

Keywords: basis, hidden component, non-negative, mutative sparseness, hierarchical representation.

1 Introduction

Hidden components detection and tracking (HCDT) is extensively utilized in data analysis and information processing. It aims to automatically explore the significant latent components of data and find an optimized representation for the data. The involved methods include latent semantic analysis (LSA), probabilistic latent semantic analysis (PLSA), latent Dirichlet allocation (LDA), non-negative matrix factorization (NMF) and non-negative sparse coding (NNSC).

In the early time, standard LSA [1] was introduced to capture and represent significant components of the data via Singular Value Decomposition. It has application in various pattern recognition problems where complex wholes can be treated as additive functions of components.

Later, PLSA [2] as well as LDA [3] showed up to inherit the ideas of LSA and enhanced the performance. Compared to LSA, PLSA adds a sounder probabilistic model meanwhile serves as a three-level Bayesian model. In LDA, the unobserved factors are assumed to have a Dirichlet prior.

To put non-negativity into constraints according to the intuitive notion, NMF [4] was introduced in 2001. Then, each observed data can take non-negative values in all latent semantic directions and the latent space in NMF doesnt need to be orthogonal.

K. Li et al. (Eds.): ICSEE 2013, CCIS 355, pp. 401–409, 2013.

In recent years, NNSC [5] regarded as a kind of neural code performs well in HCDT. Given a potentially large set of input patterns, NNSC, using a constant parameter so-called "sparseness", can find a few representative patterns combined in the right non-negative proportions and then reproduce the original input patterns.

Most methods above have good performance on decomposing multivariate data and detecting hidden components, but they still cant implement an efficient hierarchical expressiveness.

However, it is a consensus and common behavior that people always tend to find the inner relationship among the objects and then arrange them in a hierarchical structure, exp. in the patent system and library system. Thus, its desired to develop an automatic approach to learn the topics or clusters of the data, additionally the subtopics or sub-clusters in multiple layers.

In this paper, we propose a non-negative mutative-sparseness coding towards hierarchical representation. The whole organization of the data has a tree-like structure with the entire collection situated at the root level. Then, the subsequent levels of the tree function as the further expanded analysis of the data. Particularly for a sublevel, the sparseness of each data for each basis is adjusted according to the corresponding hidden components of the data in upper level. And the natural data non-negativity is also preserved. Hence, keeping the accurate reconstruction, there are connections among the data sparseness in each layer. Our experimental evaluations show that the proposed method performs excellent in document clustering and owns great efficiency of hierarchical representation.

Once again, the research motivation comes mainly from exploring bases, interaction and hierarchy of dataset. The analysis adding mutative sparseness serves to show a clearer data framework and different significance of each data, meanwhile bring forth ideas to information dissecting and representing.

2 Non-negative Mutative Sparseness Coding (NMSC)

The proposed algorithm provides a useful framework to implement the hierarchical representation of the data. There are three main steps as follows:

2.1 Learning the Basis of the Root Level

Generally, data tend to be combined with a set of sparse response on the basic elements in neural system, text information processing models, etc. Meanwhile, according to the objective law of cognition, these data and their elements should be non-negative, which has been also widely recognized.

Whats more, at the beginning of analyzing a mono-level structure or the root level in a hierarchical structure, each data shares equal importance and contribution. So the sparseness of all the data on a specific basis should be the same.

Thus, in the proposed algorithm, the root level of the data is analyzed by means of standard NNSC. NNSC can effectively find the optimal non-negative bases and then reproduce the data as the combination of sparse hidden components involving linear additive interactions.

Assume that non-negative input data X is a matrix, then linear decomposition describes the data as $X \approx AS$. The matrix A contains as its columns the basis vectors of the decomposition. The rows of matrix S contain the corresponding hidden components that give the contribution of each basis vector in the input vectors.

Establishing the objective function as follow:

$$F(A, S) = \frac{1}{2}\|X - AS\|^2 + \lambda \sum_{ij} S_{ij} \qquad (1)$$

To get the small reconstruction error with the sparseness, the function must be minimized with the constraints $\forall ij : A_{ij} \geqslant 0, S_{ij} \geqslant 0$ and $\forall i :\| a_i \|= 1$, where a_i denotes the i:th column of A. λ (≥ 0) is the sparseness-regulatory parameter to control the tradeoff between sparseness and accurate reconstruction.

Then, through constantly updating the basis matrix A and hidden components matrix S until the objective function reaching the minimum, the optimal A and S can be achieved.

Hence, in the root level, each data can be represented as linear additive combination of a few bases and then grouped into a certain cluster corresponding to the basis which gives the largest contribution to the data.

2.2 Deep Learning of the Basis for the Subsequent Level

In the root level, all the data have been clustered, but the training experiments show that there is a large size-gap among the maximum and minimum clusters. Commonly, the data with several different dimensions are still grouped into a same cluster. But the basis corresponding to the cluster cant describe all the related data especially for the ones with low projection values on the basis. As a result, these data lack adequate descriptions so that the local representation of the whole dataset is limited and not detailed. Therefore, its desired to have deeper analysis of relatively big clusters by further analyzing and expanding the corresponding bases.

Moreover, the sparseness of the response of the data based on expanded bases should not be invariant. The more intense response of the data in certain base, the more potentially the data contains the sub-level information of the base, when these data are more accurately described. Thus, to more precisely represent the data locally near the certain base, the sparseness of these data are accordingly higher.

a) First step is to expand the number of bases to make each basis composed of several sub-bases. Because these sub-bases can have more detailed description of the data especially for the ones with poor representation in the upper level.

Assume that we further analyze the which denotes the k:th basis B_k ($1 \leq k \leq r$). To keep the reconstruction of the original data, all the bases excluding B_k (number:r-1) in the root level should act as a part of B_k's sub-bases and the others are new sub-bases (number: p). Hence, there are (r+p-1) sub-bases of B_k in all.

So, the sub-basis-matrix is $\overline{A} = [\overline{A}_{n \times p}, \overline{A}_{n \times (r-1)}]$, where $\overline{A}_{n \times (r-1)}$ consists of all the columns excluding the k:th one in A. The sub-hidden-components-matrix is $\overline{S} = [\overline{S}_{p \times m} \ \overline{S}_{(r-1) \times m}]^T$, where $\overline{S}_{(r-1) \times m}$ consists of all the rows excluding the k:th one in S.

b) The basis B_k gives different contributions to each data, so the corresponding sparseness of the data should be different in the process of further analyzing B_k. In other words, the data with larger hidden component on B_k should have larger sparseness in the sublevel, which can be implemented through increasing the sparseness regulatory parameter. Then, all the data related to B_k in the upper level can be described more accurately.

Establishing objective function as follow:

$$F'(\overline{A}, \overline{S}) = \frac{1}{2} \|X - \overline{A}\ \overline{S}\|^2 + \overline{\lambda} \sum_{j=1}^{m} S_{kj} \sum_{i=1}^{r+p-1} \overline{S}_{ij} \tag{2}$$

where $\overline{\lambda}$ is the sparseness regulatory parameter. Then minimize the function under the update rules:

1) Set t=0. Iterate the steps (2)–(4) until convergence.
2)

$$\overline{A}' = \overline{A}^t - \mu(\overline{A}^t \overline{S} - X)(\overline{S})^T \tag{3}$$

Any negative value in $\overline{A}'$ is set to zero;

Rescale each column of $\overline{A}'$ to unit norm, and then set $\overline{A}^{t+1} = \overline{A}'$.

3)

$$\overline{S}^{t+1} = \overline{S}^t .* ((\overline{A}^{t+1})^T X) ./((\overline{A}^{t+1})^T (\overline{A}^{t+1}) \overline{S}^t + \Lambda) \tag{4}$$

where .* and ./ denote element-wise multiplication and division (respectively), and Λ is the sparseness regulatory parameters matrix namely

$$\Lambda = \overline{\lambda} [1 \ 1 \cdots \ 1]^T [S_{k1}, S_{k2}, \cdots, S_{km}] \tag{5}$$

4) Increment t.

Consequently, we get the optimal sub-basis matrix $\overline{A}$ and sub-hidden-components matrix $\overline{S}$ of the basis B_k.

c) Expanding and further analyzing the bases is illustrated in the following figure:

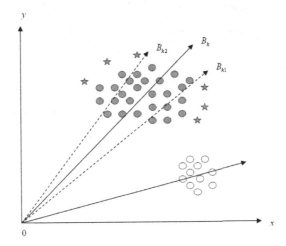

Fig. 1. Illustration of further analyzing bases in NMSC

In Figure 1, the solid arrows denote the bases in the root level while the dashed ones in the sub-level. Its obvious that all the data take non-negative directions in the space. Here, we still take the basis B_k as an example. Although B_k is the basis of the data in shade, its limited to efficiently represent several data especially for the marginal ones (stars) which have relatively low projection values on B_k. To improve the local representation of the dataset, we tend to find sub-bases B_{k1}, B_{k2}, $\cdots$ for further analyzing the data. Specifically, in the process of expanding B_k, the data nearer to B_k should have larger sparseness than the others because the data potentially contain more sub-level information of B_k. As a consequence, these sub-bases serve to have more detailed and precise description of the data in shade than only B_k .

2.3   Hierarchical Representation for the Data

In the proposed algorithm, each basis in any level if necessarily can be further analyzed and decomposed into several bases in the next level, which connects all the bases as a tree-like structure. Then, all the data tend to be correspondingly put into the tree organization. Hence, this completes the hierarchical representation of the original data.

2.4   Plate Notation for NMSC Model

In this part, we utilize the plate notation [6], [7] to show NMSC model applied in document clustering. Its graphically illustrated in Figure 2.

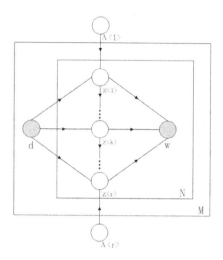

Fig. 2. Plate notation of NMSC model

$z < k >_{ij}$ is the topic for the j:th word in document i and the k:th level. w_{ij} denotes the actual word used and $\lambda < k >_i$ the sparseness regulatory parameters for further analyzing the i:th topic in the k:th level. In addition, the circles in shade indicate that the variables are observable, while the other empty circles indicating the latent variables. The directed edges imply the dependency between the variables.

According to the plate notation, its clear that any topic in a non-root level not only depends on the document corpora and sparseness regulatory parameter but also the topics of the upper level.

3   Experiment

In this section, we evaluate the proposed NMSC for the problem of document representation. We implement document clustering and compare the results with other typical clustering methods.

3.1   Data Corpora

We use the Reuters database as the document corpora. Each document in the corpora has been manually labeled with one or a few topics and also indicated which cluster it belongs to. The Reuters database contains 21578 documents and 135 topics, equally as the clusters. However, due to fair comparison with other clustering methods with assumption that each document belongs to only one cluster, weve removed the multi-topic documents and retained 9494 documents within the 51 clusters each of which contains 5 documents at least.

3.2   Document Features

The famous weighted term-frequency (tf-idf) [8] vector is used to represent the document. We collected the vocabulary set $W = \{w_1, w_2, \cdots, w_m\}$ after removing stop-words. The tf-idf vector of d_i document is defined as $X_i = [x_{1i}, x_{2i}, \cdots, x_{mi}]^T$,

$$x_{ji} = t_{ji} \cdot log(\frac{n}{idf_j}) \tag{6}$$

in which t_{ji}, idf_j, n denote the term frequency of the word w_j, the number of the whole documents containing w_j, and the number of the documents. Particularly, the words in title should generally be more important than in text. According to statistics, a word in the title has 5 times the significance of the same one in the context [9]. So during processing word frequency, all the words existing both in title and text should be taken into specific consideration. The definition of x_{ji} is formulated as $x_{ji} = 6t_{ji} \cdot log(\frac{n}{idf_j})$. Moreover, each vector X_i is also normalized into unit length. Thus, the $n \times m$ matrix X denotes the data matrix with the non-negative elements.

3.3   Evaluation

The test document data is randomly selected from the data matrix X, mixed with the documents from several clusters. For each round of test, the document feature vectors from selected k clusters are processed by the cluster methods. We evaluate the result with the labeled cluster from the ground truth provided by the original Reuters data. The accuracy of the clustering result is defined as the proportion of the documents which are partitioned into the same cluster in the ground truth:

$$accuracy = \frac{\sum\limits_{i=1}^{n} \delta(c_i, l_i)}{n} \tag{7}$$

where c_i is the index of the cluster that document d_i belongs to, and l_i is the index of the labeled cluster in the ground truth. We apply the max matching strategy or the Kuhn-Munkres algorithm as the assignment approach to find matched cluster between the result and ground truth.

3.4   Implementation and Comparisons

The proposed NMSC is much suitable to implement for the task of document clustering, due to the non-negative property of the document feature. In this test, we set up the NMSC with 3 layers and the method automatically removed the cluster containing less than 5 documents.

To evaluate the proposed NMSC, we also tested the same data with the other two famous graph-based cluster methods, Average Association (AA in short) [10] and Normalized Cut (NC in short) [11]. The graph $G = G(V, E)$ is the input of these graph-based methods, in which the vertex set $V = \{d_j\}$ is the set of the documents and the edge set $E = \{e(i,j)\}$ is the Euclidean distance between the document features. The "cut" is defined as the sum of edges between two clusters, defined as

$$cut(A, B) = \sum_{i \epsilon A, j \epsilon B} e(i, j) \tag{8}$$

The criterion of the Average Association and Normalized Cut are formulated as

$$AA = \frac{cut(A, A)}{|A|} + \frac{cut(B, B)}{|B|} \tag{9}$$

$$NC = \frac{cut(A, B)}{cut(A, V)} + \frac{cut(A, B)}{cut(B, V)} \tag{10}$$

In the formal work of Average Association, the author has proved that AA is equivalent to the LSA [1] with the K-means clustering method [10], in respect of their criterion function. Solving the "cut" problem is an NP-hard computational problem, while Shi and Malik propose an approximation method, the eigenvector-based criterion, which minimizes the normalized cut efficiently.

We also implement NMF [4] and NNSC [5] for comparison. The setup of these two methods is same as that of the NMSC. There are 50 testing rounds in total. Hence, the final accuracy is the average accuracies of all the rounds.

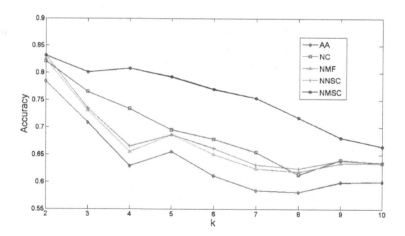

Fig. 3. Comparison of performance. The accuracy of each cluster method results in different cluster number k.

We can see from the Figure 3 that comparing with the other methods including NNSC, the NMSC performs best in the task of document clustering. Actually, the NNSC is just one layer of NMSC, whose bases are assumed to be same weighted for data representation, while NMSC can more precisely discriminate the importance of each base through discovering the hierarchical structure.

Thus, when the number of the clusters is small, the performance of the NNSC and NMSC is almost same. However, in practice, there always exist much more clusters in the data set, especially in the document corpora. From the performance shown in the Figure 3, the proposed NMSC proves more acceptable than the other methods in processing the multiple-cluster dataset.

4 Conclusions

In this paper, we have put forward non-negative mutative-sparseness coding as a useful framework to explore hidden components and hierarchical representation of data. Uncommonly, through constantly adjusting sparseness of the data in the process of further analyzing respective bases, all the data can be well described and grouped into proper clusters of multiple layers. In the experimental evaluations, NMSC embodies its great efficiency in clustering and sufficient merit in hierarchical representing documents. Then for practical application, NMSC will help to boost the development of patent system, library system, etc. In addition, towards future work, we also aim to test NMSC in other fields such as image and audio processing.

References

1. Dumais, S.T.: Latent Semantic Analysis. Annual Review of Information Science and Technology 38, 188 (2005)
2. Hofmann, T.: Probabilistic Latent Semantic Indexing. In: Proceedings of the Twenty-Second Annual International SIGIR Conference on Research and Development in Information Retrieval, SIGIR 1999 (1999)
3. Blei, D.M., Ng, A.Y., Jordan, M.I.: Latent Dirichlet Allocation. Journal of Machine Learning Research 3, 993–1022 (2003)
4. Lee, D.D., Sebastian Seung, H.: Algorithms for Non-negative Matrix Factorization, pp. 556–562. MIT Press (2001)
5. Hoyer, P.O.: Non-negative sparse coding. In: Proceedings of the 12th IEEE Workshop, pp. 557–565 (2002)
6. Ghahramani, Z.: Graphical models (Speech). Tübingen, Germany (2008) (retrieved)
7. Buntine, W.L.: Operations for Learning with Graphical Models. Journal of Artificial Intelligence Research (AI Access Foundation) 2, 159–225 (1994) ISSN: 11076-9757
8. Yuan, X.: A Clustering Algorithm for Web Document Based on Theme. Journal of Chengdu University (Natural Science Edition) 29(3) (2010)
9. Han, K.: Extract Subject from Chinese Text with Three Different Levels. Journal of Chinese Information 15(4) (2001)
10. Zha, H., Ding, C., Gu, M., He, X., Simon, H.: Spectral relaxation for k-means clustering. In: Advances in Neural Information Processing Systems, vol. 14 (2002)
11. Shi, J., Malik, J.: Normalized cuts and image segmentation. IEEE Transactions on Pattern Analysis and Machine Intelligence 22(8), 888–905 (2000)

Effects of Interruptible Load Program on Equilibrium Outcomes of Electricity Markets with Wind Power

Xuena An, Shaohua Zhang, and Xue Li

Key Laboratory of Power Station Automation Technology
Department of Automation, Shanghai University, 200072 Shanghai, China

Abstract. High wind power penetration presents a lot of challenges to the flexibility and reliability of power system operation. In this environment, various demand response (DR) programs have got much attention. As an effective measure of demand response programs, interruptible load (IL) programs have been widely used in electricity markets. This paper addresses the problem of impacts of the IL programs on the equilibrium outcomes of electricity wholesale markets with wind power. A Cournot equilibrium model of wholesale markets with wind power is presented, in which IL programs is included by a market demand model. The introduction of the IL programs leads to a non-smooth equilibrium problem. To solve this equilibrium problem, a novel solution method is proposed. Numerical examples show that IL programs can lower market price and its volatility significantly, facilitate the integration of wind power.

Keywords: Interruptible Load Programs, Cournot Competition, Equilibrium Analysis, Wind power, Electricity Markets.

1 Introduction

Under the background of the worldwide restructuring and deregulation of electric power industries and the rapid development of the intermittent renewable energy sources, DR has got massive attentions and applications, and also become one of the most important parts in the ongoing smart grid development around the world [1]-[4].

As an effective measure of DR programs, IL program aims to take advantage of flexibility in customers power consumption to reduce the expensive spinning reserve and avoid the generation capacity investment for the increasing power demand [5]-[6]. In an IL program, consumers sign an interruptible load (IL) contract with a retailer, which allows the retailer to interrupt part or all of the supply of electricity to the consumers at times of high wholesale market prices or when system reliability is jeopardized. Generally, the main terms specified in an IL contract include: the interruptible load volume that the consumers are willing to provide; the economic compensation for load interruption; and the interruption condition, for example, when the wholesale market price is higher than a certain value (hereinafter referred to as IL trigger price).

K. Li et al. (Eds.): ICSEE 2013, CCIS 355, pp. 410–416, 2013.

Due to the fact that the emerging electricity wholesale markets are more akin to oligopolistic markets than perfect competitive markets, generators have the ability to manipulate the market price by the strategic behaviors, which has a serious impact on efficiency and reliability of electricity markets. Therefore, analysis of generators strategic behaviors and market power using oligopolistic equilibrium models has become one of hot topics in electricity market fields [7]-[8]. To date, most of the related research works are based on the Cournot equilibrium model [9]-[11] and the supply function equilibrium (SFE) model [12]-[13].

It can be noted that up to now, there is few research work to address the impacts of the IL contracts on generators strategic behaviors and the equilibrium outcomes in the wholesale market with wind power. Given this background, this paper commits to address the above mentioned problem.

2   Theoretical Model

2.1   Demand Model with IL Contract

For electricity wholesale markets without IL contracts, the market demand D at a future time period t (1h) is generally be expressed by the following linear demand curve:

$$D(p) = A - bp \qquad (1)$$

where, p is the market price at time t; A and b are non-negative coefficients. b is a parameter to describe the magnitude of demand elasticity, a relatively large value of b indicates a relatively high demand elasticity. This demand curve can be shown by line segment 1 and 2 in Fig.1.

Consider that there is an IL contract in the wholesale market which can be exercised at time t. Let Q and k_0 denote the IL contract volume and the trigger price, respectively. When the market price p is higher than k_0, the customers that take part in the IL contract will be curtailed a load demand of Q, and will be paid a certain amount of economic compensation.

The market demand for the wholesale market with the IL contract can be derived as follows. For market prices below k_0, because the IL contract will not be exercised, the market demand will not be affected by the IL contract, which can be illustrated as line segment 1 in Fig. 1. For market prices greater than k_0, because the IL contract will be exercised, the market demand will be lowered by Q, which can be illustrated as line segment 4 in Fig. 1. From Fig. 1, it can be found that, when the market price p is equal to k_0, either the whole IL contract volume Q is curtailed or no load demand is interrupted, will give rise to non-smoothness in the demand curve at the point of k_0. This non-smoothness in the demand curve may lead to inexistence of the market equilibrium state. To deal with this problem, we suppose that when $p = k_0$, the IL contract can be partially exercised. That is, when $p = k_0$, an endogenous volume V is applied for interruption decision of the IL contract, which satisfies

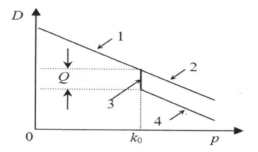

Fig. 1. Demand curve with IL contract

$0 < V < Q$. Therefore, the market demand for the wholesale market with the IL contract can be illustrated by line segment 1, 3 and 4 in Fig. 1, and can be formulated as follows:

$$D(p) = \begin{cases} A - bp, & if\ p < k_0 \\ (A - V) - bp, & if\ p = k_0 \\ (A - Q) - bp, & if\ p > k_0 \end{cases} \tag{2}$$

2.2   Cournot Equilibrium Model with IL Contract and Wind Power

Suppose that there are n conventional strategic generators and a certain number of wind power units in the wholesale market with the IL contract, and the market demand D is expressed as (2).

The conventional generator takes a form of quadratic cost function as follows:

$$C_i(q_i) = 0.5c_i q_i^2 + a_i q_i, i = 1, 2, \cdots, n \tag{3}$$

where, q_i is generator $i's$ output; c_i and a_i are cost parameters which are non-negative.

Assume that the wind power units are price-takers in the market competition. The output of wind power units in time t, q_w, is assumed to follow a normal distribution with a mean value of μ and a standard deviation of σ. The market demand D satisfies:

$$D = q_w + \sum_{i=1}^{n} q_i \tag{4}$$

Under the assumption of Cournot-type competition, the generators compete in the wholesale market by bidding their power outputs. When the constraints on generators' output and transmission capacity are ignored, generator $i's$ optimization problem in the wholesale market can be described as:

$$\underset{q_i}{Max}\, \pi_i = pq_i - C_i(q_i)$$

$$s.t. q_i = D(p) - q_w - \sum_{j=1, j \neq i}^{n} q_j \tag{5}$$

The Cournot equilibrium model for the wholesale market with the IL contract and wind power can be obtained by gathering n generators' optimization problems expressed by (5).

3   Solution Method

It can be noted from (2) that introduction of the IL contract will produce a piecewise market demand, which makes this equilibrium model difficult to solve using conventional analytic methods. In this paper, a novel solution method is proposed to solve this equilibrium problem. In this method, according to the relationships between the market price p and the trigger price k_0, three cases are considered as follows.

From (2) and (4), the corresponding demand faced by the conventional generators can be expressed as:

$$D_c = \sum_{i=1}^{N} q_i = \begin{cases} A - bp - q_w, & if \ p < k_0 \\ (A - V) - bp - q_w, & if \ p = k_0 \\ (A - Q) - bp - q_w, & if \ p > k_0 \end{cases} \quad (6)$$

The calculation process of this method can be described as:

Step 1: Assume that the market price p is less than k_0, that is $p < k_0$. The demand faced by the conventional generators can be obtained from (6). By solving the generators optimization problem, the corresponding equilibrium overcomes can be calculated.

If the equilibrium market price is less than k_0, that is, the market equilibrium satisfies the assumption that $p < k_0$, then these equilibrium outcomes are actually the solution of the equilibrium problems. Otherwise, if the equilibrium market price is greater than k_0, then turn to step 2.

Step 2: Assume that the market price p is assumed to be greater than k_0, that is $p > k_0$. Following the step 1, the corresponding market equilibrium results also can be obtained.

If the equilibrium market price is greater than k_0, and satisfies the assumption that $p > k_0$, then these equilibrium outcomes are actually the solution of the equilibrium problems. Otherwise, if the equilibrium market price is less than k_0, then turn to step 1 again. If the equilibrium market price is less than k_0 when the assumption of $p > k_0$ is used, while under the assumption of $p < k_0$, the equilibrium market price is greater than k_0, then turn to step 3.

Step 3: Assume that the market price p is assumed to be equal to k_0, and the IL contract can be partially exercised. Under the assumption of $p = k_0$, following the step 1, each generators equilibrium output can also be obtained.

4   Numerical Examples

Suppose that there are three conventional strategic generators in a wholesale market. Each generator's cost parameters are listed in Tab.1. Consider that

Table 1. Cost parameters of generators

Generator	$a_i(\$/MWh)$	$c_i(\$/MW^2h)$
1	12.0	1.0
2	10.0	1.5
3	8.0	2.0

there is an IL contract in the wholesale market which can be exercised at a future time t (1h). Let Q and k_0 denote the IL contract volume and the trigger price, respectively. In period t, assume that μ=20MW, A=200MW. The Monte Carlo simulation is employed to deal with uncertainty of the wind power.

4.1   Impacts of IL Contract Volume on Market Equilibrium

When k_0=100\$/MWh, the variations of the mean value and the standard deviation of the equilibrium market price with increasing the IL contract volume are shown in Fig.2, where the impacts of the wind power output uncertainty and the demand elasticity are considered. It can be found that introduction of the IL contract can reduce the equilibrium market price and its volatility. Especially when the uncertainty of wind power output is relatively large, this effect is more obvious. The reason is that, a higher volatility of wind power output makes it more possible to exercise the IL contract, thus the equilibrium market price and its fluctuation will be lessened more effectively. These results show that the introduction of IL programs can increase the system flexibility and is conducive to wind power integration into power systems. Furthermore, it can also be seen that, when the demand elasticity is relatively low, the generator has relatively strong ability to exert market power and raises market price. For this case, the IL program can be used to reduce the equilibrium market price and its volatility more effectively.

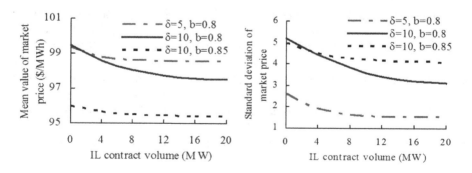

Fig. 2. Impact of IL contract volume on the mean value and the standard deviation of market price

4.2   Impacts of IL Trigger Price on Market Equilibrium

When Q=20MW, b=0.8$(MW)^2$h/\$, the variations of the mean value and the standard deviation of the equilibrium market price with increasing the IL trigger price are illustrated in Fig.3, where the impacts of the uncertainty of wind power output is considered. It can be observed that both the equilibrium market price and its volatility will be lowered with decreasing the IL trigger price. This is because the possibility to exercise the IL contract increases with decreasing the IL trigger price. Especially when the uncertainty of wind power output is relatively large, this effect is more obvious because the possibility to exercise the IL contract will be relatively large for more uncertain demand. In addition, as the IL trigger price increases, the possibility to exercise the IL contract will decrease, thus the impact of the IL contract will be weakened. These results demonstrate that the IL trigger price in the IL contract should be determined carefully to achieve the target of lowering market price and its volatility. Especially, when the standard deviation of wind power output is relatively large, a relatively low trigger price can reduce the market price and its volatility more effectively.

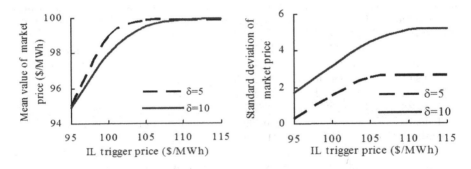

Fig. 3. Impact of IL trigger price on the mean value of equilibrium market price and on the standard deviation of market price

5   Conclusion

The impacts of IL programs on the equilibrium outcomes of electricity wholesale markets with wind power and the generators' risk are examined by proposing a Cournot equilibrium model, in which the IL programs are considered in the market demand model. The simulation results show that: (i) The introduction of the IL programs can reduce the equilibrium market price and its volatility. Especially when the uncertainty of wind power output is relatively large, this effect is more obvious. Nowadays, it is well known that the integration of wind power asks for more flexibility of power systems. This result shows that the IL

program is conducive to wind power integration into power systems. (ii) The
IL trigger price has a significant influence on the electricity market equilibrium
outcomes. A relatively low trigger price is helpful to reduce the market price and
its volatility.

Acknowledgment. This work was supported in part by the National Science
Foundation of China under Grant No. 51007052.

References

1. Farhangi, H.: The Path of the Smart Grid. IEEE Power and Energy Magazine 8(1),
   19–28 (2010)
2. Rahimi, F., Ipakchi, A.: Demand Response as a Market Resource under the Smart
   Grid Paradigm. IEEE Trans. on Smart Grid 1(1), 82–88 (2010)
3. Cappers, P., Goldman, C., Kathan, D.: Demand Response in U.S. Electricity Markets: Empirical Evidence. Energy 35(4), 1526–1535 (2010)
4. Walawalkar, R., Fernands, S., Thakur, N.: Evolution and Current Status of Demand Response (DR) in Electricity Markets: Insights From PJM and NYISO.
   Energy 35(4), 1553–1560 (2010)
5. Chen, H., Billinton, R.: Interruptible Load Analysis Using Sequential Monte Carlo
   Simulation. IEE Proceedings-C, Generation, Transmission and Distribution 148(6),
   535–539 (2001)
6. Yu, C.W., Zhang, S.H., Chung, T.S., Wong, K.P.: Modeling and Evaluation of Interruptible Load Programs in Electricity Markets. IEE Proceedings-C, Generation,
   Transmission and Distribution 152(5), 581–588 (2005)
7. Ventosa, M., Bayllo, A., Ramos, A., Rivier, M.: Electricity market modeling trends.
   Energy Policy 33(7), 897–913 (2005)
8. Helman, U.: Market Power Monitoring and Mitigation in the US Wholesale Power
   Markets. Energy 31(6-7), 877–904 (2006)
9. Li, G., Shi, J., Qu, X.: Modeling Methods for GenCo Bidding Strategy Optimization in the Liberalized Electricity Spot Market: A State-of-the-art Review. Energy 36(8), 4686–4700 (2011)
10. Chen, H., Wang, X.: Strategic Behavior and Equilibrium in Experimental
    Oligopolistic Electricity Markets. IEEE Trans. on Power Sysems 22(4), 1707–1716
    (2007)
11. Badri, A., Rashidinejad, M.: Security Constrained Optimal Bidding Strategy of
    GenCos in Day Ahead Oligopolistic Power Markets: a Cournot-based Model. In:
    Electrical Engineering, pp. 1–10 (2012)
12. Chung, T.S., Zhang, S.H., Wong, K.P., Yu, C.W., Chung, C.Y.: Strategic For-
    ward Contracting in Electricity Markets: Modelling and Analysis by Equilibrium
    Method. IEE Proceedings-C, Generation, Transmission and Distribution 151(2),
    141–149 (2004)
13. Yu, C.W., Zhang, S.H., Wang, X., Chung, T.S.: Modeling and Analysis of Strategic
    Forward Contracting in Transmission Constrained Power Markets. Electrical Power
    Systems Research 80(3), 354–361 (2010)

Two-Point Estimate Method for Probabilistic Optimal Power Flow Computation Including Wind Farms with Correlated Parameters

Xue Li, Jia Cao, and Dajun Du

Key Laboratory of Power Station Automation Technology
Department of Automation, Shanghai University, 200072 Shanghai, China

Abstract. This paper is concerned with the probabilistic optimal power flow (POPF) calculation including wind farms with correlated parameters which contains nodal injections. The two-point estimate method (2PEM) is employed to solve the POPF. Moreover, the correlation samples between nodal injections and line parameters are generated by Cholesky Factorization method. Simulation results show that 2PEM is feasible and effective to solve the POPF including wind farms with correlated parameters, while the 2PEM has higher computation precision and consume less CPU time than Monte Carlo Simulation.

Keywords: Probabilistic Optimal Power Flow, Wind Farms, Correlated Parameters, Cholesky Factorization Method, Point Estimate Method.

1 Introduction

The impacts of wind farms should be studied with the large scale wind power integrated into the power system due to the characteristics of wind as randomness, intermittent, and fluctuation. The probabilistic optimal power flow (POPF) computation including wind farms can be used to not only assess the economic impacts in steady state operation of system, but also analyze the basis of various economic behaviour under the environment of power market.

The existing probabilistic methods applied to power systems can be divided into three categories: Monte Carlo simulation (MCS) [1], point estimate method [2] and analytical method [3]-[4]. The probability description of state voltage and branch power flow can be obtained accurately by MCS [1], however, it usually consumes large computation effort. The point estimate method [2] is widely applied to the probability distribution fitting of the optimal power flow solution, which is based on the deterministic optimal power flow calculation and can calculate the statistical moment of the quantity of state efficiently. The first order second moment method [3] and Cornish-Fisher series [4] belong to the analytical method. However, the lectures above mentioned, not considering the correlation between random variables such as load and generation power, produce the impractical probabilistic results.

To solve the probabilistic power flow with correlated parameter problem, the covariance matrix transformation technique [5] is combined into the two-point

K. Li et al. (Eds.): ICSEE 2013, CCIS 355, pp. 417–423, 2013.

estimate method with nodal injection mutual correlation. Considering spatially correlated power sources and loads in [6], a probabilistic power flow model is constructed, which is then solved using an extended point estimate method. The multivariate dependent random numbers with any given distribution are generated in [7]-[8].

Motivated by the above observations, this paper investigates the POPF calculation including wind farms with correlated parameters which contain loads and generation powers. The 2PEM and MCS are used to solve the POPF with correlated parameters. Finally, to analyze the impacts of correlated parameters on the power system, the POPF with correlated parameters and with independent parameters are made on the improved 5-bus test system.

2   Probabilistic Model in Wind Power Generation System

2.1   Wind Turbine Model

In the estimation of long-term wind speed, it is usually modeled as follows [9]:

$$\varphi(v) = \frac{k}{c}(\frac{v}{c})^{k-1} \exp[-(\frac{v}{c})^k] \quad (k > 0, v > 0, c > 1) \tag{1}$$

where v represents wind speed, and $\phi(v)$ is the Weibull probability density function with the shape parameter k and the scale parameter c.

The wind turbine model can be expressed as follows:

$$P_m = \frac{1}{2}\rho A v^3 C_p \tag{2}$$

where ρ is the density of air (Kg/m^3), A is the area swept out by the turbine blades (m^2), v is the wind speed (m/s) and C_p is the dimensionless power coefficient. C_p can be expressed as a function of the blade tip speed ratio and be obtained by interpolation method.

2.2   Induction Wind Generator Model

The induction generator's equivalent circuit can be simplified into $\Gamma - type$ equivalent circuit when induction generator is used in wind farm [9]. In short, the real power injected into grid and the reactive power absorbed from the grid generated by induction wind generator can be expressed as

$$P_e = -\frac{U^2 r_2/s}{(r_2/s)^2 + x_k^2} \tag{3}$$

$$Q_e = -\frac{r_2^2 + x_k(x_k + x_m)s^2}{r_2 x_m s}P_e \tag{4}$$

where x_1 is the stator reactance, x_2 is the rotor reactance, r_2 is the rotor resistance, $x_k = x_1 + x_2$, x_m is the magnetizing reactance, s is the slip of induction machine, U is the generator voltage, respectively.

3 Probabilistic Optimal Power Flow Model for Grid-Connected Induction Wind Power System

The objective function of the POPF including wind farms is formulated as the minimization of the total fuel cost for conventional generation

$$\min \sum_{i \in S_G} (a_{2i} P_{Gi}^2 + a_{1i} P_{Gi} + a_{0i}) \tag{5}$$

where S_G is the set of power generation, P_{Gi} is the output of conventional generator's real power in i^{th} generator, a_{2i}, a_{1i} and a_{0i} are the generation cost coefficients, respectively. If wind farm is connected at i^{th} ($i \in N_w$) bus, the corresponding power flow equations for grid-connected induction wind power system are given by

$$\begin{cases} P_{ei}(e_i, f_i, s_i) - P_{Di} - P_i(e, f, t) = 0 \\ Q_{ei}(e_i, f_i, s_i) - Q_{Di} - Q_i(e, f, t) = 0 \ , i \in N_w \\ P_{mi} - P_{ei}(e_i, f_i, s_i) = 0 \end{cases} \tag{6}$$

where P_{mi} and P_{ei} are the mechanical power of wind turbine and the electrical power of wind generator at the ith bus, N_w is the set of the buses connected with wind farm, P_{Di} and Q_{Di} are real and reactive load, node powers P_i and Q_i are the function of the real part e, imaginary part f of node voltage and ratio t, respectively. Real and reactive power generations, ratios, voltage amplitudes and line currents are limited due to equipment and system constraints

$$\begin{cases} P_{Gi}^{\min} \le P_{Gi} \le P_{Gi}^{\max} i \in S_G \\ Q_{Gi}^{\min} \le Q_{Gi} \le Q_{Gi}^{\max} i \in S_G \\ t_{ij}^{\min} \le t_{ij} \le t_{ij}^{\max} (i, j) \in S_T \\ (V_i^2)^{\min} \le (e_i^2 + f_i^2) \le (V_i^2)^{\max} i \in S_N \\ I_{ij}^2 \le (I_{ij}^2)^{\max} (i, j) \in S_L \end{cases} \tag{7}$$

where S_T, S_N and S_L are the set of the transformers, the system nodes and the restricted line, respectively.

4 Cholesky Factorization Method

For N dimension random vector M in the power system, including generator outputs, and load real and reactive powers, its covariance matrix C_M is given by:

$$C_M = E(MM^T) = R \tag{8}$$

where R is a symmetric matrix. The matrix R can be obtained by the Cholesky Factorization method [10], i.e., $R = LL^T$, where the Cholesky Factor L is a lower triangular matrix. Suppose that W is an N dimension random vector with variance equal to one and independent mutually, i.e.,

$$C_W = E(WW^T) = I \tag{9}$$

where I is an identity matrix.

Using $M = LW$, the covariance matrix of M can be expressed as:

$$C_M = E(MM^T) = E(LWW^TL^T) = R \tag{10}$$

Therefore, the relevance is set in the random vector of M.

5   Two-Point Estimate Method

By path following interior point method [11], the POPF problem can be mathematically described by a set of the simplified nonlinear equations. In the 2PEM, [2], two concentrations ($K=2$) are used for each random variable in this scheme, the location and weights. Taking into account the mean and standard deviation of the bus power injection, the estimate point can be computed using the concentrations (or weights) located at the estimate point. Suppose that the correlation coefficient among all correlated random variables is set as 0.9 and its standard deviation is set as 5% of its expectation value. Then, calculate the wind turbine mechanical power and the electrical power using (2) and (3). The following algorithm is given as:

1) Set $K=1$, $E(Z_r) = 0$, $E(Z_r^j)=0$.
2) Compute the standard location $\xi_{i,k}$ and the weight $\omega_{i,k}$ of the 2m scheme.
3) Compute the estimated points $x_{i,k}(i = 1, 2, ..., m.k = 1, 2)$ located on both sides of the expectation, and obtain the sample values $(\mu_{x_1}, \mu_{x_2}, ..., x_{i,k}, ..., \mu_{x_m})$.
4) The Path Following Interior Point method [11] is used to solve the deterministic optimal power flow.
5) The slips of induction wind generator, along with the nodal voltages, are corrected at each iteration of the POPF calculation. Then, compute the estimated value Z_r of nodal voltage and nodal power.
6) Calculate the obtained results at each step using (11), and set k=k+1

$$\begin{cases} E(Z_r)^{(k+1)} = E(Z_r)^{(k)} + \omega_k \times Z_r(k) \\ E(Z_r^2)^{(k+1)} = E(Z_r^2)^{(k)} + \omega_k \times [Z_r(k)]^2 \end{cases} \tag{11}$$

7) The procedure is terminated when all the random variables are ergodic. Otherwise, repeat the point estimate iteration procedure until all the random variables are ergodic.
8) Compute the mean and standard deviation of nodal voltage and nodal power.

6   Numerical Example

The proposed method was tested on a modified 5-bus test system [3]. For the 5-bus test system, the equivalent wind farms were connected at node 1 of 5-bus test system by the two step-up transformer and double circuit transmission line, so the new nodes 6, 7 were added as shown in Fig. 1. Without specification, all

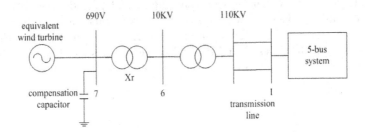

Fig. 1. The 5-bus system connected with wind farms

data were taken as per-unit value and the base value was 100MVA. A computer program was implemented in MATLAB to solve the POPF problem.

There are 20 identical wind turbines in the wind farms, each of which has the nominal capacities equal to 600 KW. The parameters of wind farms are given as follows. $\rho=1.2245Kg/m^3$, $A=1840m^2$. The identical cut-in wind speeds, cut-out wind speeds and rated wind speeds of wind farms are set as 3m/s, 20m/s, and 14m/s, respectively. The parameters of a wind generator are stator impedence $r_1 + jx_1 = 1.3883 + j16.6417$, rotor impedance $r_2 + jx_2 = 0.6217 + j18.1767$, magnetizing reactance $x_m = 591.18$. By simulating (1) with the shape parameter ($k=2$) and the scale parameter ($c=6.7703$), 10000 wind speed samples were generated. Using (2), 10000 wind power outputs were calculated.

The results from MCS were treated as the normorlized values. The average error indices [3] for 5-bus test system are shown in Table 1, where ε_e, ε_f are the average relative errors of real part and imaginary part of nodal voltage, ε_P and ε_Q, are the average relative errors of active power and reactive power of nodal power, respectively. Meanwhile, the PDF and CDF of nodal injection real power in no. 1 node with independent parameters and with correlated parameters are shown in Figs. 2-3.

Table 1. Average error indices for 5-bus test system

Average error		With independent parameters	With correlated parameters
$\varepsilon_e(\%)$	μ	0.1900	0.1425
	σ	24.1966	22.8479
$\varepsilon_f(\%)$	μ	1.1043	4.3279
	σ	7.0584	11.7695
$\varepsilon_P(\%)$	μ	1.7221	2.1980
	σ	9.9744	17.8061
$\varepsilon_Q(\%)$	μ	0.3996	0.5799
	σ	2.6794	10.6935

From the Table 1, it can be seen that the 2PEM provides good means and standard deviations, except few poor standard deviations. Meanwhile, it can also be noted that the standard deviation average errors are greater than the mean ones in almost all the cases. It indicates that the accuracy of the point estimate method deteriorates as the order of the estimated statistical moment becomes higher.

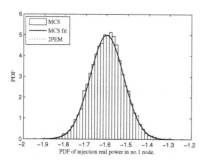 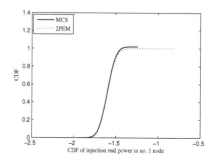

Fig. 2. PDF and CDF of injection real power in no. 1 node with independent parameters

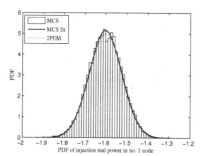 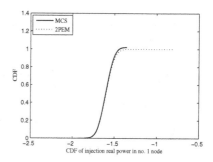

Fig. 3. PDF and CDF of injection real power in no. 1 node with correlated parameters

For the POPF with independent parameters, from the Fig. 2, it can be found that the PDF and CDF of injection real power in no. 1 node obtained by the 2PEM coincide with the results obtained by MCS basically. Therefore, the effectiveness and feasibility of the 2PEM are confirmed. For the POPF with correlated parameters, from the Fig. 3, it can be noted that the impacts of considering the correlated parameters and introducing wind farms on the power system are existed because the PDF of injection real power obtained by the 2PEM are deviated to that by MCS wherein the peak value of PDF of injection real power in no. 1 node.

7 Conclusion

2PEM combined with path following interior point method is then employed to solve the POPF including wind farms with correlated parameters. The simulation results show that: (i) The effectiveness and feasibility of the 2PEM are confirmed to solve the POPF including wind farms with correlated parameters. (ii) the superiority of the proposed method is that it has higher calculation precision and requires less execution time to solve the POPF than the MCS method dramatically. (iii) After considering the correlated parameters, some impacts on the power system occur, especially for the nodal injection real power. Thus, the POPF including wind farms with correlated parameters can be used as an efficacious analytical tool for the researchers to address long term studies such as transmission congestion and reliability analysis in the power system.

Acknowledgment. This work was supported in part by the National Science Foundation of China under Grant No. 51007052 and in part by the innovation fund project for graduate students of Shanghai University No. SHUCX120091.

References

1. Zhang, H., Li, P.: Probabilistic Analysis for Optimal Power Flow Under Uncertainty. IET Generation, Transmission & Distribution 4(5), 553–561 (2010)
2. Morales, J.M., Perez-Ruiz, J.: Point Estimate Schemes to Solve the Probabilistic Power Flow. IEEE Transactions on Power Systems 22(4), 1594–1601 (2007)
3. Li, X., Li, Y.Z., Zhang, S.H.: Analysis of Probabilistic Optimal Power Flow Taking Account of the Variation of Load Power. IEEE Transactions on Power Systems 23(3), 992–999 (2008)
4. Chemozhukov, V.: Rearranging Edgeworth-Cornish-Fisher Expansions. Institute for Fiscal Studies and Massachusetts Institute of Technology 42(2), 419–435 (2010)
5. Li, X., Li, Y.Z., Li, H.Y.: Comparison and Analysis of Several Probabilistic Power Flow Algorithms. Proceedings of the CSU-EPSA 21(3), 12–17 (2009)
6. Morales, J.M., Baringo, L., Conejo, A.J., Minguze, R.: Probabilistic Power Flow with Correlated Wind Sources. IET Generation, Transmission, & Distribution 4(5), 641–651 (2010)
7. Usaola, J.: Probabilistic Load Flow with Correlated Wind Power Injections. Electric Power Systems Research 80(5), 528–536 (2010)
8. Papaefthymiou, G., Pinson, P.: Modeling of Spatial Dependence in Wind Power Forecast Uncertainty. In: Probabilistic Methods Applied to Power Systems (PMAPS) International Conference, Mayaguez, pp. 1–9 (2008)
9. Li, X., Du, D.J., Pei, J.X., Menhas, M.I.: Probabilistic Load Flow Calculation with Latin Hypercube Sampling Applied to Grid-Connected Induction Wind Power System. Transactions of the Institute of Measurement and Control, 1–10 (2011), doi:10.1177/0142331211410101
10. Morales, J.M., Conejo, A.J., Perez-Ruiz, J.: Simulating the impact of Wind Production on Locational Marginal Prices. IEEE Transactions on Power Systems 26(2), 820–828 (2011)
11. Li, X., Cao, J., Du, D.J.: Comparison of Levenberg-Marquardt Method and Path Following Interior Point Method for the Solution of Optimal Power Flow Problem. International Journal of Emerging Electric Power Systems 13(3), 1–21 (2012)

An Efficient Two-Stage Gene Selection Method for Microarray Data

Dajun Du[1,2], Kang Li[2], and Jing Deng[2]

[1] Key Laboratory of Power Station Automation Technology
Department of Automation, Shanghai University, 200072 Shanghai, China
ddj@shu.edu.cn
[2] School of Electronics, Electrical Engineering and Computer Science,
Queen's University Belfast, United Kingdom
k.li@ee.qub.ac.uk, jdeng01@qub.ac.uk

Abstract. Gene selection is a key issue in the analysis of microarray data with small samples and variant correlation. The main objective of this paper is to select the most informative genes from thousands of genes with strong correlation. This is achieved by proposing an efficient two-stage gene selection (TSGS) algorithm. In this algorithm, the L_2-norm penalty are firstly introduced to achieve the grouping effect for the highly correlated genes. To overcome the small samples problem, the augmented data technique is then used to produce an augmented data set. Finally, by using the recently proposed two-stage algorithm, the most informative genes can be selected effectively. Simulation results confirm its effectiveness of the proposed approach in comparison with the popular Elastic Net method.

Keywords: Gene selection, microarray data, small samples, two-stepwise selection method, variant correlation

1 Introduction

DNA microarray technology has been widely employed to obtain microarray data which provides useful information for extracting disease-relevant genes, diagnosis, and classification of disease, etc. [1]. This measurement inevitably involves destroying the actual system (or cells), which means that sample sizes are small [2] and many important genes can be highly correlated. The number of genes is significantly greater than the number of samples, while only a small number of the thousands of genes show strong correlation with a certain phenotype [3]. Therefore, the gene selection is very crucial for disease diagnosis and treatment, biological experiment and decision, etc.

Discriminant analysis of microarray data can be referred to as feature selection in machine learning [4]. According to the way of calculating the feature evaluation index, the existing feature classification methods can be classified into three categories: filters, wrappers, and embedded methods. The filter approach [5] is widely used based on gene ranking. However, the drawback is that such a selection procedure is independent of the specific required prediction/classification

K. Li et al. (Eds.): ICSEE 2013, CCIS 355, pp. 424–432, 2013.
© Springer-Verlag Berlin Heidelberg 2013

task. The wrapper method [6] usually consists of the search procedure and the evaluation criterion. However, exhaustive search of all subsets is too expensive to implement from a high-dimensional feature space. Unlike the filter and wrapper that separate the variable selection and training processes, the embedded methods like the boosting method [7] incorporated feature selection into the construction procedure of the classifier or regression model.

Recently, some new embedded learning approaches have been proposed to achieve grouping effect by introducing the penalty term into the cost function. The most popular algorithm is the Elastic Net [8], which can encourage a grouping effect, i.e., the system variables (genes/regressors) can be naturally grouped together according to regulatory pathways where the within-group correlations are very high. In fact, the Elastic Net method is a forward selection method. It is well known that all previously selected regressors remain fixed and some insignificant regressors can not be removed from the model later. This probably leads the Elastic Net to miss a good model. Therefore, further effort still need to been made to improve the prediction performance and to reduce model size.

In this paper, an efficient two-stage gene selection (TSGS) method is proposed, which can solve the small samples and variant correlation problems. The paper is organized as follows. Section 2 gives some preliminaries. Section 3 presents the proposed two-stage gene selection method. Simulation results are presented in Section 4, followed by concluding remarks in Section 5.

2 Preliminaries

Suppose a set of data samples, denoted as $D_N = \{(x_i, y_i), i = 1, \ldots, N\}$, where $x_i = [x_{i1}, \ldots, x_{iM}]^T$ is the input vector, and y_i is output. Let $y = [y(1), \ldots, y(N)]^T$ be the response and $X = [x_1^T; x_2^T; \ldots; x_N^T] = [x_{(1)}, x_{(2)}, \ldots, x_{(M)}]$ be model matrix, where $x_{(j)} = [x_{1j}, \ldots, x_{Nj}]^T, j = 1, \ldots, M$ represent the predictors (i.e., gene or candidate regressors). After a location and scale transformation procedure, the input and response variables are centred (mean=0) to remove the intercept and also the input vectors are standardized, i.e.

$$\sum_{i=1}^{N} y_i = 0, \quad \sum_{i=1}^{N} x_{ij} = 0, \sum_{i=1}^{N} x_{ij}^2 = 1, j = 1, \ldots, M. \tag{1}$$

If these data are microarray gene expression data, then $x_{(j)}$ represents the jth gene, x_i represents the expression levels of M genes of the ith sample tissue and y_i may represent the tumor type. The gene selection problem can be approximated by a linear-in-the-parameters model of the form

$$y = X\beta + \varXi, \tag{2}$$

where $\beta = [\beta_1, \ldots, \beta_M]^T \in \Re^M$ is the estimated coefficients, $\varXi = [\varepsilon(1), \ldots, \varepsilon(N)]^T \in \Re^N$ is residual.

Different cost functions, often involving a trade-off between model complexity and training accuracy, lead to alternative architectures. Therefore, the core difference of learning methods focus on the optimisation of the different cost functions. According to the "sum-of-squared errors + penalty" criterion, these optimisation problems can be formulated into the following form as

$$\hat{\beta} = \arg\min_{\beta}\{\|y - X\beta\|^2 + f(\lambda, \beta)\}, \tag{3}$$

where $f(\lambda, \beta)$ is usually L_1-norm penalty function or L_2-norm penalty function or Elastic Net penalty function (both L_1-norm penalty and L_2-norm penalty).

The two-stage stepwise selection method [9] only consider the optimisation of the sum-of-squared errors, which can not select the gene with high correlation. To extend the gene selection ability of the recently proposed two-stage stepwise selection, the L_2-norm penalty is added into the cost function as follows

$$J(\beta, \lambda_2) = \|y - X\beta\|^2 + \lambda_2\|\beta\|^2, \tag{4}$$

where λ_2 is the regularisation parameter, and $\|\beta\|^2 = \sum_{j=1}^{M} \beta_j^2$.

The estimator $\hat{\beta}$ is the minimizer of (4)

$$\hat{\beta} = \arg\min\{J(\beta, \lambda_2)\} \tag{5}$$

3   Two-Stage Gene Selection Method

3.1   The Grouping Effect of the Ridge Penalty

Qualitatively speaking, a regression method exhibits the grouping effect if the regression coefficients of a group of highly correlated variables tend to be equal (or a change of sign if negatively correlated) [8]. In fact, the (4) is the ridge optimisation problem. The L_2 penalty in ridge regularisation can provide the grouping effect as shown in the following.

After the regression matrix X is standardised, then obviously

$$X^T X = \begin{bmatrix} 1 & \rho_{12} & \cdots & \rho_{1M} \\ * & 1 & \cdots & \rho_{2M} \\ \vdots & \vdots & \ddots & \vdots \\ * & * & \cdots & 1 \end{bmatrix} \tag{6}$$

where $\rho_{i,j}$ is the sample correlation between i^{th} regressor and j^{th} regressor, '$*$' represents the symmetrical structure. The ridge estimator is expressed by

$$\hat{\beta} = (X^T X + \lambda_2 I)^{-1} X^T y \tag{7}$$

Theorem 1: Suppose that the response y is centred, the regression matrix X are standardised, and $\hat{\beta}$ is the solution of (4). If $\hat{\beta}_i \hat{\beta}_j \neq 0$, then

$$\left\|\hat{\beta}_i - \hat{\beta}_j\right\| \leq \frac{\|y\|}{\lambda_2}\sqrt{2(1 - \rho_{ij})} \tag{8}$$

where $\rho_{i,j}$ represents the correlation between i^{th} and j^{th} regressor.

Proof. The subgradient of the objective function (4) with repect to $\hat{\beta}$ satisfies

$$\frac{\partial J(\beta, \lambda_2)}{\partial \beta_k}\bigg|_{\beta=\hat{\beta}} = 0, \qquad \beta_k \neq 0 \tag{9}$$

For $\hat{\beta}_i \hat{\beta}_j \neq 0$, it follows from (9) that

$$\hat{\beta}_i - \hat{\beta}_j = \frac{1}{\lambda_2}(x_{(i)}^T - x_{(j)}^T)(y - X\hat{\beta}) \tag{10}$$

From (4) and (5), we have

$$\left\| y - X\hat{\beta} \right\|^2 \leq J(\lambda_2, \hat{\beta}) \leq J(\lambda_2, \beta = 0) = \|y\|^2 \tag{11}$$

Since X are standardized, it can be easily obtained that

$$\left\| x_i^T - x_j^T \right\|^2 = \left\| x_{(i)} \right\|^2 + \left\| x_{(j)} \right\|^2 - 2x_{(i)}^T x_{(j)} = 2(1 - \rho_{ij}) \tag{12}$$

Using (11) and (12), (10) can be re-written as

$$\left\| \hat{\beta}_i - \hat{\beta}_j \right\| \leq \frac{\|y\|}{\lambda_2}\sqrt{2(1 - \rho_{ij})} \tag{13}$$

This completes the proof.

Theorem 1 describes the difference between the coefficient paths of the ith and j^{th} regressor. If the ith and j^{th} regressor is highly correlated, the regression method will assign almost identical coefficients (only a change of sign if negatively correlated). Therefore, Theorem 1 provides a quantitative description for the grouping effect of the Ridge penalty.

However, like most variable selection methods, the regularised two-stage stepwise selection method [9] is still unable to select more than N regressors in the $M \gg N$ scenario. The augmented data technique provides an effective method by supplementing a fictitious set of data points taken according to an orthogonal experiment.

3.2 Overcoming $M \gg N$ Using Augmented Data Technique

An augmented data set method for regularised regression problems was introduced as follows.

Theorem 2 [10]: The ridge estimator is equivalent to a least squares estimator when the actual data are supplemented by a fictitious set of data points taken according to an orthogonal experiment H_k; the response y being set to zero for each of these supplementary data points.

According to Theorem 2, since H_k is an orthogonal matrix, obviously the matrix $H_k^T H_k$ is diagonal. If the orthogonal columns have also been constructed correctly, this diagonal matrix will be equivalent to a scalar multiplied by the identity matrix I_M. For any value λ_2, the matrix may always be scaled so that $H_k^T H_k = \lambda_2 I_p$. Therefore, for the data set (X, y) and a scalar λ_2, an augmented data set (Φ, Y) with $N + M$ observations and M predictors can be defined by

$$\Phi_{(N+M) \times M} = \begin{bmatrix} X \\ \sqrt{\lambda_2} I_M \end{bmatrix} \quad and \quad Y_{(N+M)} = \begin{bmatrix} y \\ 0 \end{bmatrix}. \tag{14}$$

where $\Phi \overset{\Delta}{=} [\phi_1, \ldots, \phi_M]$. This augmented data set can now be fed into the original two-stage stepwise selection algorithm to form a two-stage gene selection procedure. The least square estimate the parameters is given by

$$\hat{\theta} \overset{\Delta}{=} \arg\min_{\theta} \{(Y - \Phi\theta)^T (Y - \Phi\theta)\} = ((\Phi)^T \Phi)^{-1} (\Phi)^T Y. \tag{15}$$

It is found that such an augmented data technique not only integrates Ridge regularisation directly into the regression matrix, but also allows the following two-stage stepwise selection method to overcome the $M \gg N$ problem. It is noted that the augmented data technique does not change the correlation between the ith and j^{th} regressor, $i, j = 1, \ldots, M$. The augmented data set can then be solved by the following two-stage stepwise algorithm [9].

3.3   Two-Stage Stepwise Selection Method

Stepwise selection is the recommended subset selection technique owing to its superior performance [11]. By contrast, the recently proposed two-stage selection algorithm [9], which includes a forward selection stage and a second backward refinement stage, provides an efficient path.

Forward Recursive Selection - First Stage. The forward selection stage selects the regressors based on their contributions to maximizing the model error reduction ratio, one regressor at a time. The selection procedure continues until some termination criterion is met or a desired model size is reached.

Model Refinement - Second Stage. The above forward stage generates a model, however, forward selection stage is subject to the constraint that all previously selected regressors remain fixed and cannot be removed from the model later. To overcome this deficiency, each previously selected term needs to be checked again and the model is refined. This review is repeated until all the selected model terms are more significant than those remaining in the candidate pool. Finally, a satisfactory model is produced.

3.4   Complete Algorithm

The complete algorithm can be summarized as follows.

Step 1 Initialization: Set a small parameter set Λ including positive values for λ_2, e.g., $\Lambda = \{0.01, 0.1, 1, 10, 100\}$, and the iteration index $I = 1$. Set the forward selection step (*i.e.*, S) to a positive integer.

Step 2 Model constuction:

(A) The parameter λ_2 picks a grid value from Λ, then the augmented data set (Φ, Y) is generated according to (14).

(C) Forward selection: At each step, the net contribution for all remaining candidate model terms are computed, and find the one that produces the maximum contribution. The selection procedure continues until some termination criterion is met or a desired model size is reached. An initial model with n regressors is produced.

(D) Backward model refinement: Each previously selected regressor in the initial model is shifted to the nth position and compared with all remaining candidate terms. The shifting and comparison procedures are repeated until no insignificant term remains in the selected model. A satisfactory model is finally constructed.

(E) If K-fold cross-validation (CV) is used as a termination criteria, then for each of K experiments, $K - 1$ folds and 1 fold of the augmented data set are used for training and validation, respectively. After Step 2 (B)-(D) are operated K times, an n-unit model can be finally determined by K-fold CV.

Step 3 Determining the λ_2 and the corresponding model:

(A) The procedure is monitored and terminated when λ_2 picks the final element from Λ. Otherwise, set $I = I + 1$, and go to step 2.

(B) The λ_2 can be chosen by the one giving the smallest CV error, and the corresponding model is produced.

4 Simulation

Arthritis is a form of joint disorder that involves inflammation of one or more joints. The Arthritis data set [12] consists of rheumatoid arthritis (RA) and osteoarthritis (OA) types. RA is a systemic disease characterized by an aggressive infiltration of the synovium, which degrades cartilage and bone. However, OA does not display these histological features, but degradative proteases are nevertheless produced in the synovium. Although OA is the most common type of arthritis, RA is recognized as the most crippling or disabling type of arthritis. Therefore, the classification of these two clinically distinct forms of arthritis is an important issue.

The arthritis data describe the expression of 755 genes in 7 OA and 24 RA samples. The correlations among all 755 genes were shown as Fig. 1. It is obviously seen that there exist some genes with high correlation. To verify the ability of overcoming the small sample of proposed TSGS approach, all 31 data samples was used as a training set and the initial parameters of $S = 100$ and $\lambda_2 = 0.5$ were set. Model fitting and tuning parameter selection by 5-fold cross-validation (CV) were operated on the training data. The solution paths of the parameter estimates are shown in Fig. 2, where 42 genes (> 31) were selected by CV. This means that the proposed TSGS method can solve the small sample problem.

Furthermore, the correlations of all the selected 42 genes are shown in Fig. 3, some highly correlated genes can be chosen, i.e., $\{672, 320\}$, $\{44, 186\}$, $\{93, 75\}$, $\{616, 204\}$ and $\{529, 513\}$. Fig. 4 shows the solution paths of the 320th and 672th gene. It can be seen again that as a new highly correlated gene was added to the model, it immediately adjusted the previous parameter estimates so that the correlated genes shared almost equal values. This indicates that the proposed TSGS method can solve the gene correlation problem.

Finally, to compare the performance with the popular Elastic Net method, the training data set randomly selected 4 OA and 14 RA samples, and the remaining 13 samples were use to test. The entire process was repeated 50 times, leading to 50 different training and testing data sets. The initial parameters $S = 100$ and a small parameter set $\Lambda = [0.5, 1, 5, 10, 50, 100]$ for λ_2 were set, respectively. The proposed TSGS and Elastic Net methods were applied to the 50 training and testing data sets, and 50 runs were performed. The statistical results, including model size, training error and testing error are shown in Table 1. It is obviously seen that the proposed method produced not only the more compact model but also the better classification performance.

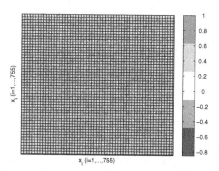

Fig. 1. The correlations among All 755 genes

Table 1. Statistical performance of two algorithms

	Method	Arthritis
Elastic Net	Model size	46.62 ± 23.74
	Training error	0.44 ± 1.11
	Testing error	1.56 ± 1.05
TSGS	Model size	27.32 ± 22.90
	Training error	0.48 ± 1.05
	Testing error	1.14 ± 0.94

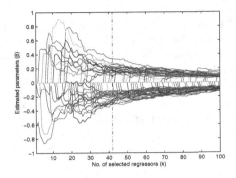

Fig. 2. Solution paths of TSGS for 100 genes

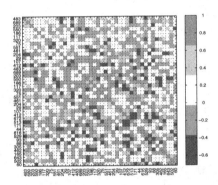

Fig. 3. The correlations among all the selected 42 genes

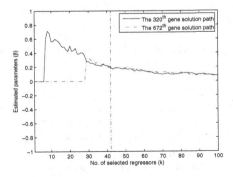

Fig. 4. Solution paths of the 320th and 672th gene

5   Conclusion

The paper has proposed a two-stage gene selection algorithm for microarray data with small samples and variant correlation. The L_2-norm penalty are firstly introduced to achieve the grouping effect for the highly correlated genes. By using the augmented data technique, an augmented data set can be then produced. The most informative genes can be selected effectively by the proposed two-stage algorithm. Compared with the popular Elastic Net method, the proposed TSGS method demonstrates the better results. The identified gene clusters may provide a chance for the exploratory analysis of microarray data.

Acknowledgment. The authors would like to thank the Research Councils UK under grant EP/G042594/1, the National Science Foundation of China (61074032, 60834002, 51007052, 61104089), Science and Technology Commission of Shanghai Municipality (11ZR1413100), the innovation fund project for Shanghai University.

References

1. Golub, T.R., et al.: Molecular classification of cancer: class discovery and class prediction by gene expression monitoring. Science 286, 531–537 (1999)
2. Guyon, I., Elisseeff, A.: An introduction to variable and feature selection. Journal of Machine Learning Research 3, 1157–1182 (2003)
3. Liu, B., Wan, C., Wang, L.: An efficient semi-unsupervised gene selecttion method via spectra biclustering. IEEE Transactions on Nanobioscience 5(2), 110–114 (2006)
4. Kohavi, R., John, G.H.: Wrappers for feature subset selection. Artificial Intelligence 97(1-2), 273–324 (1997)
5. Cai, R., Hao, Z., Yang, X., Wen, W.: An efficient gene selection algorithm based on mutual information. Neurocomputing 72, 991–999 (2009)
6. Zhou, X., Mao, K.Z.: LS bound based gene selection for DNA micorarray data. Bioinformatics 21(8), 1559–1564 (2005)
7. Freund, Y., Schapire, R.: A dicision-theoretic generalization of on-line learning and an application to boosting. Journal of Computer and System Sciences 55, 119–139 (1997)
8. Zou, H., Hastie, T.: Regularization and variable selection via the elastic net. J.R. Statist. Soc.B 67(2), 301–320 (2005)
9. Li, K., Peng, J.X., Bai, E.W.: A two-stage algorithm for identification of nonlinear dynamic systems. Automatica 42(7), 1189–1197 (2006)
10. Marquardt, D.W.: Generalized inverses, ridge regression, biased linerar estimation, and nonlinear estimation. Technometrics 12(3), 591–612 (1970)
11. Nelles, O.: Nonlinear system identification. Springer (2001)
12. Sha, N., Vannucci, M., Brown, P., Trower, M., Amphlett, G.: Gene selection in arthritis classification with large-scale microarray expression profiles. Comparative and Functional Genomics 4, 171–181 (2003)

Investigation of AIMD Based Charging Strategies for EVs Connected to a Low-Voltage Distribution Network

Mingming Liu and Seán McLoone

Department of Electronic Engineering, National University of Ireland, Maynooth
Maynooth, Ireland
mliu@eeng.nuim.ie, s.mcloone@ieee.org

Abstract. In this paper we consider charging strategies that mitigate the impact of domestic charging of EVs on low-voltage distribution networks and which seek to reduce peak power by responding to time-of-day pricing. The strategies are based on the distributed Additive Increase and Multiplicative Decrease (AIMD) charging algorithms proposed in [5]. The strategies are evaluated using simulations conducted on a custom OpenDSS-Matlab platform for a typical low voltage residential feeder network. Results show that by using AIMD based smart charging 50% EV penetration can be accommodated on our test network, compared to only 10% with uncontrolled charging, without needing to reinforce existing network infrastructure.

Keywords: EV charging, AIMD, Smart Grid, Distributed algorithm.

1 Introduction

In the near future, with an increasing number of EVs plugging into the grid in residential areas it is likely that coincident uncontrolled charging will overload local distribution networks and substantially increase peak power requirements. It is therefore essential that careful consideration is given to developing smart grid infrastructure and charging strategies to mitigate the impact of the roll out of EVs on the grid. EV charging strategies have been the focus of considerable research effort in recent years [1]-[4]. Clement-Nyns et al. [4], propose a coordinated charging method is to minimize power losses and maximize the main grid load factor. In Richardson et al. [2] a technique based on linear programming to determine the optimal charging rate is developed in order to maximize the total power that can be delivered to EVs while meeting distribution network constraints. In [1] a coordinated charging algorithm using both quadratic and dynamic programming is employed to shift the EV loads to off-peak times while minimizing the power losses for both deterministic and stochastic data. In [6] a transportation micro-simulation is employed to secure power system operation using a multi-agent system (MAS) to coordinate EV charging behavior. In [8], to maximize a customer's own utility, a simple adaption strategy based on price

K. Li et al. (Eds.): ICSEE 2013, CCIS 355, pp. 433–441, 2013.

feedback is effectively used to solve the distributed EV charging problem in the smart grid. Most recently, Stüdli et al. [5] propose charging strategies based on Additive Increase and Multiplicative Decrease (AIMD) algorithms that can be implemented in a decentralized fashion to maximize power utilization by EVs while achieving a fair allocation of power across customers. In this paper we investigate the performance of AIMD based charging strategies for EVs connected to a low-voltage distribution network with regard to mitigating the impact of EVs on the grid from the perspective of transformer loading levels and voltage profiles. In particular, we propose an extension of the AIMD charging strategy that seeks to reduce peak power by responding to time-of-day pricing. The strategies are evaluated using simulations conducted on a custom OpenDSS-Matlab platform for a typical low voltage residential feeder network.

2   Methodology

2.1   Assumptions

We make several assumptions in investigating the impact of EV charging on a LV distribution network, most of which are consistent with previous studies in [2], [3], [5] and [6]. The assumptions are as follows:(i) All EV batteries have a capacity of 20 kWh; (ii) Each EV charger is connected to a standard household outlet at 230V; (iii) The maximum power output from the EV home charger cannot exceed 3.7kW; (iv) Each EV has the ability to adapt its charge rate in real-time and continuously; (v) The initial state of charge (SOC) energy request for each EV ranges from 5kWh to 15kWh uniformly; and (vi) Power flow is unidirectional from grid to vehicle.

In order to implement smart charging strategies in practice, several specific assumptions are also required in relation to the communication and sensing capabilities of the smart grid infrastructure, namely:

1. Each EV charging point is equipped with a communication device and is able to receive broadcast signals from a local server.
2. Each EV charging point is able to detect its line voltage in real-time.
3. Each EV charging point is able to send the voltage signal back to the local server and regulate its own charge rate by commands from the server.
4. A centralized server is installed in the substation and is able to sense the available resource and broadcast signals to the local servers.

2.2   Distributed AIMD Algorithm

The basic idea of AIMD was originally applied in the context of decentralized congestion control in communication networks [7]. In [5] Stüdli et al. proposed applying AIMD to EV charging problems and investigated a number of practical scenarios. In this paper, the framework we assume is consistent with the domestic charging scenario demonstrated in [5]. The basic decentralized AIMD algorithm for EV charging is summarized as follows:

while *battery not charged* **do**
  if $P(k) < \overline{P}(k)$ **then**
    | $c_i(k+1) = c_i(k) + \alpha.\Delta T$
  else
    generate uniform random number, p **if** $p < p_i$ **then**
      | $c_i(k+1) = \beta^{(1)}.c_i(k)$
    else
      | $c_i(k+1) = \beta^{(2)}.c_i(k)$
    end
  end
end

Here, $c_i(k)$ is the charge rate of the ith EV, $P(k)$ is the total power demanded by all the EVs connected and $\overline{P}(k)$ maximum power available to charge EVs at the kth time instant. α is the additive constant value in kW/s, $\beta^{(1)}$, $\beta^{(2)}$ are the multiplicative constants, which are selected at random with probability p_i and ΔT is the time interval between EV charge rate updates. During operation each active EV charge point additively increases its charge rate until a capacity event occurs at which point it applies a multiplicative decrease to the charge rate. A capacity event occurs when the power $P(k)$ demanded by the active EV charger points exceeds the maximum available power $\overline{P}(k)$. Here $P(k)$ is computed as:

$$P(k) = \sum_{i=1}^{N(k)} c_i(k), \tag{1}$$

where $N(k)$ is the number of active chargers at the kth time instant. The $P(k) < \overline{P}(k)$ capacity event condition is monitored by a central monitoring station (server) which broadcasts a message to the charge points when events occur. As disscussed in [5], this algorithm guarantees an equitable 'average' distribution of the power if each charge point chooses the same α, β and p, while requiring a minimum of communication infrastructure.

2.3 Formation of the Smart Charging Strategy

In this section a new charging strategy is formulated based on the decentralized AIMD method. The objective is to achieve benefits for both utilities and customers. In order to do this we incorporate power system constraints on voltages and load balance into the AIMD implementation so that all EVs can share the maximum amount of available power fairly while ensuring that the distribution network continues to operate within acceptable limits. In addition, we modulate the available power signal in response to a varying electricity price [11] to affect a shift in EV loads away from periods of high demand, thereby reducing peak-power capacity requirements and ultimately the cost to the consumer.

The infrastructure needed to support the proposed strategy consists of EV charging points, local monitoring stations at the distribution transformers and a central monitoring station at the main sub-substation. Each charge point must be capable of running the AIMD algorithm, measure its own socket voltage $V_i(k)$, and maintain bidirectional communication with the local monitoring station for its residential area. The role of each local monitoring station is to: (i) receive voltage capacity event signals from the EV chargers in its local residential area and generation capacity broadcasts from the central monitoring station; (ii) detect local infrastructure capacity events, and (iii) broadcast capacity event information to the charge points in its area. The role of the central monitoring station is to monitor available power in real-time and broadcast generation capacity events via the local monitoring stations to the charge points.

With this infrastructure in place the basic AIMD algorithm running on each charge point is modified to respond to all capacity events (voltage, infrastructure and generation) and to transmit a voltage event message to the local monitoring station when the voltage level drops below a minimum acceptable level, V_{min}. The pseudo code for the algorithm is as follows:

```
while battery not charged do
    if capacity event then
        generate uniform random number, p if  p < p_i then
            | c_i(k + 1) = β^(1).c_i(k)
        else
            | c_i(k + 1) = β^(2).c_i(k)
        end
    else
        | c_i(k + 1) = c_i(k) + α.ΔT
    end
    if V_i(k) < V_min then
        | transmit voltage event message
    end
end
```

The local monitoring station for a given residential area broadcasts a capacity event signal to the active EV charge points in its area if any of the following conditions are satisfied: (i) A generation capacity event is broadcast by the main substation; (ii) A voltage event message is transmitted by any of the charge points in its residential area; or (iii) A local infrastructure constraint violation is detected (e.g. transformer overload).

The central monitoring station is responsible for determining the power available and broadcasting a generation capacity event when this is exceeded. The total instantaneous power consumption is given by

$$P(k) = \sum_{j=1}^{N} h_j(k) + \sum_{i=1}^{N(k)} c_i(k), \tag{2}$$

where $h_j(k)$ represents the non-EV power consumption for the jth house on the distribution network at time k, $c_i(k)$ is the charge rate of the ith active EV charge point, N is the number of houses on the distribution network and $N(k)$ is the number of active charge points. The instantaneous available power is computed as

$$\overline{P}(k) = P_{rated} - \lambda - \Delta(k). \tag{3}$$

Here P_{rated} (kVA) is the maximum capacity that can be drawn from the substation and is the lesser of the available generation capacity or the substation rating. λ (kVA) is a constant 'safety margin' for secure operation. $\Delta(k)$ (kVA) is a time varying factor introduced to create an artificial reduction in available power at times of high electricity prices and is computed as

$$\Delta(k) = (E(k) - E_{min}).\xi. \tag{4}$$

In (4) ξ is a constant tuning parameter, $E(k)$ (cent/kWh) is the Time-of-Use(TOU) price [11] at time k and E_{min} is the minimum TOU price during the day.

In the AIMD EV charging study in [5] EV charging was considered independently of other household power consumption with $P(k)$ computed according to (1). This results in a significant communication overhead as the central substation cannot distinguish EV charge point power usage from other residential power consumption, hence this information must be communicated continuously to the central monitoring station by the active EV charge points. In our formulation $P(k)$ and $\overline{P}(k)$ are defined in terms of overall residential power consumption levels, hence $P(k)$ can be directly sensed at the central substation.

3   Simulation Platform

To evaluate the performance of the proposed AIMD charging strategy a test distribution network incorporating EVs is simulated based on a typical LV residential feeder layout. A simplified schematic diagram for the test network is given in Fig. 1. In our simulations, the voltage is set at 1.05pu at the source end of the external grid. A 2MVA distribution substation is connected to the external grid to bring the voltage level to 10kV. This substation feeds three distribution transformers serving residential areas. Each distribution transformer is connected by an unbalanced 336 MCM ACSR transmission line of different length (modeled as Pi-Equivalent circuits). Both household loads and EV charging loads are connected at the secondary side of each distribution transformer. As illustrated in Fig.1, the household loads with EV charging points are separated into three phases, and the number of houses connected to each phase is indicated in parenthesis. Non-EV charging loads are lumped together using balanced three phase modeling. The distance between each house connected to a given phase is randomly chosen between 10-50m. To simulate EV charging with this network a custom OpenDSS-Matlab platform was employed. OpenDSS [9], an open source electric power Distribution System Simulator, was used to simulate the test network and calculate its power flow distribution and voltage profiles at each sample instant. A master programme was developed in Matlab to simulate the operation of network over a period of time, with the following steps performed at each sample instant:

(a) simulate EVs connecting and disconnecting to charge points;
(b) compute the instantaneous charge rates according to the AIMD algorithm described in the previous section;
(c) generate updated OpenDSS simulation parameters;
(d) call the OpenDSS software to simulate the current state of the distribution network.

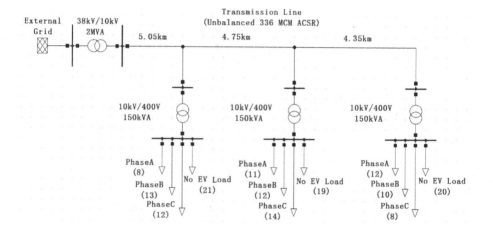

Fig. 1. Schematic diagram of the distribution network

4   Case Study

Utility companies hope that in the short to medium term (10-20 years) smart charging strategies will enable them to accommodate the extra loads represented by EV charging without needing to upgrade their distribution network infrastructure. To predict the impact of EV charging on the grid we assume a maximum penetration of EVs of 50% in the medium term and simulate the performance of both uncontrolled and smart charging under these circumstances for typical Irish winter grid loads. Residential power consumption winter profiles were generated based on residential customer smart meter electricity trial data provided by the *Commission for Energy Regulation* (CER) in Ireland [12].

For our AIMD algorithm implementation E_{min} and $E(k)$ are set in accordance with [11], $\xi=12$, P_{rated} is set to 450kVA and $\lambda=50$kVA. V_{min} is selected as 0.9pu, the minimum acceptable voltage level in Ireland, and the capacity limit of each of the distribution transformers is set to 150kVA. Charging is performed in accordance with the assumptions set out in Section 2.1. We also assume that each EV is automatically disconnected from the grid when it is fully charged. Updates are performed every 5 minutes (i.e. $\Delta T = 300$s).

From Irish traffic survey data [10] it can be concluded that the majority of commuters arrive home between 4pm-8pm each day. To capture this in our simulation we generate EV home arrival times from a normal distribution centered at 6 pm with a standard deviation of 1 hour. It should be noted that this falls within the period normally associated with peak-power on the Irish grid (5-7pm), hence unregulated EV charging has the potential to substantially increase peak-power on the grid. Fig. 2(a) shows the voltage profile of the minimum voltage level in the distribution network for three different scenarios: (i) no EVs on the grid; (ii) uncontrolled charging of EVs; and (iii) charging of EVs with our proposed AIMD smart charging algorithm. The corresponding power flows at the substation are plotted in Fig. 2(b). As expected the impact of uncontrolled EV charging is to increase the voltage drops significantly. The Minimum non-EV voltage on all buses was 0.92 pu during peak-periods, but with uncontrolled EV charging coinciding with peak-power, bus voltages drop to 0.87 pu, leading to voltage issues on the network. In addition, transformers are overloaded with demanded power exceeding the available power by 65%. Thus, for our test distribution network uncontrolled charging at 50% EV penetration cannot be supported. Simulations conducted for different EV penetration levels (not included) show that the maximum level that can be sustained under these conditions is 10%. In contrast, with AIMD smart charging voltage sags are substantially reduced (0.93pu compared to 0.87pu) and capacity and infrastructure constraints are maintained while making best use of available power.

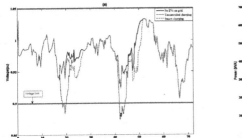 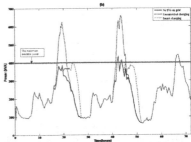

Fig. 2. Comparison of uncontrolled and AIMD smart charging of EVs for a typical LV network over a 72 hour period in mid-winter (0 = midnight): (a) Minimum voltage on the distribution network; (b) substation power flow

The daily price variation signal $E(k)$, set according to [11], is plotted in Fig. 3(a). The impact of the inclusion of this term to modulate the available power signal in the AIMD algorithm is clearly evident when comparing the voltage sags for uncontrolled and smart EV charging. In the latter the significant voltage sags have been postponed until after the peak-time (around 9 pm) reflecting a corresponding shift in the EV charging load. This is further highlighted in Fig. 3(b), which shows a comparison of the EV load profile obtained using AIMD with and without the inclusion of the price modulated available power term $(\Delta(k))$ in (3)).

Table 1 provides a summary comparison of the different charging strategies considered in the paper. From the table it can be seen that *SmartP* (i.e. price-adjusted AIMD) EV charging provides the best overall performance with the lowest line voltage drops, smallest maximum power requirements and the cheapest charging rate (40-50% less than with uncontrolled charging).

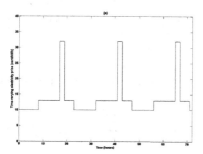 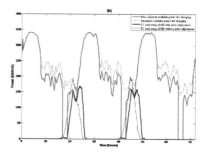

Fig. 3. (a) Plot of the TOU electricity price signal $E(k)$; (b) Comparison of EV loads obtained using AIMD smart charging with and without price adjusted available power.

Table 1. Comparison of all charging scenarios considered

Scenarios	Mini Voltage(p.u)	Max Load[1] (%)	Energy[2] (kWh)	Cost(cents/kWh)
NoEV	0.92	108	-	-
Unctrl	0.87	165	10.15	19.70
Smart[3]	0.91	118	10.15	12.69
SmartP[4]	0.92	109	10.15	10.90

5   Conclusions

In this paper, a novel distributed smart EV algorithm has been proposed for managing EV charging on a low-voltage residential distribution network. The algorithm, which is based on the AIMD EV charging algorithms proposed in [5], is designed to take account of capacity, infrastructure and voltage constraints on the network and encourage EV charging at off-peak times. Using a simulation of a representative low-voltage residential distribution network with 50% EV penetration the proposed algorithm has been benchmarked against uncontrolled charging and shown to effectively mitigate the impact of EV charging on the grid. Our results show that for the scenarios considered the proposed AIMD charging strategy is able to comfortably support up to 50% EV

---

[1] Maximum loading as a percentage of the sub-station rating (400kVA).
[2] Average energy per EV per day.
[3] AIMD smart charging without price adjustment.
[4] AIMD smart charging with price adjustment.

penetration without requiring strengthening of the distribution network infrastructure. Furthermore, by taking into account TOU pricing a significant reduction in the cost of EV charging can be achieved for the customer. Thus, the proposed AIMD based charging strategy has the potential to provide significant benefits to both EV owners and to utility companies.

Acknowledgments. The authors would like to thank the Irish Social Science Data Archive (ISSDA) for providing access to the CER Smart Metering Project data. The authors also gratefully acknowledge funding for this research provided by NUI Maynooth (Doctoral Teaching Scholarships Programme) and Science Foundation Ireland (grant number 09/SRC/E1780).

References

1. Clement-Nyns, K., Haesen, E., Driesen, J.: Analysis of the impact of plug-in hybrid electric vehicles on residential distribution grids by using quadratic and dynamic programming. In: Proc. EVS24 Int. Battery, Hybrid and Fuel Cell Electric Vehicle Symposium, Stavanger, Norway (May 2009)
2. Richardson, P., Flynn, D., Keane, A.: Optimal Charging of Electric Vehicles in Low-Voltage Distribution Systems. IEEE Trans. Power Systems 27(1), 268–279 (2012)
3. Richardson, P., Flynn, D., Keane, A.: Impact assessment of varying penetrations of electric vehicles on low voltage distribution systems. In: 2010 IEEE Power and Energy Society General Meeting, July 25-29, pp. 1–6 (2010)
4. Clement-Nyns, K., Haesen, E., Driesen, J.: The Impact of Charging Plug-In Hybrid Electric Vehicles on a Residential Distribution Grid. IEEE Trans. Power Systems 25(1), 371–380 (2010)
5. Stüdli, S., Crisostomi, E., Middleton, R., Shorten, R.: A flexible distributed framework for realising electric and plug-in-hybrid vehicle charging policies. International Journal of Control 85(8), 1130–1145 (2012)
6. Galus, M.D., Waraich, R.A., Andersson, G.: Predictive, distributed, hierarchical charging control of PHEVs in the distribution system of a large urban area incorporating a multi agent transportation simulation. In: Proc. 17th Power Systems Computations Conference (PSCC), Stockholm, Sweden (2011)
7. Shorten, R., Wirth, F., Leith, D.: A positive systems model of TCP-like congestion control: asymptotic results. IEEE/ACM Transactions on Networking 14(3), 616–629 (2006)
8. Fan, Z.: A Distributed Demand Response Algorithm and its Application to PHEV Charging in Smart Grids. Preprint, to appear in IEEE Trans. Smart Grid (2012)
9. Open DSS Manual. Electric Power Research Institute (July 2010), http://sourceforge.net/projects/electricdss/
10. CSO: National Travel Survey 2009. Central Statistics Office (2009), http://www.cso.ie/en/media/csoie/releasespublications/documents/transport/2009/nattravel09.pdf
11. SEAI: Time-varying Electricity Price. Sustainable Energy Authority of Ireland (August 2012), http://www.seai.ie/Renewables/Smart_Grids/The_Smart_Grid_for_the_Consumer/Home_Consumer/Smart_Meter/Time_varying_electricity_price/
12. CER: Electricity Customer Behaviour Trial. Commission for Energy Regulation (2012), www.ucd.ie/issda/data/commissionforenergyregulation/

A Fast Market Competition Approach to Power System Unit Commitment Using Expert Systems

Lijun Xu[1,2], Kang Li[2], and Minrui Fei[1]

[1] Shanghai Key Laboratory of Power Station Automation Technology,
School of Mechatronical Engineering and Automation, Shanghai University, Shanghai, China
[2] School of Electronics, Electrical Engineering and Computer Science
Queen's University Belfast, University Road, Belfast, BT9 5AH, UK
ruby_mickey@sina.com, k.li@qub.ac.uk, mrfei@staff.shu.edu.cn

Abstract. This paper proposes a novel two-layer hierarchical market competition algorithm (THMCA) combined with expert system for the unit commitment(UC) problem in power systems. Two hierarchical population individuals are defined in the algorithm, namely the holding companies and the subsidiary companies which altogether form conglomerates. Market competitions among these conglomerates lead to the convergence to a monopoly at the end, resulting in an optimal solution of the above problem. In the meanwhile, expert system is used to produce several expert rules for heavy constraints handling not only in the pre-scheduling process and in the THMCA process as well, ensuring that the positions of all companies are feasible and near-optimal solutions to the UC problem. The algorithm is shown to have a fast execution speed for UC application and the comparison simulation results on a power system with up to 100 generating units have demonstrated the effectiveness on cost reduction of the proposed method.

1 Introduction

The unit commitment (UC) [1-5] is a complex combinatorial optimization problem to determine the power generation schedule of units in order to satisfy load demand at minimum cost in power systems. A variety of constraints on the generating units must be satisfied, including the time-dependent ones. Therefore, to solve the nonlinear, non-convex, large scale, mixed integer UC problem is a challenging task. It should also be noted that to solve this problem can be computationally expensive for large power systems.

To tackle this issue, several intelligent optimization methods can be adopted, for example genetic algorithm (GA) [6], particle swarm optimization (PSO) [7], clone evolutionary algorithm [8], bacterial foraging (BF) [9] and shuffled frog leaping algorithm (SFLA) [10] etc. GA is one of the early proposed evolutionary algorithms which has found many successful applications. However, it is computationally expensive, and the convergence cannot be guaranteed. The comparison results [7] show that the PSO is more efficient than GA. But unfortunately, in solving the UC problem by PSO, many particles in a randomly created swarm are not feasible solutions because of their violation of constraints, and especially to meet the minimum up/down time limits. Bacterial foraging (BF) algorithm is a feature selection method based on a heuristic search

K. Li et al. (Eds.): ICSEE 2013, CCIS 355, pp. 442–450, 2013.

strategy with fast computing speed. But a large storage capacity is required to complete the computation. Shuffled frog leaping algorithm (SFLA) combines the benefits of both the genetic-based memetic algorithm and the social behavior-based PSO algorithm. However, most of these methods suffer from the curse of dimension when applied to a modern large-scale power system with heavy constraints. Moreover, these methods commonly get stuck at a local optimum rather than reaching the global optimum. Even small percentage reduction in fuel costs typically leads to considerable savings for electric utilities. In this paper, a new fast approach called two-layer hierarchical market competition algorithm (THMCA) combined with expert system is proposed to efficiently solve the constrained UC problem.

2   Formulation of the Unit Commitment (UC) Problem

The UC problem is to determine the start-up and shut-down schedules and outputs of generation units to meet the dynamic demand load during the scheduling horizon at a minimal total cost. In general, the fuel cost(FC) of the ith generator is described as:

$$FC_i^t(P_i^t) = a_i + b_i P_i^t + c_i(P_i^t)^2, \quad i = 1, 2, \cdots, N_{unit} \tag{1}$$

where a_i, b_i, and c_i are fuel cost coefficients of the ith unit, N_{unit} represents the number of units, t is the index for time intervals, P_i^t is the power output of the ith unit.

The total operating cost is expressed as the sum of fuel costs, start-up costs, and shutdown costs of the generating units:

$$TFC = \sum_{t=1}^{T} \sum_{i=1}^{N_{unit}} FC_i^t(P_i^t) u_i^t + SUC_i^t(1 - u_i^{t-1})u_i^t + SDC_i^t u_i^{t-1}(1 - u_i^t) \tag{2}$$

$$SUC_i^t = \begin{cases} SUC_{ihot} & if \ T_i^0 \in [MDT_i, \ MDT_i + T_{icold}] \\ SUC_{icold} & if \ -T_i^0 \geq MDT_i + T_{icold} \end{cases} \tag{3}$$

where SUC_i^t and SDC_i^t are start up and shut down costs respectively. u_i^t is the status of the ith unit at time t. SUC_{ihot} is the hot start-up cost, SUC_{icold} is the cold start-up cost and T_{icold} is the cold start time.

$$u_i^t = 1 \quad if \ online \quad or \ u_i^t = 0 \quad if \ offline \tag{4}$$

The units and system constraints of UC problem, which must be satisfied during the UC scheduling period, are as follows:

The ith online unit generation limits:

$$P_{imin}^t \leq P_i^t \leq P_{imax}^t \tag{5}$$

where P_{imin}^t and P_{imax}^t are minimum and maximum output power of ith online unit.

For security requirements and economic reasons, some units are assigned with the must-run status:

$$u_i^t = 1 \quad for \quad t = [t_{i,begin}^{run}, t_{i,end}^{run}] \tag{6}$$

where $t_{i,begin}^{run}$ and $t_{i,end}^{run}$ are beginning and ending time of ith online must-run unit.

For maintenance requirements, some units are assigned with the must-off status:

$$u_i^t = 0 \qquad for \quad t = [t_{i,begin}^{off}, t_{i,end}^{off}] \tag{7}$$

where $t_{i,begin}^{off}$ and $t_{i,end}^{off}$ are beginning and ending time of ith online must-off unit.

Ramp up/down rates for online units:

$$P_i^t - P_i^{t-1} \leq UR_i \tag{8}$$

$$P_i^{t-1} - P_i^t \leq DR_i \tag{9}$$

where DR_i and UR_i are down and up ramp rate limits of ith unit.

System power balance:

$$\sum_{i=1}^{N_{unit}} P_i^t u_i^t = P_D^t, \quad t = 1, 2, \cdots, T \tag{10}$$

where P_D^t is the system load demand at time t.

The spinning reserve of the power system:

$$\sum_{i=1}^{N_{unit}} \hat{P}_{i\max}^t \geq P_D^t + P_R^t \tag{11}$$

where $\hat{P}_{i\max}^t$ is obtained by $\hat{P}_{i\max}^t = \min\{P_{i\max}^t, P_i^{t-1} + \tau UR_i\}$ with $\tau = 10\ min$, which is 10 minutes maximum active power of the ith unit with up ramp rate limit. And P_R^t is the system reserve at time t.

Minimum up/down time:

$$T_i^c \geq MUT_i \qquad if \quad T_i^c \geq 0 \tag{12}$$

$$T_i^c \leq -MDT_i \qquad if \quad T_i^c < 0 \tag{13}$$

where T_i^c represents the continuous ON/OFF status duration of the cth cycle of the ith unit. MUT_i and MDT_i are minimal up time and minimal down time respectively.

3　Two-Layer Hierarchical Market Competition Algorithm (THMCA)

The two-layer hierarchical market competition algorithm (THMCA) is inspired by competitions among enterprises in economic activities. The THMCA begins with an initial population called perfect competition companies. Some of the best companies that have the best objective fitness function values are selected to become the holding companies. The rest become the subsidiary companies which are then divided among the holding companies based on their power. The power of a holding company is positively proportional to its fitness value. The holding companies and their subsidiary companies form different conglomerates. Then subsidiary companies move toward their relevant holding companies and the position of the holding companies will be updated if necessary. In the next step, the market competition among the conglomerates begins, and the weak conglomerates are eliminated. This is called monopolistic competition procedure. Then the oligopoly procedure begins. The market competition will gradually lead to an increase in the power of strong conglomerates (oligopolies) and a decrease in the power of weaker ones. Finally the weak conglomerates which could not to improve their performance will collapse. These competitions among the conglomerates will cause all the companies to converge to a state called monopoly where only one conglomerate exits in the market and all the other companies become subsidiary companies of this holding company.

To begin with, an initial population called perfect competition companies is created. In a D-dimensional problem, the position of the i-th company is defined as $Company_i = [x_{i,1}, x_{i,2}, \cdots, x_{i,D}]$, $i = 1, 2, \cdots, N_c$, where N_c is total number of competition companies. The cost function of the i-th company can be defined as:

$$f_i = cost_i(x_{i,1}, x_{i,2}, \cdots, x_{i,D}) \tag{14}$$

$N^{holding}$ of the most powerful competition companies are selected as the holding companies, which form different conglomerates. The remaining N^{sub} are the subsidiary companies of these holding companies. Theoretically, the normalized cost of the n-th holding company can be defined as:

$$F_n^{holding} = \max_i \{f_i^{holding}\} - f_n^{holding} \tag{15}$$

where $f_n^{holding}$ is the cost of the n-th holding company.

The normalized power of each holding company is defined as:

$$P_n^{holding} = \left\| F_n^{holding} / \sum_{n=1}^{N^{holding}} F_n^{holding} \right\| \tag{16}$$

Then the initial number of subsidiary companies of a conglomerate becomes

$$N_n^{sub} = round\{P_n^{holding} \times N^{sub}, 0\} \tag{17}$$

where N_n^{sub} is the initial number of subsidiary companies of the n-th conglomerate. For each holding company, N_n^{sub} of the subsidiary companies are randomly selected and allocated. These subsidiary companies along with their holding company form the conglomerate.

Then the subsidiary companies start to move toward their relevant holding companies. The positions of the subsidiary companies of the n-th conglomerate are updated.

$$SUB_{n,i} = sub_{n,i} + [rand() \times \omega(holding_n - sub_{n,i})]/\cos\theta \tag{18}$$

where $sub_{n,i}$ is the position of the i-th subsidiary company of the n-th holding company, $rand()$ is a random number between 0 and 1, ω is a weight factor, and $holding_n$ is the position of the n-th holding company. To search different points around the holding company, a random amount of deviation is added to the direction of movement. The movement of a subsidiary company toward its relevant holding company at its new direction θ is a random angle between $-\varepsilon$ and ε, where $\varepsilon > 0$ is the parameter that adjusts the deviation from the original direction.

However, a subsidiary company in a conglomerate may reach a position with lower cost than its holding company. In such case, the positions of the holding company will be replaced by the higher one. The rest will move toward the new holding company.

The total power of a conglomerate depends on both the power of the holding company and the power of its subsidiary companies. But the holding company has larger weights. This total power is defined by the weighted cost of two hierarchical companies:

$$power_n^{cong} = [p_n^{cong}]^{-1} = [f_n^{holding} + \tau \overline{f_{n,i}^{sub}}]^{-1} \tag{19}$$

where p_n^{cong} is the total cost of the n-th conglomerate and $0 < \tau < 1$. Cost $\overline{f_{n,i}^{sub}}$ is the geometry mean of i-th subsidiary company in n-th conglomerate. In fact, τ represents the role of the subsidiary companies in determining the total power of a conglomerate.

The market competition among conglomerates begins and all the conglomerates try to take possession of the subsidiary companies of other conglomerates. This competition is modeled by picking some of the weakest subsidiary companies of the weakest conglomerates and making a competition among all conglomerates to possess these subsidiary companies. Each of the conglomerates will have a likelihood of taking possession of these subsidiary companies based on its total power; therefore, powerful conglomerates have greater chance to possess subsidiary companies. The possession probability of each conglomerate must be found. The normalized total cost of each conglomerate is calculated as:

$$P_n^{cong} = \max_i \{p_i^{cong}\} - p_n^{cong} \tag{20}$$

where P_n^{cong} is the normalized total cost of the n-th conglomerate.

The possession probability of each conglomerate is given by

$$pp_n^{cong} = \| P_n^{cong} / \sum_{n=1}^{N^{holding}} P_n^{cong} \| \tag{21}$$

where pp_n^{cong} is the possession probability of the n-th conglomerate.

Finally, vector **Dif** is formed :

$$\mathbf{Dif} = \left(PP_1^{cong} - rand_1, PP_2^{cong} - rand_2, \cdots, PP_{N^{holding}}^{cong} - rand_{N^{holding}} \right) \tag{22}$$

where $rand_1, rand_2, \cdots, rand_{N^{holding}}$ are random numbers between 0 and 1.

The mentioned subsidiary companies will be given to an conglomerate which has the maximum relevant index in **Dif** vector.

The powerless conglomerates will collapse in the market competition. Different criteria can be defined for collapse mechanism. In this paper, a conglomerate is assumed to be collapsed when it loses all of its subsidiary companies.

After the market competitions, all the conglomerates will collapse except the most powerful one and all companies under their possession become subsidiary companies of this conglomerate. All the subsidiary companies also have the same positions and the same fitness. In such case, the algorithm stops.

4   Implementation of THMCA for UC Problem Using Expert System

This work develops an expert system, which comprises six expert rules (ER1-ER6), for satisfying all constraints. The expert system is applied in both the pre-scheduling

Table 1. Constraints of the 10-unit system for daily scheduling

Units	Constraints
1	$u_1^t = 1$ for $t = 1$ to 24 (must-run)
2	$MUT_2 = 8$ initial status $T_2^0 = 1$ (already online for an hour)
3	$MDT_3 = 6$ initial status $T_3^0 = -2$ (already offline for 2 hours)
4	$u_4^t = 1$ for $t = 6$ to 20 (must-run)
6	$u_6^t = 0$ for $t = 1$ to 7 (must-off)
10	$u_{10}^t = 0$ for $t = 1$ to 24 (must-off)

process and the THMCA process to ensure that the positions of all companies are feasible and near-optimal solutions for the UC problem.

4.1   Pre-scheduling Using Expert Rules(ER) to Generate Initial Competition Companies

Pre-scheduling Using to Determine the Initial Status of Units. In this paper, pre-scheduling expert rules are developed to guarantee not only that it can satisfy some constraints in advance, but also that it can reduce the problem dimensions.

ER1: For a must-run unit, assign online status "1" between $t = [t_{i,begin}^{run}, t_{i,end}^{run}]$.

ER2: For a must-off unit, assign offline status "0" between $t = [t_{i,begin}^{off}, t_{i,end}^{off}]$.

ER3: Examine if the initial status T_i^0 satisfies the minimum up/down time constraints (12) and (13). If not, continue the initial status u_i^0 until it does.

To demonstrate the pre-scheduling using expert rules, consider a popular test system with 10 units over a 24 hour scheduling horizon. The constraints of the units are given in Table 1.

Competition Company Position Definition. The position of a competition company in integer-coded THMCA for a UC problem is composed of a sequence of online or offline cycle durations of the units during the scheduling horizon. A positive integer in the $Company_i$ vector represents duration of continuous online status of that unit, while a negative one represents duration of continuous offline status. According to a general daily load profile with two peaks, the maximum number of online or offline cycles for peak load to be units is 5. Therefore, the number of scheduling cycles is set 5 and a competition company in THMCA implementation for daily scheduling is a vector with $5 \times N_{unit}$ elements. The remaining blanks represented by "*" in the $Company$ vector are filled with randomly generated integer numbers according to (5).The encoding of a company comprising online or offline cycle duration of units and the unit commitment schedule is $Company$=[24,0,0,0,0,7,*,*,*,*,-4,*,*,*,*,15,*,*,*,*,*,*,*,*,-7,*,*,*,*,*,*,*,*,*,*,*,*,*,*,*,*,-24,0,0,0,0].

4.2   Constraints Handling Using Expert System

System Constraints. Combining the system power balance equality (10) and spinning reserve requirement inequality (11) with the generation limits inequality (5) enables the coupling constraints to be integrated and modified as shown in (23) and (24):

$$\sum_{i=1}^{N_{unit}} P_{i\min}^t \leq P_D^t \tag{23}$$

$$\sum_{i=1}^{N_{unit}} P_{i\max}^t \geq P_D^t + P_R^t \tag{24}$$

ER4 (Excess capacity rule): For $\sum_{i=1}^{N_{unit}} P_{i\min}^t > P_D^t$, assign offline status "0" to online unit with the highest cost until constraint (23) is satisfied.

ER5 (Insufficient capacity rule): For $\sum_{i=1}^{N_{unit}} P_{i\max}^t < P_D^t + P_R^t$, assign online status "1" to online unit with the lowest cost until constraint (24) is satisfied.

Minimum Up/Down Time Constraints. For the minimum up/down time constraints during the THMCA process, ER3 is applied the same as pre-scheduling.

Ramp Rate Constraints. Combining the ramp rate limit inequalities (8) and (9) with the generation limit inequality (5) yields the generation limits and the ramp rate limits of unit in hour as (25), which can be determined in the THMCA process.

ER6 (Ramp rate rule):

$$\max(P_{i\min}^t, P_i^{t-1} + DR_i) \leq P_i^t \leq \min(P_{i\max}^t, P_i^{t-1} + UR_i) \tag{25}$$

5   Simulation Studies

The proposed THMCA using expert system was tested on a popular power system with Num = 10, 20, 50 and 100 generating units. The required data for ten generating units is the same in [10]. For power system with 20 generating units, the data of the ten units was duplicated and the total load demand was multiplied by two. For the problem with more units, the data was scaled appropriately.

The optimal THMCA parameters for ten generating units which were chosen after several runs are given as $N^{holding} = 10$, $N^{sub} = 200$, $\tau = 0.2$, $\omega = 2$. For power system with more units, the same parameters were utilized, except for $N^{holding}$ and N^{sub} which increase correspondingly. Another parameter to be selected is $\theta \in (45°, 90°)$, which adjusts the deviation from the original direction. These values were found suitable to produce good solutions in terms of the processing time and the quality of the solutions.

From the results of 10 simulation runs, it is found that the optimal solution can be obtained after 8-th to 10-th market competition interactions. Fig.1 shows the convergence of the iteration for the 10-unit power system. This indeed verifies the high convergence rate of the algorithm.

For the comparison purpose among the THMCA and other optimization methods, all the simulation experiments used the same basic parameter settings.

Table 2 lists the solution of UC for 10-unit system obtained by THMCA using ES, THMCA without ES, the shuffled frog leaping algorithm (SFLA) [10], BF [9], GA [6], hybrid quantum clone evolutionary algorithm (HQCE) [8], and quantum inspired PSO (Q-PSO) [7]. It is obvious that THMCA using ES has satisfactory results in comparison with other methods, especially for the execution time.

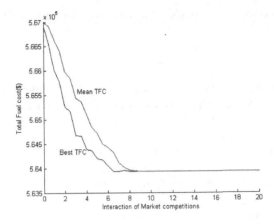

Fig. 1. 10-runs convergence of market competition algorithm in 10-unit power system

Table 2. Comparison of costs and average execution time of different methods for 10-unit system

Method	SUC	FC	Best TFC	Mean TFC	Mean Time(s)
			Costs($)		
GA	4090	561735.00	565825.00	565889	220
HQCE	4090	559856.30	563938.00	563952	19
PSO	2095	562899.00	565804.00	565868	24
QPSO	4090	559852.30	563942.30	563950	18
BF	2100	562742.00	564842.00	564866	20
SLFA	4090	559847.70	563937.70	563942	19
THMCA without ER	4090	559847.40	563937.40	563940	18
THMCA with ER	4090	559846.60	563936.60	563938	11

Table 3. Comparison of costs and average execution time of different methods for 10 to 100 units

Method	TFC($(Mean Time(s)))			
	Unit			
	10	20	50	100
GA	565825(220)	1126243(733)	2680085(4065)	5627437(15733)
HQCE	563938(19)	1123370(53)	2673694(142)	5625937(477)
PSO	565804(24)	1126148(69)	2679703(186)	5627510(624)
QPSO	563942(18)	1125510(51)	2673392(134)	5609085(449)
BF	564842(20)	1124892(34)	2674028(86)	5611514(282)
SLFA	563937(19)	1123261(33)	2673458(87)	5624526(289)
THMCA without ER	563937(18)	1122904(31)	2672610(83)	5622742(282)
THMCA with ER	563936(11)	1122872(17)	2672332(45)	5622457(156)

The best comparison results of total fuel cost and mean execution time for power system with up to 100 generating units are shown in table 3. The execution time of THMCA with ES increases almost linearly with the size of the UC scheduling problem. The overall execution time obtained by THMCA with ES is less than that of other methods. Our fast method also provide large savings of total fuel cost for large scale power systems.

6   Conclusions

UC in a power system is a constrained nonlinear optimization problem. A new optimization algorithm namely two-layer hierarchical market competition algorithm (THMCA) has been proposed to solve the problem effectively. The proposed method has been tested and compared with a few alternatives. Simulation results show that the computational time and total fuel cost of THMCA using expert system are less than other algorithms such as SFLA, GA, BF, quantum-inspired PSO, and hybrid quantum clone evolutionary algorithm. However, the performance of the THMCA also depends on certain parameters selected. Future research includes developing more efficient algorithms and addressing uncertainties in the power system.

Acknowledgment. This work was financially supported by National Natural Science Foundation of China under Grants No.61074032,and 61273040. Project of Science and Technology Commission of Shanghai Municipality under Grants No.10JC1405000 and No. 11ZR1413100, and UK-China Science Bridge Project under RCUK grant EP/G042594/1.

References

1. Wang, Q., Guan, Y., Wang, J.: A Chance-Constrained Two-Stage Stochastic Program for Unit Commitment With Uncertain Wind Power Output. IEEE Transactions on Power Systems 27(1), 206–215 (2012)
2. Bertsimas, D., Litvinov, E., Sun, X., Zhao, J., Zheng, T.: Adaptive Robust Optimization for the Security Constrained Unit Commitment Problem. IEEE Transactions on Power Systems PP(99), 1–12 (2012)
3. Daneshi, H., Srivastava, A.: Security-constrained unit commitment with wind generation and compressed air energy storage. IET Generation, Transmission and Distribution 6(2), 167–175 (2012)
4. Jiang, R., Wang, J., Guan, Y.: Robust Unit Commitment With Wind Power and Pumped Storage Hydro. IEEE Transactions on Power Systems 27(2), 800–810 (2012)
5. Makarov, Y., Etingov, P., Ma, J., Huang, Z., Subbarao, K.: Incorporating Uncertainty of Wind Power Generation Forecast Into Power System Operation, Dispatch, and Unit Commitment Procedures. IEEE Transactions on Sustainable Energy 2(4), 433–442 (2011)
6. Chen, Q., Zhong, Y., Zhang, X.: A pseudo genetic algorithm. Neural Comput. & Applic. 19(1), 77–83 (2010)
7. Jeong, Y., Park, J., Jang, S., Lee, K.: A New Quantum-Inspired Binary PSO: Application to Unit Commitment Problems for Power Systems. IEEE Transactions on Power Systems 25(3), 1486–1495 (2010)
8. Xu, L., Fei, M.: A Hybrid Quantum Clone Evolutionary Algorithm-Based Scheduling Optimization in a Networked Learning Control System. In: Chinese Control and Decision Conference, CCDC 2010, pp. 3632–3637 (2010)
9. Eslamian, M., Hosseinian, S.H., Vahidi, B.: Bacterial foraging-based solution to the unit-commitment problem. IEEE Trans. on Power Syst. 24(3), 1478–1488 (2009)
10. Ebrahimi, J., Hosseinian, S.H., Gharehpetian, G.B.: Unit Commitment Problem Solution Using Shuffled Frog Leaping Algorithm. IEEE Transactions on Power Systems 26(2), 1–9 (2011)

Spatial and Temporal Variability Analysis in Rainfall Using Standardized Precipitation Index for the Fuhe Basin, China

Rongfang Li[1,3], Lijun Cheng[1,4], Yongsheng Ding[1,2,*],
Yunxiang Chen[3], and K. Khorasani[4]

[1] College of Information Sciences and Technology, Donghua University
Shanghai, 201620, P.R. China
[2] Engineering Research Center of Digitized Textile & Fashion Technology
Ministry of Education, Donghua University
Shanghai, 201620, P.R. China
[3] Jiangxi Provincial Institute of Water Sciences, Nanchang, Jiangxi, P.R. China
[4] Department of Electrical & Computer Engineering,
Concordia University, Montreal, Quebec, H3G1M8, Canada
ysding@dhu.edu.cn

Abstract. In this paper, the standard precipitation index is investigated to study the rainfall characteristics in the Fuhe basin in China. Our study highlights some important and critical issues by analyzing the daily precipitation data of 17 hydrological stations spanning the years 1970 to 2008. Specifically, we have considered the precipitation distribution, the droughts and floods occurrence frequency, and the spatial distribution characteristics in the Fuhe basin by the year, the month and the day. The study provides a basic rationale in favor of appropriate water management, and the inhabitant's coordination and management, and drought and flood mitigation in the Fuhe basin in Jiangxi, China.

Keywords: Rainfall Analysis, Standard Precipitation Index (SPI), Flood and Drought Analysis, Fuhe basin.

1 Introduction

Extensive body of literature has now appeared that argue that global climate changes have led to great variations in the precipitation and the atmospheric moisture [1],[2]. There seems to be little doubt that the fluctuations of natural atmospheric circulation and greenhouse gases [3] are influencing the complex patterns of natural rainfall both spatially and temporally. These have also caused the precipitation intensity incensement and the frequency reduction of the annual average rainfall leading to droughts and floods that occur in the world more frequently in recent years [1]. This is particularly the case in the subtropical regions of the Fuhe basin in the Jiangxi province, China. In the years 1998 and

---

* Corresponding author.

K. Li et al. (Eds.): ICSEE 2013, CCIS 355, pp. 451–459, 2013.

2010, the floods caused by the continuous torrential rain [4] have nevertheless resulted in multi-defense riverbank breaches in the Fuhe basin. In July 2009, April 2011 and January 2012, the region had to face the extreme drought [5], [6]. Drought and flood have become the primary natural disaster in Jiangxi and have seriously impacted the industrial and agricultural production in this region. Analyzing the precipitation characteristics and its evolution patterns can provide a theoretical basis for drought and flood management and mitigation in the context of climate change in the Fu river basin.

A large number of experiments have demonstrated that the Standard Precipitation Index (SPI) enjoys a strong sensitivity property and is capable of accurately capturing the characteristics of the rainfall distribution with time variations [7]-[9]. The SPI can be employed and is indeed flexible and suitable for precipitation estimation of both dry and wet spells at different hydrological levels and locations [10].

In this paper, a 10-days scale SPI is used to study the rainfall characteristics in the Fuhe basin for the first-time in the literature. The objective of this study is mainly to analyze the precipitation distribution by year, month, day and the changes in the SPI characteristics in the Fuhe basin from the years 1970 to 2008, which can provide the basis for the prediction of droughts and floods in this region.

The remainder of the paper is organized as follows. The study area and the data base are introduced and described in Section 2. The basic SPI metric is reviewed in Section 3. In Section 4, the results and discussion are provided. Finally, concluding remarks and future work are provided in Section 5.

2   Study Area and Its Database

For this study, the historical daily precipitation (mm) time series data sets are collected from the years 1970 to 2008 in 17 hydrological stations which cover a duration of 39 years (468 months and 14244 days). Fig. 1 shows the schematic diagram of the hydrological stations spatial distribution in the Fuhe basin where the 17 hydrological stations are marked. Agricultural production is quite dominant in the Fuhe basin. Large-scale agricultural irrigation areas are concentrated mainly in the S2, S3, S4, S5, and S7 regions around the Fu River in the basin as shown in Fig. 1.

3   Standard Precipitation Index Metric

The Standard Precipitation Index (SPI), which was developed by McKee et al. in 1993, provides a metric for the degrees of wetness and aridity by evaluating the precipitation for any period of interest (e.g., days, weeks, months, seasons, years). In the time scale of interest, a gamma function is used as a probability density function (pdf) to fit the distribution of the observed data, and then the probability is calculated and the probability is transformed to an SPI index value by using an inverse normal (Gaussian) function. For further details on the specifics of the procedure the reader can refer to the reference [11].

Associated with the SPI index, positive values indicate greater than the mean for wet conditions and negative values indicate less than the mean for drought. Normally, part of the SPI range is arbitrarily split into 7 level conditions [11]. The specific partitions are shown in Table 1, where the drought and the flood categories correspond to specific classification ranges of $|SPI| > 1$. Following the SPI calculation the process is smoothened by using a moving average window. In other words, for the SPI-10 days calculation, the 10th day SPI is based on the mean precipitation from the first day to the 10-th day in the same period, the 11th day SPI is based on the mean precipitation from the second day to the 11-th day in the same period, etc. In the present study, the value of the SPI-10 days is calculated as shown in Fig. 2.

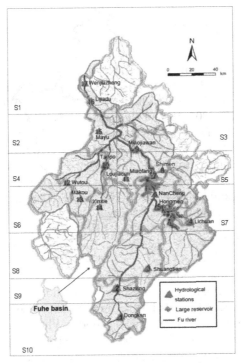

Fig. 1. The schematic diagram of the spatial hydrological stations distribution in the Fuhe basin

4   Results and Discussions

4.1   Years Trends and Analysis

Fig. 3 shows the annual average rainfall distribution from the year 1970 to the year 2008 in 17 hydrometric stations over the Fu river basin. It can be seen that

Table 1. Drought category based on the SPI index

SPI Value	Category
>2.0	Extremely wet
1.5-1.99	Very wet
1.0-1.49	Moderately wet
-0.99-0.99	Near normal
-1.0-1.49	Moderately dry
-1.5-1.99	Very dry

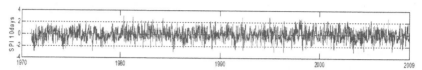

Fig. 2. The standard precipitation indices (SPIs) series based on the average rainfall over the Fu river basin from Jan 1970. to Dec 2008 for a total of 468 months and 14244 days.

the inter-annual variability of the annual rainfall is significant. Specifically, the years 1980, 1997, 1998, 2002 are wet years and have significant rainfalls, and there are local floods or large area floods in these years. Whereas in the years 1971, 1973, 1978, 2003, 2007, the Fu river basin was subject to severe droughts. By using a simple linear regression in time and the water quantity, one can obtain the expression $y = 9.9589x - 18130$ for the precipitation and years, where x denotes the year ranging from 1970 to 2008 and y denotes the annual rainfall as shown by the red line in Fig. 3. Given that the slope is positive this indicates that the rainfall is increasing year by year in the Fuhe basin, and the overall trend of the rainfall is gradually increasing.

According to the classification of the SPI index (Table 1), using the 10-day smoothing moving average, the rainfall frequency within the normal range $(-1 <= SPI <= 1)$, drought $(SPI < -1)$, wet $(SPI > 1)$ from the years 1970 to 2008 are calculated and shown in Fig. 4. It can be seen that the wet label has been the dominant position since 1978, and specifically in 1986-1989, 1990-1992, 1997-2002, 2005 floods have occurred. The years 1998 and 2002 were the extremely wet periods with significant floods. Whereas the years 1970-1979 were the relatively dry concentrated periods, especially in 1971 and 1973 where a sustained and extremely drought period has occurred. Following these years in 1986, 1996, 2003 and 2008, droughts have occurred again.

In addition, a linear regression is also used to analyze the normal year frequency trends. We have obtained the following linear regression equation as $frequency = -0.00046796year + 1.6699$ with a norm of the residuals being 0.33434, where year is to be selected from 1970 to 2008. It can be seen from Fig. 4 that the linear regression normal level curve presents a slight downward trend, which indicates that the normal rainfall turn lower with time and the drought and flood in recent years have became more intense and the fluctuation of the weather have become larger than prior years.

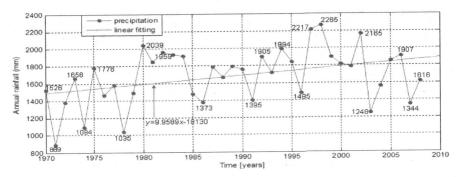

Fig. 3. Average precipitation of 17 hydrometric stations and the associated linear trend in the Fu river basin during the years 1970-2008

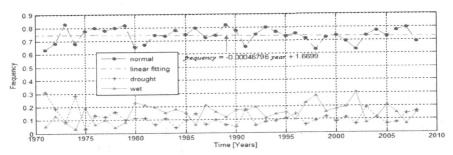

Fig. 4. Frequency of the SPI 10-days of drought and flood in the years 1970-2008 based on the average rainfall over the Fu river basin

To summarize, the overall precipitation amount is increasing gradually every year from 1970 to 2008, and at the same time the precipitation spatial fluctuations are increasing year by year in the Fuhe basin. The frequency of floods is more intensive than the droughts. The basin's average drought time accounts for 11.34% whereas the SPI-10 days is less than -1 and the basin's flood average time accounts for 14.77% during 39 years from 1970 to 2008 whereas SPI-10 days is larger than 1, and the normal state of the basin where SPI-10 days lie in [-1,1],that is not dry and not wet is 73.89 %.

4.2   Months Trends and Analysis

Fig. 5 shows the average monthly rainfall of the Fuhe basin. As can be seen the month of June has the largest rainfall whereas the month of December has the lowest, and the precipitation in December accounts for only 15% precipitation of the month of June. Rainfall is concentrated in the months of April to June, accounting for 46.85% of the annual rainfall. However, in the months of July, August and September, the agricultural water usage peaks, and the precipitation sharply declines, which have accounted for the 20.27% of the annual rainfall.

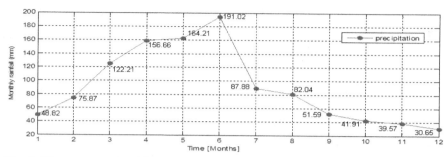

Fig. 5. The average monthly rainfall of the 17 hydrometric stations in the Fuhe basin

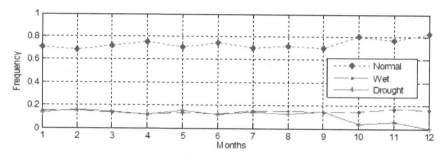

Fig. 6. Drought monthly frequency changes by using the SPI-10 days smoothing sliding analysis for the years 1970-2008 based on the average rainfall in the Fu river basin

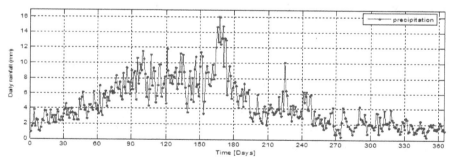

Fig. 7. The average daily rainfall statistics in the 17 hydrometric stations of the Fu basin

In order to capture the rainfall change trends and regularity, the absolute and relative change trends of the rainfall using the SPI-10 days are examined and the details are shown in Fig. 6 where the definitions of normal, drought and flood are the same as that in Fig. 4. One can observe that flood and drought signs have appeared in each month in the Fuhe basin and the natural disasters have never stopped, and there are less abnormal events that occurred in the month of October, and the normal frequency is nearly 0.8. From October to December, there was a significant decrease in the relative frequency of dryness as

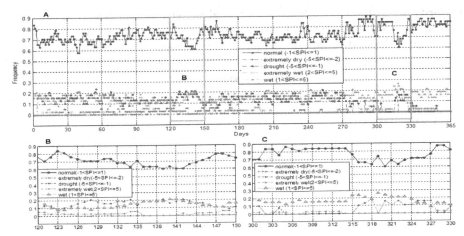

Fig. 8. Daily average drought and flood frequency fluctuations by using the SPI-10 days with a total of 39 years average from Jan. 1, 1970-Dec. 31, 2008 in the Fuhe basin

compared to the average year precipitation and the difference may be caused by the maximum rainfall season in the months of April-June. During these periods, the water leads to a strong water surface and uplifts the Fu river level. However, when the month of July arrived, the river basin water is close to its saturation, although the precipitation has sharply dropped. Moreover, if the rainfall slightly increases, then flood occurs and the phenomenon would lead to a number of floods that is more than that of droughts in the following several months.

4.3 Days Trends and Analysis

The precipitation information of temporal and spatial can further be refined to specific days. In fact, the daily time and spatial rainfall distributions can be regarded as a map of the rain of the Fuhe basin, which can guide other production activities. Fig. 7 shows the average daily rainfall distribution. It can be seen that the overall daily rainfall distribution is quite uneven, and the largest daily rainfall exists in the 167-th day of 16.082 mm, whereas at the beginning and at the end of the year, the precipitation has reached its lowest point close to 0.5mm. On the other hand, the amount of rainfall is not proportional to the precipitation spatial fluctuations.

Fig. 8 depicts the spatial average daily rainfall changes relative to the rainfall mean. The frequency of the abnormal rainfall is fluctuating around 30%. However, the rainfall fluctuations around the 146-th day show a downward slope which reaches the normal frequency at its lowest point as shown in Fig. 8B. Note that the normal precipitation frequency is only about 60% and the flood and the drought appear continuously during the period and the extreme droughts and floods outbreak frequently. These phenomena can be also found during the 300th-330th days, as shown in Fig. 8C. In addition, one can observe that the extremely dry and the wet occur frequently in the 10th-200th days, namely the majority are in the first half of the year and the spatial and temporal distribution of the water resources are extremely uneven. The main reason is that the basin suffers from the affection of the western pacific subtropical high pressure during this period every year, and the rainfall amount and the rain area depend completely on the atmospheric circulations.

5   Conclusion

In this paper, the SPI index is used to study the rainfall characteristics in the Fuhe basin from the years 1970 to 2008. Some of the most important conclusions that are drawn from this study are as follows:

(1) The overall amount of rainfall is increasing gradually from the 1970 to 2008 in the Fuhe basin, and at the same time the spatial fluctuations of the rainfall is increasing year to year. The basin's average drought time accounts for 11.34% and the flooding average time accounts for 14.77% during the past 39 years. The frequency of the floods is becoming more intensive than the droughts.

(2) From the months of April to June, annual rainfall reaches its maximum, but in the months of October-December, it remains at its lowest level. Droughts and floods have never stopped in a year, especially in the first half of the year. The frequency of droughts and floods is high. From October to December, the drought relative frequency shows a significant decrease as compared to the average year precipitation.

(3) Overall the daily rainfall distribution is quite uneven, and the frequency of the abnormal daily rainfall is around 30%, in which the largest rainfall fluctuations occur around the 146-th day and the 300-330th days in the year.

This study also provides an overall and general understanding of the rainfall fluctuations behavior and characteristics in the Fuhe basin as well as some important information and characteristics of droughts and floods in the Fuhe basin. There are several important directions that are worth investigating in future. First, one can analyze the impact area of the droughts and the floods and the affected region breadth. It is necessary to investigate how to migrate the disasters according to the precipitation characteristics. Secondly, the multi-characteristics, multi-linear interactive relationships can be analyzed and their joint distribution characteristics in the droughts and floods investigated. One might also expect that such interactions could result in higher droughts and floods accuracy predictions.

Acknowledgments. This work was supported in part by the Key Project of the National Nature Science Foundation of China (No. 61134009), the National Nature Science Foundation of China (No. 60975059), Specialized Research Fund for the Doctoral Program of Higher Education from Ministry of Education of China (No. 20090075110002), Specialized Research Fund for Shanghai Leading Talents, Project of the Shanghai Committee of Science and Technology (Nos. 11XD1400100, 11JC1400200, 10JC1400200).

References

1. Xu, X., Du, Y.G., Tang, J.P., Wang, Y.: Variations of temperature and precipitation extremes in recent two decades over China. Atmospheric Research 101(1-2), 143–154 (2011)
2. Kundzewicz, Z.W., Mata, L.J., Arnell, N.W., Doll, P., Jimenez, B., Miller, K., Oki, T., Sen, Z., Shiklomanov, I.: The implications of projected climate change for freshwater resources and their management. Hydrological Sciences Journal 53(1), 3–10 (2008)
3. Dore, M.H.I.: Climate change and changes in global precipitation patterns: What do we know? Environment International 31(8), 1167–1181 (2005)
4. Chen, P., Chen, X.: Spatio-temporal variation of flood vulnerability at the Poyang Lake ecological economic zone. Jiangxi Province, China. Water Sci. Technol. 65(7), 1332–1340 (2012)
5. Zhang, X.P., Li, R.F., Lei, S., Fu, Q., Wang, X.X.: The Study of dynamic monitor of rice drought in Jiangxi Province with remote sensing. Procedia Environmental Sciences 10(Part B), 1847–1853 (2011) In: 3rd Annu. Conf. Environmental Science and Information Application Technology, ESIAT (2011)
6. Xie, D.M., Yang, Y., Deng, H.B., Fang, Y., Fan, Z.W.: A study on the hydrological characters in the five river-catchments in Jiangxi Province. Acta Agriculturae Universitatis Jiangxiensis 31(2), 364–369 (2009)
7. Elsa, E.M., Ana, A.P., Luis, S.P., Joao, T.M.: Analysis of SPI drought class transitions using loglinear models. Journal of Hydrology 331(1-2), 349–359 (2006)
8. Duggins, J., Williams, M., Kim, D.Y., Smith, E.: Change point detection in SPI transition probabilities. Journal of Hydrology 388(3-4), 456–463 (2010)
9. Li, W., Kerrylee, R., Joanne, L., Neil, S.: The impacts of river regulation and water diversion on the hydrological drought characteristics in the Lower Murrumbidgee River, Australia. Journal of Hydrology 405(3-4), 382–391 (2011)
10. Khan, S., Gabriel, H.F., Rana, T.: Standard precipitation index to track drought and assess impact of rainfall on watertables in irrigation areas. Irrig. Drainage Syst. 22(2), 159–177 (2008)
11. McKee, T.B., Doesken, N.J., Kleist, J.: The relationship of drought frequency and duration to time steps. In: 8th Conference on Applied Climatology, Anaheim, California, January 17-22, pp. 179–184 (1993)

Material Identification of Particles in Space-Borne Electronic Equipments Based on Principal Component Analysis

Guofu Zhai[1], Jinbao Chen[1], Shujuan Wang[1], Kang Li[2], and Long Zhang[2]

[1] School of Electrical Engineering and Automation, Harbin Institute of Technology,
150001, Harbin, P.R. China
cjb851118@yahoo.com.cn
[2] School of Electronic, Electrical Engineering and Computer Science,
Queen's University Belfast, Belfast, UK
kli_qub@hotmail.com

Abstract. The existence of loose particle left inside the space-borne electronic equipments is one of the main factors affecting the reliability of the whole system. It is important to identify the particle material for analyzing their sources. The conventional material identification algorithms mainly rely on frequency and wavelet domain features. However, these features are usually overlapped and redundant, resulting in unsatisfactory material identification accuracy. The main objective of this paper is to improve the accuracy of material identification. The principal component analysis (PCA) is employed to reselect the nine features extracted from time and frequency domains, leading to six less correlated principal components. The reselected principal components are used for material identification using support vector machines (SVM). The experimental results show that this new method can effectively distinguish the type of materials including wire, aluminum and tin particles.

Keywords: Loose particles, Material identification, Principal component analysis, Support vector machine.

1 Introduction

The space-borne electronic equipments are widely used in communication, remote control and scientific experiments in a satellite system. Their reliability is vital to the success of the mission and safety of personnel and equipments. However, due to their complex structure and production process, loose particles may be left inside, such as the wire pieces, aluminum scraps and tin dregs. The loose particles can be freed and collide randomly, which is caused by vibration and shock. This can lead to component malfunction, system breakdown, or even aerospace catastrophes. Therefore, it is critical to investigate the loose particle detection and identification technologies for the development of aerospace industry [1].

K. Li et al. (Eds.): ICSEE 2013, CCIS 355, pp. 460–468, 2013.

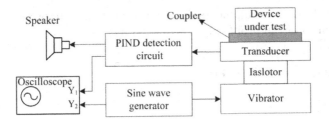

Fig. 1. PIND system of MIL-STD-883E method

The Particle Impact Noise Detection (PIND) test is a traditional screening technique which is used to detect free particles in hermetical components and specified in MIL-STD-883E standard [2]. Fig. 1 represents the structure of PIND system. A vibrator generates a series of shocks and vibrations. These shocks can free the particles and vibrations make them to collide with the component's walls. The collided energy is captured by the acoustic sensors and outputted in the forms of sounds and voltage, which can be used to estimate the existence of loose particles by watching the waveforms on an oscilloscope or listening to the sound from a speaker.

The PIND test is a typical method for detecting the presence of loose particles inside the electromagnetic relay. However, due to the many subjective factors of operators, wrong conclusions are easily drawn, and the accuracy is only about 50% [3]. Therefore, some researchers have proposed several approaches of the automatic detection and material identification of particles. Considering the coefficient of restitution and quality of particles, dynamical mathematical model is established for particles collision within aerospace relays. The best vibration condition is derived, and it indicates that output power of the particle is in proportional to the vibration acceleration and the amplitude of velocity of vibrator [4] [5]. For the relays, material identification has also been studied by analyzing the acoustic signal pulse duration, the spectrum shape factor and the linear prediction coefficients as the features [6]. The disturbance signal energy distribution vectors in vibration acceleration are used to classify tin, glass and rubber particle based on back propagation (BP) neural networks [7]. In addition, a material identification method is presented by employing wavelet and neural networks for aerospace power [8].

However, the features directly obtained from frequency and wavelet domains depend on experience. This can result in overlapping and redundancy. As a result, the actual performance of the particle impact acoustic method based on direct features selection will be seriously affected for large and complex space-borne electronic equipments.

To solve this problem, nine features of impact acoustic signals are selected from time domain and frequency domain. PCA is then used for further feature extraction, dimension reduction and de-noising. The first six principal components are used as inputs to a SVM model. This led to the proposal of a novel particle material identification method based on PCA and multi-SVM for space-borne electronic equipments. Experimental results are presented, to verify the effectiveness of the proposed classification method.

2   Experimental Set-Up

2.1   Test System

The data used in this work were collected from PIND automatic detection system as shown in Fig. 2. With a high-power vibrator, vibrations make the particles to collide with the walls of the space-borne electronic equipment. The collided energy is released in the form of elastic wave propagating along the walls. Four acoustic sensors mounted on the walls convert the collision signal into electronic signal. After amplification and data acquisition, the electronic signals are recorded by a computer. The broad bandwidth acoustic sensors were previously calibrated using an absolute capacitive transducer. The sensor is followed by a 10 kHz high pass filter and a main amplifier. The system bandwidth is from 10 kHz to 200 kHz with the gain of 60 dB. With four channels data acquisition card, acoustic emission (AE) signal is digitized at a sampling rate of 500 kHz before being stored on disk. This system satisfies the measurement of particle collision signals.

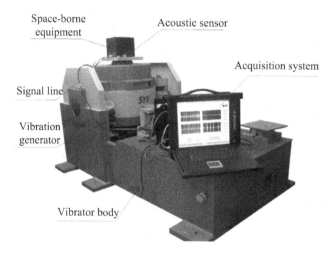

Fig. 2. Photo of the experimental system

2.2   Test Condition

The experimental objective is a rectangular space-borne electronic equipment with a size of $150mm \times 120mm \times 100mm$ in dimensions and the wall thickness is $2mm$. In order to carry out the experiments, three different material particles including wire pieces, aluminum scraps and tin dregs were prepared. The weight of particle ranges from 0.5 mg to 10 mg, and the shape is close to sphere. According to GJB-65B-99 and MIL-STD-883E standards and device-level mechanical environmental routine test conditions, the vibration frequency was set as 40 Hz, and acceleration $5g(1g = 9.8m/s^2)$.

3   Feature Extraction Based on PCA

In traditional methods, the features directly obtained from different material particle collision signals could lead to overlapping patterns. In order to reduce the redundant information, PCA was performed for further feature extraction.

3.1   Basis of PCA

As a widely used statistical technique, PCA has been employed to reduce the dimensionality of problems and to transform interdependent coordinates into significant and independent ones [9]. PCA is to convert a set of correlated variables into a set of values of linearly uncorrelated variables called principal components. This transformation is defined in such a way that the first principal component has the largest possible variance, and each succeeding component in turn has the highest variance possible under the constraint that it be orthogonal to the preceding components.

In this work, the PCA is performed with the following eigenvector algorithm:

(1) normalizing the original data X and calculating the covariance matrix C_x;

(2) by eigenvector decomposition of the C_x, the eigenvectors $U_i(i=1, 2, \cdots, K)$ and corresponding eigenvalues λ_i are sorted in descending order;

(3) obtaining the principle components(PCs) PC_i (i=1, 2, $\cdots$, K) by projecting X onto the resulting eigenvectors U_i;

(4) to ensure the integrity and dimension reduction of the information, selecting the first N principal components according to the eigenvalues λ_i in descending order.

Define R_N as the accumulative contribution rate of the first N principal components with respect to the whole principal components, and according to the threshold of R_N, the first N principal components are chosen as feature vectors.

3.2   Feature Extraction

The identification accuracy depends on the features that are sensitive to the fault. There are overlapping and noise for different features, and some features may even weaken the detection ability. A variety of time domain and frequency domain features have been proposed including acoustic waveform parameters and power spectrum.

In this paper, pulse duration time, energy, zero-crossing rate (ZCR), amplitude divided by duration time and rise time divided duration time are used as the features in time domain. Pulse duration time and energy measurements can be directly linked to the physical parameters of the material. Moreover, ZCR is defined as a measure of the number of times within a given period that the amplitude of the acoustic signal passes through a value of zero [10]. ZCR is an important parameter for acoustic signal classification. It is computed as follows:

$$Z_i = \sum_{m=1}^{n-1} sign[x_i(m-1) * x_i(m)] .$$ (1)

where n is the number of samples in the ith pulse.

The power spectrum of the signal reflects the signal frequency energy distribution. When the frequency components change, the power spectrum of the centroid position will also change. In other words, the power spectrum disperses or concentrates as the frequency components increases or decreases. Hence, the power spectrum of the centroid and the spectral energy distribution can reflect the frequency domain characterics. In this paper, the spectrum centroid (FC), mean square frequency (MSF), spectrum variance (VF) and peak frequency are chosen as the frequency domain features. They are defined as follows:

$$FC = \frac{\sum_{i=N_1}^{N_2} f_i \cdot PSD_i}{\sum_{i=N_1}^{N_2} PSD_i} .$$ (2)

$$MSF = \frac{\sum_{i=N_1}^{N_2} f_i^2 \cdot PSD_i}{\sum_{i=N_1}^{N_2} PSD_i} .$$ (3)

$$VF = \frac{\sum_{i=N_1}^{N_2} (f_i - FC)^2 \cdot PSD_i}{\sum_{i=N_1}^{N_2} PSD_i} = MSF - FC^2 .$$ (4)

f_i is the magnitude of the ith frequency, PSD_i is the power spectral amplitude of the ith frequency.

All the aforementioned features in both time and frequency domain were used as an initial variable pool, and then the PCA was used to further select representative features from this pool. The single and accumulative contribution rate of principal components obtained by PCA is shown in Fig. 3. In this study, the R_N threshold of 0.75 leads to the number of principal components $N = 2$ for the clustering analysis.

By comparison with wire, aluminum (Al) and tin particles as in the Fig. 4, it is feasible to distinguish the different material particles using PCA. For the wire particles, material hardness is lower than that of tin and aluminum. However, it is difficult to classify tin and aluminum due to the overlapping patterns.

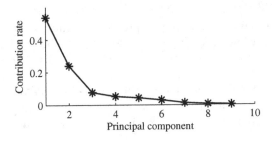

Fig. 3. Contribution rate of principal components

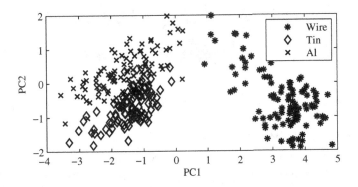

Fig. 4. Clustering results from the PCA for different material particles

4   Classification Using SVM

In order to enhance the accuracy, the data mining method of SVM combined with PCA were used to identify the particle material.

4.1   Basis of SVM

SVM is a machine learning method using small samples. Based on structural risk minimization principle, SVM minimizes the empirical risk and Vapnik-Chervonenkis (VC) dimension simultaneously. SVMs can efficiently perform non-linear classification problems by implicitly mapping their inputs into high dimensional feature spaces using the Kernel tricks [11].

The conventional SVMs are designed for binary classification problems. In order to train the SVM, a serial of training samples including positive and negative samples are needed. SVM aims to find a linear or nonlinear hyper-plane with maximum margin to separate the positive and negative examples from the training samples.

4.2 SVM Design

The conventional classifier for binary problems cannot be used to deal with multi-class classifications directly. In this research, the binary tree (BT) method was adopted to identify the varieties of particle materials. The SVM-BT is able to employ a coarse-to-fine strategy that makes coarse classes easy to differentiate. According to the above analysis, using one SVM (SVM1) is first achieved classification with a high accuracy on two coarse classes, namely the metallic material (tin and aluminum) and nonmetallic material (wire), and then go through the next level of classification on tin and aluminum classes using another SVM (SVM2), as shown in Fig.5.

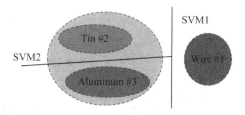

Fig. 5. Multi-class SVM design for particle material identification

4.3 Classification Result

In this study, particle impact signals obtained in the laboratory with the aforementioned three types of materials are used, and the data are split into one training data set and one test data set. The training data contains 250 samples of class 1 (wire), 250 samples of class 2 (tin) and 250 samples of class 3 (aluminum) from 0.5mg to 10mg particle impact signals. These data were all randomly selected to provide enough information for the training model.

In order to obtain the input parameters, especially the first N principal components which has most significant effect on the classification accuracy, SVM2 in Fig. 5 was chosen as the object to be optimized due to the difficulty in accurately distinguishing between tin and aluminum. Thus, by analyzing the classification performance of SVM2 with different number of input vectors after PCA, the number of support vectors (SVs), the different margins, and the error rate are shown in Table 1. It was found that $N = 6$ ($R_N = 0.978$) is the best choice that will produce the SVM model with better generalization performance.

The classification results of the SVM classifiers on the test data are presented in Table 2. With the first six principal components features, the accuracy rate of classifying the wire particle class is 94.8%, 90.8% for the aluminum and 90.0% for the tin. As listed in table 3, it is also found that the accuracy rate with nine features directly is 91.2%, 82.0% and 84.0% for wire, aluminum and tin particles, respectively. This indicates that the method to combine the PCA is more effective than directly using SVMs. Statistically, misjudgements exist between a small mass (0.5-2 mg) wire particle and a large mass (8-10 mg) tin or aluminum particle, and between tin and aluminum in the same mass. This overlapping patterns are caused by different size of the particles.

Table 1. SVM output with different top number of principal components

PCs	SVs	Margin	Error rate
1	496(99.2%)	0.000001	45.4%
2	424(84.8%)	0.000001	38.8%
3	446(35.7%)	0.000067	27.3%
4	52(10.4%)	0.001007	17.6%
5	46(9.2%)	0.008764	13.1%
6	**31(6.2%)**	**0.087357**	**10.2%**
7	37(7.4%)	0.087546	11.3%
8	37(7.4%)	0.087822	13.6%
9	39(7.8%)	0.088081	14.2%

Table 2. SVM output using PCA features

	Classifier output			
	Wire	Al	Tin	Accuracy
Wire	237	5	8	94.8%
Al	5	227	18	90.8%
Tin	4	225	21	90.0%

Table 3. SVM output using features directly

	Classifier output			
	Wire	Al	Tin	Accuracy
Wire	228	13	9	91.2%
Al	13	205	32	82.0%
Tin	10	210	30	84.0%

5 Conclusion

In this paper, a material identification method based on principal component analysis has been investigated for the particle remainders on space-borne electronic equipments. Some valuable conclusions have been drawn as follows:

(1) Nine features have been identified in the time and frequency domains. PCA is then used for further feature extraction.

(2) The first six principal components are used as the inputs to the SVM model and they give a better classification performance than directly using the nine features to build the SVM model. The accuracy of particle material identification is above 90% in the experiments.

(3) In view of the dimensionality reduction and noise reduction capabilities of PCA and generalization ability of SVM, the proposed particle material identification method can be extended to other sealed devices.

References

1. Zhang, H., Wang, S.J., Zhai, G.F.: Research on Particle Impact Noise Detection Standards. IEEE Transactions on Aerospace and Electronic Systems 44(2), 808–814 (2008)
2. MIL-STD-883G, Department of Defense Test Method Standard Microcircuits (2006)
3. Gao, H.L., Zhang, H., Wang, S.J.: Research on Auto-detection for Remainder Particles of Aerospace Relay Based on Wavelet Analysis. Chinese Journal of Aeronautics 20(1), 75–80 (2007)
4. Zhang, H., Wang, S.J., Zhai, G.F.: Research on Vibration Test Conditions in Particle Impact Noise Detection. In: 4th International Symposium on Instrumentation Science and Technology, Part 1-2(48), pp. 46–53 (2006)
5. Zhang, H., Wang, S.J., Zhai, G.F.: Dynamic model of particle impact noise detection. In: 2nd International Conference on Reliability of Electrical Products and Electrical Contacts, pp. 301–304 (2007)
6. Wang, S.J., Gao, H.L., Zhai, G.F.: Research on Feature Extraction of Remnant Particles of Aerospace Relays. Chinese Journal of Aeronautics 20(6), 253–259 (2007)
7. Zhai, G.F., Wang, S.C., Wang, S.J.: Classification of Remainder Material for Aerospace Relay Based on Wavelet Transform. Transactions of China Electrotechnical Society 24(5), 52–59 (2009) (in Chinese)
8. Wang, S.J., Chen, R., Zhang, L.: Detection and Material Identification of Loose Particles Inside the Aerospace Power Supply via Stochastic Resonance and LVQ Network. Transaction of the Institute of Measurement and Control, 1–9 (2011)
9. James, D., Scott, S.M., O'Hare, W.T., Ali, Z., Rowell, F.J.: Classification of Fresh Edible Oils Using a Coated Piezoelectric Sensor Array-based Electronic Nose with Soft Computing Approach for Pattern Recognition. Transaction of the Institute of Measurement and Control 26(1), 3–18 (2011)
10. Downie, J.: The Scientific Evaluation of Music Information Retrieval Systems: Foundations and Future. Computer Music Journal 28(2), 12–33 (2004)
11. Li, X.H., Shen, L.S., Li, H.Q.: Estimation of Crowd Density Based on Wavelet and Support Vector Machine. Transaction of the Institute of Measurement and Control 28(3), 299–308 (2006)

Research and Application of Multiple Model Predictive Control in Ultra-supercritical Boiler-Turbine System

Hengfeng Tian[1,2], Weiwu Yan[1,2,*], Guoliang Wang[1,2], Shihe Chen[3]
Xi Zhang[3], Yong Hu[1,2], and Nan Li[1,2]

[1] Department of Automation, Shanghai Jiao Tong University, Shanghai, 200240
[2] Key Laboratory of System Control and Information Processing,
Ministry of Education of China, Shanghai, 200240
tianhengfeng1987@126.com, yanwwsjtu@sjtu.edu.cn
[3] Guangdong Electric Power Research Institute, Guangzhou 510600

Abstract. Based on analyzing the characteristics of Ultra-supercritical unit, this paper introduced a multiple model MCPC (Multivariable Constrained Predictive Control) structure with three inputs and three outputs for coordination control of Ultra-supercritical unit. In the structure, double-layer structure of optimization was used to obtain good steady and dynamic performance, and piecewise linear models at the different operating points of Ultra-supercritical unit were used to deal with nonlinearity. In the real-time simulation, nonlinear model of 1000MW Ultra-supercritical unit in [1] was considered. Finally, the result of real-time simulation was given in the paper.

Keywords: Ultra-supercritical Unit, Multivariable Constrained Predictive Control, Coordination control, Multiple Model.

1 Introduction

In China, electric power is mainly generated by coal power plant due to the primary energy structure. As the largest user of coal, Electric power industry has to improve coal utilization efficiency. Ultra-supercritical unit technology has good inheritance and is easy to be implemented in large-scale. So among the clean coal power generation technologies, combining Ultra-supercritical power generation technology with efficient flue gas purification technology is the most feasible way to achieve efficiently producing clean power in large-scale. If the Ultra-supercritical unit capacity ratio increases to 20% in the next ten years, Chinese average coal consumption of thermal power generating units for power supply will drop about 20g/kWh [2]. Improving Ultra-supercritical unit proportion and optimization of Ultra-supercritical unit control level is an important direction of promoting the coal-fired power generation technology.

---

* This research was supported by National Natural Science Foundation of China (Grant Number: 60974119).

K. Li et al. (Eds.): ICSEE 2013, CCIS 355, pp. 469–476, 2013.

Coordination control system of Ultra-supercritical unit is the core of thermal power unit control system [3]. However, it needs to solve the following difficulty: (1) Coupling between boiler and turbine. (2) Strong nonlinearity. (3) Large delay. Due to the above control difficulty, the parameters of the coordination control system based on PID controller suffer from a large fluctuation in the process with changing load. For Model predictive control can effectively deal with the coupling and time-delay problems, it has been applied successfully in chemical industry and many other fields. It also began to be applied to the power plant control. J. A. Rovnak and R. Corlis [4] discussed the DMC method in the control of generating units in both theoretical and practical. U. C. Moon and K. Y. Lee [5] presented an adaptive DMC method for traditional unit and used a three order model as a simplified model. For superheater outlet temperature and reheater outlet temperature control problem, U. Moon and W. Kim [6] built a 44 input output model, and presented the simulation results. For the system with strong nonlinearity, Model predictive control based on a single model can't guarantee a good performance. Multiple model method has the characteristics of intelligent control and can combine the classical modeling, control method with advanced control ideas. Goodwin and Narendra developed the stable multiple model adaptive control. M.Y.Fu developed direct multiple model adaptive control method [7]. Z. Binder constructed multiple model controller based on the probability weighted form. Narendra developed adaptive control of discrete-time systems using multiple models, which belonged to indirect multiple model control [8]. However, there aren't many researches about controller based on MCPC for the coordination control of Ultra-supercritical unit.

The multiple model predictive control method is explored to solve the nonlinear control problems for the control of process with changing load in this paper. The paper is organized as following. A kind of Ultra-supercritical unit coordination control system with three inputs and three outputs is analyzed in section 2. In section 3, the multiple models MCPC with double-layer structure is introduced and several groups of step response models at different load point are established. Section 4 presented the simulation results. Conclusions are given in section 5.

2   Ultra-supercritical Unit Coordination Control System Analysis

This paper simplifies the coordination control system to a multivariable control system with three inputs and three outputs. The three input variables were the fuel flow (Kg/s), the governor valve opening (%) and the water flow (Kg/s). And the three output variables were the steam pressure (MPa), load (MW), and steam temperature (°C). Their interaction relations were shown in Figure 1 [9]. Coordinated control system of Ultra-supercritical unit is a typical nonlinear system. Once Load differs by 10%, the model will have a large deviation. Step disturbance can be applied at different load work point to establish several groups of step response models of the system. The input output relationship of the

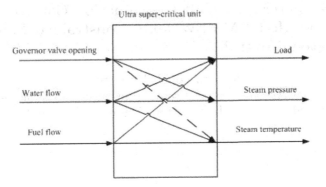

Fig. 1. The relationship between inputs and outputs of Ultra-supercritical unit

system can be expressed as the following form:

$$Y_{k+1|k} = Y_{k+1|k-1} + A\Delta U_k \tag{1}$$

For different step response models, the Dynamic matrix A is different. The predicted output is

$$Y_M(k) = [\tilde{y}_M(k+1|k), \tilde{y}_M(k+2|k), ..., \tilde{y}_M(k+P|k)]^T \in \Re^{mP \times 1}. \tag{2}$$

Zero input response vector is

$$Y_0(k) = [\tilde{y}_0(k+1|k), \tilde{y}_0(k+2|k), ..., \tilde{y}_0(k+P|k)]^T \in \Re^{mP \times 1}. \tag{3}$$

Incremental control vector is

$$\Delta U(k) = [\Delta u(k|k), \Delta u(k+1|k), ..., \Delta u(k+M-1|k)]^T \in \Re^{nM \times 1}. \tag{4}$$

Dynamic matrix A is defined as

$$A = \begin{pmatrix} A^{(1)} & \cdots & & 0 \\ \vdots & \ddots & & \\ A^{(M)} & \cdots & & A^{(1)} \\ \vdots & & & \vdots \\ A^{(P)} & \cdots & & A^{(P-M+1)} \end{pmatrix} \in \Re^{mP \times nM}. \tag{5}$$

The K moment dynamic matrix $^{(k)}$ is defined as

$$A^{(k)} = \begin{pmatrix} a_{1,1}^{(k)} & a_{2,1}^{(k)} & \cdots & a_{n,1}^{(k)} \\ a_{1,2}^{(k)} & a_{2,2}^{(k)} & \cdots & a_{n,2}^{(k)} \\ \vdots & \vdots & \ddots & \vdots \\ a_{1,m}^{(k)} & a_{2,m}^{(k)} & \cdots & a_{n,m}^{(k)} \end{pmatrix} \in \Re^{m \times n}. \tag{6}$$

3   Ultra-supercritical Unit Coordination Control System Multiple Model Multivariable Constrained Model Predictive Control

The multiple models MCPC with double-layer structure is introduced in the section. It is mainly composed of steady optimization, dynamic optimization and multiple models policy. The structure diagram is shown in figure 2.

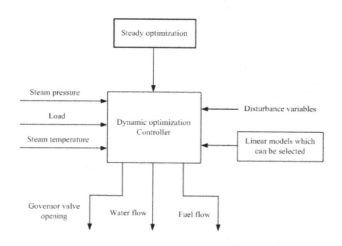

Fig. 2. structure diagram of multiple models MCPC system

Steady optimization was generally used for the purpose of the economic benefits. In order to obtain better economic benefits, steady optimization integrates the Ultra-supercritical unit load and energy consumption into optimization problems. It takes the constraint conditions of the inputs and outputs into consideration and pushes the input and output values of steady state as far as possible to the optimal value of meeting the constraints. Steady optimization can be written as follows

$$
\min_u f(u, y, c)
$$
$$
hi(u, y, \alpha) = 0, i = 1, ..., n_e,
$$
$$
gj(u, y, \beta) \leq 0, j = 1, ..., n_{in}
$$

(7)

In which, α, β are constant, u,y optimization variables, c the constant in target function, n_e the equality constraint number and n_{in} the inequality constraint number. The steady-state model used in steady optimization is the linearized model of the nonlinear system.

The problem of degrees of freedom needs to be considered in steady optimization. When the degrees of freedom (degrees of freedom=effective free MV number-controlled CV number) is insufficient, how to choose to release and relax the soft constraints and hard constraints is an important problem [6].

Two kinds of constraints corresponding to setpoint control and zone control are used in the multiple models MCPC. Setpoint control requires controlled variables or manipulated variables to maintain a given value. Its form is as follows:

$$CV_i = kc_i$$
$$MV_j = km_j. \tag{8}$$

The optimization process emphasizes the concept of degree of freedom. When a variable is constrained by setpoint control, the degree of freedom will be reduced. Therefore, CV zone control is often used to replace setpoint control. Zone control of CV and MV requires that corresponding CV or MV should be maintained within certain limits. Its form is as follows:

$$kc_{\min,i} \le CV_i \le kc_{\max,i}$$
$$km_{\min,j} \le MV_j \le km_{\max,j}. \tag{9}$$

Dynamic optimization is based on steady optimization value and use DMC to accomplish the rolling optimization process with constraints:

$$\min_{\Delta U_k} J_k = \|W_k - Y_k\|_Q^2 + \|\Delta U_k\|_R^2 + \|U_k - U_{IRV,k}\|_V^2 \tag{10}$$
$$s.t. C\Delta U_k \le b,$$

in which W_k is the expected value of CV, Y_k the value of CV, ΔU_k the incremental control vector, U_k the value of MV and $U_{IRV,K}$ the MV steady-state optimal value which constitutes the soft constraint of MV variable. Matrix V is the soft constraint weighted matrix. Matrix C and vector b integrate all the hard constraints.

When the Ultra-supercritical load differs by 10%, models of Ultra-supercritical unit have a large deviation. Multiple models policy was used to deal with the nonlinearity of models. The piecewise step response models of Ultra-supercritical unit at different load point are built at the load work point 600MW, 700MW, 800MW, 900MW, 1000MW respectively. They are shown in figure 3:

In the multiple models policy, only outputs of two piecewise models which are adjacent to the current load were weighted for final output of MCPC controller. The weight is a linear function of the distance from the current load to the selected model's load. Other models which are not near the current load are ignored.

4   Simulation Results

In simulation, the nonlinear model of 1000MW Ultra-supercritical unit in [1] was used. The multiple models MCPC is used to control the model in real-time

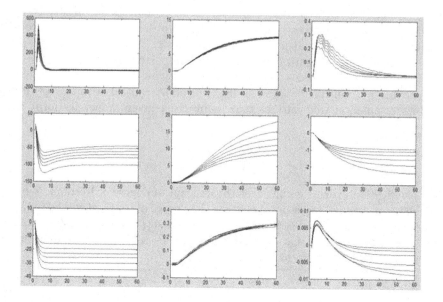

Fig. 3. piecewise step response models at different load point of Ultra-supercritical unit

simulation environment. MV1, MV2, MV3 represents the governor valve opening, the fuel flow and the water flow respectively. CV1, CV2, CV3 represents the load, the steam temperature and the steam pressure respectively. Load is constrained by setpoint control while steam temperature and steam pressure is constrained by zone control. Figure 4 and Figure 5 showed the manipulated variables and controlled variables in simulation when load changed from 1000MW to 900MW. In Figure 4, the red curve, orange curve and blue curve represented the governor valve opening, the water flow, the fuel flow respectively. In Figure 5, the red curve, orange curve, blue curve represented the load, steam pressure and steam temperature respectively. Figure 6 was the figure of change of load. Figure 7 was the figure of change of the steam temperature. Figure 8 and Figure 9 showed manipulated variables and controlled variables respectively when the load changed from 600MW to 700MW.

The results show that the value of CV could rapidly track the expected value of CV in the process of changing load and the setpoint control was accurate in the steady working condition. The load could be changed by the rate of 20-30MW per minute. Steam pressure smoothly rose and dropped following load. Steam temperature fluctuated in the range of 1 degrees around of setpoint 599. The system could guarantee good static and dynamic performance. So the multiple models MCPC with double-layer structure is a promising method for coordination control of Ultra-supercritical unit.

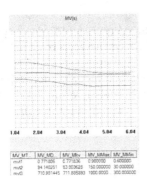 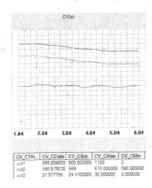

Fig. 4. manipulated variable results when load changed from 1000MW to 900MW

Fig. 5. controlled variable results when load changed from 1000MW to 900MW

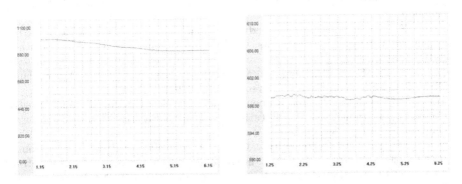

Fig. 6. load results when load changed from 1000MW to 900MW

Fig. 7. steam temperature results when load changed from 1000MW to 900MW

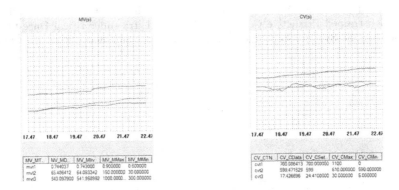

Fig. 8. manipulated variable results when load changed from 600MW to 700MW

Fig. 9. controlled variable results when load changed from 600MW to 700MW

5   Conclusions

This paper introduced a MCPC structure with three inputs and three outputs for coordination control of Ultra-supercritical unit. Piecewise step response models at the different operating points of Ultra-supercritical unit were built in order to deal with nonlinearity. In the real-time simulation, nonlinear model of 1000MW Ultra-supercritical unit in [1] was used. The result of real-time simulation showed the proposed control method could guarantee good steady and dynamic performance.

References

1. Liang, Q., Zeng, D., Liu, J., Yan, S., Xue, Y.: Modeling and Simulation Analysis for 1000MW Ultra-supercritical unit. North China Electric Power University 39(11), 1889–1892 (2011) (in Chinese)
2. Yang, X.: Thermal process automatic control. Tsinghua University Press (2000) (in Chinese)
3. Xiong, X., Liu, J.: The research of advanced control method of coordinated control system for Ultra-supercritical units. Control and Computer Engineering Institute (2011) (in Chinese)
4. Rovnak, J.A., Corlis, R.: Dynamic Matrix based Control of Fossil Power Plant. IEEE Transactions on Energy Conversion 6(2), 320–326 (1991)
5. Moon, U.C., Lee, K.Y.: Step-Response Model Development for Dynamic Matrix Control of a Drum-Type Boiler–Turbine System. IEEE Transactions on Energy Conversion 24(2), 423–430 (2009)
6. Moon, U., Kim, W.: Temperature Control of Ultra-supercritical Once-through Boiler–turbine System Using Multi-input Multi-output Dynamic Matrix Control. Journal of Electrical Engineering & Technology 6(3), 423–430 (2011)
7. Fu, M.Y.: Minimum switching control for adaptive tracking. In: Proceedings of the IEEE Conference on Decision and Control, December 11-13, vol. 4 (1996)
8. Narendra, K.S., Xiang, C.: Adaptive control of discrete-time systems using multiple models. IEEE Transactions on Automatic Control 45(9), 1669–1686 (2000)
9. Wang, G.-L., Yan, W.-W., Chen, S.-H., Zhang, X., Shao, H.-H.: Study on Multivariable Constrained Predictive Control of 1000MW Ultra-supercritical Once-through Boiler-turbine System. In: Chinese Process Control Conference (2012) (in Chinese)

Control Structure Selection
for Vapor Compression Refrigeration Cycle

Xiaohong Yin*, Wenjian Cai, Shaoyuan Li, and Xudong Ding

Department of Automation, Shanghai Jiao Tong University, Shanghai 200240,
P.R. China
School of Electrical and Electronic Engineering, Nanyang Technological University,
Singapore 639798
School of Information and Electrical Engineering, Shandong Jianzhu University, Jinan
250101, China
{yxh1985,syli}@sjtu.edu.cn, {ewjcai,XDDING}@ntu.edu.sg

Abstract. A control structure selection criterion which can be used to
evaluate the control performance of different control structures for the
vapor compression refrigeration cycle is proposed in this paper. The calculation results of the proposed criterion based on the different reduction
models are utilized to determine the optimized control model structure.
The effectiveness of the criterion is verified by the control effects of the
model predictive control (MPC) controllers which are designed based
on different model structures. The response of the different controllers
applied on the actual vapor compression refrigeration system indicate
that the best model structure is in consistent with the one obtained by
the proposed structure selection criterion which is a trade-off between
computation complexity and control performance.

Keywords: Dynamic modeling, Model reduction, Approximation, Multiple time-scale system, Singular perturbation.

1 Introduction

The ever-increasing demand for accurate mathematical modeling for simulation
and controller design leads to models of high complexity, arising from high-order
and/or complex nonlinearities. However, complex, nonlinear and multi-variable
models, for instance in the heating, ventilating, and air-conditioning (HVAC)
system, present big challenges for the controller design. In order to make a trade-
off between the modeling accuracy and the convenience in controller design,
control structure selection plays an increasingly important role over the last
two decades[1]. Vapor compression refrigeration cycle system is an important
and necessary part of HVAC system, which is composed of primary components,
such as compressor, condenser, expansion valve, and evaporator. The system is a
high dimensional system that has obviously nonlinear thermodynamic coupling
characteristics and time-varying dynamics characteristics. As a result, direct

---

* Corresponding author.

K. Li et al. (Eds.): ICSEE 2013, CCIS 355, pp. 477–485, 2013.
© Springer-Verlag Berlin Heidelberg 2013

numerical simulation in such a large scale system becomes an intractable task. Through model reduction to choose appropriate control structure is an approach to overcome this problem[2].

Model reduction methods have become increasingly popular in recent years. There are a number of mathematical formulations and systematic strategies proposed for the model reduction of the HVAC system. One of the most common model reduction schemes is balanced truncation which was first introduced by Mullis and Roberts (1976)[3], and later Moore (1981) applied it in the systems and control literature. Furthermore, this approach was extended by He[4] to the controller design of vapor compression cycle. Unfortunately no validation results were presented to prove the accuracy of the reduced order model. Another popular model reduction method is the balanced residualization introduced by Fernando K and Nicholson H[5]. A 4th-order model for the transcritical vapor compression system was proposed by Rasmussen[6] using residuatization principles and dimensional hankel singular value perturbation method to remove the redundant mass balance state. The model preserved the physical meaning of the dynamic states but still presented some complex challenges for the controller design and implementation. Based on Laguerre polynomials, Wang[7] proposed new methods for model reduction of coupled systems in the time domain. By defining projection matrices according to laguerre coefficients, reduced order coupled systems are generated to match a desired number of these coefficients. The new methods retained the stability of coupled systems, but didn't implement on an actual HVAC system.

In this paper, a structure selection criterion, which can be used to choose the optimal control structure and evaluate the control performance of different control structures, is proposed. The remainder of the paper is organized as follows. Section 2 details the dynamic model of the vapor compression cycle system. Section 3 proposes a structure selection criterion which is used to evaluate the performance of different control structures and choose the optimal simplify model. Section 4 presents the experimental results which justify some of the conclusions discussed in the previous sections. Section 5 summarizes the main conclusions.

2   Dynamic Modeling and Model Linearization

A typical single vapor compression cycle system is shown in Figures 1. Using the lumped-parameter and moving-boundary method, the dynamic model of each component of vapor compression cycle was derived by B.P. Rasmussen and is brief listed below[6].

1) Compressor: The dynamics of the compressor is considered to be much faster than those of heat exchangers, therefore, its mass flow rate can be modeled as a static component

$$\dot{m}_k = \omega_k V_k \rho_k \left(1 + C_k + D_k \left(\frac{P_{ko}}{P_{ki}} \right) \right) \tag{1}$$

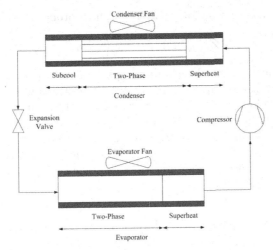

Fig. 1. Vapor compression refrigeration cycle. The system has four components: a compressor, a condenser, an expansion valve, and an evaporator.

where $\dot{m}_k$ is the mass flow rate of the refrigerant through the compressor, ω_k is the motor shaft speed, V_k is the effective displacement volume of the compressor, C_k and D_k are volumetric efficiency coefficients for the compressor, n is the polytropic coefficient, P_{ki} and P_{ko} are the inlet pressure and outlet pressure across the compressor, respectively.

2) Condenser: According to the state of refrigerant, the condenser can be divided into three sections: a subcooled liquid section, a two-phase section and a superheated vapor section. The condenser model has 7 states and 5 inputs. It can be expressed by a non-linear state space form shown below:

$$Z_c(x_c, u_c) \cdot \dot{x}_c = f_c(x_c, u_c) \tag{2}$$

where the state variables are: length of the two condensation regions L_{c1} and L_{c2}; refrigerant pressure P_c; refrigerant outlet enthalpy h_{co}; the wall temperatures in the three regions T_{cw1}, T_{cw2}, and T_{cw3}, respectively, the input variables are mass flow rate of the inlet and outlet, $\dot{m}_{ci}$ and $\dot{m}_{co}$; refrigerant inlet enthalpy h_{ci}; air temperature T_{ca} and air mass flow rate $\dot{m}_{ca}$; respectively.

3) Expansion valve: The expansion valve is also modeled as a static component; its mass flow rate can be calculated from the orifice equation

$$\dot{m}_v = C_v A_v \left[\rho_v(P_{vi} - P_{vo})\right]^n \tag{3}$$

where $\dot{m}_v$ is the mass flow rate of refrigerant through expansion valve, C_v is the orifice coefficient, A_v is the opening area, ρ_v is the refrigerant density. P_{vi} and P_{vo} are the inlet and outlet pressure across the expansion valve, respectively.

4) Evaporator: Similar to the condenser model, the evaporator can be divided into two regions, i.e., a two-phase region with a mean void fraction, and a

superheated region. It can be formulated by a fifth order non-linear model.

$$Z_e(x_e, u_e) \cdot \dot{x}_e = f_e(x_e, u_e) \tag{4}$$

where the state variables are length of two phase flow L_{e1}; refrigerant pressure P_e; refrigerant outlet enthalpy h_{eo}; the wall temperatures in the saturated and the superheated region, T_{ew1} and T_{ew2}, respectively; the input variables are mass flow rate of the inlet and outlet, $\dot{m}_{ei}$ and $\dot{m}_{eo}$, refrigerant inlet enthalpy h_{ei}, air temperature T_{ea} and air mass flow rate $\dot{m}_{ea}$, respectively.

The composite model of whole vapor compression refrigerant cycle can be obtained by appropriately combining the component models according to the relations between the variables.

3   Dynamic Modeling and Model Linearization

For the purposes of high control accuracy and simple calculation, the optimized control structure needs to be chosen for advancing towards further studies. A structure selection criterion, which is used to evaluate the performance of different control structures and choose the optimal simplify model, is proposed in the following section.

Firstly, a twelfth-order linear state-space model obtained from the model linearization considering the steady-state point of operation. After designing a continuous process at steady state for given operating conditions, control structure selection is an important part of process control. A popular control structure selection method is the singular perturbation model reduction method[8] by which the model is decoupled into two parts: the fast one and the slow one. This algorithm can to some extent simplify the model, but it does not show whether it is the optimized control structure for controller design.

In this section the evaporator pressure and the superheat of evaporator are chose in the proposed criterion for controller design of this system. The overall system model is firstly reduced from 12th-order to reduced-order models using singular perturbation method, and then giving random varying inputs to the full-order model and the same varying inputs to the reduced-order models, then the structure selection criterion is defined as follows:

$$J(i) = (\bar{P}_e(i) - P_e(i))^2 + (\bar{T}_{e\_sh}(i) - T_{e\_sh}(i))^2 i = 1, ..., Z \tag{5}$$

where $\bar{P}_e$ and $\bar{T}_{e\_sh}$ are separately the values of the evaporator pressure and the superheat of evaporator computed by the full-order model, P_e and T_{e_sh} are separately the values of the evaporator pressure and the superheat of evaporator computed by the reduced-order model. Z is the whole simulation time. This criterion evaluates the deviations between the full-order model and the reduced-order model in each sample point i.

Denote the cost value J in (5) of the reduced-order model with dimension n as J_n, the value of J_n at time i is obtained from the reduced-order model with dimension n, and denote it as $J_n(i)$. Similar to $J_n(i)$, the value of J_n at time $i+1$

is obtained from the reduced-order model with dimension n, denoted as $J_n(i+1)$. Then we can define difference over the whole working time Z, as follows:

$$\Delta J_n = \sum_{i=1}^{Z-1} \left[(J_n(i+1) - J_n(i))^2 / J_n^2(i) \right] \qquad (6)$$

This criterion is to ensure a smooth command profile of the output variables. The structure selection criterion aims to find the optimal structure which minimizes the sum of quadratic partial variances of the evaporator pressure and the superheat of evaporator. With different controller dimension n, different ΔJ_n can be computed. The n-order control structure with minimum ΔJ_n should be chose as the optimized control structure.

Thus, the structure selection criterion algorithm performs the following steps:

Step 1: The overall system model is reduced first from 12th-order to a particular low order model using singular perturbation method, and then further reduced in descending order from that reduced-order model to 2nd-order model;

Step 2: Giving same random various inputs to the full-order model and the reduced-order models, respectively;

Step 3: Compute the criterion J between the reduced-order models and the full-order model as described in (5) at each sampling time;

Step 4: Calculate the differences ΔJ_n of the reduced-order models expressed as (6) over the whole simulation time;

Step 5: Compare the values of ΔJ_n and choose the model which minimums ΔJ_n as the optimized control structure.

4   Experiment Results

4.1   Simulation Results of the Control Structure Selection

To verify that the control structure selection can choose optimized control structure effectively, a comparison study has been carried out. The model has been firstly linearised around a steady state operating point indicated in Table 1. Then with the singular perturbation method, the system model is reduced to 4th-order model while which is still complicated to control, therefore, according to the step 1 of structure selection criterion, the model is further reduced to 3rd-order and 2nd-order model, and the ΔJ_n among the 4th-order, 3rd-order and 2nd-order control structures are compared.

Through the simulation, the structure selection criterion is calculated shown in Table 2 which indicates that the 3rd-order control structure has the minimum structure selection criterion. According to the structure selection criterion in (5)-(6), the 3rd-order control structure is thus finally chosen for optimal control model structure, convenient to the controller design of the refrigeration cycle system.

Table 1. The steady state operating conditions for the model linearization

State variables	Units	Value
Condensing pressure	kPa	933
Evaporating pressure	kPa	342
Air temperature at the condenser inlet	°C	29
Air temperature at the evaporator inlet	°C	23
Air mass flow rate through the condenser	kg/s	0.45
Air mass flow rate through the evaporator	kg/s	0.19
Compressor speed	rpm	1100
EEV Opening	%	11.8

Table 2. The Structure Selection Criterion

Control structure	2I2O	3I3O	4I4O
Selection	40.3048	14.3525	15.3394

4.2   Experimental Results of the MPC Controllers

In this section, MIMO predictive controllers are designed and the control performances are compared with the conclusion drawn from the optimized control structure chosen in the previous section.

MPC is a control algorithm which computes a sequence of control inputs based on an explicit prediction of outputs within some future horizon. The application of MPC in HVAC systems can be found in the research of Xu[9] and Matthew S. Elliott[10]. In order to define how well the predicted process tracks the set-point, an objective function $J(k)$ for the predictive control needs to be optimized as follows:

$$minJ(k) = J_y(k) + J_{\Delta u}(k) \tag{7}$$

$$J_y(k) = \sum_{i=1}^{P}\sum_{j=1}^{n_y}\{q_j\,[\omega_j\,(k+i|k) - y_j\,(k+i|k)]\}^2 \tag{8}$$

$$J_{\Delta u}(k) = \sum_{i=0}^{M-1}\sum_{j=1}^{n_{mv}}\{\lambda_j\Delta u_j\,(k+i|k)\}^2 \tag{9}$$

(8) computes the weighted sum of squared deviations for the deviation of the outputs from the setpoints. (9) computes the weighted sum of squared deviations for incremental manipulated variables. Where k is the current sampling interval, $k+i$ is the future sampling interval, P is the prediction horizon, n_y is the number of plant outputs, q_j and λ_j are the weight of output and input j, $\omega_j(k+i)$ is the desired output at instant $k+i$, $y_j(k+i)$ is the actual output at instant $k+i$, M is the control horizon, n_{mv} is the number of the inputs.

Three MPC controllers have been designed based on three different control structures and the performances of them have been compared in terms of reference tracking and disturbance rejection. To verify the robustness of controllers, the changing ambient temperature as the disturbance is added as Figure 2.

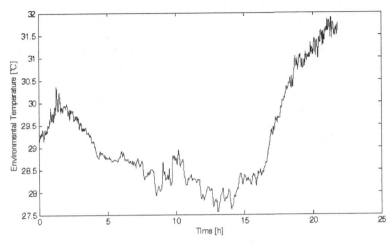

Fig. 2. The disturbance. It shows the changes of actual external environment temperature during 24 hours from 2 pm to 2 pm of the next day in Singapore

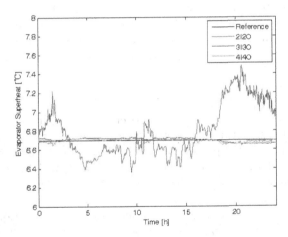

Fig. 3. The superheat of evaporator. It shows the superheat outputs of the three controllers due to the change in the disturbance.

In the three MPC controllers, weights of 1 are placed on the evaporator pressure and superheat, and rate weights of 0.1 and 0.01 are placed on rpm and expansion valve, respectively. Moreover, rate weights of 0.001 are placed on the $\dot{m}_{c-ai}$ in 3I3O MPC controller, and rate weights of 0.01 are placed on the $\dot{m}_{c-ai}$ and $\dot{m}_{e-ai}$ in 4I4O MPC controller. A control interval of 1 seconds are used, with a control horizon of 15 intervals and a prediction horizon of 50 intervals. The results for disturbance rejection for different model control structures are shown in figure 3 and figure 4.

Figure 3 and 4 show the comparison of superheat and evaporator pressure among the three controllers, where the 3I3O controller has better performance in

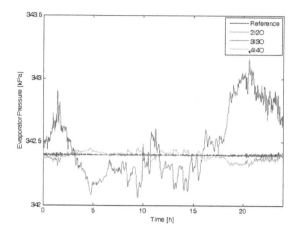

Fig. 4. The pressure of evaporator

resisting disturbance and strong tracking than the 2I2O controller and 4I4O controller, which indicate that MPC controller based on 3I3O control structure not only drives the system towards the references satisfactorily but also rejects the disturbances provoked by the changes on the temperature variable, which is consistent with the conclusion obtained by the proposed structure selection criterion.

5   Conclusion

This paper presents a structure selection criterion to simplify model and select the optimal control structure for advancing towards further studies. To validate its effectiveness, a comparison study has been carried out. According to three low-order models, three controllers based on multivariable model predictive control strategy are designed respectively, and the comparative results of the control performance indicate that the optimal model structure is consistent with the one obtained by the proposed structure selection criterion. Thus, the proposed method of choosing optimal control structure for the refrigeration cycle is a good trade-off which not only decreases the complexity of the computation, but also improves the reliable control performance.

Acknowledgments. The work was funded by National Research Foundation of Singapore: NRF2008EWT-CERP002-010. The other project partners are also acknowledged.

References

1. Antoulas, A.C.: An overview of approximation methods for large-scale dynamical systems. Annual Reviews in Control 29, 181–190 (2005)
2. An, Y., Gu, C.Q.: Model reduction for large-scale dynamical systems via equality constrained least squares. Journal of Computational and Applied Mathematics 234, 2420–2431 (2010)

3. Mullis, C.T., Roberts, R.A.: Synthesis of minimum roundoff noise fixed-point digital-filters. IEEE Transactions on Circuits and Systems 23, 551–562 (1976)
4. He, X.D., Liu, S., Asada, H.H.: Modeling of vapor compression cycles for multivariable feedback control of HVAC systems. Journal of Dynamic Systems Measurement and Control-Transactions of the ASME 119, 183–191 (1997)
5. Fernando, K.V., Nicholson, H.: Singular perturbational model-reduction of balanced systems. IEEE Transactions on Automatic Control 27, 466–468 (1982)
6. Rasmussen, B.P., Alleyne, A.G.: Control-oriented modeling of transcritical vapor compression systems. Journal of Dynamic Systems Measurement and Control-Transactions of the ASME 126, 54–64 (2004)
7. Wang, X.L., Jiang, Y.L.: Model order reduction methods for coupled systems in the time domain using Laguerre polynomials. Computers and Mathematics with Applications 62, 3241–3250 (2011)
8. Ding, X.D., Jia, L., Cai, W.J., Wen, C.Y.: Dynamic modeling and model order reduction for vapor compression cycle. In: 2008 Chinese Control and Decision Conference, China, pp. 1718–1723 (2008)
9. Xu, X.H., Wang, S.W., Huang, G.S.: Robust MPC for temperature control of air-conditioning systems concerning on constraints and multitype uncertainties. Building Services Engineering Research and Technology 31, 39–55 (2010)
10. Elliott, M.S., Rasmussen, B.P.: Model-based predictive control of a multi-evaporator vapor compression cooling cycle. In: 2008 American Control Conference, USA, vols. 1-12, pp. 1463–1468 (2008)

Dynamic Tensegrity Based Cooperative Control of Uninhabited Vehicles

SookYen Lau and Wasif Naeem

School of Electronics, Electrical Engineering and Computer Science
Queen's University Belfast, University Road, Belfast BT7 1NN, UK
{slau02,w.naeem}@qub.ac.uk

Abstract. A new formation control methodology is presented in this paper. The proposed technique is modelled by using the concept of cross-tensegrity structures. The main task is to regulate the desired formation of a group of vehicles and to perform point-to-point manoeuvring in the plane. The position of the controlled vehicles in the formation changes with respect to the admissible tendon forces by varying the lengths of bars in the dynamic tensegrity structure modelling. This change of bars' dimensions for geometric transformation is not possible in the application of tensegrity concept in the physical structural engineering. It has been demonstrated that this control method allows more flexibility over a wide range of different shape switching tasks using the predictable tendon control forces under the limited communication's range. The proposed approach is also scalable to any number of pairs of autonomous vehicles in the formation.

1 Introduction

Multi-vehicle formation control can be characterised by the range and orientation angles between the craft which allows all of them to move effectively as a whole to perform a cooperative task. This coordination in multiple vehicles has become important in both theory and practical applications. Vehicles formation control has been widely performed in land, marine and aerial applications such as transportation, search and rescue, exploration, surveillance and microsatellite clusters [CW05]. The most recognised advantages of this concept include wide sensing area coverage, vast object transportation and energy conservation due to the reduction of friction in each vehicle, to name a few.

The problems in formation control often encompass formation achieving, formation keeping and formation transforming whilst carrying out a task. Many methodologies have been proposed to solve these problems, such as implicit polynomial (IP) [EU10], elliptic Fourier descriptors (EFD) [EU10] and Lyapunov function [LX05] which have been developed to address the shape dynamics of the ensembles of vehicles.

In this paper, the multi-vehicle formation control problem is solved using the concept of tensegrity structures. These structures consist of strings (in tension) and bars (in compression) [SO09], where the strings are attached to the ends of the bars so called as node. This concept has been borrowed from the field of architectural engineering and mechanical structures.

K. Li et al. (Eds.): ICSEE 2013, CCIS 355, pp. 486–495, 2013.
© Springer-Verlag Berlin Heidelberg 2013

In [LN12], a formation control strategy was proposed which was used to drive the dynamic group of vehicles into a specified formation with control forces that are represented by admissible tendon forces in tensegrity structures. The formation controller that was proposed in [LN12] has been improved with an advanced formation changing performance and cascaded with the previous tendon controller that has the function of maintaining the formation's geometry under the disturbance condition. Here, attention is paid to the development of virtual tensegrity-based formation control algorithms for vehicle's motion systems.

A virtual tensegrity structure is used to describe the entire formation as a single rigid/solid tensegrity configuration that is invariant under translation and rotation of the structure. The desired motion is assigned to move the virtual tensegrity structure as a whole in a plane. In this dynamic tensegrity-based formation control, the position of each vehicle in the group can be controlled by varying the length of the virtual bars. Note that this change in bars' length is not possible in the use of tensegrity concept in architecture and mechanical structures. The ratio of bars' lengths is used to control the admissible tendon forces (control forces) that are applied on the vehicles in the formation. The overall formation control system here is formed by three main considerations: vehicles formation geometry that is modelled by a virtual tensegrity configuration, communications topology that is represented by strings and bars of the tensegrity structure, and the interaction control algorithm.

In the remainder of this paper, Section 2 outlines the benefit of tensegrity's properties. Formation's problem formulation is defined in Section 3 whilst formation controller design will be explained in Section 4. Simulation results are shown in Section 5 to demonstrate the formation achieving, formation maintaining and dynamic switching between different shapes formation while manoeuvring on the plane. Concluding remarks are made in Section 6.

2 Tensegrity Structures

Fuller [Ful62] first used the word tensegrity as a contraction of tensional integrity and the first tensegrity structure was built by the artist Kenneth Snelson [Sne65]. In biology, animal skeletal system for smooth locomotion has proved that the tensegrity is a fundamental building architecture of life [EP5]. In architectural engineering, geometric arrangement in these structures can sustain tension and compression, hence make the buildings responsive to natural environmental disturbances such as earthquakes and winds [SO09].

The stability and rigidity of tensegrity structures have been proven by energy function in mathematics [Con82]. This has motivated the development of tensegrity framework in the design and analysis of static and dynamic systems to achieve shape formation control and other engineering functions. The absence of physical connections between rigid members in the structure allows more flexibility over a wide range of different shapes using the predictable tendon force response. This great impact of flexible and deployable control in tensegrity can be used as a solution for the problem of geometry changing in formation control.

The formation controller in this paper is designed to perform formation changing based on the ratio for the length of the virtual bars. The requirement for geometric rotation on the plane can also be achieved by varying the formation's orientation angles. The controller is designed to ensure: (a) the vehicles in formation can respond under the limited communication length, (b) the movement of controlled vehicles are predictable over a wide range of different formations, and (c) the flexible control can be applied to all the vehicles in the formation changing.

3   Problem Formulation and Definition

Coordination architecture and communications topology are key factors in the implementation of formation control. Here, a centralized control architecture is applied, where all the vehicles in the formation are controlled relative to a central virtual leader. The advantage of a virtual leader is that it eliminates the possibility of leader breakdown.

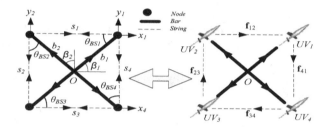

Fig. 1. Vehicles communication topology referred to a cross-tensegrity structure

The topology of the formation is modelled by a virtual cross-tensegrity structure, as shown in Figure 1, where the nodes are defined as vehicles and edges are represented by strings/elastic springs (s) and bars (b). In this figure, the virtual leader is represented by the cross point of the bars, O. The edges correspond to communication topology and the tendon control force between the vehicles. The direction of communication between the vehicles and the tendon forces are represented as uni-directional as shown in Figure 1. $\mathbf{f}_{ij}$ represents the tendon control force that is exerted on i^{th} vehicle according to the jth vehicle. A tendon force controller has been designed in [LN12] by assuming a spring with elastic characteristics which experiences the properties of both spring and string. The same controller will be employed here for controlling and maintaining the shape of a group of vehicles.

In addition to the controller and communication topology, a formation definition is needed to define the location and orientation of the vehicle's formation in the plane. The position vector of ith vehicle is defined as $P_i = [P_{ix}, P_{iy}, P_{iz}]$ where the relative distance between any two vehicles (i^{th} and j^{th}) in a group of N vehicles formation is given as:

$$-r_{ij} = r_{ji} = P_j - P_i = [P_{jx}, P_{jy}, P_{jz}] - [P_{ix}, P_{iy}, P_{iz}] \tag{1}$$

Where $i, j = 1, \ldots, N$, $i \neq j$. Hence, the complete motion definition can be described in a three dimensional Euclidean space that consists of relative distance (r_{ji}) and attitude (ψ_i, θ_i, ϕ_i) of all the vehicles in the formation, where ϕ_i, θ_i, ψ_i, are the roll, pitch and yaw angles respectively.

3.1   Tensegrity Structure Definition

In the tensegrity structure, let n be the number of nodes in a structure given by the three-dimensional vectors, $\mathbf{n}_i$, where $i = 1,...,n$. The position vector of the ith node in the structure is defined as $\mathbf{n}_i = [n_{ix}, n_{iy}, n_{iz}]$. Hence, the configuration of the entire tensegrity structure can be described as the node vector, $\mathbf{n} = [\mathbf{n}_1 \ ... \ \mathbf{n}_n]^T$. Let $\mathbf{m}$ be the vector that describes the member (edge) in the structure that connecting any two nodes. In cross tensegrity structures such as one shown in Figure 1, members are represented by strings/elastic springs (s_1, s_2, s_3, s_4) and bars (b_1, b_2), hence $\mathbf{m}$ can be written as:

$$\mathbf{m} = \begin{bmatrix} \mathbf{m}_s \ \mathbf{m}_b \end{bmatrix}^T = \begin{bmatrix} s_1 \ s_2 \ s_3 \ s_4 \ b_1 \ b_2 \end{bmatrix}^T \in \mathfrak{R}^{3m} \qquad (2)$$

The synchronisation of vehicles' positions in the formation requires the consideration of Equation 3. This kinematic Equation defines the position of four vehicles (represented by nodes, n_1, n_2, n_3, n_4 in the tensegrity structure) according to the virtual leader (represented by the cross point of bars, O in the tensegrity structure) in the plane by referring to the virtual leader's fixed body axes frame (X_B, Y_B, Z_B).

$$\begin{aligned}
P_1 &= \tfrac{l_{b1}}{2}[cos(\beta_1 + \psi_v)cos(\alpha_1), sin(\beta_1 + \psi_v)cos(\alpha_1), sin(\alpha_1)] + P_v \\
P_2 &= \tfrac{l_{b2}}{2}[sin(-\beta_2 + \psi_v)cos(\alpha_2), cos(-\beta_2 + \psi_v)cos(\alpha_2), sin(\alpha_2)] + P_v \\
P_3 &= \tfrac{l_{b1}}{2}[-cos(\beta_1 + \psi_v)cos(\alpha_1), -sin(\beta_1 + \psi_v)cos(\alpha_1), -sin(\alpha_1)] + P_v \\
P_4 &= \tfrac{l_{b2}}{2}[-sin(-\beta_2 + \psi_v)cos(\alpha_2), -cos(-\beta_2 + \psi_v)cos(\alpha_2), -sin(\alpha_2)] + P_v
\end{aligned} \qquad (3)$$

Where ψ_v is the heading angle of the virtual leader while P_v is the position of leader in the plane. β_1 is the angle between the bar, b_1 to the horizontal body axis of the virtual leader (X_B) while β_2 is the angle between the bar, b_2 and the virtual leader's vertical body axis (Y_B). α_1 is the angle between $X_B Y_B$-body axes frame to bar, b_1 while α_2 is the angle between $X_B Y_B$-body axes frame to the bar, b_2. l_b represents the total length of the bar while θ_{BSi} is the bar-string angle between any given bar and the corresponding string (s_i) in Equation 3 and its use in formation control will be elaborated further in Section 4 (A).

4   Formation Control Methodology

In this section, a centralised tensegrity-based formation control system is described which makes use of the tendon forces in formation changing. The use of autopilot in the formation's manoeuvring task is also explained.

4.1   Formation Controller Design

The formation control method should be flexible so that the shape changes can be efficiently carried out to adopt the changes in the unknown environment. Here, the formation control is achieved by maintaining the distances ($l_b/2$) between the nominated pair of vehicle and the virtual leader by using the concept of cross-tensegrity as depicted in Figure 1 and Equation 3.

It is well known that a large system can be easily managed by forming the sub-systems with lower dimensionality within the complex system. Hence, the individual subsystems can be analysed and solved efficiently [YNIL12]. Here, each subsystem is formed by a follower vehicle (represented by the node) and a virtual leader (represented by cross bars' point, O) with an interspacing distance of $l_b/2$ (half the bar's length). Since all the vehicles are referring to the same virtual leader, a centralised formation control method is implemented in the system that consisting of n subsystems, where n is also represent the number of vehicles in the formation. Note that four such subsystems are needed in the setup of the complete formation so that the modelling of cross-tensegrity structure can be applied in the formation control.

For each subsystem in the formation, an individual tendon controller is designed to synchronise a pair of vehicles according to the virtual leader as shown in Figure 2a. A direct communication link is established between the two vehicles whereas the controller output is the tendon force which will be elaborated more in the following section. The signal measured by the controller is the relative distance between the two synchronised vehicles which is denoted by r_{ij}. The position of a vehicle in the formation is then controlled by the applied tendon force with respect to its designated neighbour vehicle by specifying their equilibrium interspacing distance, $l_{tensegrity}$. The position of the virtual leader is used by all the vehicles in the formation to calculate the current interspacing which the autopilot is required to regulate during the manoeuvring task.

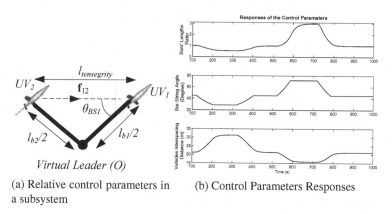

(a) Relative control parameters in a subsystem

(b) Control Parameters Responses

Fig. 2. Different Shape Transformation

The ratio of bars' lengths, l_{b2}/l_{b1} and the bar-string angle, θ_{BS} are the two key control parameters in estimating the vehicles' equilibrium distance ($l_{tensegrity}$). For example, UV_1 makes its decision to move according to the tendon force, $\mathbf{f}_{12}$ that is applied on it with respect to its neighbour, UV_2 in order to maintain their equilibrium interspacing distance, $l_{tensegrity}$. This tendon force is dependent on the relative distance between the nominated pair of vehicles, r_{ij}.

Hence, by varying the ratio of l_{b2}/l_{b1}, the bar-string angle (θ_{BS1}) changes and the new equilibrium interspacing distance ($l_{tensegrity}$) between the controlled vehicles can be obtained as well. Figure 2b shows the response of the interspacing distance between UV_1 and UV_2 changes according to the two key control parameters, l_{b2}/l_{b1} and θ_{BS1}.

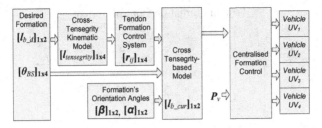

Fig. 3. Block Diagram of the cross-tensegrity based centralised formation control

In the demonstration, both of the bars' length were set at 30m in the beginning. Due to the inherent geometry properties of the cross-tensegrity structure in the formation control, the bar-string angle, θ_{BS1} and the vehicles' interspacing distance, r_{21} were initially at 45 degree and 21.2m, respectively. The l_{b1} was then dilated from 30m to 55m at time 120s for a period of 200s. The reference input (l_{b1}) is taken to be a ramp signal with 0.3 (m/s) slope in order to avoid any sudden changes to the formation. Note that the change of l_{b1} causes the ratio to reduce from 1 to 0.5. The reduction of the ratio also reduces the bar-string angle (45 to 29 degree) but increases the distance between the vehicles from 21.2m to 31.3m.

The l_{b1} was then contracted from 30m to 10m at time 520s for the same period. This time the ratio value increased to 3, which also increased the angle of θ_{BS1} to 71.6 degree but reduced the interspacing of the UV_1 and UV_2 to 15.8m. Hence, a wide variation of formation shape can be converged by regulating the ratio of bars' length, l_{b2}/l_{b1}. Figure 3 depicts the complete setup of the formation control system.

The matrix parameters are defined as $[\mathbf{l}_{b\_d}]_{1\times2} = [l_{b1\_d}, l_{b2\_d}]$; $[\mathbf{l}_{tensegrity}]_{1\times4} = [l_{s1\_d}, l_{s2\_d}, l_{s3\_d}, l_{s4\_d}]$; $[\theta_{BS}]_{1\times4} = [\theta_{BS1}, \theta_{BS2}, \theta_{BS3}, \theta_{BS4}]$; $[\mathbf{r}_{ij}]_{1\times4} = [s_1, s_2, s_3, s_4]$; $[\alpha]_{1\times2} = [\alpha_1, \alpha_2]$; $[\beta]_{1\times2} = [\beta_1, \beta_2]$ and $[\mathbf{P}_v] = [P_{v_x}, P_{v_y}, P_{v_z}]$. Where l_{b1_d} is desired length of the bar, b_1; l_{s1_d} is desired length of the string, s_1 and l_{b_cur} is the current length of the bar during formation changing. Note that the ratio is used to perform the formation's shape changing while formation's rotation can be achieved by regulating the formation's bearings, β and α. The formation rotation task will be demonstrate in the Section V. The extension to obstacle avoidance using the rotation formation technique is possible but is not cover here.

4.2 Tendon Controller

As mentioned earlier, the tendon force, $\mathbf{f}_{ij}$ was applied on vehicle (i^{th}) with respect to its neighbour vehicle (j^{th}). This force was designed to have a much larger elastic limit compare to its proportional limit as defined mathematically in Equation 4 in [LN12]:

$$\mathbf{f}_{ij} = \begin{cases} K \ln \frac{r_{ij}}{l_{tensegrity}} & \text{if } 0 < r_{ij} \leq l_{ultimate} \\ K \exp(-\frac{r_{ij}-l_{tensegrity}}{l_{break}}) & \text{if } l_{ultimate} < r_{ij} \leq l_{break} \\ 0 & \text{if } l_{LF} > l_{break} \end{cases} \tag{4}$$

Where $K = K_1 \alpha_{ij} \omega_{ij}$. And K_1 is a gain parameter that is proportional to the disturbance force, $\mathbf{d}_f$ and adapt the tendon controller to external disturbances, $K_1 \propto \mathbf{d}_f$. ω_{ij}

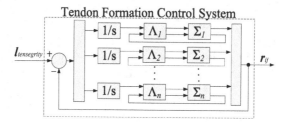

Fig. 4. Block diagram of tensegrity-based formation closed-loop system. Subsystems and tendon controllers are denoted by Σ and Λ, respectively. $l_{tensegrity}$ and r_{ij} are both vector quantities.

is defined as stress [Con82] and α_{ij} is a signed scalar parameter that determines the attracting ($\alpha_{ij} > 0$) or repelling ($\alpha_{ij} < 0$) force that is exerted on the ith vehicle with respect to its neighbour j^{th} vehicle [LN12].

The parameter, $l_{tensegrity}$ is the equilibrium tensegrity length in the structure, which in control terms, it is the desired distance between i^{th} and j^{th} vehicles. r_{ij} is the current relative distance between the two vehicles, $l_{ultimate}$ is the maximum distance between the controlled pair of vehicles, where ultimate tensile strength (attracting or positive force) that is applied to the ith vehicle increases. After this point, this attracting force starts to reduce. This is done in order to reduce the rebound force that will occur on the ith vehicle if the disturbance is suddenly removed.

l_{break} is the maximum length of the string; the string is fractured at this point if the disturbance force continues to be added to the string. In formation control terms, l_{break} is the maximum communication length between the vehicles. The control force, $\mathbf{f}_{ij}$, is equal to zero at this point to give up a straying vehicle rather than trying to apply more force on it to pull it back to the formation. This vehicle might collide with the other vehicles in the formation when the disturbance force is suddenly removed due to the large restoring force.

In the cross tensegrity-based formation control, there will be n applied tendon forces and $n(n-1)/2$ communication links in a formation containing n vehicles shown as block diagram in Figure 4.

4.3   Nomoto Model/Autopilot Design

In this paper, all the vehicles are restricted to two-dimensional motion, hence the parameters of roll and pitch angles (ϕ and θ) can be eliminated. For the simplification in designing the controller, all the vehicles in the formation are assumed to have the same dynamics which can be represented by a linear first order Nomoto model given by Equation 5.

$$T\ddot{\psi} + \dot{\psi} = K_n\delta \tag{5}$$

whose transfer function is:

$$\frac{\psi}{\delta}(s) = \frac{K_n}{s(1+Ts)} \tag{6}$$

Where T is the time constant in the system and K_n is the gain that can be uniquely determined from the input rudder angle (δ) and the output heading angle (ψ). The values of K_n and T are chosen to be 0.049 and 17.78 for simulation purposes [TC99].

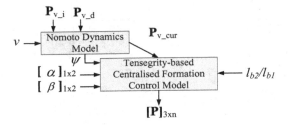

Fig. 5. Block diagram of the formation control model

A suitable PD heading controller was developed which was tuned heuristically. This controller regulates the desired heading ($\psi(t)$) of the vehicles which maintains the trajectory of the vehicle. It must also have the function of performing the change of heading without excessive oscillations and in the minimum possible time.

Figure 5 shows the complete formation control setup, where $\mathbf{P}_{3\times n} = [(P_{1x}, P_{1y}, P_{1z}),$ $(P_{2x}, P_{2y}, P_{2z}), ..., (P_{nx}, P_{ny}, P_{nz})]$ is the position vector of the controlled unmanned vehicles in the formation. The task was to drive the virtual leader with desired heading angle, ψ and velocity, v from one way-point (P_{v-i}) to the next (P_{v-d}). While the controlled vehicles in the formation will synchronise their positions with the virtual leader to perform the formation manoeuvring task on the plane.

5   Implementation and Simulations

The objective in this section is to perform the formation achieving; to regulate the inter-UV spacing between the vehicles within the prescribed communications range; to perform the shape transformation and carrying out the manoeuvring task in a plane. All the parameters in the tendon-driven system [LN12] which is shown in Figure 4 are considered to be unity.

The key control parameters are; the length of bars (l_{b1} and l_{b2}) which are assumed to be of the same length of 30m at equilibrium for the formation; and the formation's orientation angles (β_1 and β_2) are set at 45 degree according to the virtual leader's body X_B- and Y_B- axis, respectively. The length of the bars and the formation's orientation angles were changed individually as well as together for simulation purposes to demonstrate the effectiveness of the proposed strategy.

The length of the bar (b_1), l_{b1} was changed twice at 120s and 520s for a period of 200s. The formation changing performance when the length of bar b, is varied is simulated in Figure 6. From the controller responses shown in Figure 6c, it can be seen that the vehicles (UV_1 & UV_3) experienced negative (repelling) forces at 120s when the bar's length was dilated from 30m to 55m. Note that the controller responded to the bar's variations because of its dependent on particular string's length. The bar's length, l_{b1} was returned to its original length of 30m at 320s with a positive attracting force. The control forces for the subsequent bar's length contraction performance, l_{b1} varied from 30m to 10m at 520s can be expected as shown in Figure 6c.

Figure 7 shows simulation results for different shape transformation tasks. Figure 7a depicted the formation in performing turning manoeuvre while its bar (b_2) was dilated

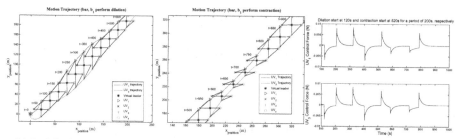

(a) b_1 Dilation Performance   (b) b_1 Contraction Performance(c) Control Forces Performance

Fig. 6. Varying of bar's length, l_{b1} for shape changing performance

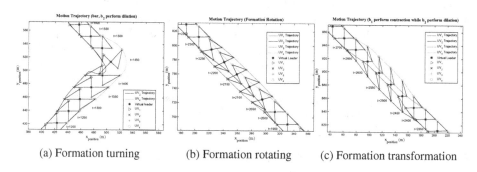

(a) Formation turning   (b) Formation rotating   (c) Formation transformation

Fig. 7. Different Shape Transformation

from 30m to 55m for a period of 200s. The formation rotation is also performed in this work which is shown in Figure 7b. This rotation of the formation is accomplished by increasing the value of β_1 and β_2 angles from 45 to 65 degree respectively at time 2000s for a period of 200s. This may useful for obstacle avoidance purposes. Finally, an overall shape transformation has been performed in Figure 7c by contracting b_1 and dilating the b_2 simultaneously at 2370s. Note that there is no crossover or any collisions between any of the vehicles in the formation.

6   Conclusion

In this paper, a new formation control methodology using the concept of cross-tensegrity structure has been proposed. In this dynamic tensegrity-based formation control, the position of the controlled vehicles in the formation changes with respect to bars' length. This virtual change for shape transformation is not possible in the application of tensegrity concept in architecture and mechanical structures. The presented approach allows more flexibility over a wide range of different shape switching using the predictable tendon response under the prescribed communication's range. The proposed method is also scalable to any number of pair of autonomous vehicles in the formation and is only limited by the communication bandwidth. Extension to 3D motion is possible using this technique but is not covered here due to the lack of space. Shape changing with avoidance manipulability can be improved for obstacles avoidance. The modelling of tensegrity structure under the formation control is required to develop more advanced mathematical machinery that can be used to analyse the stability of the formation configuration.

References

[CW05] Chen, Y.Q., Wang, Z.M.: Formation control: a review and a new consideration. In: Proceedings of IEEE/RSJ International Conference on Intelligent Robots and Systems, Edmonton, Canada, August 2-6, pp. 3181–3186 (2005)

[Con82] Connelly, R.: Rigidity and energy. Inventiones Mathematicae, 11–33 (1982)

[EP5] Ekeberg, O., Pearson, K.: Computer simulation of stepping in the hind legs of the cat: An examination of mechanisms regulating the stance-to-swing transition. Journal of Neurophysiology 94(6), 4256–4268 (2005)

[EU10] Esin, Y.H., Ünel, M.: Formation Control of Nonholonomic Mobile Robots Using Implicit Polynomials and Elliptic Fourier Descriptors. Turkish Journal of Electrical Engineering and Computer Sciences 18(5), 765–780 (2010)

[Ful62] Fuller, R.B.: Tensile-integrity Structures. U.S. Patent 3, 063, 521 (1962)

[LN12] Lau, S.Y., Naeem, W.: Tensegrity-Based Formation Control of Unmanned Vehicles. In: Proceedings of UKACC International Conference on Control, Cardiff, UK, September 1-3, pp. 1–6 (2012)

[LX05] Li, X.H., Xiao, J.Z.: Formation Control in Leader-Follower Motion Using Direct Lyapunov Method. International Journal of Intelligent Control and Systems 10(3), 244–250 (2005)

[SO09] Skelton, R.E., de Oliveira, M.C.: Tensegrity Systems. Springer, New York (2009)

[Sne65] Snelson, K.: Continuous Tension, Discontinuous Compression Structures. U.S. Patent 3, 169, 611 (1965)

[TC99] Tzeng, C.Y., Chen, J.F.: Fundamental properties of linear ship steering dynamic models. Journal of Marine Science and Technology 7(2), 78–88 (1999)

[YNIL12] Yang, A., Naeem, W., Irwin, G.W., Li, K.: Novel decentralised formation control for unmanned vehicles. In: Proceedings of IEEE Intelligent Vehicles Symposium, IV 2012, Madrid, Spain, June 3-7, pp. 13–18 (2012)

Application of Multivariable Model Prediction Control to Ultra-supercritical Unit

Shihe Chen[1], Xi Zhang[1,*], Guoliang Wang[2,3], and Weiwu Yan[2,3]

[1] Guangdong Electric Power Research Institute, Guangzhou 510600
[2] Department of Automation, Shanghai Jiao Tong University, Shanghai, 200240
[3] Key Laboratory of System Control and Information Processing,
Ministry of Education of China, Shanghai, 200240
zhangx.sjtu@gmail.com, yanwwsjtu@sjtu.edu.cn

Abstract. In this paper, development of ultra supercritical unit control is summarized. Based on analyzing the control difficulties and the input-output relationship of Ultra-supercritical Units, a model predictive control scheme for Ultra-supercritical Unit is proposed. The input variables are fuel flow, turbine valve opening and water flow and output variables are load, steam temperatureteam pressure. The algorithm and implementation method are also given in details.

Keywords: Ultra-supercritical units, Multivariable Model Predictive Control.

1 Introduction

Electric power is mainly generated by coal power plant due to the primary energy structure in China. According to the electric power development plan of 2020, Chinese installed generating capacity will increase from the current 0.8 billion kilowatts to 0.9 billion kilowatts in 2020, among which coal power plants take up 75% of the total capacity. As the biggest consuming customer of coal, electric power industry should improve utilization efficiency of coal as well as enhance the production efficiency of coal power plant to adapt to the future requirement of energy saving, emission reduction, low-carbon and sustainable development. The international coal-fired power-generating technology has two developing tendencies. With the mature coal gasification technology in coal chemical industry, the first one uses the integrated gasification combined cycle (IGCC) technology to realize high-efficiency low polluting power generation. The other one improves efficiency through increasing the steam parameter of conventional power-generating sets, which means using the supercritical units and ultra-supercritical units. The higher the steam parameters of ultra-supercritical units are, the higher the thermal efficiency is. The main steam pressure of ultra-supercritical units is 25 31MPa, while the temperature of main steam and reheated steam is 580 610. Not only 2% 3% higher than the thermal efficiency of

* This research was supported by National Natural Science Foundation of China (Grant Number: 60974119).

K. Li et al. (Eds.): ICSEE 2013, CCIS 355, pp. 496–504, 2013.

subcritical units, but also the thermal efficiency of ultra-supercritical is 4% higher than that of the supercritical units, as shown in figure 1 (IEA "Focus on Clean Coal", 2006). IGCC technology has promising prospect to improve efficiency, although the problem of high investment caused by relatively complex system still need to be solved. Ultra-supercritical unit technology has good inheritance and is easy to be implemented in large-scale. Ultra-supercritical units have good performance as well as subcritical units in various of the reliability, availability, heat mobility and unit life. Among all kinds of clean coal power generation technology, the combination of ultra-supercritical power-generating technology and efficient gas purification technology is the most feasible way to realize the large-scale production of high-efficient and clean power-generating in a short period of time. If the proportion of ultra-supercritical units capacity can be raised to 20% in the following ten years, the average coal consumption rate of fossil power plants can be reduced about 20g/kWh. It means about 0.36 billion tons of coal in 10 years, equivalent to 0.2 billion tons of CO2 emission, can be saved [1]. Therefore, increasing optimization and control level of ultra-supercritical fossil power plants are important aspects in boosting the development of fire coal power-generating technology.

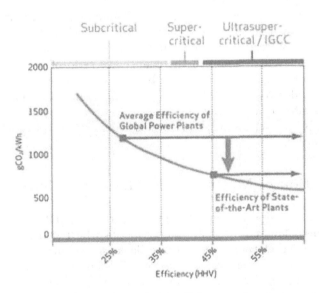

Fig. 1. Comparison of unit efficiency and emissions

2   The Difficulties of Ultra-supercritical Fossil Power Plant Control

The high operating parameters value (pressure and temperature) of ultra-supercritical units and the once-through boiler that needs extensive pitch peaks have raised higher performance demands for the control system design of supercritical units. Due to complex dynamic characteristics, the Ultra-supercritical units have the following control difficulties[2,3,4]:

(1) Coupling between turbine and boiler. Without the buffering of steam drum, dynamic characteristics of the ultra-supercritical boiler are influenced by the terminal resistance. The turbine valve opening on the one hand controlled the turbine power, and on the other hand it influenced directly the characteristics of terminal resistance of the boiler. This is different from the situation of the drum boiler.

(2) Strong non-linearity. Ultra-supercritical units have complex controlled characteristics. As the load changes, the dynamic characteristics parameters of units change greatly. Most ultra-supercritical units are operated in voltage transformation, even sometimes units will be operated under subcritical pressure. With the large difference between the steam properties of supercritical and subcritical and migration of evaporation zone (phase transformation points) under different combustion ratios, strong non-linearity and variable parameter characteristics are presented in ultra-supercritical units, which makes it more difficult to be controlled than normal units.

(3) Delay in dynamic characteristics. One of the important points for ultra-supercritical units control lies in the control of main steam and reheated steam temperature. Steam temperature is generally controlled by the coarse tuning of ratio of fuel to water and the first and second level desuperheating water. Because of its delayed response to the changes of ratio of fuel to water, the outlet temperature cannot be used as feedback quantity of ratio of fuel to water regulation. In order to increase the response speed and precision of ratio of fuel to water, intermediate point temperature or enthalpy is usually used as feedback variable of control loop to realize the regulation and control of main steam temperature.

3   Application of Multivariable Model Predictive Control

The main steam temperature control of ultra-supercritical unit is a typical object with large time delay and dynamic characteristics which change greatly along with change of load. Currently, most controllers used for main steam temperature control of ultra-supercritical unit are still PID controllers or adaptive PID controllers with adaptive measures added on original controllers. The nature of delay of PID regulation decides the existence of various disadvantages in main steam tempearture control system based on PID. However, Model predictive control (MPC) is suitable to solve the time delay and multivariable control problems. MPC was firstly used in the slow process like chemical industry. Richalet proposed MHPC (model heuristic predictive control) or MAC (model algorithm control) in 1976 and 1978 and applied it in process control [5,6,7]. Cutler et al. presented DMC (dynamic matrix control) In 1980 [8] and Garcia et al presented QDMC (quadratic dynamic matrix control)[9] in 1986. Later in 1987, Clarke et al. proposed the GPC (generalized predictive control)[10,11] and Yuan Pu

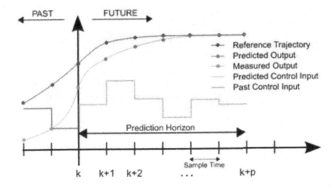

Fig. 2. Predictive Control Schematic

presented the SSPC (state space predictive control) or SFPC (state feedback predictive control) in 1993 [12]. All of these are called model-predictive control algorithm, which is the general name of control algorithm based on models. Generally speaking, predictive control algorithm is constituted by three parts which are predictive model, rolling optimization and feedback correction, as shown in figure 2: Because of its successful applications in many areas, predictive control is started to be applied in fossil power plant. Literature [13] discusses the application of DMC in the control of power-generating units on both theory and practical, in which power, steam pressure and steam temperature are used as manipulated variables. The DMC control method is used in steam temperature in [14]. For the controlled object with single input and single output, the control results are better than traditional PID. In [15], the simulation result of reheated steam temperature controlled by DMC combining with state feedback is given.[16] presented a self-adaptive DMC method of traditional unit and used the third-order model as the simplified model. In [17], for the superheater and reheater temperature control, a 44 input output model was built and the simulation results were presented. In the paper, step response was applied to establish dynamic response matrix, and the manipulated variables were obtained through online optimization.

4   System Analysis of Ultra-supercritical Units Control

The structure model of coordinated control system object of ultra-supercritical units can be simplified into a three-input and three-output system, with the input being the fuel quantity M(%), turbine valve opening T(%) and feed water flow W (%), and the output being pressure before the turbine P(MPa), unit load Ne(MW), outlet steam temperature of separator θ (°C) or enthalpy (KJ/Kg). The relationship of the input variables and output variables is shown in figure 3. The solid line represents strong correlation and the dotted line represent weak correlation[18]. In the range of certain working conditions, the model of the unit

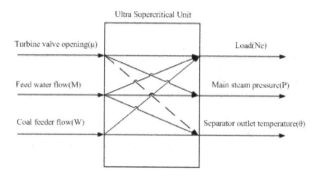

Fig. 3. Input-output relationship of the thermal power units

can be similar to linear model..The input-output step response relationship can be expressed as the following form:

$$Y_{k+1|k} = Y_{k+1|k-1} + A\Delta U_k \tag{1}$$

in which, the model prediction output is:

$$Y_M(k) = [\tilde{y}_M(k + 1|k), \tilde{y}_M(k + 2|k), ..., \tilde{y}_M(k + P|k)]^T \in \Re^{mP \times 1}. \tag{2}$$

The zero input response vector is:

$$Y_0(k) = [\tilde{y}_0(k + 1|k), \tilde{y}_0(k + 2|k), ..., \tilde{y}_0(k + P|k)]^T \in \Re^{mP \times 1}. \tag{3}$$

And the control incremental vector is:

$$\Delta U(k) = [\Delta u(k|k), \Delta u(k + 1|k), ..., \Delta u(k + M - 1|k)]^T \in \Re^{nM \times 1}. \tag{4}$$

The dynamic matrix A can be defined as:

$$A = \begin{pmatrix} A^{(1)} & \cdots & 0 \\ \vdots & \ddots & \\ A^{(M)} & \cdots & A^{(1)} \\ \vdots & & \vdots \\ A^{(P)} & \cdots & A^{(P-M+1)} \end{pmatrix} \in \Re^{mP \times nM}. \tag{5}$$

The process of predictive control accomplishing constrained rolling optimization is

$$\min_{\Delta U_k} J_k = \|W_k - Y_k\|_Q^2 + \|\Delta U_k\|_R^2 + \|U_k - U_{IRV,k}\|_V^2 \tag{6}$$
$$s.t. C\Delta U_k \leq b,$$

After the extreme value of optimized index J is calculated, it can be seen that the control variable at the current k moment is:

$$\Delta U(k) = (A^T Q A + R)^{-1} A^T Q [W(k) - Y_0(k)] \tag{7}$$

in which A is the dynamic matrix of unit predictive model, W the setpoint value vector, Q the deviation weight matrix and R the control weight matrix. Deviation weight matrix Q represents the error control degree in P future time domains and control matrix R characterizes the constraint degree for control increments. In actual, control strategies are usually made from the whole economic indicator of the system (comprehensively considering the output and control energy, the production materials and savings). The combination of steady-state optimization and dynamic optimization of the system is realized in [19,20] by overall linear optimization and rolling optimization conducted by changeable constraint control. Usually on the basis of equation (3), the manipulated variables of k instant exerted by predictive control on the system is the first group manipulated variables of the optimization. In the ultra-supercritical fossil power plant predictive control, the coal feed flow, feed water flow and valve opening bias value calculated through predictive control algorithm are added to the coal master control of the coordinated control loop, and along with the correction of feed water master control and turbine master control on dynamic feedforward loop, the ideal control effects are therefore realized.

5   The Implementation of Model-Prediction Predictive Control Scheme

The operation of the control scheme is divided into two phases which are the simulation phase and the field conduction phase.

a) Test in simulation system: with noise added in the system, the load varied (referring to the track) and set value changed in order to test reliability of the algorithm and the software.

b) Test in field: Comparing simulation, controlled variables are sampled from plant and the algorithm is implemented based on actual set value of the setpoints. The error should be calculated and curve plotted for synthetical assessment. Under this stage, the setting of the operating points does not enter the actual system.

c) Operation of actual system: After finishing the two tests and assessing results strictly, the control system is put into real operation by setting real setpoint and manipulated variable into Real-time environment.

Specific steps:

Step 1: Prophase design of the controller. Before using the predictive controller, designers should be familiar with relevant knowledge of the process and have more discussions with operators and technicians to complete the prophase design of the controller. During the implementation of the program, designers may need to communicate and discuss with technicians to determine the manipulated variables, disturbance variables and controlled variables of the predictive controller.

Step 2: Preliminary checkout of the process data, In the preliminary checkout of the process data, designers need to make sure whether the variable signals of the predictive controller are normal. The breakdown of any transmitters and valve related to variables need to be repaired to ensure the safe and effective operation of the controller.

Step 3: Test of the process data.If variables which are relevant to the controller are normal, further test should be conducted on the controlled object. Test is the process of performing disturbance experiment on every input variable and at the same time recording and collecting data. The test of process object is very crucial. Accurate test will shorten the time of the test run and ensure the long-term and stable operation of the controller.

Step 4: Identification of the process model.After the data test, the data obtained from the object test can be used to obtain the transfer function matrix through system identification.

Step 5: Building controller configuration files and simulation and adjustment for controller off-line.With the identified model, upper and lower limits of the disturbance variables, controlled variables, and manipulated variables are determined using the simulation model in simulation. After evaluation of the controller's performance, parameters can be further adjusted in order to reach expected performance.

Step 6: Online test of controller. The online test of the predictive controller runs firstly in test mode to test whether the software is operating normally as well as examine the accuracy of the model. Under this mode, the controller will accomplish all algorithms whose output however will not be added to the controlled object. All upper and lower limits of control variables will be fixed within the range very close to the current set value. Then predictive controller is started meanwhile the performance of the controller is monitored and adjustments will be made if necessary.

Step 7: Maintenance of the predictive controller.Maintenance is needed to ensure the optimal performance for any kind of predictive controller. An important measure often taken is to detect the upper and lower limits of manipulated and controlled variables. Another important measure is update costs for steady state variable.

6 Conclusions

Based on analyzing the control difficulties of ultra-supercritical unit and the input-output relationships of ultra-supercritical units, this paper introduce a predictive control for ultra-supercritical unit with input variables fuel flow, turbine valve opening and water flow respectively and output variables load and steam temperature, steam pressure respectively. Also, the specific relationship between the scheme and basic control layer DCS and detailed implementation steps are given.

References

1. Yang, X.Y.: Automatic control of thermal process. Tsinghua University Press, Beijing (2000) (in Chinese)
2. Xi, Y.G.: Model Predictive Control. National Defence Industry Press, Beijing (1993) (in Chinese)
3. Wang, D.J.: Optimal control theory of H2 and H_∞. Harbin University of Technology Press, Harbin (2000) (in Chinese)
4. Xiao, B.L.: Development Progress of Automation and Information Technologies for Domestic Power Plants. Journal of Chinese Society of Power Engineering 31(8), 611–618 (2011) (in Chinese)
5. Xi, Y.G.: Model Predictive Control. National Defence Industry Press, Beijing (1993) (in Chinese)
6. Wang, D.J.: Optimal control theory of H2 and H_∞. Harbin University of Technology Press, Harbin (2000) (in Chinese)
7. Richalet, J., Rault, A., Testud, J.L., Papon, J.: Algorithmic control of industrial processes. In: Proceedings of the Fourth IFAC Symposium on Identification and System Parameter Estimation, pp. 1119–1167 (1976)
8. Cutler, C.R., Ramaker, B.L.: Dynamic matrix control, a computer algorithm. In: Proceedings of Automatic Control Conference, San Francisco, CA (1980)
9. Garcia, C.E., Morshedi, A.M.: Quadratic programming solution of dynamic matrix control (QDMC). Chemical Engineering Communications 46, 73–78 (1986)
10. Clarke, D.W., Mohtadi, C., Tuffs, P.S.: Generalized predictive control. Part1: The basic algorithms. Automatica 23, 137–148 (1987)
11. Clarke, D.W., Mohtadi, C., Tuffs, P.S.: Generalized predictive control. Part2: Extensions and interpretations. Automatica 23, 149–160 (1987)
12. Yuan, P., Zuo, X., Zheng, H.T.: State variable feedback predictive control. Automatica Sinica 19, 569–577 (1993)
13. Rovnak, J.A., Corlis, R.: Dynamic Matrix based Control of Fossil Power Plant. IEEE Transactions on Energy Conversion 6(2), 320–326 (1991)
14. Sanchez, L.A., Arroyo, F.G., Villavicencio, R.A.: Dynamic Matrix Control of Steam Temperature in Fossil Power Plant. In: IFAC Control of Power Plants and Power Systems, Cancun, Mexico, pp. 275–280 (1995)
15. Hua, Z., Hua, H., Lu, J., Zhang, T.: Research and application of a new predictive control based on state feedback theory in power plant control system. In: IEEE Congress on Evolutionary Computation, pp. 4378–4385 (2007)
16. Moon, U.C., Lee, K.Y.: Step-Response Model Development for Dynamic Matrix Control of a Drum-Type Boiler–Turbine System. IEEE Transactions on Energy Conversion 24(2), 423–430 (2009)

17. Moon, U., Kim, W.: Temperature Control of Ultrasupercritical Once-through Boiler–turbine System Using Multi-input Multi-output Dynamic Matrix Control. Journal of Electrical Engineering & Technology 6(3), 423–430 (2011)
18. Wang, Z.S., Xia, M., Zhao, S.L., et al.: Coordinated Control Strategy Analysis and Optimization of Beilun 1 000 MW USC Unit. Electric Power Construction 31(1), 87–94 (2010)
19. Xu, C.: Research and Applications of Multi-variable Constrained Model Predictive Control Algorithm and Software. Shanghai Jiaotong University (2002) (in Chinese)
20. Joe Qin, S., Badgwell, T.A.: A survey of industrial model predictive control technology. Control Engineering Practice 11, 733–764 (2003)

Research on Test Conditions of Sealed-Relay Remainders Based on High-Speed Video Technology

Guotao Wang[1], Shujuan Wang[1], Kang Li[2], Pengfei Niu[1], and Shicheng Wang[3]

[1] School of Electrical Engineering and Automation, Harbin Institute of Technology,
Harbin, China
[2] School of Electronics, Electrical Engineering and Computer Science,
Queen's University Belfast, BelfastBT9 5AH, UK
[3] Aviation College, PLA, China
{star5892,wsc_lce}@163.com, kli_qub@hotmail.com

Abstract. Particle Impact Noise Detection (PIND) is an important method for detecting remainders within space relays. During the PIND process of sealed relays, the movement of remainders often cannot be observed directly. A new method to analyze the movement of remainders based on high-speed video technology is therefore proposed in this paper. The PIND process in special transparent cavities is studied, and the parameters to describe the kinetic movements of remainders are calculated based on image processing and data analysis. Then the PIND test conditions are explored with this method. Experimental results show that this method can capture the motion state of remainders in real time and calculate the remainder motion parameters effectively. This method offers an intuitive and reliable means of verifying the PIND test conditions.

Keywords: High-speed video technology, movement analysis method, reminders, Particle Shock Noise Detection.

1 Introduction

The space relay is a kind of sealed relays which are often used for signal transmission, control execution and power distributions in spacesystems. Its reliability is vitalin assuringthe reliability of the whole system[1, 2]. However, remainders, such as solder sinters, metal dusts, colophony and aqua sealsin the relayscan causemalfunctioning, short-circuit or open-circuit, which potentially leads to major space flight accidents[3-5]. The Particle Impact Noise Detection (PIND) is an effective method to detect the existence of remainders in a space relay. Figure 1 shows the diagram of the PIND system. The PIND system generates a series of shocks and vibrations. The shocks and vibrations activate the particles to impact on the relay shell. The energyof the collision is transformed to sound and voltage signals, which areused to estimate the existance of remainders. Figure 2 shows the typical remainder movement in a sealed relay.

The widely employed PIND method nowadays may fail from time to time in industrial applications. Test conditions are claimed to a key factor. Inappropriate test conditions may not be able to free the remainders which lead to a wrong result, or

K. Li et al. (Eds.): ICSEE 2013, CCIS 355, pp. 505–514, 2012.

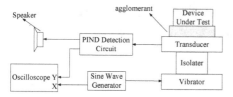

Fig. 1. Schematic diagram of PIND system

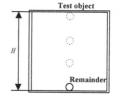

Fig. 2. Remaindermovement in sealed cavity relay

cause too much stress on test objects which leads to potential hazards. Therefore, more effective test conditions of PIND should be explored to improve the detection accuracy.

The existing researches on PIND are mainly focused on the detection systems and algorithms. McCullough discussed the detection effect of different test conditions, but there was no theoretical analysis [6]. Du showed that vibration acceleration and excessive shocks could lead to the fatigue fracture of the internal structure of integrated circuits [7]. Zhang carried out a detailed analysis of the PIND mechanics, and proposed a dynamic model of remainders movement. However the verification was mainly based on simulations, not experiments [8-11]. In summary, little has been done on the movement of remainders and PIND test conditions.

The nature of the research on remainder test conditions is to select appropriate conditions which lead to the strongest collision between remainders and the shell. Obviously, when other conditions are fixed, the faster the initial relative velocity is, the stronger the collision is, and the generated output signal will be stronger accordingly.The initial relative velocity between remainders and test objects can be used as a basis for judging the merits of the test conditions.

The shell of sealed relays is often made of metal alloys, so the remainder movement can not be directly observed.Previous studies of the test conditions were only based on system output signal analysis or modeling derivations, the visual inspection were impossible and the detailed remainder movement could not be studied experimentally due to the restriction of experimental conditions. Further, the remainder motion is a collision process with high frequency, thereforethe motion parameters are vital in test condition researches.. Given that ordinary photographic equipment can not obtainclear images, therefore, a movement-analysis methodof remainders based on high-speed video technology isproposed in this paper.

2   Movement-Analysis Methods of Remainders Based on High-Speed Video Technology

2.1   The Overall Scheme Design

The movement of remainders in relays during the tests is a complex collision and vibration process. The research on the test conditions for relays with complex internal structures is very difficult. The popular method is to consider the relay as a rectangular cavity and the remainders as balls with a collision energy recovery coefficient, and the test conditions are based on the movement of the remainders. Although this method cannot accurately simulate the movement of remainders inside relays, the relationships between test conditions and the output signals can be investigated and used as a basis for improving the PIND test conditions. Therefore, the movement of remainders in transparent rectangular cavities and relays with a transparent shell is researched in this paper. In addition, there exist a wide range of remainders and test objects, therefore, accurate estimation of motion parameters under all different conditions will not feasible, and the research on test conditions should be based on statistics. In summary, the proposed analysis method of remainder movements requires:

a) Direct observation of the motion process of remainders in transparent cavity and relays with a transparent shell, and the remainders are supposed to be larger than 1mm in diameter;

b) Record of sufficient experimental images continuously and automatic preprocessing is necessary;

c)Automatic calculation of the motion parameters of remainders and test objects.

Based on these, a movement-analysis system is developed based on high-speed video technology. The system uses an existing testing platform of remainders and a sealed transparent cavity. The system can pre-process the image of remainders with appropriate image enhancement algorithms, calculate the location of remainders with reference to the inner wall of the cavity; calculate the motion parameters of remainders and test objects based on kinematic principle and plot the trajectory of parameter changes. The overall diagram is shown in Figure 3. Here, the York phantom v7.3high-speed camera with a storage space of 8GB is used in the analysis system.

2.2   Image Processing of Remainder Movements

1)   Pre-processing and Filtering

According to preliminary experiments, the parameters of the camera were set as follows: the resolution was 256×64, the pixel size was 22μm; the frame rate was 100000fps; the time of inter frame and exposure were set to 10μs and 8μs respectively to ensure that the whole process of remainder motion can be captured. The ratio of pixel size ρ is 46pixels/mm according to calibration.

In order to reduce the influence of color difference during image processing and enhance image quality, the Gamma correction and grayscale processing and enhancement are needed [12, 13]. The image pre-processing procedure of the system is shown in Figure 4.

The noise will inevitably affect the image quality during the image acquisition process, and image enhancement techniques are used for denoising. Three factors, including calculation efficiency, filtering effect and edge effects, are important in image enhancement. Therefore, six filtering methods were chosen in the research, including median filter, Sobel operator, Prewitt operator, contrast enhancement filters, adaptive filtering and Laplace Gaussian operator. The original grayscale images were processed with the six methods and the results are compared as in Figure 5.From Figure 5, we can find that the images processed with the median filter have the clearest edge and less noise points. The median filtering algorithm denoises only after isolating the noise point, so the edge information is retained. Blur edges and numerous noise points are the salient features of remainder movement images. Follow-up contrast tests also indicate that the filtering effect is better improved when the comparing area is 2×2 rather than 3×3.

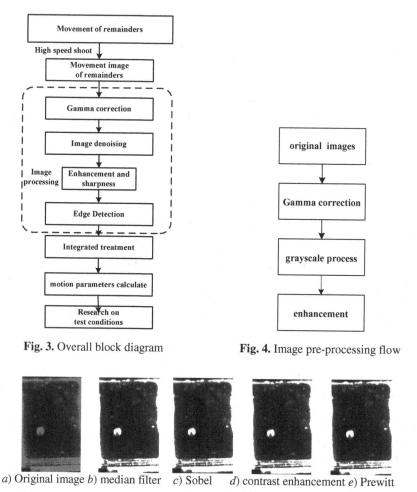

Fig. 3. Overall block diagram   Fig. 4. Image pre-processing flow

a) Original image *b)* median filter   *c)* Sobel   *d)* contrast enhancement *e)* Prewitt

Fig. 5. The original image and filtered images

2) Edge Detection of Remainder Image

The edge detection of remainder images is made on the filtered images in this system. The edge detection algorithm should have the following features:

a) The detection effect on spherical graph is important almost remainders used are spherical.

b) The detection effects on salient points of the graphic edge are important as the edge detection is the basis for the follow-up centroid extraction algorithm and the positions of salient points are very important.

c) The calculation speed must be considered as a large number of data points are used in the edge detection.

In this paper, five edge detection algorithms were chosen, including Roberts operator, Sobel operator, Prewitt operator, Log operator and Canny operator. Filtered images were processed with the above five methods and the results were compared as shown in Figure 6.

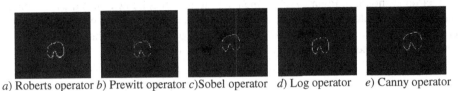

a) Roberts operator *b)* Prewitt operator *c)*Sobel operator   *d)* Log operator   *e)* Canny operator

Fig. 6. The edge detected images

From figure 6, we can find that the image processed with the Roberts operator is better in terms of continuity and has the clearest outline. Roberts operator finds the edge based on local difference method, and its advantages include high positioning accuracy and high calculation speed, and thus the Roberts operator is chosen as the image edge detection algorithm.

3)   The Estimation of Motion Parameters

The location extraction of remainders is to find the location range in a grayscale image. Grayscale digital image can be transformed from RGB three-dimensional space to a two-dimensional gray-scale matrix. The location range can be found in the two-dimensional matrix with a set threshold.

A typical grayscale two-dimensional matrix of remainders image is shown in Table 1.The highlighted numbers in the table produces the contour of the remainder. Generally speaking, grayscale values of the remainder and noise points are larger than that of backgrounds. We have done a number of comparative experiments and found that the remainders regional intensity ranges from 101 to 143, while majority of the other values are less than 100.Therefore, the threshold is set to 100.When the grayscale value is less than 100, the system will set the value to 0; while greater than 100, set to 255.This procedure generates a new grayscale image.

Table 1. Typical remainders grayscale value matrix

23	13	24	34	54	65	23	23
22	2	32	*109*	*106*	67	45	41
12	46	*110*	*120*	*108*	*104*	34	45
66	21	*109*	*105*	*107*	*121*	45	34
16	24	33	*116*	*111*	26	16	111
34	11	34	56	23	67	32	43

After the contour of the remainders is found, the geometric center of remainders can be calculated with extremes algorithm, and then the radius *r* of the remainder and the vertical axis of centroid *y* can be calculated.

The location extraction of the cavity wall is one of the difficulties in the proposed system. The scattering of light during high-speed shooting can cause significant interference; the uneven of the cavity surface can also influence the extraction effect. We have done a number of comparative experiments and choose the Hough transform as the straight-line extraction method. The Hough transform can convert the image from image space to Hough parameter space, and extract straight lines based on the Point-line duality between the two spaces [14-15]. The Typical result of cavity wall extraction is shown in Figure 7.The time interval of shot is known, thus the movement parameters include displacement, velocity and acceleration can be calculated by comparing two adjacent-remainder moving images. The calculating process of remainder movement parameters is shown in Figure 8.

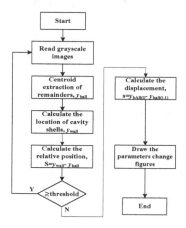

Fig. 7. cavity wall extraction

Fig. 8. Calculating process of remainder movement parameters

2.3   The Performance Validation of the Proposed Method

The main function of the proposed method is to locate the remainders and test objects and calculate the motion parameters. The performance can be verified by comparing the automatic processing results and manual measurements. Assume the diameter of remainder is R and the distance between remainders and sealed cavity is h. We carried out the validation of these two parameters. The results are shown in Table 2 and 3.

Table 2. Performance verification of R (mm)

measured values / actual values	1	2	3	4	5	6	7	8	9	10
1.0 (mm)	1.02	0.96	1.03	1.02	0.99	0.98	1.02	1.00	1.00	1.03
1.5 (mm)	1.50	1.53	1.52	1.48	1.50	1.49	1.52	1.51	1.50	1.46
2.5 (mm)	2.55	2.56	2.48	2.50	2.43	2.55	2.46	2.53	2.44	2.53

Table 3. Performance verification of h (mm)

measured values / actual values	1	2	3	4	5	6	7	8	9	10
9.7 (mm)	9.72	9.75	9.66	9.67	9.69	9.69	9.73	9.70	9.68	9.71
4.3 (mm)	4.35	4.32	4.33	4.27	4.31	4.30	4.30	4.33	4.28	4.27
1.5 (mm)	1.66	1.56	1.62	1.42	1.58	1.46	1.70	1.52	1.41	1.63

From the experimental results, the calculation of R is almost accurate; the error is less than 0.06mm. The calculation of h is comparatively accurate when $h>R$, the error is less than 0.05mm, but when $h\leq R$, the calculation error may more than 0.2mm as the remainder will interfere the location extraction of the cavity in this case.

3   Study of Remainder Test Conditions

The movement of remainders in a sealed cavity is affected by the height of the cavity, vibration acceleration, vibration frequency, the features of the remainders and energy coefficient of restitution between remainders and the cavity, etc. Due to the gravity, collision between remainders and the bottom wall is always stronger than that of the upper wall, so collision between remainders and the bottom wall is mainly concerned in experimental studies. In the following, the average relative speed before collision between remainders and the bottom wall is denoted as $y_{average}$; the height of the cavity as H, the diameter of remainder as R, the test vibration frequency as f, the test acceleration peak value of vibration as a_m, and further we assume the upward direction is the negative orientation.

3.1   Study of Relationships between Test Conditions and Output Signals Based on Transparent Cavities

1) Experiments with Varying Frequency

Experiments with varying frequency were carried out. Bearing Steel balls were selected to be remainders, for which R is 1.5mm and 1mm respectively; the dimension of transparent cavities are 30mm × 30mm × 40mm and 20mm × 20mm × 20mm respectively. Remainders are put into transparent cavities and tested with varying frequency. The bandwidth is between 10Hz and 50Hz; f was step changed by 10Hz each time; and a_m is 5g; We have captured 200 collision images for each combination of test conditions, remove images with $h\leq R$, and then select 50 images with a collision between remainders and the bottom wall for movement analysis.

It was found that when f is set below 20Hz, the vibration generator may become unstable; if above 110Hz, the bounce height of remainders is less than its own diameter due to small amplitude of the vibration table, which is disadvantageous for detection. Therefore, the actual range of f is chosen between 20Hz and 110Hz. The $y_{average}$ results of experiments with varying frequency are shown in Table 4, where the experimental results of approximate periodic collision vibration are recorded in italic.

Table 4. $y_{average}$ results of experiments with varying frequency (m / s)

f(Hz) H,R(mm)	20	30	40	50	60	70	80	90	100	110
40,1.5	0.929	0.772	*1.227*	0.630	0.465	0.401	0.727	0.279	0.227	0.128
40,1	0.897	*2.925*	*1.173*	0.595	*0.681*	0.398	0.353	0.430	0.332	0.105
20, 1.5	1.078	*1.326*	0.521	0.687	*0.998*	0.485	0.399	0.336	0.315	0.147
20, 1	1.002	*1.357*	*0.496*	0.607	0.603	0.456	0.368	0.316	0.296	0.129

The experimental results show that:

a) Frequency of the test conditions has a great influence on the test results. If other conditions are fixed, the $y_{average}$ reduces obviously while the f increases, which verified the theory in [9] and [10]. The reason is the acceleration time of remainders and test objects gradually decreases with the increase of vibration frequency.

b) The remainder sometimes performs approximate periodic collision vibration in the sealed cavity. The phenomenon of approximate periodic collision vibration has a great influence on the test results, which may double the $y_{average}$ and improve the success rate of detection effectively.

c) The phenomenon of approximate periodic collision vibration does not always improve the $y_{average}$; it may make the $y_{average}$ less than the value of adjacent high-frequency conditions with small probability.

2) Experiments of Varying Acceleration

Varying acceleration experiments were also performed. A 1.5mm diameter bearing steel ball was selected to be a remainder; the size of transparent cavities are 40mm×40mm×70mm and 20mm × 20mm × 20mm respectively; the range of a_m was set between 2g and 10g and a was step changed by 1g each time. In order to minimize the interference of approximate periodic collision vibration, f is fixed at 50Hz; and likewise, 50 images are selected. The $y_{average}$ of the experiments with varying acceleration are shown in Table 5.

Table 5. $y_{average}$ results of experiments with varying acceleration (m / s)

a(g) H(mm)	2	3	4	5	6	7	8	9	10
70	0.033	0.041	0.083	0.119	0.187	0.227	*0.289*	0.283	0.290
20	0.223	0.483	0.629	*0.687*	0.691	0.717	0.763	0.706	0.720

The above three experiments indicate that:

a) Acceleration peak value of the test conditions has a great influence on the test results. If other conditions are fixed, the $y_{average}$ increases while a_m increases. The reason is that the acceleration time of remainders and test objects can obtain a higher speed during the same accelerating time.

b) The speed of remainders and test objects is very low while a_m is low, which is disadvantageous for detection.

c) If other conditions are fixed, there is a threshold in the acceleration peak value, *if a_m is* lower than the threshold, increasing a_m can improve $y_{average}$ significantly; *if a_m is* higher than the threshold, the effect of increasing a_m is significantly reduced. In the table, the thresholds and corresponding $y_{average}$ are recorded in italic.

It should be noted that remainders may be confined inside the relay for a number of reasons. If a_m is very low, it may not activate remainders [16].

3.2   The Improved Test Conditions for Relays with a Transparent Shell

Test conditions were also researched for relays with a transparent shell. Bearing Steel balls were selected as remainders, r are 1.5mm and 1mm respectively and the sizes of two relays are 30mm × 30mm × 40mm and 20mm × 20mm × 20mm. Remainders were put in relays and tested with varying frequency and acceleration conditions; and likewise, 50 images were selected.

According to the experimental results and analysis, it is recommended that the test conditions of PIND for sealed relay could be improved if the following principles are adopted:

a) The test frequency range should be between 20Hz and 90Hz.

b) The step value should be reduced in the low frequency range (20Hz-40Hz), and more tests within this frequency range should be attempted.

c) In order to overcome the binding, a_m higher than the standard ones should be used if the test objects are not damaged.

d) For the test objects with large height dimension, a_m higher than the standard ones or *f* lower than standard ones should be used to improve the success rate of detection.

4   Conclusions

Test conditions of sealed-relay remainders have been studied based on high-speed video technology. The main conclusions are as follows:

a) A movement-analysis method of remainders based on high-speed video technology have been developed; corresponding image processing algorithms and motion parameters extraction algorithms have been designed. The motion parameters of remainders and test objects can be extracted automatically using this method.

b) Based on the method, it is found that the detection effectiveness is inversely proportional to the test frequency, and proportional to the acceleration peak value; the phenomenon of approximate periodic collision vibration of remainders and thresholds in the test acceleration peak value are found.

c) According to the experimental results, suggestions for improvement of PIND test conditions have been proposed, which can improve the success rate of detection of remainders in sealed relays.

References

[1] Ding, M.S.: Reliability analysis of digital relay. In: 8th IEEE International Conference on Developments in Power System Protection, The Netherlands, vol. 1, pp. 268–271 (2004)

[2] Phil, R.: Testing techniques to improve relay reliability. EE: Evaluation Engineering 44(4), 44–48 (2005)

[3] Zheng, N.C.: Reliability analysis and countermeasures of relays in the control system. Journal of Rocket Propulsion 31(4), 58–62 (2005)

[4] Ingmar, H., Marc, K., Armin, W.: Channel adaptive scheduling for cooperative relay networks. In: 2004 IEEE 60th Vehicular Technology Conference, vol. 60(4), pp. 2784–2788 (2004)

[5] Li, D.: The implementing inspection for space product superabundance and its prevent and control standards. Aerospace Standardization (1), 17–20 (2006)

[6] McCullough, R.E.: Hermeticity and Particle Impact Noise Test Techniques. In: 14th Annual Proceedings of Reliability Physics, vol. (8), pp. 256–262 (1996)

[7] Du, Y., Lv, D., Pan, W.L.: Study of particle impact noise detection ⟍PIND⟋. Electronic Product Reliability and Environmental Testing~(1), 34--39 (2005)

[8] Zhang, H., Wang, S.J.: Research on Particle Impact Noise Detection Standards. IEEE Transactions on Aerospace and Electronic Systems 44(2), 808–814 (2008)

[9] Zhang, H., Wang, S.J., Zhai, G.F.: Dynamic model of particle impact noise detection. In: IEEE IECON, pp. 2577–2581 (2004)

[10] Zhang, H., Wang, S.J., Zhai, G.F.: Test conditions discussion of particle impact noise detection for space relay. In: IEEE IECON, pp. 2566–2572 (2004)

[11] Zhang, H., Wang, S.J., Zhai, G.F.: One-period stability analysis of particle impact noise detection for space relay reminders. Acta Aeronautica ET Astronautica Sinica 26(3), 362–366 (2005)

[12] Arnold-Bos, A., Malkasse, J.P., Kervern, G.: A preprocessing framework for Automatic underwater images denoising. In: European Conference on Propagation and Systems (March 2005)

[13] Sendur, L., Seselsnick, I.W.: Bivariate shrinkage functions for wavelet-based denoising exploiting interscale dependency. IEEE Trans. on Signal Processing (November 2002)

[14] Duda, R.O., Hart, P.E.: Use the Hough transform to detect lines and curves in Pictures. Comm., Assoc. Comput. 15, 11–15 (1972)

[15] Rong, C.L., Tsai: Gray-Scale Hough Transformfor Thick line Detection in Gray-Scale Images. Pattern Recognition 28(5), 647–661 (1995)

[16] Wang, G., Wang, S., Li, K.: A Study on the Activation Condition of Vibration Tests for Particle Impact Noise Detection of Space Relay Remainders. In: ICIEA 2011, pp. 2610–2614 (2011)

Maximum Power Point Tracking
for Photovoltaic System
Using Model Predictive Control

Chao Ma, Ning Li, and Shaoyuan Li*

Department of Automation, Shanghai Jiao Tong University
Key Laboratory of System Control and Information Processing
Ministry of Education of China, Shanghai 200240
ning_li@sjtu.edu.cn

Abstract. In this paper, T-G-P model is built to find maximum power point according to light intensity and temperature, making it easier and more clearly for photovoltaic system to track the MPP. A predictive controller considering constraints for safe operation is designed. The simulation results show that the system can track MPP quickly, accurately and effectively.

Keywords: PV system, MPPT, T-G-P model, MPC.

1 Introduction

In the recent decades, Photovoltaic (PV) power generation system has been paid more and more attention. To make output power of photovoltaic system most, energy converter should track Maximum Power Point (MPP) which is very important in order to ensure the efficient operation of the solar panel. Because of solar panels' non-linear current-voltage characteristics, only one particular operating point is the MPP (Fig.1). In addition, the MPP changes with temperature and light intensity, making the tracking control of the MPP a complicated problem.

To overcome these problems, many tracking control strategies have been proposed. One of the most widely used methods is perturbation and observation [1]. This method uses an algorithm known as "hill climbing" and can be implemented by applying a disturbance to the reference voltage signal of the solar panel. P&O method is simple and suitable for changeable conditions, nevertheless a larger disturbance will lead to a higher value of oscillation amplitude [2]. Another conventional approach, PI controller, requires sufficient time for the system to reach steady state operation increasing the time interval between two

* This work was supported by the National Science Foundation of China (60825302, 60934007,61074061), the High Technology Research and Development Program of China (2011AA040901), the Key Project of Shanghai Science and Technology Commission (10JC1403400), and Key Project of China's Ministry of Railways (J2011J004).

K. Li et al. (Eds.): ICSEE 2013, CCIS 355, pp. 515–523, 2013.

successive reference outputs form the MPPT and hence deteriorating dynamic performance [3,4]. Fuzzy Logic Control (FLC) has been used in MPPT, for this method is appropriate for non-linear control. However, FLC with fixed parameters is inadequate in application where the operating conditions change in a wide range and the available expert knowledge is not reliable. Adaptive Fuzzy Logic Control can solve this problem because it can re-adjust the fuzzy parameters to obtain optimum performance, but the computation cost is much higher than conventional FLC [5].

In this paper, according to the uniqueness of MPP in specific conditions and the PV system's non-linear characteristics, the T-G-P model is discussed. The predictive controller is designed based on the PV energy converter mixed logic dynamic model using hybrid system theory. At last, the simulation results verify the validity of the model and the effectiveness of the proposed controller.

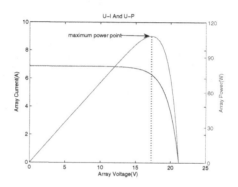

Fig. 1. U-I and U-P characteristics of photovoltaic cell

2   Photovoltaic Cell and T-G-P Model

The physical structure of a photovoltaic cell is similar to that of a diode in which the p-n junction is subjected to sun exposure. When illumination and environment keep invariable, photovoltaic cell is a non-linear DC power. Photovoltaic cell's mathematic model could be given as follows [6]:

$$I = I_{PV} - I_O[\exp(\frac{V + IR}{aV_T}) - 1] - (\frac{V + IR_s}{R_p}) \tag{1}$$

where I_{pv} is the current generated by the incidence of light; I_o is the reverse saturation current; a is the impact factor of the diode(usually 3-5); R_s is the series resistance; R_p is the parallel resistance; q is the electron charge, which has constant value of $1.6 \times 10^{-19}C$; T is the temperature of the p-n junction in q; k is the Bolzmann constant ($1.38 \times 10^{-23}J/K$).

I_{pv} exhibits a linear relationship with light intensity and also varies with temperature. I_o is also temperature-dependent [7]. Fig.1 shows that the MPP is unique in specific conditions. That means different light intensity and temperature will lead to different MPP.

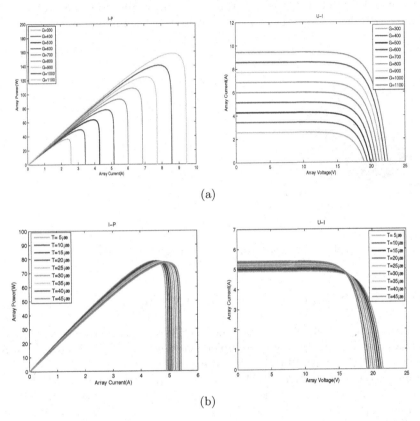

(a)

(b)

Fig. 2. I-P and V-I curves under different light intensity(a)and temperature(b)

Fig.2 shows the I-P and V-I characteristics for different values of light intensity and temperature. As shown in Fig.2 (a) the maximum power increases with increasing light intensity. On the other hand, Fig.2 (b) shows that the maximum power is little influenced by the sharp increase of temperature in natural environment, and also indicates that the output voltage exhibits an inverse relation with temperature.

Consider the uniqueness of the MPP in different atmospheric conditions, the T-G-P model which presents the effects of temperature and light intensity on MPP is discussed. Fig.3 reveals that the change of light intensity exerts a greater influence on the MPP compared with that of temperature. By using the tool named 1stOpt15PRO, a T-G-P model could be expressed by a function: $P = f(T, G)$.

In this work, the fitting function is:$P = (c1 + c2 * G + c3 * G^{(}2) + c4 * T)/(1 + c5 * T)$. In which c1=-0.1754, c2=0.1132, c3=4.0779E-5, c4=0.0032, c5=0.000235. After measuring the value of temperature and light intensity, the MPP in that condition could be calculate through this function. In addition, the PV output voltage at MPP can be regarded as constant in every sampling time.

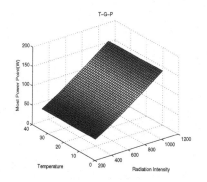

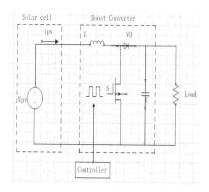

Fig. 3. The effects of T and G on MPP   **Fig. 4.** Photovoltaic system

3   Modeling of Photovoltaic Energy Converter

Photovoltaic system consists of solar power, DC-DC converter, and controller. Fig.4 shows the photovoltaic system implemented in this work. Power switch S could realize MPPT control by changing duty cycle.

The main control objective in this paper is to drive the switch with a duty cycle to make the PV output current equals to its reference. Two equations of are built when the switch is on and off. For output current and converter output voltage should be controlled, i and v_o are selected as state variables. The sampling period is set the same with PWM period T_s, the system switches between the following two subsystems:

When S is on:

$$\begin{cases} \dot{x} = A_1 x + B_1 u \\ y = Cx \end{cases} \quad kT_s \leq T < kT_s + t_{on} \tag{2}$$

When S is off:

$$\begin{cases} \dot{x} = A_2 x + B_2 u \\ y = Cx \end{cases} \quad kT_s + t_{on} \leq T < (k+1)T_s \tag{3}$$

In (2) and (3): $A_1 = \begin{pmatrix} 0 & 0 \\ 0 & -\frac{1}{RC} \end{pmatrix}, A_2 = \begin{pmatrix} 0 & -\frac{1}{L} \\ \frac{1}{C} & -\frac{1}{RC} \end{pmatrix}, B_1 = B_2 = \begin{pmatrix} \frac{1}{L} \\ 0 \end{pmatrix}, C = (1\ 0), x = (i\ v_0), t_{on}$ means the time when the switch is on in one PWM period. In one PWM period T_s, output current is controlled by changing duty cycle, $0 \le d(k) = t_{on}/T_s \le 1$.

In order to formulate an adequate boost converter model, which is of fundamental importance for the subsequent derivation and implementation of the optimal control problem, it is necessary to construct the discrete model. To describe interaction between the switching dynamics and continuous dynamics, the period of length T_s is divided in sub-periods of duration $\tau_s = T_s/N$ with $N \ge 2$. N subinterval could be divided into three categories of state: (i) the system in state on; (ii) the system in state off; (iii) the system is in transition state. $\xi(n)$ is state of every sub-interval τ_s, $n \in (0, 1, \cdots, N-1)$, and $\xi(0) = x(k), \xi(N) = x(k+1)$, define binary variables δ_n as $\delta_n = 1 \leftrightarrow d(k) \ge n/N, n = 0, 1, \cdots, N-1$, in which $\delta_n = 1$ means the switch S is on at the time of $n\delta_s$, paper [8] has introduced the detailed information.

4   Model Predictive Control for MPPT

Mixed Logic Dynamic (MLD) formulation captures the associated hybrid features and allows the definition of the optimal control problem in a convenient way. By introducing auxiliary logical variable and continuous variable, the state variable can be expressed as linear state equations with constraints can be expressed as inequation:

$$x(k+1) = A'x(k) + b'd(k) + R\delta(k) + Gz(k) \tag{4}$$

$$E_1 d(k) + E_2 \delta(k) + E_3 z(k) \le E_4 x(k) + E_5 \tag{5}$$

which will act as predictive model and constraints respectively. For MLD model, paper [9] has given a detailed introduction.

The main control objective in this work is to regulate the output current to its reference I^*. Let $\Delta d(k)$ indicates the absolute value of the difference between two consecutive duty cycle. This term is introduced in order to reduce the presence of unwanted chattering in the input when the system has almost reached stationary conditions. The objective of system optimization is finding the optimal control sequence $D^* = [d^*(k), d^*(k+1), \cdots, d^*(M-1)]$, which makes the system output track expected reference trace and the performance function (4) least.

$$\min_{D,x,d} = \sum_{k=0}^{P} ||y(k+i/k) - y_r(i)||_{Qr}^2 + \sum_{k=0}^{M} ||\Delta d(k+i-1/k)||_{Qm}^2 \tag{6}$$

P is prediction horizon, M is control horizon, $y(k+i/k)$ is the predicted output at instant k, y_r is reference signal, $||x||_Q^2 = x^T Q x, Q = Q^T \ge 0, Q_y$ is weighting coefficient to penalize output signal error and Q_m is a weighting coefficient to penalize big changes in input signal. For the numerical optimization to be tractable, the predictive state $x(k + i/k)$ should be replaced by equivalent relations in terms of design variables i.e. manipulated variable-d, auxiliary logic variable-δ and auxiliary continuous variable-z. Then the predictive output could be expressed as

$$y(k + i/k) = \zeta + \Omega\gamma(k + i/k) \tag{7}$$

Where $r(k + i/k) = [d(k), \cdots, d(k + M - 1), \delta(k), \cdots, \delta(k + M - 1), z(k), \cdots]$, $\zeta = C(A', \cdots, A'^P)^T x(k)$,

$$\Omega = C \begin{pmatrix} B' & \cdots & 0 & R & \cdots & 0 & G & \cdots & 0 \\ A'B' & \cdots & 0 & A'R & \cdots & 0 & A'G & \cdots & 0 \\ \vdots & \vdots & \vdots & \vdots & \vdots & \vdots & \vdots & \vdots & \vdots \\ A'^{P-1}B' & \cdots & A'^{P-M}B' & A'^{P-1}R & \cdots & A'^{P-M}R & A'^{P-1}G & \cdots & A'^{P-M}G \end{pmatrix}$$

By substitution of $y(k + i/k)$ into performance function of the system J, an easy-to-tract form of optimization problem is obtained as follows:

$$\min_{\gamma} \quad \gamma^T H\gamma + 2f^T\gamma$$

$$s.t. \quad \begin{cases} (5) \\ 0 \leq d(k + i/k) \leq 1 \\ y_{min} \leq y(k + i/k) \leq y_{max} \end{cases} \tag{8}$$

Where $H = \Omega^T Q_y \Omega + Q_m$, $f = 2\zeta^T Q_y \zeta - 2d_0^T Q_m$. This control problem could be converted to solving the corresponding Mixed Integer Quadratic Programming (MIQP) problem with matlab function miqp.m. The optimal control sequence is $\gamma_t^{*M-1} = [d_t^*(0), \cdots, d_t^*(M - 1), \delta_t^*(0), \cdots, \delta_t^*(M - 1), z_t^*(0), \cdots]$ at the time of t. Only $d^*(0)$ is required when controlled the system. At the time of $t = t + 1$, repeat the above steps to realize rolling optimization.

In this paper, light intensity, temperature and output voltage of photovoltaic cell should be measured in every sampling period in order to readjust the reference current I*. In the predictive control strategy, aforementioned procedures should be performed at each time step. The numerical optimization is performed and the first element of computed control sequence is applied to the system. During this procedure, constraints based on components restriction, such as the diode reverse current, limit the output current within I_{max}; duty cycle $d(k)$ should be valued between 0 and 1, and the volatility of $d(k)$ should be reduced, avoiding the damage to the power switch and unnecessary energy loss caused by frequent operation.

5   Simulation Results

In this work, the hybrid system toolbox is used for simulation. The predictive horizon P=3, the control horizon M=2, $Q_y = 1$, $Q_m = 0.01$, Ts=50μs, number

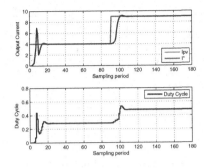

Fig. 5. The Current tracking and Duty Cycle

of the sub-interval N=2. As Fig.5 shows, though oscillates at first, the output current could track the reference current I*, even I* fluctuates form 4A to 9A suddenly.

In order to validate the proposed control methodology, simulation and experiment based on the data of light intensity and temperature were carried out on the photovoltaic system of SJTU showed in Fig.6.Fig.7 shown the Light intensity and temperature measured every 5mins form 7.a.m. to 17.p.m on July.21 2012.

Fig. 6. Photovoltaic system owned by Department of Automation, SJTU

Based on the data of light intensity and temperature, MPP could be calculated by the proposed T-G-P model. After measuring the PV output voltage (it changes form 14.8V to 17.2V very slowly), which can be regarded as constant in one sampling time, the MPPT reference output current I* is clear. In Fig.8, though the charge-discharge of energy-storage elements may lead to the oscillation of output current (the blue line) during the regulation procedure, the output current could track the reference current (the red line) well, more importantly, the output power (Pout, the blue line) follows the MPP (the red line) perfectly.

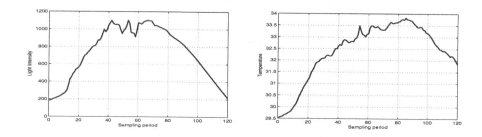

Fig. 7. G and T changes form 7.a.m. to 17.p.m. on 7.21.2012

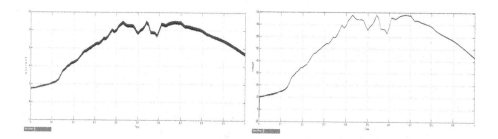

Fig. 8. The output current and output power tracked their reference respectively

At last, the operating point trajectory should be compared with the I-P and U-I curves to test whether the PV system is working on the MPP. As the Fig.9 shows, the PV system works on the theoretical MPP even though the environment keeps changing.

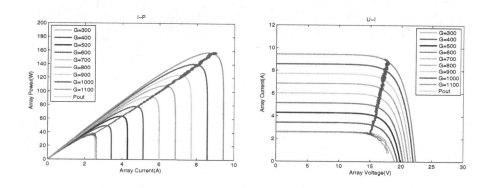

Fig. 9. Operation points on I-P and U-I curves

6   Conclusions

This work analyzed the characteristics of the PV system. Considering the uniqueness of the MPP in different conditions, the T-G-P model was built. Based on this model, the reference current could be worked out by the data of the light intensity, temperature and output voltage in a sampling period. The boost converter model was built and the predictive controller was designed. The simulation results showed the effectiveness of the controller and the accuracy of maximum power point tracking for PV system.

References

1. Hua, C., Shen, C.: Comparative study of peak power tracking techniques for solar storage system. In: IEEE Appl. Power Electron. Conf. Exposition Proc., vol. 2, pp. 679–683 (1998)
2. Hussein, K.H., Muta, I., Hoshino, T., Osakada, M.: Maximum photovoltaic power tracking: an algorithm for rapidly changing atmospheric conditions. IEEE Proc. Generation Transmission Distribution 142(1), 59–64 (1995)
3. Pandey, A., Dasgupta, N., Mukerjee, A.K.: High-performance algorithms for drift avoidance and fast tracking in solar MPPT system. IEEE Transactions on Energy Conversion 23, 681–689 (2010)
4. Ropp, M.E., Gonzalez, S.: Development of a MATLAB/Simulink model of a single-phase grid-connected photovoltaic system. IEEE Transactions on Energy Conversion 25, 207–216 (2010)
5. Suttichai, P., Yosanai, S.: Maximum power point tracing using adaptive fuzzy logic control for grid-connected photovoltaic system. Renewable Energy 30, 1771–1788 (2005)
6. Wang, W., Li, S.: A real-time modeling of photovoltaic array. In: The 23rd China Process Control Conference (2012)
7. Mohammad, H.M., Ali Reza, R.: A hybrid maximum power point tracking method for photovoltaic systems. Solar Energy 85, 2965–2976 (2011)
8. Giovanni Beccuti, A., Morari, M.: Optimal Control of the Boost dc-dc Converter. In: The 44th IEEE Conference on Decision and Control, December 12-15 (2005)
9. Chen, P., Ren, S.: Hybrid Dynamic Modeling and MPPT Control for photovoltaic Power Generation System. In: The 3rd International Conference on Measuring Technology and Mechatronics Automation (2011)

Fault Diagnosis for Power System Transmission Line Based on PCA and SVMs

Yuanjun Guo, Kang Li, and Xueqin Liu

School of Electronics, Electrical Engineering and Computer Science,
Queen's University of Belfast, Belfast, BT9 5AH, UK
{yguo01,k.li,xueqin.liu}@qub.ac.uk

Abstract. This paper presents the application of a fault detection method based on the principal component analysis (PCA) and support vector machine (SVM) for the detection and classification of faults in power system transmission lines. Consider that the data may be huge with a number of strongly correlated variables, method which incorporates both the principal component analysis (PCA) and support vector machine (SVM) is proposed. This algorithm has two stages. The first stage involves the use of the PCA to reduce the dimensionality as well as to find violating point of the signals according to the confidential limit. The features of each fault extracted from the data are used in the second stage to construct SVM networks. The second stage is to use pattern recognition method to distinguish the phase of the faulty situation. The proposed scheme is able to solve the problems encountered in traditional magnitude and frequency based methods. The benefits of this improvement are demonstrated.

Keywords: Fault Diagnosis, Transmission Line Faults, Principal-Component Analysis (PCA), support vector machine (SVM).

1 Introduction

In power systems, the transmission line is the vital link between the electricity power production and usage. To find the accurate fault location in the transmission line based on the measurement of the currents and voltages is of great importance, since a more accurate location results in the minimization of the amount of time spent by the line repair crews in searching for the fault [1].

Since the modern power system is well equipped with advanced measurement and protection instruments, a huge amount of data has been collected with greater dimensionality and quantity. Applications of statistical monitoring techniques could be useful in extracting and interpreting process information from massive data sets in order to discriminate between power system normal or faulty states. At the same time, pattern recognition techniques could also be used to distinguish which phase of the power system is faulty. Consequently, various multivariable statistical process control algorithms have been investigated and implemented in power systems recently [2].

K. Li et al. (Eds.): ICSEE 2013, CCIS 355, pp. 524–532, 2013.
© Springer-Verlag Berlin Heidelberg 2013

The wide use of control charts for multivariate quality control situations is based on the correlations between all the process variables. One approach that has proved particularly powerful is the use of Principal Component Analysis (PCA) in combination with T^2 and Q charts. PCA divides data information into the significant patterns, such as linear tendencies or directions in model subspace, and the uncertainties, such as noises or outliers located in residual subspace [3].

For the use of PCA techniques to fault detection and identification, Venkat has summarized a comprehensive and systematic review about the process diagnosis [4]. Recently, Zheng utilized fault diagnosis method in power system using multiclass least square support vector machines classifiers and proved the effectiveness on the basis of experiments.

Despite the reported progress in power system fault detection over the past few years, two major problems still exist in the fast detection of the transmission lines. One problem is that there are a lot of data collected from online monitoring systems, which are multivariate and correlated. The other one is that there are various types of faults in transmission lines. For real-time fault diagnosis in transmission lines, the two problems should be considered simultaneously. This is a problem to balance the real time implementation and the accuracy.

In this paper, a fault diagnosis approach based on PCA and SVM is proposed to tackle the problem. In the proposed approach, a feature extraction algorithm based on PCA is used to reduce the dimensionality. The PCA monitoring scheme and the SVM are utilized to pinpoint the fault inception point and implement fault recognition, respectively. The experimental results show that the proposed approach is capable of detecting and recognizing the faults effectively.

This paper is organized as follows. Section 2 presents the preliminaries for the proposed approach, followed by the description of the monitoring scheme in Section 3. The application study for various fault situations are given in Section 4, as well as the discussion of the implementation of the algorithms. Finally, the conclusion is given in Section 5.

2   Preliminaries

2.1   Principal Component Analysis

Let the data in the training set, consisting of m observation variables and N observations for each variable, be stacked into a matrix $\mathbf{X} \in \mathbb{R}^{N \times m}$, given by

$$\mathbf{X} = \begin{bmatrix} x_{11} & x_{12} & \cdots & x_{1m} \\ x_{21} & x_{22} & \cdots & x_{2m} \\ \vdots & \vdots & \cdots & \vdots \\ x_{N1} & x_{N2} & \cdots & x_{Nm} \end{bmatrix} \tag{1}$$

PCA decompose the data matrix into a score matrix $T \in \mathbb{R}^{N \times k}$ and a loading matrix $P \in \mathbb{R}^{m \times k}$ ($k \leq m$ is the number of retained principal components) [3].

Then the sample covariance matrix of the training set is equal to

$$S = \frac{X^T X}{N - 1} \tag{2}$$

An eigenvalue decomposition of the matrix S is

$$S = V \Lambda V^T \tag{3}$$

which reveals the correlation structure for the covariance matrix, where Λ is a diagonal and V is orthogonal. The Principal Components (PCs) are represented by the loading and score vectors so that PCA can decompose the observation matrix $\mathbf{X}$ as:

$$\mathbf{X} = t_1 p_1^T + t_2 p_2^T + ... + t_k p_k^T + E = \mathbf{T}\mathbf{P}^T + \mathbf{E} \tag{4}$$

$t_i, i = 1, ..., k$ are vectors, also named scores of the data, which contain information on how the samples are related to each other. $p_i, i = 1, ..., k$ are the loadings, also are the eigenvectors of the covariance matrix, $\mathbf{E}$ is the residual matrix. The important statistic for PCA monitoring is given by Hotelling's T^2, which is the sum of normalized squared scores defined as [5],

$$T_i^2 = t_i \lambda^{-1} t_i^T = x_i p \lambda^{-1} p^T x_i^T \tag{5}$$

where t_i is the ith row of k score vectors from PCA model, and λ^{-1} is a diagonal matrix containing the inverse eigenvalues associated with the k eigenvectors. The equation represents the distance in principal component model subspace, which shows that T^2 is a measure of the variation within the training data set.

However, monitoring the output variable via T^2 based on the first k PCs is not sufficient. If a totally new type of special event occurs which was not present in the reference data used to develop the PCA model, the new observations X_{new} will move off the plane. Such new events can be detected by computing the squared prediction errors (SPE) of the residuals of new observations [6].

$$SPE_{X_{new}} = \sum_{i=1}^{m} (X_{new,i} - \hat{X}_{new,i})^2 \tag{6}$$

This statistic is referred to as the Q-statistic, or distance to the model. It represents the squared perpendicular distance of a new observation from the plane. When the process is normal, this value should be small. Upper control limit for this statistic can be computed from training data set. Q statistic indicates how well each sample conforms to the PCA model, it is a measure of the amount of variation in each sample not captured by the k principal components retained in the model.

2.2   Support Vector Machine

SVM has become an increasingly popular technique for machine learning activities including classification, regression, and outlier detection [7]. The idea of using SVMs for separating two classes is to find support vectors, which refers to a representative of training data points, to define the bounding planes, in which the margin between the both planes is maximized [8].

SVM maps the input vectors into some high dimensional feature space through nonlinear methods. In this space a linear decision surface is constructed with special properties that ensure high generalization ability of the network. The number of support vectors increases with the complexity of the problem.

3   The proposed Fault Detection Scheme

This paper introduces a monitoring scheme that extracts the features of different type of faults and find the fault inception point and fault type using PCA-SVM.

3.1   PCA Monitoring Scheme

In PCA, Hotelling's T^2 and Q statistics are used to realize monitoring of the processes. To compute PCA using the covariance method, several steps are introduced:

1. Calculate the empirical mean and the empirical mean along each dimension.
2. Calculate the deviations from the mean.
3. Compute the eigenvalue and eigenvectors of the covariance matrix.
4. Calculate the cumulative energy content for each eigenvector and chose the basic vectors.

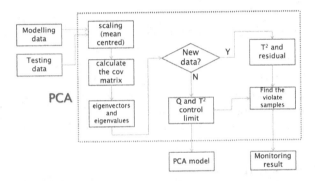

Fig. 1. Diagram of PCA monitoring scheme

After establishing the PCA model from the training data, the control limit of the T^2 and Q statistics can also be determined, the new data set can be computed and get the new T^2 and Q value and compared with the PCA model and the statistics, as shown in Fig.1.

Fig.1 shows the diagram of PCA algorithm when used in monitoring. In general, the conventional PCA-based fault detection methods use T^2 and SPE or Q statistics. The confidence limits for these indicators are calculated based on the assumption of the multivariate normality of observations and errors, respectively.

3.2   SVM Classification

For data generated from different type of faults, PCA can reduce the dimension of the huge data significantly. During the process of PCA, an intermediate result matrix is produced, consisting of the eigenvalues of the covariance matrix of the data sets. These elements are uncorrelated and independent, thus can be considered as the main features for detecting the certain fault type. Therefore, at the same time with data dimensionality reduction, features extracted by PCA can be used to build the corresponding feature data-base.

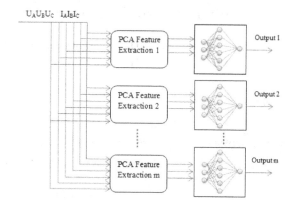

$U_A U_B U_C \quad I_A I_B I_C$

Fig. 2. Diagram of SVM network with features as input

With these advantages, it is possible to use SVM to process new data with the data-base to identify the type of fault, such as the phase-to-phase, phase-to-ground or three phases fault. The classification scheme is shown in Fig.2.

As shown in Fig.2, the signals with multiple variables as the input are preprocessed to generate the uncorrelated and independent eigenvalues by PCA based on the data collected under different situations. The featured eigenvalues are then used as the input of the SVM network to classify the faults.

4   Implementation

The proposed method was applied to the fault diagnosis in the power system transmission line as shown in the following.

4.1   Faults in Transmission Line

The simulation was carried out using SimPowerSystems package in Matlab, which described a three-phase, 60Hz, 735kV power system transmitting power from a power plant through 600 km transmission line. The transmission line was split into two 300km lines connected between three buses. To simulate all the situations, different type of faults were considered by different combinations of source impedances. For each condition, the fault simulation was carried out by changing the parameters in the fault breaker and the data was collected respectively.

4.2   Simulation Results

Two data sets were both recorded from the simulated transmission line corresponding to fault free and faulty conditions. The fault free data set described normal power transmission behaviour and included a total of 4000 samples. The faulty condition consisted seven groups of data represented different type of

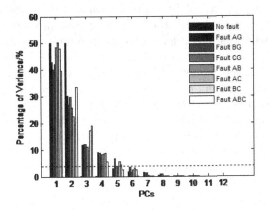

Fig. 3. Percentage of variance captured by PCs

faults as described previously, and were used for testing the performance of the proposed PCA and SVM method.

In the first stage, a PCA monitoring model was produced using the recorded fault-free data set. First, the number of PCs that can describe the variance of data was determined. Fig.3 shows the percentage of variance captured by each PC with a dashed line representing 5% of the variance limit. The variance captured by some PCs which was less than 5% can be considered as the irrelevant information. This PCA approach thus compressed the variation incorporated in the 12 process variables to 5 PCs, which were capable of capturing 97.3% in average for seven fault groups.

After the PCA model was established, a new data set with a certain kind of fault was estimated using the model. The confidential control limits for T^2 and Q statistics were 95%, and the monitoring chart was plotted in Fig.4.

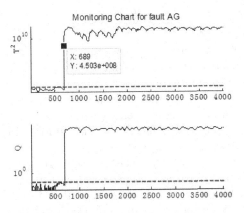

Fig. 4. Monitoring chart for faulty signals

In Fig.4, the sample point of 689 can be considered as the violating fault inception point as it violated both the T^2 and Q control limits. In the section before point 689, which is also referred to the pre-fault signal, the calculated results of T^2 and Q statistics are both under the control limit. The section after the inception point is named post-fault signal, with high values of statistic results. Then the unusual event can be detected by these high values of statistics. A effective set of multiple variate control charts is therefore a T^2 chart on the five dominant orthogonal PCs plus a SPE or Q chart. Fig.4 also indicates that these statistics respond sharply to the abrupt changes in the voltages and currents caused by transmission line faults.

For a verification of the PCA model working under noisy condition, the model based on the noisy signals was identified next. In order to provide a fair comparison, the same five PCs were retained in the model. Noise was generated based on the random normal distribution, with 5% to 30% of the signal amplitude as the fluctuation range at the interval of 5%. Fig.5 presents the entire situation under the 10%(left part with logged ordinate) and 15%(right part with linear ordinate)noisy condition. This indicates that the value of T^2 decrease with the increasing amplitude of the added noise. Some spikes appeared when the noise amplitude was bigger than 20% of the signal amplitude. The details were given in Table.1. In general, the error caused by different range of random noises were under 1% which was acceptable. These have led to a conclusion that the statistics control limit work well in the charts under the noise condition less than 30% except some fluctuations added.

Table 1. Table of PCA monitoring error under different noise condition

Noise Condition	Inception Point	PCA Result	Error
5%	689	703	0.3499%
10%	689	702	0.3249%
15%	689	714	0.6248%
20%	689	712	0.5749%
25%	689	728	0.9748%
30%	689	735	1.1497%

From the monitoring scheme, it is clear that PCA is capable of finding the accurate fault inception point. Due to that PCA modelling only based on the normal data set, any type of disturbance would result in a change in the covariance structure of the observation data, which will be detected by the high value of statistic charts. Therefore, the modelling results are able to find the violate point, and are independent of the fault type. In the following part, the focus is on the feature extraction according to the faults. The diagonal elements of matrix S (intermediate result of PCA) consists of the eigenvalues of the covariance matrix of the data sets, which are uncorrelated and independent, thus can be considered as the main features of the certain fault type.

The second stage of monitoring scheme is to use SVM to classify the fault type. The SVMs were set up using supporting vector machine toolbox in MATLAB.

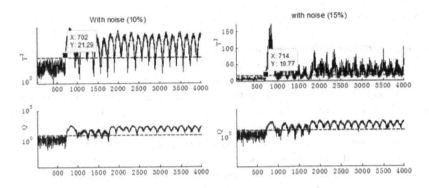

Fig. 5. Monitoring chart for noisy faulty signals(left 10%, right 15%)

To set up a certain kind of SVM according to the fault type, the input contains Training data matrix and Group data. For the training data, each row corresponds to an observation, and each column corresponds to a feature. As for the Group variable, each element of Group specifies the type of the corresponding row of Training data. In this simulation, it consisted of logical number 1 and 0, which indicated belong to this certain type of fault and not belong to, respectively. For the fault type, AG, BG and CG represented the phase-to-ground fault, AB, BC, AC were the two-phases faults, ABC was the three phases fault while N stands for normal condition with no fault. The radial basis function with the default scaling factor of 1 was chosen as the kernel function to map the training data into kernel space. Primary rough results has been simulated as shown in Table.2.

Table 2. Table of SVM classification result

Fault Type	SVMs							
	AG	BG	CG	AB	BC	AC	ABC	N
AG	1,0	1,0	1,0	1,0	1,0	1,0	0,0	0,0
BG	1,0	1,0	1,0	1,0	1,0	1,0	0,0	0,0
CG	1,0	1,0	1,0	1,0	1,0	1,0	0,0	0,0
AB	0,0	0,0	0,0	1,0	1,0	1,0	0,0	0,0
BC	0,0	0,0	0,0	1,0	1,0	1,0	0,0	0,0
AC	0,0	0,0	0,0	1,0	1,0	1,0	0,0	0,0
ABC	0,0	0,0	0,0	0,0	0,0	0,0	1,0	0,0
N	0,0	0,0	0,0	0,0	0,0	0,0	0,0	1,0

From the classification results, it can be seen that the performance of SVMs were not satisfactory. Without optimal structure design and parameter choices, SVMs can only correctly classify phase-to-ground, phase-to-phase and three phases faults. The future work include to develop improved algorithm with optimized parameters and selected features as the inputs for SVM.

5   Conclusion

This paper demonstrated that the proposed PCA-SVM approach is a powerful tool for monitoring power system transmission process. This methods is capable of capturing the relationship between the recorded variables from the data, and providing confidential limit charts for the violate fault points. It also helps to extract the features of the faulty signal under different faulty situations, which are used as the the inputs of the SVMs to classify these faults correctly.

The work however has also raised some important issues with respect to the implementation of the combination method: (1) the monitoring statistics based on linear PCA are not sufficient to identify the nonlinear relations of the process variables, therefore, it is necessary to develop nonlinear and dynamic extensions for the nonlinear and dynamic system; (2) parameters of SVMs in this paper is not optimized or selected to achieve precise classifications. So the work in the near future is to solve these problems and to improve the PCA-SVM methods.

Acknowledgment. This work was financially supported by RCUK including Engineering and Physical Sciences Research Council (EPSRC) under grant EP/G042594/1, and partly supported by the China Scholarship Council, and National Natural Science Foundation of China under Grants 61271347 and 61273040.

Reference

[1] Salat, R., Osowski, S.: Accurate fault location in the power transmission line using support vector machine approach. IEEE Transactions on Power Systems 19(2), 979–986 (2004)

[2] Frank, P.M.: Fault diagnosis in dynamic systems using analytical and knowledge-based redundancy: A survey and some new results. Automatica 26(3), 459–474 (1990)

[3] Jolliffe, I.T., MyiLibrary: Principal component analysis, vol. 2. Wiley Online Library (2002)

[4] Venkatasubramanian, V., Rengaswamy, R., Yin, K., Kavuri, S.N.: A review of process fault detection and diagnosis:: Part i: Quantitative model-based methods. Computers & Chemical Engineering 27(3), 293–311 (2003)

[5] Dryden, I.L., Mardia, K.V.: Statistical shape analysis, vol. 4. John Wiley & Sons, New York (1998)

[6] Kresta, J.V., Macgregor, J.F., Marlin, T.E.: Multivariate statistical monitoring of process operating performance. The Canadian Journal of Chemical Engineering 69(1), 35–47 (1991)

[7] Suykens, J.A.K., Vandewalle, J.: Least squares support vector machine classifiers. Neural Processing Letters 9(3), 293–300 (1999)

[8] Jakkula, V.: Tutorial on support vector machine (svm). School of EECS, Washington State University (2006)

Energy Consumption Analysis for a Single Screw Extruder

Jing Deng[1], Kang Li[2], Eileen Harkin-Jones[1], Mark Price[1],
Nayeem Karnachi[1], and Minrui Fei[3]

[1] School of Mechanical & Aerospace Engineering
Queen's University Belfast, Belfast, BT9 5AH, UK
[2] School of Electronics, Electrical Engineering and Computer Science,
Queen's University Belfast, Belfast, BT9 5AH, UK
[3] School of Mechatronic Engineering and Automation,
Shanghai University, Shanghai 200072, China

Abstract. Polymer extrusion is regarded as an energy intensive production process, the real-time monitoring of both thermal energy and motor drive energy consumption becomes necessary for the development of energy efficient management system. The use of power meter is a simple and easy way to achieve this, however the cost sometimes can be high. Mathematical models based on the process settings provide an affordable alternative, but the resultant models cannot be easily extended to other extruders with different geometry. In this paper, simple and accurate energy real-time monitoring methods are developed for the analysis of energy consumption of the thermal heating and motor drive respectively. This is achieved by looking inside the controller, and use the control variables to calculate the power consumption. The developed methods are then adopted to study the effects of operating settings on the energy efficiency. These include the barrel heating temperature, water cooling temperature, and screw speed. The experimental results on Killion KTS-100 extruder show that the barrel heating temperature has a negative effect on energy efficiency, while the water cooling setting affects the energy efficiency positively but insignificantly. Undoubtedly, screw speed has the most significant effect on energy efficiency.

1 Introduction

Due to the low cost, super mechanical characteristics and ease of processing, polymers are becoming increasingly prevalent as basic materials in many industrial applications. According to the British Plastics Federation (BPF), the UK polymer industry is made up of 7,400 companies which produced around 2.5 million tonnes of plastics in 2011 with a turnover of £19 billion [1]. This number is still increasing as more traditional materials, such as wood, concrete and metal, are substituted by polymer.

The lightweight of plastics can help to reduce energy consumption in transportation and its related area. For instance, the Airbus A380 is 20% carbon fibre reinforced and burns 12% less fuel per seat, and for the Boeing 787 Dreamliner, the lighter but strong fuselage gives a 30% reduction in maintenance cost and 20% in fuel [1]. The processing of plastics, however, is energy intensive. In UK, the electricity bill for this purpose

K. Li et al. (Eds.): ICSEE 2013, CCIS 355, pp. 533–540, 2013.
© Springer-Verlag Berlin Heidelberg 2013

amounts to 380 million per annum. Thus, a reduction in electricity usage of 10% would result in savings of £38 million per annum and a significant reduction in environmental burden.

The extruder motor is obviously an energy intensive component which consumes around 1/3 of the total energy. Its associated power factor is always an critical issues for the plastic processing companies as lower power factor may lead to an undesired penalty (or surcharge). The on-line monitoring of motor power consumption therefore becomes necessary for the investigation of energy efficiency. Such energy monitoring can also provide useful information on the melt stability and the quality of final product. The use of power meter (e.g. HIOKI 3169-21) is of course the easiest way to monitor the motor power consumption, including the apparent power, active power, reactive power and the power factor. However, the installation of power meters for each extruder involves a big cost and disruptions to the production line. Mathematical models based on the process settings seems to be an affordable alternative for such purpose [2,3]. However, the developed models highly depend on the geometry of the extruder and the materials being processed. It is difficult to use the same model on a different machine without re-training. Fortunately, most of the motor drive provides essential information which can be used to theoretically calculate the power consumption. For a DC motor, these variables can be the rotational speed and armature current. In this paper, a simple method will be first presented for on-line monitoring of the DC motor power consumption in a single screw extruder.

Thermal heating is another big consumer of energy in polymer processing. The majority of thermal energy is used to keep the temperature of extruder barrel and die at a specific level. Thus, real-time monitoring of thermal energy also plays a key role in energy efficiency optimization. Due to the same issue in motor power monitoring, power meter and mathematical model are not suitable for this purpose. A more accurate and reliable method based on temperature controller outputs will be given in this paper.

Based on the real-time monitoring of thermal energy and motor energy consumption, the effects of operating settings on total energy efficiency in polymer extrusion can then be carried out. It is well-known that the increase of screw speed leads to lower specific energy consumption (SEC) [4]. However, the material residence time is reduced at higher screw speed, leading to possible poor melt quality. An optimal screw speed therefore needs to be properly identified. According to literature research, the temperature settings of each heating zone have different effects on the total energy consumption. However, the temperature of solid conveying zone near the feed area was found to have the most significant influence [5]. Additionally, the temperature of water cooling not only affects the extruder energy efficiency, but also determines the chiller power consumption. In this paper, the aforementioned operation variables will be considered in the analysis of extruder energy efficiency, the experimental results will provide useful information for industry to incorporate the energy management system, and further reduce energy cost in the plastic production line.

2 Fundamentals of Polymer Extrusion Process

Due to the poor heat conduction, polymer materials are usually processed through extrusion. Single screw and twin screw extruder are the most common types being widely used. In such process, the plastic granules is pushed by the screw from the feed zone to the die, and the heat generated by the shear stress and barrel heating helps to melt the granules.

Melt pressure near the screw tip is known to be proportional to the screw speed, it is also slightly affected by the melt temperature and screw geometry, as well as the material being processed. In industry, this pressure is use as the main indicator of melt quality. Unstable melt pressure would cause fluctuations on the throughput, which results in variations on the quality of final product. However, previous research has shown that the melt viscosity is probably the best indicator of melt quality [6].

The melt viscosity can be described as the resistance of material to flow, and is derived from the ratio of shear stress and shear rate of the flow as shown in equation (1)

$$\eta = \frac{\tau}{\dot{\gamma}} \tag{1}$$

where η represents the viscosity, τ is the shear stress, and $\dot{\gamma}$ denotes the shear rate. Additionally, shear stress is determined by the pressure drop in a slit die or capillary die, while the shear rate is proportional to the volumetric flow rate through the die. For a slit die, the viscosity can be calculated by [7]

$$\eta = \frac{\Delta P W H^2}{12 L Q} \frac{3n}{2n + 1} \tag{2}$$

where n is the power law index under the operating conditions, ΔP denotes the pressure drop along the slit die, W and H are the slit width and height, L is the length between the two pressure points, and finally Q represents the volumetric flow rate. According to [8], throughput Q is related to the melt pressure and screw speed. It shows that for a low density polyethylene (LDPE), the throughput can be approximated by a polynomial model with the order up to 2, and the model fit error can be less than 3%.

3   Online Energy Monitoring

The real-time monitoring of power consumption at each component is desirable for optimizing the overall energy efficiency. The usage of power meter is the easiest way to achieve this purpose, but the cost is sometimes too high. Mathematical models based on process settings can provide satisfactory accuracy in power monitoring, but it is difficult to apply the same model to other extruder with different geometry and processing materials. In this section, a simple method based on the controllers of thermal heating and motor drive will be presented. All methods were developed and verified on a Killion KTS-100 single screw extruder located at Queen's University Belfast.

3.1   Monitoring of Heating and Cooling Energy Consumption

The Killion KTS-100 extruder has separate temperature controllers (Eurotherm 808) for each heating zone. The displacement contactors AFM215-303 are used to regulate the heating and cooling. As a result, the pulse width modulation (PWM) is incorporated to implement the PID (proportional-integral-derivative) control. Moreover, zone 1-3 (solid conveying, melting, and metering) are fitted with air fan cooling, their controller output ranges from -100 to 100 instead of $[0, 100]$ for heating only control.

Each temperature controller also supports RS-422 serial communication which is then converted to RS-232 to be connected with computer. Several variables, such as setting point, measured value, and controller outputs, can be directly read or write through this communication. Therefore, by taking the controller outputs and their associated heating and cooling power, their energy consumption can be easily calculated as

$$P_{thermal} = \sum_{i=1}^{5} p_i u_i \tag{3}$$

$$p_i = \begin{cases} p_{heating} & u_i \geqslant 0 \\ p_{cooling} & u_i < 0 \end{cases} \tag{4}$$

where $p_i, (i = 1, \cdots 5)$ denotes the ith heating or cooling power, and u_i represents the ith controller output.

By using the proposed methods, it is also possible to monitor the energy consumption at different heating zones. According to the recorded data, zone 1 consumes nearly half of the total thermal energy. This can be caused by the plastic granules absorbing heat energy when passing through zone 1. However, due to the high heat conduction of metal, a significant amount of energy is wasted in cancelling the zone 1 heating and feed area water cooling. The installation of heat isolation plate between zone 1 and feed section should help to significantly reduce the overall thermal energy consumption.

3.2   Monitoring of Motor Energy Consumption

The Killion KTS-100 extruder is fitted with Eurotherm 512C motor drive and a tacho meter. The motor controller also utilizes PID control algorithm implemented through PWM. For the ease of configuration, the 512C controller also provides several terminals which can be used to read or write the motor status through analogue or digital signals. These include the screw speed, setpoint ramp, tacho feedback, torque/current limit, current meter output, start/stop, and so on.

For a DC motor, the rotational speed is known to be proportional to the motor armature voltage while the screw torque is proportional to the motor armature current. This relationship can be summarised as

$$V_a = R_a I + E_b \tag{5}$$
$$E_b = K_v \omega \tag{6}$$
$$T = K_m I \tag{7}$$
$$V_a = R_a I + K_v \omega \tag{8}$$

where V_a and R_a denote the armature voltage and current (or motor supply voltage and current), E_b is known as back EMF (Electro Motive Force), ω represents the motor rotational speed, and T denotes the torque, finally, K_v and K_m are motor specific parameters which can be identified through the measurements of $V_a, I_a, T,$ and ω.

Conventionally, the power consumption can be simply obtained through the production of armature voltage and current. However, as PWM regulation is adopted, the voltage and current change frequently at each PWM cycle. This change causes a phase shift between the voltage and current, leading to a low power factor of this motor drive system. An additional power meter attached to the motor drive power supply can verify such effect. As a result, the DC motor consumes more energy than it actually required to drive the screw. The recorded data also shows a clear relationship between the power factor and screw speed. Thus, a full representation of the motor apparent power consumption can be obtained. The resultant real-time monitoring of motor energy consumption is then used for the investigation of optimal operating conditions discussed later.

4   Effects of Process Settings on Thermal Energy Consumption

As thermal heating consumes nearly 2/3 of the total energy usage, it is desirable to study the effects of operation settings on the thermal energy consumption. Basically, the barrel and die heating temperatures, feed area water cooling temperature, and screw speed are the main adjustable variables. The following will investigate their effects on extrusion thermal energy consumption. All experiments (see table 1) were carried out on the Killion KTS-100 sincle screw extruder, and the material been processed is low density polyethylene (LDPE 2102TN00W, MFI 2.5 g/10min, density 921 kg/m^3) from SABIC.

Table 1. Experimental settings of each test on the single screw extruder

DOE	Zone 1	Zone 2	Zone 3	Screw Speed	Water Cooling
A	150°C	160°C	170°C	40	25°C
B	160°C	170°C	180°C	40	25°C
C	170°C	180°C	190°C	40	25°C
D	170°C	180°C	190°C	40	15°C
E	170°C	180°C	190°C	40	40°C
F	170°C	180°C	190°C	20	25°C
G	170°C	180°C	190°C	60	25°C

4.1   Heating Temperature Settings

The processing window of polymer melting temperature can be 50°C or even more for some materials. Too low of the heating temperature would cause the plastic granules not properly melted and more energy consumption from the motor. By contrast, higher heating temperature increases the amount of energy lost to environment. Thus an optimal heating temperature setting not only saves the energy, but also improves the melt quality.

For numerical analysis, the experimental heating temperatures were then set at three different levels: A (low), B (medium), and C (high) as shown in table 1. Each trial lasts around 90 minutes in which the first 30 minutes was allocated for the machine to reach its equilibrium point, and the data was recorded for the rest 60 minutes. The resultant total thermal energy consumption (from zone 1 to adapter), total extruder active energy consumption, motor drive active energy consumption, extruder power factor, motor drive power factor, ratio of zone 1 energy consumption to the total thermal energy usage, overall specific energy consumption, thermal specific energy consumption, and motor specific energy consumption are provided in table 2 for each trail.

Table 2. Effects of barrel temperature settings on the energy consumption (data was recorded for 50 minutes, the sampling rate was 10Hz)

	DOE A	DOE B	DOE C
Barrel temperature setting	low	medium	high
Total thermal energy (kwh)	0.808	0.910	1.000
Extruder active energy (kwh)	1.259	1.315	1.381
Motor drive active energy (kwh)	0.485	0.451	0.431
Extruder power factor	0.569	0.592	0.612
Motor drive power factor	0.419	0.412	0.406
Zone 1 vs total thermal	0.610	0.607	0.657
Total SEC (kwh/kg)	0.692	0.720	0.757
Thermal SEC (kwh/kg)	0.444	0.498	0.548
Motor SEC (kwh/kg)	0.266	0.247	0.236

The thermal energy was calculated from the temperature controllers, and the total value is a little higher than measuring from a power meter. This is caused by the lower sampling rate (1Hz) of the power meter, which is not capable of capturing the frequent changes of barrel heating and cooling. Therefore, the sum of thermal energy and motor active energy is slightly higher than the extruder active energy consumption. From table 2, higher barrel temperature settings lead to higher thermal energy consumption but lower motor active energy consumption. However, the total specific energy is increased at higher barrel heating temperature. This suggests that reasonable lower barrel heating is preferred without determinately affecting the melt quality.

4.2   Feed Area Cooling

The feed area cooling temperature setting not only affects the extruder energy consumption, but also determines the chiller energy usage. It has been studied that increasing the flow temperature by 4°C will decrease chiller operating costs by 10% [9]. The selection of chiller temperature setting should also consider the atmosphere temperature in order to save the cooling energy. For this KTS-100 single screw extruder, increasing water temperatures reduces energy usage in both thermal and motor drive (table 3). However, the power factor slightly decreased. As the effects are small compared to barrel temperature settings, it might be reasonable to put more attention on chiller operating cost instead of the extrusion while adjusting the water temperature settings.

Table 3. Effects of water cooling temperature settings on the energy consumption (data was recorded for 50 minutes, the sampling rate was 10Hz)

	DOE D	DOE C	DOE E
Water cooling temperature	low	medium	high
Total thermal energy (kwh)	1.020	1.000	0.961
Extruder active energy (kwh)	1.403	1.381	1.345
Motor drive active energy (kwh)	0.435	0.431	0.427
Extruder power factor	0.616	0.612	0.606
Motor drive power factor	0.408	0.406	0.405
Zone 1 vs total thermal	0.667	0.657	0.644
Total SEC (kwh/kg)	0.772	0.757	0.736
Thermal SEC (kwh/kg)	0.561	0.548	0.525
Motor SEC (kwh/kg)	0.239	0.236	0.234

4.3   Screw Speed

There is no doubt that higher screw speed leads to lower energy consumption. According to the results in table 4, screw speed has the most significant impact on thermal specific energy consumption (SEC). By contrast, the screw speed has small effect on the motor SEC. A small decrease on power factor can also be observed as the screw speed increased. However, the melt quality at higher screw speed need to be further investigated.

Table 4. Effects of screw speed on the energy consumption (data was recorded for 50 minutes, the sampling rate was 10Hz)

	DOE F	DOE C	DOE G
Screw speed	20	40	60
Total thermal energy (kwh)	0.994	1.000	1.048
Extruder active energy (kwh)	1.185	1.381	1.626
Motor drive active energy (kwh)	0.224	0.431	0.624
Extruder power factor	0.640	0.612	0.637
Motor drive power factor	0.301	0.406	0.500
Zone 1 vs total thermal	0.630	0.657	0.695
Total SEC (kwh/kg)	1.325	0.757	0.583
Thermal SEC (kwh/kg)	1.111	0.548	0.376
Motor SEC (kwh/kg)	0.251	0.236	0.234

5   Conclusions and Future Works

Simple and efficient energy monitoring methods have been developed in this paper. The new methods can not only monitor the thermal energy usage profile along the extruder, but also provide accurate and reliable monitoring of both thermal energy and motor drive energy consumption independent of extruder geometry and material. The effects of barrel temperature settings, water cooling, and screw speed on energy efficiency were then investigated based on the proposed monitoring methods. It has been found that lower barrel heating temperature, higher water cooling temperature, and higher screw will lead to lower specific energy consumption. However, the screw speed has the most significant impact on overall energy consumption.

High energy efficiency does not mean better melt quality, the effects of barrel heating, water cooling, and screw speed on the melt viscosity need to be further investigated to provide substantial guidance on operating point optimization. Heuristic methods, such as particle swarm optimization (PSO) or differential evolution (DE), can be adopted to find optimal process settings under specific constraints.

Acknowledgments. This work was financially supported by Engineering and Physical Sciences Research Council (EPSRC) under grant number EP/G059489/1 and partly supported by EPSRC grant EP/F021070/1 and EP/G042594/1, and Science and Technology Commission of Shanghai Municipality (11ZR1413100).

References

1. Davis, P.: Challenges for the british plastics industry. Technical report, British Plastics Federation (2011)
2. Lai, E., Yu, D.: Modeling of the plasticating process in a single-screw extruder: A fast-track approach. Polymer Engineering & Science 40(5), 1074–1084 (2000)
3. Abeykoon, C., McAfee, M., Li, K., Martin, P.J., Deng, J., Kelly, A.L.: Modelling the Effects of Operating Conditions on Motor Power Consumption in Single Screw Extrusion. In: Li, K., Fei, M., Jia, L., Irwin, G.W. (eds.) LSMS 2010 and ICSEE 2010, Part II. LNCS, vol. 6329, pp. 9–20. Springer, Heidelberg (2010)
4. Kelly, A., Brown, E., Coates, P.: The effect of screw geometry on melt temperature profile in single screw extrusion. Polymer Engineering & Science 46(12), 1706–1714 (2006)
5. Rasid, R., Wood, A.: Effect of process variables on melt temperature profiles in extrusion process using single screw plastics extruder. Plastics, Rubber and Composites 32(5), 187–192 (2003)
6. Cogswell, F.: Polymer melt rheology: a guide for industrial practice, p. 178. George Godwin Limited (1981)
7. McAfee, M.: Soft Sensor for Viscosity Control of Polymer Extrusion. PhD thesis, Queen's University Belfast (2005)
8. Chiu, S., Pong, S.: In-line viscosity control in an extrusion process with a fuzzy gain scheduled pid controller. Journal of Applied Polymer Science 74(3), 541–555 (1999)
9. Kent, R.: Energy management in plastics processing. Plastics, Rubber and Composites 37(2-4), 96–104 (2008)

Intelligent Control of Energy-Saving Power Generation System

Zhiyuan Zhang, Guoqing Zhang, and Zhizhong Guo

Department of Electrical Engineering, Harbin Institute of Technology,
Harbin 150001, China

Abstract. Highway power generation system which is environmentally friendly and sustainable provides an innovative method of energy conversion. It is also as a kind of city science and technology innovation , which has the characteristics of environmental protection and sustainable utilization. Making full use of vehicle impact speed control humps, we design a new kind of highway speed control humps combined with solar electric generation system integration. Developing green energy, energy saving and environment protection can be achieved.

Keywords: Highway power generation, Speed control humps, Electrical energy.

1   Introduction

The issue of energy shortage is widely concerned. To date, many methods of electricity generation have been proposed, such as coal-fired power, hydroelectric power, nuclear power generation and solar power. However, power equipments contain complicated structure and thus are extremely expensive. All over the world is now confronting the energy-deficient situation. Effective use of the various energy deserves much attention. There's traffic in the streets from dawn till dusk. To make full use of the vehicle impacting deceleration with energy, we designed a speed control humps with the integration of united solar power generation system. Man have always explored new energy [1-2]. Road power generation system in cities provides a green method for energy conversion by [3] improving speed control humps [4].

The system is formed by speed control humps with a vibration power generating device[5], solar energy installations, energy storage and control devices, and light installation, etc. During rush hours, when the cars press the speed control humps, kinetic energy can be converted to electrical energy. During bottom periods, this system is supplied by solar energy. As shown in Figure 1, this three-stage speed control hump consists of three parts: two fixed slope-shaped units and intermediate mover which can move up and down. When cars go across and press the speed control humps, both sides of the slope shape decelerate the car; when the wheels reach the middle part of speed control humps, motor vehicle pressure it with movement of the permanent magnet which is fixed in the mover. Fixed coil cutting magnetic field generates the electromotive force and then generated alternating current is transformed

K. Li et al. (Eds.): ICSEE 2013, CCIS 355, pp. 541–548, 2012.
© Springer-Verlag Berlin Heidelberg 2012

into direct current through rectifier devices. During bottom periods, the solar energy devices can convert solar energy into electricity, charge the battery. Small amount of electricity is provided for speed control humps with alarm lamp, while the rest for road lighting.

Fig. 1. The mock-ups of *the speed control hump*

2   Method

2.1   Speed Control Humps

Speed control hump which is used for deceleration is divided into three parts. Left view of the speed control hump is shown in Figure 2, in which 1 is the surface of the road. 2 is the fixed speed control hump's inclined plane, 3 is the speed control hump with it middle part connected with the iron yoke, which is the core of the linear motor. 4 is the speed control hump's back inclined plane, 5 is a yoke of iron connected with the speed control hump, which is the mover, 6 is the base of the device, and 7 is a coil.

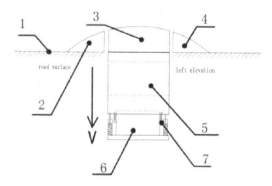

Fig. 2. The speed control hump's schematic diagram

2.2   Linear Generator

When cars pass the speed control hump, the speed control hump moves down because of the gravity of the cars. Due to the movement of the motor rotor, the permanent magnet fixed in the motor and coil fixed in the following take place relative displacement, thereby cutting the lines of magnetic induction generate electricity. As shown in Figure 3, when the car press the speed control hump, the front sloping offsets part of horizontal force. So that when the cars passed the middle part of speed control hump, the linear generator's mover moves downwards with the permanent magnet. Detailed structure of the linear generator is showed in Figure 4, Figure 5 and Figure 6.

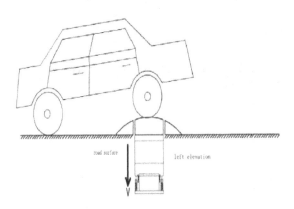

Fig. 3. Automobile passing by the speed control hump

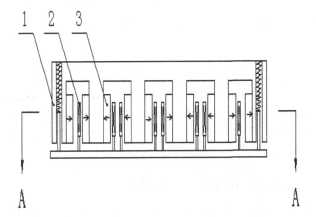

Fig. 4. Linear generator's schematic structure

Fig. 5. The actual manufacture of linear generator

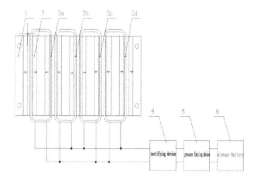

Fig. 6. The vertical view of linear generator

In Figure 4,1represents the iron yoke which is a part of the speed control hump. The motor is connected to the middle part of speed control hump. 2 is coils, namely the stator fixed in the bottom panel. 3 is the permanent magnet, and the arrows represent the relative direction of motion of the magnetic field line direction. The motor rotor and the stator are connected by a spring. The motor rotor and the stator can move vertically rather than horizontally.

2.3   Rectifier Unit and Pressure Limiting Device

Because the speed control hump is a linear reciprocating generator, the system generates alternating current. The bridge rectifier device turns direct current into alternating current. Changing into the direct current, this current restores the battery to a charged condition.

2.4   Solar Energy Installations

When there is less traffic, we introduce solar power and the speed control hump to generate electricity for battery charging, through which reliability of this system is improved

2.5   Storage Device

As the cars couldn't be continuously passing the speed control hump, electricity generated by this device is not continuous. Using energy storage devices, however, electricity could be continually collected for energy supplement all day long.

2.6   Novelty and Uniqueness

The novel structure of the speed control hump avoids the drawbacks of the level of damage caused by horizontal impact.

As shown in Figure 5, we propose a creative idea to establish "the linear power generator" type for electricity generation. The linear generator has the advantages of high energy per unit.

People who live in remote areas rely on energy grid supplied, "the linear power generator" type for generate electricity support system is better than long-distance transmission lines with the power consumption on the transmission lines.

On the basis of the device, we develop a new type of self-luminous warning the speed control hump. Traditional speed control hump is coated with reflective paint. Only when the car with the headlights light up pass the speed control hump can drivers see the deceleration zone. In our design the self-luminous LED lights installed on the surface of the speed control hump is charged by electricity generated by speed control hump, and it can also serve as warnings. Even in the night, bicycles with little lighting can also see the speed control hump.

In addition, after several modifications, the speed control hump can also be applied to the other places including pavements.

This project presents a new environmentally friendly method which is based on the linear generator character.

3   Experiments and Results

The device of permanent magnetic materials is made of NdFeB. Assume the road width L=6m, the width of the speed control hump w=0.3m, height h_1=0.5m, mover's trip h=0.25m. Estimate the weight of motor rotor NdFeB m=6kg. Assume the vehicle weight M=1.2t, the average tension of a spring T= 86N. Assume each straight-line power generation unit, the thickness of NdFeB l_1=5mm, NdFeB coil airthickness l_2 = 2mm, coil thickness l_3=5mm, the NdFeB iron yoke thickness l_4 =15mm.The thickness of enameled wire is 0.96mm. Assume the resistance is24.2 Ω / km. We can set up a group of power generation groupsN_1=80. The number of coil turns in each linear generator group N_2=100. Assume the magnetic induction B= 1.4T. Assume ampere force is constant force and the ampere force variable is F. Each generation group suffered the ampere force is F_1. Assume the linear motor rotor motion as uniformly accelerated motion. Assume the acceleration is a, the average velocity is v. Assume each generation group device output voltage is U_1 and each generating set's output current is I_1. The total output current is I. Assume each turn of coil length is l and each turn of the coil resistance is R.

$$l = 2 \times (2l_1 + 2l_2 + 2l_3 + l_4 + w) \tag{1}$$

$$R = \frac{l \times 14.2\Omega/km}{1000} \tag{2}$$

$$Mg - F - N_1T = N_1ma \tag{3}$$

$$F = N_1F_1 \tag{4}$$

$$F_1 = BI_1wN_2 \tag{5}$$

$$I_1 = \frac{U_1}{N_2R} \tag{6}$$

$$U_1 = BwvN_2 \tag{7}$$

$$2ah = (\frac{v}{2})^2 \tag{8}$$

Plugging those numbers into equation, we get U_1= 2.38273V, the speed control hump device output voltage is U=190.618V. Assume the circuit loading R_1=18 Ω ,the output power P = 2018.6W.

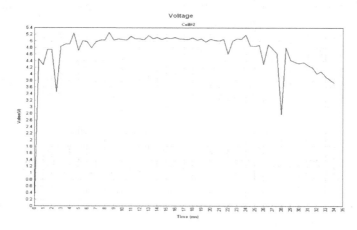

Fig. 7. Simulation results of each coil generating electromotive force

After developing the model, we use Infolytica Magnet to simulate magnetic field distribution of this model. It can be effectively solved through simulation practice. And this design method is a better solution of solving the optimization problem of the efficiency of electricity generation. As is shown in the Figure 7, the curve shows the fluctuation of simulated results analysis with a temporary change in voltage at different times.

Our three-section structure of the speed control hump with both sides of the slope shaped protects the core power generation equipment. When the cars pass the road, the front sloping in the structure of the speed control hump reminds the driver to decelerate. It reduces horizontal momentum; so that the force to power generation core part of the speed control hump is in the vertical direction all the time. The speed control hump with a power core part works as follows: when cars pass through the middle part of the three-section deceleration zone, the mover, along with permanent magnet, is pressed due to the gravity of cars. The coil is fixed, so the movement of the magnetic field magnetic induction lines cuts the coil which creates an alternating current according to Ampere's Law. While four groups of generators are demonstrated in Figure 5, according to the actual width of the road, we can increase the number of groups of straight line generators and maximize the energy harvested. According to right-hand screw rule, the level of face-to-face setting two sets of permanent magnets is the horizontally magnetic field lines. In the direction of magnetic back-to-back two sets of permanent magnets is on the contrary. As shown in Figure 6, when the permanent magnet moves up and down, the magnetic field of the adjacent coils 2a and 2b changes contrary. The coil 2c and 2d of the magnetic field changes in the opposite direction. According to right-handed rule, the direction of the current in the coil 2a and 2b is contrary; 2c and 2d generating current direction is opposite; Coil 2a and 2c is in the same current direction, the same to 2b and 2d. In other words, two adjacent coil current in the opposite direction, between adjacent two coil current in the same direction. The principles of rectifier operation were: the electric bridge rectifier device is electronic devices which rectify alternating current to direct current and charge the battery. In figure 6, 4 is a rectifier device, 5 is a pressure limiting device, and 6 is the battery.

4   Conclusions

In our design, we introduce synchronous operation linear generator, three-section speed control hump and solar panels. Compared to traditional rotary generators, our system has internal advantages including high energy conversion efficiency, since it energy consumption caused by mechanical transmission device can be reduced. Moreover, using linear generator is much more promising and reliable because gears, which could be abrasive and fragile especially when large load is applied, are eliminated. In addition, the stability of generating capacity can be guaranteed with solar energy convertor, and the electricity generated by the rectifier can serve as supplement for street lights in remote areas. Considering that speed control humps create no noise and prove to be safe, energy storage and generation of them have great potential and commercial values around the country in the near future.

References

[1] Gao, F.: Environmental Protection of Power System-Vibration Power Generation. Energy Research and Utilization 8(15), 44–45 (2011)
[2] Wu, L., Zheng, X., Yang, Y., et al.: Based on Linear Motor "octopus" type wave energy generating set. Journal of Chongqing Jiaotong University 31(1), 150–152 (2012)
[3] Han, Y., Shan, C., et al.: Decelerating band deceleration principle and applications. Road Traffic and Safety 9(6), 16–20 (2009)
[4] Hou, C., Peng, W., Jin, X., et al.: Automobile T decelerating Band Kinetics Response Analysis. Machine Design and Research 26(2), 95–99 (2010)
[5] Wu, H., Guo, Q., Ye, Y., et al.: Summary of Generating System Controlling Technique for PM Linear Motor. Micromotors 44(12), 87–91 (2011)

A Decision Support Framework for Collision Avoidance of Unmanned Maritime Vehicles

Mamun Abu-Tair and Wasif Naeem[*]

The Energy and Intelligent Control Cluster
School of Electronics, Electrical Engineering and Computer Science, Queen's University
Belfast, Belfast, UK.
{m.abu-tair,w.naeem}@ qub.ac.uk

Abstract. Recently there has been considerable interest in the development of Unmanned Surface Vehicles (USVs) due to the increasing demand in a number of maritime applications. One of the main challenges for unmanned (and even manned) vessels is detecting and avoiding the (static and dynamic) obstacles that may appear during a mission. This paper presents a practical solution for obstacle detection and avoidance mainly for USVs. More importantly, an integrated decision support framework is proposed which provides risk assessment as an integrated feature to the path planner. The hardware platform consists of a high definition video camera and a laser range sensor mounted on a pan and tilt device to detect obstacles in the vicinity of the ship. The performance of the proposed system is extensively investigated in a number of virtual maritime environment scenarios. The results reveal that the proposed system is able to detect multiple dynamic obstacles. Additionally, the proposed system provides a real-time visual interface (similar to radar screen) for the captain which includes recommended actions to avoid the obstacles in accordance with marine rules of the road.

1 Introduction

Unmanned surface vehicles (USVs) are routinely being deployed in applications such as remote sensing, surveillance, coast patrolling and providing navigation and communication support to unmanned underwater vehicles (UUVs). In many instances, they are remotely operated to perform a specific mission in open or confined waters. The intelligence of these vehicles primarily resides in the navigation, guidance and control (NGC) systems design. Ideally, the vehicle needs to operate without any human intervention. This means that the vessel's on-board control system must be self-reliant and able to maintain and supervise each on-board component. Having said that, even with the most advanced NGC design the craft cannot be fully autonomous without the presence of an obstacle detection and avoidance (ODA) system [3]. Studies have shown that, in manned vessels, more than 60% of casualties

---

[*] Corresponding author.

K. Li et al. (Eds.): ICSEE 2013, CCIS 355, pp. 549–557, 2012.
© Springer-Verlag Berlin Heidelberg 2012

at sea are caused by collisions [5]. In addition, it has been found that human error is a major contributing factor to those incidents. This could be due to an ever-decreasing number of crew members adding more responsibilities per person. In the first three months of 2012 alone, a number of worldwide maritime incidents occur including the grounding of Costa Concordia and the collision between a cargo vessel and a passenger ferry in Belfast Lough [6]. It appears from initial investigations that most, if not all, of these incidents occurred as a result of human error and were therefore preventable. This provides a sound motivation in developing autonomous collision detection and avoidance systems for both manned and unmanned vehicles. For uninhabited craft in particular, this cannot be overlooked as their collision with other manned ships could endanger human lives. Hence a human operator is always required to maintain a constant lookout for any potential obstacles which is costly.

UUVs usually operate underwater and therefore do not pose a direct threat to the ambient surface traffic. Even so, there is a great deal of research in sonar-based ODA strategies developed for UUVs as compared to USVs which is primarily due to recent surge in underwater exploration.

The autonomy of an unmanned vehicle depends on the design of a reliable NGC system. Of these, the navigation system acquires and processes data so that the guidance system can generate appropriate trajectories to be followed by the vehicle. A well-designed control system or autopilot tracks the reference commands as closely as possible. The most common form of guidance law used in unmanned vehicles (marine or airborne) is the line-of-sight (LOS) guidance [7]. In this method, a LOS angle is formed and followed between the vehicle's current position and the target location. Several other guidance laws are also based on this methodology. In the absence of a collision detection system, the unmanned vehicle will follow the reference path regardless of the presence of any intermediate objects. This could lead to catastrophe as the vehicle may run into an obstacle thus damaging the on-board components and in the worst case, sink it. The presence of an on-board ODA system is thus extremely important for the vehicle to become self-sufficient.

Unfortunately, the marine research community has been mainly focussed on advanced navigation and control systems design and little attention is paid to the area of collision avoidance. The usual way adopted to work around this problem is by human intervention [3] through a radio control channel or a wireless link thus adding to the operating cost in the form of a manned support boat. As a consequence, the usability and extent of the vessel is severely constrained. In [1], it is argued that although USVs provide an excellent platform for fast experimentation and development of guidance and control algorithms, their use is limited due to the lack of a reliable ODA system.

Motivated by the need of designing a reliable ODA system for USVs, this paper describes such an experimental platform that can improve the USVs' efficiency and safety. The proposed system employs a high definition video camera and a laser sensor mounted on a pan and tilt device to provide the NGC system of the USVs with a visual reference of the surrounding area. In addition, the proposed system is integrated with a risk assessment module which identifies any potential threat and take/recommend suitable actions in order to alleviate it.

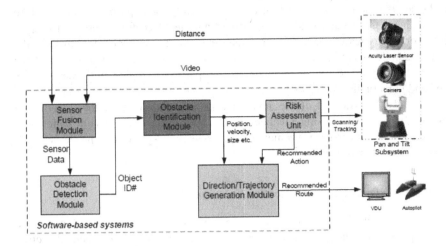

Fig. 1. Individual elements of the proposed obstacle detection and avoidance (ODA) system

The rest of the paper is organised as follows: Section 2 describes the proposed ODA system whereas Section 3 provides a full description of the simulation and experimental environment and scenarios which are used to investigate the performance of the proposed approach. The results are discussed in Section 4 whilst Section 5 concludes this study.

2   The Proposed ODA System

As mentioned earlier, a reliable ODA system is a vital element for a fully autonomous craft. For manned vehicles, obstacle detection is generally accomplished by trained crew members with the aid of a marine radar together with Automatic Radar Plotting Aid (ARPA) as well as manual lookout. For most existing USVs, a trained operator with the help of an onboard camera remotely monitors the state of the vehicle and safely navigates it around the obstacles. The proposed system is able to rectify this problem by having detection and path planning modules onboard the USV which require minimal or no intervention from a human operator. In the following, the individual elements of the proposed ODA system depicted in Figure 1 are described.

2.1   Pan and Tilt Scanner

The proposed system employs a pan and tilt device for (a) to increase the scanning range of the system and (b) to act as a stabiliser for the high definition video camera and the laser sensor mounted on it.

Three different scanning positions have been chosen to around the USV covering a 180 region. It was deemed sufficient to scan the front and sides of the USV as the path planner is reactive and therefore only close-range or pop-up obstacles will be dealt with here. Observations are made, stored and analysed for a period of time in each sector periodically according to predefined parameters or according to the USV's speed. However, if the risk assessment unit identifies any threat in one of the sector, it can advise the sensor to focus on that particular region until the threat is cleared and the situation is resolved.

Additionally, the tilt part of the device works as a stabiliser for the system. Tilt control is carried out by using feedback from a wave detection system which will minimise the effect of wave movements on the detection system and hence keep the camera and laser sensor stable.

2.2   Obstacle Detection/Identification Subsystems

The camera and the laser sensor provide the obstacle detection subsystem with a complete picture of the surrounding. For video feedback, the obstacle detection subsystem employs the well-known background subtraction technique to extract the information regarding any obstacle in the vicinity of the USV. The basic algorithm has been modified to ignore the sea waves which are also dynamic in nature. In addition to the camera system which is used to detect the presence of the obstacle, the laser sensor can provide accurate range measurements. By using the camera and the laser sensor this subsystem will extract the location, heading angle (if dynamic) and the size of the obstacle and transmit this information to the motion planner.

2.3   Direction/Trajectory Generation Module

Based on the information provided by the detection/identification system, the path planner module will provide the USV with the recommended route until the mission is accomplished and the USV reached the target.   This module is capable of generating paths that are compliant with the standard collision regulations (COLREGs) on prevention of collision at sea [8] defined by the International Maritime Organisation (IMO) [2]. Since the aim of this paper is to focus on detecting and identifying the risk of static and dynamic obstacles. Therefore the implementing of this module is not included in this paper.

2.4   Risk Assessment Unit

In addition to the above modules, the ODA system carries out its own risk assessment. Due to the fact that the detected obstacles may be dynamic, the ODA needs to assess the threat of these obstacles in a periodical manner. To accomplish this task the risk assessment unit is linked to the direction/trajectory module to advise a change in direction if there is any approaching threat. Additionally, it provides feedback to the pan and tilt subsystem if it needs further scanning of a specific region where a threat is believed to be approaching from. In order to evaluate the performance of the risk assessment unit, a computer simulator has been designed to simulate the direction/trajectory generation module.

2.5   Prediction Module

In terms of scanning, since there is only one active region at a time, the non-active sectors are dealt with using a prediction module to estimate the position of the obstacles (particularly the dynamic ones). The prediction module employs a linear technique to estimate the positions of the obstacles which are not currently being observed by the vision sensor. The risk assessment unit will continue to use the estimated position of the obstacles until it receives their actual updated positions when the system re-scans that particular region.

In the next section, the hardware and software environments used in the proposed approach are explained followed by results.

3   Experimental and Simulation Environments

The proposed system has been implemented using a Googoltech pan and tilt platform [4], a Microsoft HD video camera for object detection and identification and an Acuity laser sensor [9] for accurate range measurements. Associated with these hardware devices, a wide range of software libraries have been used to control and analyse the acquired data. The Googoltech C\C++ library has been utilised to control the pan and tilt. Additionally, the Afrog.net computer vision library has been used to implement the obstacle detection/identification subsystem.

For testing purpose, a number of maritime scenarios have been created using the commercially available Virtual Sailor Simulator [10]. Indoor experiments have been carried out in a visualisation lab at Queen's University Belfast equipped with multiple projectors and screens to emulate a panoramic view of the maritime environment. This setup has two major disadvantages. The first issue is the lack of depth information whereas the other is that the platform remains static which is not practical. In practice, a laser range sensor will be employed to obtain accurate depth information. In order to display the obstacles' information to the USV operator or to the captain of the manned ship, a virtual map is developed using the .Net environment. This is shown in Fig. 2 where the depth information is obtained from a previously stored file (in simulations) or from a laser range sensor (in practice).

The following section discusses the results of the experiment and simulation of the proposed system in detail.

4   Results

The hardware-in-the-loop simulation experiments aim to investigate the performance of the proposed ODA system in terms of detecting and identifying objects and to assist the risk assessment. The results illustrate a scenario of two dynamic ships with paths which are orthogonal to own ship's heading. Both the vessels are travelling opposite to each other at a different range to the own ship. Hence the own ship, according to the maritime rules, becomes a give-way vessel to the ship on its starboard whereas it will act as the stand-on vehicle to the ship on its port-side. The risk assessment unit determines this accordingly and may issue a recommended action to the captain/autopilot in addition to providing feedback to the path planner.

Fig. 2. Detecting multiple dynamic ships by the obstacle detection and identification subsystem.

4.1   Detection of Obstacles

Fig. 2 shows a snapshot of one of the simulation scenarios. As shown, the obstacle detection and identification subsystem has successfully detected the two cruisers present in the scene in real-time. This is shown in the form of rectangles or bounding boxes around the target vessels. Also note that the improved background subtraction technique ignores the dynamic waves in the scene which otherwise be detected as several moving objects. Once an obstacle is detected and its states estimated, the information is displayed on the virtual map to assist the remote human operator or the captain of a manned ship. This virtual mapping system also displays suitable actions that are either being taken or recommended by the risk assessment module. The following subsection will further elaborate on this.

4.2   Multiple Dynamic Obstacles Avoidance Scenario

The pan and tilt platform is employed to periodically scan in the vicinity of the USV. This provides an alternative to a radar but at a lower cost. Moreover, the system is able to detect close-range obstacles which may pose a direct threat to own ship for which reactive path planning may become necessary. Fig. 3 depicts the processed data on the virtual map of the environment developed as part of the proposed system. This is analogous to a radar screen as explained earlier. However, this virtual map only displays the information around the USV in a 180o span and continually update as the

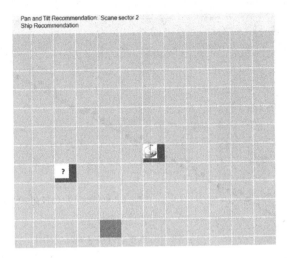

Fig. 3. A snapshot of the virtual map showing the recommend action for the pan and tilt subsystem

vessel manoeuvres. In the figure, the grey box indicates own ship's position. There are two additional icons showing a ship and a query box. The ship icon represents the actual location of a ship whilst being observed. On the other hand, the query box is the estimated position of the second ship which cannot be seen because of its existence in the non-active scanning region. This means that the proposed system

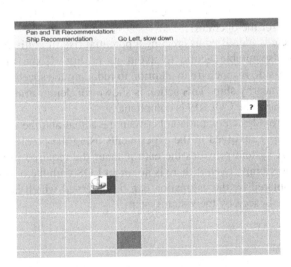

Fig. 4. A snapshot of the virtual map showing the recommended action for the ship control subsystem

always keeps track of any previously detected obstacles and predict their positions even if they are not visible directly by the camera. In addition to the scanning, the risk assessment unit of the proposed system guides the scanner to further investigate a

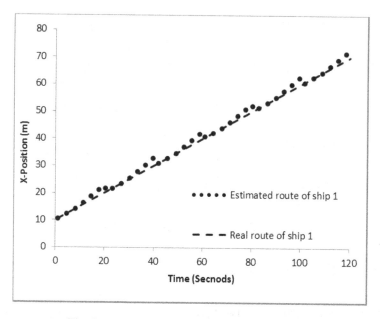

Fig. 5. The real and estimated route of target ship 1

specific region to assist in determining potential collision threats in that sector. Fig. 3 illustrates this scenario in which the risk assessment unit requests the pan and tilt device to re-scan sector 2. The request can be seen at the top of the virtual map and will be highlighted for the operator.

Once the obstacle is identified as creating a potential risk of collision, the proposed system recommends suitable action to avoid it. In Fig. 4, the proposed system recommends a port-side manoeuvre in addition to reducing the speed in order to avoid the collision of the nearby ship. This action is shown for demonstration purpose only and may not necessarily be COLREGs compliant.

Finally, Fig. 5 provides a comparison between the actual and the estimated relative positions of the ship 1 involved in the experiments using the prediction system. It is clear that the ODA system provides the path planner the required information of all the obstacles in the vicinity even if there is no direct LOS with the obstacle. However, some error accumulated by the estimation can be observed which is corrected when the obstacle is directly seen by the vision system.

5   Conclusion

The provisioning of obstacle detection and avoidance system of unmanned surface vehicles is a challenging research issue. This paper highlights the development of such an experimental platform. The key contributions of this paper are twofold: (1) proposition of a practical vision-based ODA module and (2) conducting the extensive performance evaluation of the proposed system in virtual dynamic maritime environments. The proposed system used a combination of a pan and tilt device, camera and a laser range sensor to detect and identify potential threats to the USV. In addition, a virtual map is created which is analogous to a radar screen providing up-to-date information of the state of objects in the vicinity. The results have shown that the proposed system is able to detect multiple dynamic obstacles. Additionally, the proposed system is shown to have the ability to handle and manage the risks that may occur during a particular USV mission.

Acknowledgments. The authors would like to acknowledgement the Engineering and Physical Sciences Research Council, UK for the financial support. EPSRC Grant Reference No. EP/I003347/1

References

1. Caccia, M.: Autonomous surface craft: Prototypes and basic research issues. In: 14th Mediterranean Conference on Control and Automation, MED 2006, Ancona, Italy, pp. 1–6 (2006)
2. Commandant, U. C. G. International regulations for prevention of collisions at sea, 1972 (72 COLREGS) (1999)
3. Corfield, S.J., Young, J.M.: Advances in Unmanned Marine Vehicles. In: Game Changing Technology for Naval Operations, vol. 69, ch. 12, pp. 311–326. The Institution of Electrical Engineers (2006)
4. Googoltech, http://www.googoltech.com/ (Date accessed, March 15, 2012)
5. Jingsong, Z., Price, W.G., Wilson, P.A.: Automatic collision avoidance systems: Towards 21st century. Department of Ship Science 1(1) (May 2008)
6. MAIB, Marine Accidents Investigation Branch, http://www.maib.gov.uk/ (Date accessed, March 15, 2012)
7. Naeem, W., Sutton, W., Ahmad, S.M., Burns, R.S.: A review of guidance laws applicable to unmanned underwater vehicles. Journal of Navigation 56(1), 15–29 (2003)
8. Naeem, W., Irwin, G.W.: Evasive Decision Making in Uninhabited Maritime Vehicles. In: Proceedings IFAC World Congress, Milan, Italy, August 28-September 2, pp. 12833–12838 (2011)
9. Schmitt Industries Ltd., Acuity distance measurement sensor, AccuRange 3000, http://www.acuitylaser.com/ (Date accessed, March 16, 2012)
10. Virtual Sailor, http://www.hangsim.com (Date accessed, March 16, 2012)

Author Index